TIGER CHECK

TIGER CHECK

Automating the US Air Force Fighter Pilot in Air-to-Air Combat, 1950–1980

Steven A. Fino

Johns Hopkins University Press
Baltimore

The views expressed in this document are those of the author and do not reflect the official policy or position of the United States Air Force, Department of Defense, or the US Government.

Printed in the United States of America on acid-free paper
9 8 7 6 5 4 3 2 1

Johns Hopkins University Press
2715 North Charles Street
Baltimore, Maryland 21218-4363
www.press.jhu.edu

Library of Congress Cataloging-in-Publication Data

Names: Fino, Steven A., 1974– author.
Title: Tiger check : automating the US Air Force fighter pilot in air-to-air combat, 1950–1980 / Steven A. Fino.
Other titles: Automating the US Air Force fighter pilot in air-to-air combat, 1950–1980
Description: Baltimore : Johns Hopkins University Press, [2017] | Includes bibliographical references and index.
Identifiers: LCCN 2016052578 | ISBN 9781421423272 (hardcover : alk. paper) | ISBN 9781421423289 (electronic)
Subjects: LCSH: Fighter planes—United States—History—20th century. | Fighter pilots—United States—History—20th century. | Airplanes, Military—Armament—United States—History. | United States. Air Force—Weapons systems—Technological innovations.
Classification: LCC UG1242.F5 F542 2017 | DDC 623.74/64—dc23
LC record available at https://lccn.loc.gov/2016052578

A catalog record for this book is available from the British Library.

Special discounts are available for bulk purchases of this book. For more information, please contact Special Sales at 410-516-6936 or specialsales@press.jhu.edu.

Johns Hopkins University Press uses environmentally friendly book materials, including recycled text paper that is composed of at least 30 percent post-consumer waste, whenever possible.

CONTENTS

ACKNOWLEDGMENTS

A few years ago at an annual meeting of the Society for Military History, I approached a colleague to congratulate him on his recent publication. His new book was based on his dissertation, which I was using to inform my own research. He thanked me, and then he shared with me how laborious the dissertation-to-book process had been. His candid confession caught me off guard. But he was right.

Fortunately, I've had the benefit of an amazing team cheering me on. P. J. Springer offered early encouragement for the project. Roe Smith introduced me to Johns Hopkins University Press, and Bob Brugger in turn provided me with my first lessons on how to transform a dissertation into a publishable book. After he left, Elizabeth Demers and her outstanding colleagues patiently guided me through the entire process. Brian MacDonald's keen eye was invaluable during the copy-edit stage. And special thanks also go to an anonymous reviewer who provided critical insights on the manuscript. It was a true pleasure working with and learning from them all. Whatever errors remain, they are mine alone.

Of course, this book would not have materialized without the extraordinary support I received while at MIT and in the years since. David Fahrenkrug and Tom Ehrhard initiated the process by persuading the powers that be in the Air Force to send me off to school. While at MIT, my committee of David Mindell, Sheila Widnall, and Owen Coté; colleagues like Joy Rankin, John Tylko, Stephen Zoepf, J. C. Ryan, and Bill Young; professors like Roe Smith; and the Engineering Systems Division staff, especially Beth Milnes, were invaluable. Ray O'Mara, Tim Schultz, Stephen Wright, John Terino, Case Cunningham, Robin Kapavik, Meg Martin, William Fry, and Reid Rasmussen have been stalwart friends and supporters throughout.

For offering unfettered access to their shelves of historical records, I'm indebted to the incredible staff at the Air Force Historical Research Agency at Maxwell Air Force Base, including Sylvester Jackson, Maranda Gilmore, and Leeander Morris; Brett Stolle at the National Museum of the USAF and Ray Ortensie at the Air Force Materiel Command History Office, both at Wright-Patterson Air Force Base; Jerry White, the Air Force historian at Nellis Air Force Base; and Debbie Douglas at the MIT Museum. Several former Air Force leaders and air combat veterans also graciously shared their time with me, providing the firsthand accounts and personal experiences that were essential to my research, including Larry Welch, Jasper Welch Jr., Jeff Cliver, Dick Anderegg, Tom Sokol, Jere Wallace, Bill Sparks, and Stephen "Shad" Dvorchak.

To Keith Ferris, Stephen Doyle, Barbara Stevens (for her late husband, Bob), and Robert Longo who agreed to let me use their amazing art to help tell my story, thank you.

Finally, to my wonderful wife and our three beautiful children, your boundless love continues to amaze. It has been a wild ride, filled with far too many cross-country moves and deployments. I'm not sure when or where it will end, but at least we're all still smiling. It is because of your tireless support that this book exists, and that is why it is dedicated to you.

ACRONYMS AND ABBREVIATIONS

AAA	antiaircraft artillery
AB	air base
ACEVAL	Air Combat Evaluation
ACM	air combat maneuvers
ACMI	air combat maneuvering instrumentation
ACT	air combat tactics
ADC	Air Defense Command
AFB	Air Force Base
AFHRA	Air Force Historical Research Agency
AI	airborne interceptor / airborne interception
AIM	air intercept missile
AIMVAL	Air Intercept Missile Evaluation
AMRAAM	advanced medium-range air-to-air missile
AOA	angle of attack
ASE	allowable steering error
ATC	Air Training Command
CAP	combat air patrol
CAS	control augmentation system
CC	central computer
CSAF	chief of staff of the air force
DTIC	Defense Technical Information Center
ECM	electronic countermeasures
EID	electronic identification
FEAF	Far East Air Forces
FIG	Fighter-Interceptor Group
FIW	Fighter-Interceptor Wing
GAR	guided air rocket
GIB	guy in back
HOTAS	hands on throttle and stick
HUD	head's up display
INS	inertial navigation system
IR	infrared
LCG	lead-computing gunsight
mil	milliradian

MiG	Mikoyan-Gurevich
MIT	Massachusetts Institute of Technology
MRM	medium-range missile
MSFRIC	Muir S. Fairchild Research Information Center
NARA	National Archives and Records Administration
NKAF	North Korean Air Force
NMUSAF	National Museum of the US Air Force
NVAF	North Vietnamese Air Force
NVN	North Vietnam
PACAF	Pacific Air Forces
PD	pulse-Doppler
RIO	radar intercept officer
ROE	rules of engagement
RP	Route Pack
RPA	remotely piloted aircraft
RTAFB	Royal Thai Air Force Base
RWR	radar warning receiver
SAC	Strategic Air Command
SAM	surface-to-air missile
SARH	semi-active radar homing
SEA	Southeast Asia
SOR	Specific Operational Requirement
SRM	short-range missile
SVN	South Vietnam
TAC	Tactical Air Command
TD	target designator
TDC	target designator control
TFS	Tactical Fighter Squadron
TFW	Tactical Fighter Wing
TFX	Tactical Fighter Experimental
TISEO	Target Identification System, Electro-optical
TWS	track-while-scan
USAF	United States Air Force
Vc	closing velocity
VID	visual identification
VSD	vertical situation display
WSO	weapons system operator

TIGER CHECK

CHAPTER 1

Introduction

The duty of the fighter pilot is to patrol his area of the sky and shoot down any enemy fighters in that area. Anything else is rubbish.

MANFRED VON RICHTHOFEN, circa 1918

Soaring high above the muck and filth–filled subterranean world of trench warfare, the fighter pilots of the First World War captured the attention of the Western world. Soldiers would crane their necks skyward and gaze awestruck as battles unfolded only a few thousand feet above them. With buzzing planes tracing tight circles among the clouds, alternately diving low and zooming high, onlookers were often blissfully unaware of the life-or-death struggle that was occurring in the warring pilots' cockpits. Only a sudden "snap," the sight of a plane crumpling, and then the sound of it striking the earth with an ignoble "thud" would remind the observers that the aerial demonstrations were no barnstorming display manufactured simply for their entertainment. On the home front, the reality of this novel form of combat was even further removed, with peoples' imaginations fueled by sensationalized reporting of gallant aviators and always-thrilling aerial victories. Whole nations quickly developed a love affair with their own favorite ace-sons: America had Eddie Rickenbacker; France, Roland Garros and Georges Guynemer; Britain, Albert Ball. But among them all, only Germany's Manfred von Richthofen transcended national borders.

Born to an aristocratic family, an excellent gymnast-athlete, and an avid hunter, Richthofen was the prototypical fighter pilot ace of the Great War. Educated, disciplined, fiercely competitive, and a good marksman to boot, Richthofen first enlisted as a cavalry officer. However, with trench warfare obviating the centuries-old need for horse-mounted soldiers, he became disenchanted and thereafter set his sights on gaining admission into Germany's fledgling air service. His request was honored, and the young Richthofen soon began flying combat missions near the Eastern Front. In the summer of 1916, while still searching for his first aerial victory, Richthofen caught the attention of Germany's leading ace at the time, Oswald Boelcke. A fruitful apprenticeship emerged, but it was cut short on October 28, 1916, when Boelcke plummeted from the skies following a midair collision with another German pilot in the unit, Erwin Böhme.[1] Albeit brief, Richthofen prospered immensely from the personal tutelage Boelcke provided through his *Dicta*, the essential precepts for air combat.[2] When Richthofen later assumed command of the Jagdstaffel (or Jasta) 11, he earned a reputation not only as a supremely talented aviator but also as a respected leader. Following the "bloody April" 1917 aerial battles near Arras, in which Richthofen's pilots accounted for nearly a third of all aerial victories over the British Royal Flying Corps, Richthofen was promoted to Rittmeister and given command of the new Jagdgeschwader 1. Richthofen's legendary "Flying Circus" was born, and its leader, now known for flying auspiciously painted red aircraft, achieved worldwide renown as the Red Baron.[3]

Richthofen was typically "dispassionate" in the reports he filed following his aerial victories. His account of his seventy-fifth victory, which occurred on April 2, 1918, shortly after he took command of the Flying Circus, was no different. It offers a valuable description of what air combat entailed during the later stages of the First World War:

> About 12:30pm I attacked a British R.E. [aircraft] at an altitude of 800 metres above the woods at Moreuil, [and] just below the clouds. As the adversary did not see me until very late, I managed to approach within 50 metres of him. I fired from ten metres' distance until it began to burn. When the flames shot out, I was only five metres away from him. I could see the pilot and observer twisting out of their aeroplane [seats] to escape the flames. The machine did not burn in the air, but gradually burned [on the way] down. It fell out of control to the ground, where it exploded and burned to ashes.[4]

After the mission, Richthofen was awarded the Order of the Red Eagle Third Class with Crown and Swords, an honor normally reserved for regimental commanders celebrating strategic victories. Richthofen, ever the "modest, open, sincere person" in the words of one fellow pilot, noted the irony: "[There was] never a strategic victory. Every air battle, no matter how big it is, always ends up in individual combats."[5]

Before his death, Richthofen sent at least eighty Allied planes tumbling to the ground, along with a good number of their crews. Yet, despite his hand in the deaths of many of their comrades, Allied pilots still routinely toasted the famed German flyer. One London journalist observed in 1918 when talking with a group of British aviators that, although all professed a strong desire to be the one who brought the Baron down, "there is not one who would not equally gladly have shaken hands with [Richthofen] had he been brought down without being killed or who would not so have shaken hands [with Richthofen] if brought down by him."[6] When Richthofen was finally felled behind Allied lines on April 21, 1918, most likely not by another pilot but instead by a round fired from a machine gun on the ground, Allied troops conducted a solemn military funeral, complete with rifle salute, for their "gallant and worthy foe."[7]

The iconic Red Baron lives on today, battling Snoopy, jamming to the tunes of Led Zeppelin, and rolling out tasty frozen pizzas.[8] Richthofen, more so than any other pilot, came to symbolize the mythical flying ace, a chivalric and noble knight who rose like a phoenix from the morass of mechanistic warfare during the Great War to restore humanity's faith in itself. Unlike those countless muddied soldiers climbing out of trenches and racing to their deaths at the sound of a starter's gun, a pilot like Richthofen survived and captured glory through individual cunning and skill.[9] Widely popularized stories of aviators pausing in battle to acknowledge each other's flying acumen or dropping wreaths on the wreckage of a fallen foe similarly rekindled in the citizenry notions of long-lost battlefield gallantry and honor.[10] And, in contrast to the chateau-bound, medal-encrusted, white-haired generals who measured victory in yards of No Man's Land captured, the dashing youths clad in leather jackets and silk scarfs who took to the heavens had an irreducibly simple mission—find the enemy and force him from the sky. "Anything else is rubbish," Richthofen purportedly once remarked.[11] Even heads of state celebrated the aces as the "Cavalry of the clouds. . . . They are the knighthood of the war, without fear and without reproach."[12]

Richthofen's iconic red triplane, the Fokker Dr.I, had long since vanished from the sky by the time of the Korean War nearly thirty-five years later, but the image of the Flying Knight, as well as many of his tactics, persisted. Patrolling the skies along the North Korean-Manchurian border known as MiG Alley, F-86 Sabre pilots of the fledgling US Air Force (USAF) clamored at the chance to down an enemy MiG jet fighter. Earlier, Richthofen had implored his Jasta pilots to have "the courage to fly in close to the enemy before opening fire." It was a lesson adopted from Boelcke's *Dicta*, "the secret of aerial victory."[13] Now in Korea, flying jet-powered aircraft at velocities approaching the speed of sound, some of which were equipped with a new radar-ranged, lead-computing gunsight that promised longer-range firing opportunities against enemy aircraft, American Sabre pilots were instructed to do the same. Leaders of the 4th Fighter-Interceptor Group (FIG) implored their pilots, " 'When in doubt, attack'—be a determined, aggressive killer—thats [*sic*] what you're flying a fighter for."[14] In another Sabre unit in Korea, the 51st FIG, pilots were told, " 'Every man's a Tiger!' The aggressive spirit must always be foremost if the pilot and his organization are to enjoy a successful career in the fighter business."[15]

Like most young Sabre pilots flying in MiG Alley, First Lieutenant Douglas Evans "sure wanted to become an ace" when he set sail from San Francisco destined for war.[16] The youngest son of the decorated Marine colonel Francis T. Evans Sr., the fourth Marine Corps officer to complete aviation training and the first to perform a loop-to-loop in a seaplane, Doug Evans had raced off to enlist in the US Army Air Forces the day he turned eighteen. But it was already 1943, and after his nineteen-month flight-training program, the freshly winged Evans joined scores of other new pilots wilting in the Texas heat waiting for an invitation to the Big War that never came. He was discharged and went back to school. After stints flying for the Reserves and the Air National Guard, and numerous letters to the Pentagon, Evans managed to score a coveted F-86 assignment to Korea, arriving aboard the USS *General John Pope* in mid-1951. After spending a few months learning the intricacies of jet combat while flying as a wingman alongside a more experienced flight leader, Evans finally got the opportunity to lead others into combat.[17]

One afternoon in early November 1951, Evans was leading Green flight, a formation of four F-86 Sabres. He was accompanied by three other four-ship Sabre formations: Red, White, and Blue flights. All were collectively assigned the task of seeking out and destroying enemy MiG fighters that might try to disrupt the Allied aircraft hurtling bombs at the Chinese Communist troops below. Approaching the Yalu River border that divided North Korea from the Communist sanctuary in Manchuria, the radio in Evans's "Spartan" cockpit erupted with chatter. Another group of Sabres slightly ahead of Evans's group had spotted twelve MiGs rocketing across the river. Moments later, a member of White flight called out eight more MiGs that were trying to sneak around the right side of Evans's group. Evans in turn spied four additional enemy fighters approaching from the front-left side. Then, the leader of Blue flight chimed in that there were six more MiGs approaching from the left. Evans later recalled, "Everybody went into a general dogfight, Sabres and MiGs churning around everywhere I looked. . . . It seemed like no matter who I started after, there were always others coming at me to keep my eyes from my gunsight—a very exasperating situation."[18]

Throughout the ensuing melee, Evans's wingman, flying as Green 2, performed admirably, maintaining a steady formation position scarcely two hundred feet alongside Evans.[19] The intent behind this "fighting wing" position was that with the wingman so near the flight lead, he could easily cover the flight lead's tail, which in turn would allow the leader to focus his full attention on spotting a MiG, maneuvering into position, and prosecuting a successful gun attack. "The flight and element leaders are the 'shooters,'" the 51st FIG's manual on air tactics decreed, "and the wingmen are the 'lookers.'"[20] The arrangement, however, was slightly counterintuitive. Shooting down an enemy aircraft was in some ways an easier assignment than trying to fly alongside someone who was trying to score a kill against a wild and unpredictable target. Remembering his first missions in Korea, Evans remarked, "Flying as a wingman in a wild dogfight is the toughest combat job there is."[21]

At one point during the swirling dervish, Evans and his wingman ended up directly above a MiG, "close enough to make out the pilot through his canopy and notice that some of the red paint had peeled off the nose of his plane." But before Evans could descend into position on the MiG's tail, he noticed "two more swept-wing jobs coming right up our tails."[22] Evans bent his jet around to thwart the attack, but then he spied twelve distant specks, rapidly growing, on the horizon—more MiGs. After another series of high-G, heart-pounding maneuvers, Evans and his wingman ended up behind two of the MiGs in a steep, climbing right turn. Evans explained:

> In the first moments I closed slightly, concentrated on the [MiG] wingman, and began hammering away, but my blasted gunsight was out of whack . . . so I was forced to guess range and lead. . . . I hit him with two or three flashes out of four or five bursts, but we had gotten into such a steep spiral it was tough to maintain any steady aim. . . . I was steadily losing ground until it seemed like we were falling backwards. My controls were mushy from the low airspeed, so I checked our tails before a last squirt and rolled over and down—talk about frustrated, and those two MiGs, going right up like elevators, disappeared through the overcast [clouds].[23]

Despite missing out on the kill, Evans took pride in the fact that "this hunting is fair, and the game has an equal opportunity of blowing your head off if you flub your vigilance or goof a maneuver. These counter-moves of opposing planes trying to kill each other is the damndest game of chess; you hardly have time to realize how dangerous it is."[24]

In many ways, Evans's and his fellow Sabre pilots' experiences over MiG Alley were comparable to Richthofen's over the Western Front almost thirty-five years earlier. In contrast, when Air Force first lieutenant David "Abby" Sveden Jr. launched in an F-15C Eagle from Incirlik Air Base (AB), Turkey, on the afternoon of January 19, 1991, he took to the skies armed with a vastly different conception of air combat. Granted, there were some similarities. For example, Sveden's task that afternoon was similar to that of the earlier fighter pilots: seek out and destroy enemy aircraft. And Sveden, like Evans before him, flew in a formation with other fighter aircraft. But rather than being tucked alongside his flight members as Evans's flight mates had been, Sveden was flying more than two miles away from them in a formation known as "tactical." The responsibilities were different, too. In this formation, each pilot shared responsibility for monitoring the formation's vulnerable six o'clock position while simultaneously scanning the sky ahead with the F-15's powerful radar. If anyone, including a wingman, detected a target, he was expected to immediately initiate the electronic identification (EID) process, communicate the target's presence to other members of the formation using agreed-upon code words, and, if appropriate, maneuver into position to launch a high-speed missile at the target.[25]

Sveden, a youngster on his first F-15 assignment who had been in the Air Force for only four years, was flying in the number four position in Rambo flight on the January 19 mission.[26] Along with another four-ship of F-15s in Conan flight, Sveden's formation was tasked with escorting more than sixty aircraft to attack targets near Kirkuk in northern Iraq. As the Eagles turned southward from their staging point in eastern

Turkey and crossed over the Iraqi border, the F-15 pilots were alerted to the possible presence of Iraqi fighter jets in the area. The Eagles continued south. At a distance of forty miles from an enemy airfield, several of the Eagle pilots observed a small green "brick," a synthetically generated radar target, appear about halfway up their four-inch-square radar scope, offset slightly toward the right side. The position of the radar target roughly corresponded to the last reported position of the enemy fighters. Sveden quickly commanded a radar lock and started the EID process. After a few moments, he knew the radar targets to be enemy fighters, and he reported that fact to his flight mates using the radio. As the distance to the enemy targets collapsed to twenty miles, the F-15 pilots observed the single "brick" split into two, which indicated the presence of at least two aircraft. The Rambo flight lead, Captain Steve "Gunga" Dingee, directed the other F-15s in Conan flight to continue south with the strike package; Rambo, he declared, was turning west to intercept the enemy fighters.[27]

The F-15s in Rambo flight locked on to the radar targets according to the prebriefed targeting plan. Two short radio calls later and the Eagle pilots were sure that they had targeted and properly "sorted" the two separate aircraft: Sveden was locked on to a target at nine-thousand-feet altitude; Sveden's element leader, Captain Dan "Spyro" Prather in Rambo 3, was locked on to a target a couple thousand feet lower at seven thousand feet.[28] From fifteen miles away, the Eagles already knew where their foes were, how many there were, what altitude they were flying at, what type of aircraft they were flying, and which Eagle was going to shoot which target when the time came.

The turn to the west toward the Iraqi fighters had pushed Sveden and Prather slightly out in front of the Eagles' formation. At a distance of ten miles to the enemy aircraft, the designated missile-firing range, Sveden pressed the red pickle button on his control stick, and one of his four, twelve-foot-long, radar-guided AIM-7 Sparrow air intercept missiles rocketed off toward the target. Seconds later, Prather launched a missile at his target. Sveden glanced back inside his cockpit and checked his 360-degree radar warning receiver (RWR) to confirm that no enemy was targeting him. Sveden and Prather, still about ten thousand feet above their unsuspecting prey, pushed forward on their control sticks and started to descend. They peered through the small target designator (TD) box in their head's up display (HUD), looking for their targets. At about eight-miles range, a small black dot appeared in the center of each pilot's TD box.[29] Sveden and Prather fired a second missile at each target from a distance of about four miles. Prather recalled:

> About this time, the first missile hit the first guy. They were [both hit] within a couple of seconds of each other, but I think Abby's missile hit his guy first. . . . It made a pretty big fireball, and you could see this fireball descending rapidly from 8,000 or 10,000 feet down toward the Tigris River. Almost instantly, the second [target was hit], making a big round blossoming orange-black fireball. I couldn't really see any debris falling from it; it just engulfed the whole thing. Those [hits]

were from our first missiles. Our second missiles weren't even required, but I remember seeing them guide through the fireballs and exploding when they got near the debris.[30]

Sveden and Prather's experience of launching a missile at an enemy target before ever visually seeing it was similar to about half of the USAF Eagle's thirty-four kills in Desert Storm.[31] One Eagle pilot who eventually racked up three kills during the war, Captain Rhory "Hoser" Draeger, described his victories: "For me, at least up to that point, it was the highlight of my career. After working all those years for that time and that place, it all came together. It worked out quite nicely. . . . It was something special."[32]

But for Draeger, Sveden, Prather, and the other Eagle pilots, it was also something vastly different from Evans's experience of air combat in an F-86 Sabre. Evans, as Richthofen had before him, needed to rely on keen eyesight to gain the advantage that accrued to the pilot who first sighted his adversary; Sveden and his fellow Eagle drivers used the power of electromagnetic energy emanating from the radars installed in their aircraft to detect their foes from more than forty miles away. For Richthofen and Evans, victory demanded the courage to close within a quarter mile before hammering away with their aircraft's machine guns, and the stick-and-rudder skills to gain and maintain that offensive position even if the enemy started to react defensively. Sveden only needed to turn his aircraft about sixty degrees to the right, use the middle finger of his left hand to lock the unsuspecting Iraqi aircraft, glance at his RWR to determine if he was at risk of being shot down, and then push the pickle button with his right thumb to send a nearly four-hundred-pound missile on its way to obliterate his adversary. In light of their radically different experiences of air combat, how can Evans and Sveden both be labeled "fighter pilots"?

Long before Desert Storm, while American Sabres and Soviet MiGs were still tangling with each other in the skies over Korea, the editors at *Life* magazine in February 1952 published an eight-page, illustrated spread depicting the remarkable transformation in "US military aviation from World War I to the present." Looking toward the future, the article foretold of fighters that may soon reach "speeds of more than 1,000 mph" and "rockets which may soon replace machine guns." Moreover, it predicted that the new armament would be aimed automatically by an "electronic 'brain'" carried aloft in the small fighters. Having surveyed these amazing technological advances looming on the horizon, the *Life* editors declared, "Fighters are rapidly becoming planes flown by men more like scientists than pilots."[33]

Indeed, aerial warfare in the ensuing three decades would undergo a remarkable transformation, from the close-fought, swirling melees of Evans's time in Korea to the immediate precursor to the video game–like war Sveden experienced in Desert Storm. Countless texts have charted the development of the novel air combat artifacts that emerged during the 1950s, '60s, and '70s. Regrettably, many of these "great machine" narratives cast these advancements as obvious, almost-linear milestones toward faster-flying and more-complex aircraft armed with longer-range and more-powerful weapons systems (figure 1.1).[34] Although a few "great machine" texts successfully avoid the trap of technological determinism as they unpack the black boxes of specific aviation technologies, notably Ron Westrum's study of the Sidewinder guided missile and Glenn Bugos's analysis of the F-4 Phantom II, most fighter aviation histories overlook a critical component of their respective technologies—that of the individual pilot tasked with operating the new equipment. It is as if, by opening an artifact's black box, authors felt it necessary to seal the operators of that technology in a black box all their own.[35] Consequently, the fascinating intersection between the human pilot and his machine that the editors at *Life* alluded to in 1952 has gone largely unexplored.[36]

Even the numerous biographies, autobiographies, and cultural histories of fighter pilots often fail to address the dynamic relationship that exists between pilot and machine in air combat. Far too frequently, these types of texts ignore the human-machine interaction in favor of developing a "great pilot" narrative. The reified fighter pilot, possessing a unique set of timeless skills, commands center stage in these accounts, and often the only pages devoted to detailing his interactions with his aircraft are those which describe how the heroic pilot compensated for a sudden mechanical breakdown or technological failing. Put simply, a pilot's survival and success, according to these sources, depend on his possessing the "right stuff."[37] Moreover, that "right stuff" has remained essentially unchanged over time: traits that were valuable for the Red Baron, such as good eyesight, excellent hand-eye coordination, and a daredevil-like aggressiveness that bordered on recklessness, remain the critical components of being a successful fighter pilot today, regardless of the dramatic changes in air combat technology.[38]

The history of air combat within the US Air Force from 1950 to 1980 belies these banal "great machine" or "great pilot" narratives. Studying the Air Force's three primary air combat aircraft that collectively define this period—the F-86E Sabre of the 1950s, the F-4C Phantom II of the '60s, and the F-15A Eagle of the '70s—and the actions of the men who flew them reveals a continually evolving relationship between the fighter pilot and his equipment.[39] Each of these aircraft possessed weapons systems that were a generational improvement over their immediate predecessors.[40] In each case, the pilots flying these aircraft were confronted with new gadgets and increasing automation designed to make them more lethal in air-to-air combat and, at the same time, less dependent on the innate, physical skills that had defined successful fighter pilots in decades past. Occasionally, the fighter pilots expressed their concern that the new equipment would transform air combat from the celebrated close-range aerial duel between two lone chevaliers into an assassin's duty whose outcome depended solely on missiles launched according to procedures-driven checklists. Lured into the service with notions of becoming "America's Knights of the Sky" (figure 1.2),

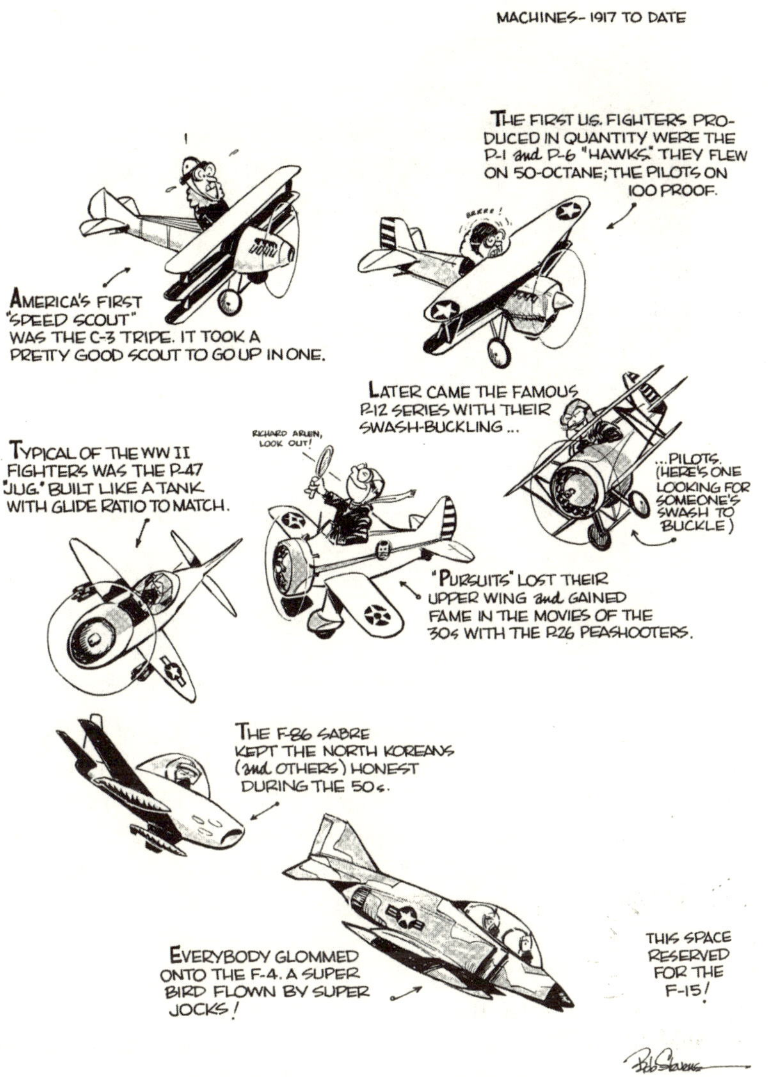

FIGURE 1.1. The "Natural" Progression of Fighter Aircraft. Bob Stevens's "There I Was" cartoon captured the pilots' enthusiasm for their recently announced F-15 Eagle, successor to the F-4 Phantom "super bird flown by super jocks." Stevens's cartoon also reflects the common tendency to portray fighter aircraft development as a natural, almost linear progression toward ever faster- and higher-flying machines. *Air Force*, May 1973. Copyright Bob Stevens. Reprinted by permission.

FIGURE 1.2. US Air Force Recruiting Advertisement, 1954. The image of the fighter pilot as a heroic knight has been a centerpiece of the fighter pilot myth since its inception during the First World War. Originally constructed as a powerful retort to the mass, mechanistic slaughter occurring in the trenches, the claimed ancestral heritage linking the solitary aviator with the "Knights of Arthur's train" was frequently pushed to the forefront, as in this advertisement. *Air Force,* May 1954. USAF.

the fighter pilots worried that their new equipment would transform them into just another "black box" in their aircraft (figure 1.3). Once seated in the confines of their new cockpits, the fighter pilots were forced to adapt their relationships with their equipment. They also had to adjust their relationships with their fellow pilots. As they did, they tried to make both sets of relationships—personal and machine—conform to preconceived notions of what it meant to be a fighter pilot. A fascinating coevolution of human and machine emerged over the ensuing decades as the skills, tasks, and responsibilities of each were renegotiated and reallocated according to the changing requirements of air combat and the powerful myth of the fighter pilot.

For example, the F-86E Sabre's new A-1C(M) radar-ranging gunsight would, Air Force engineers boasted, finally allow "the pilot to direct his full attention to the selected target," forever eliminating the need for him to apply "Kentucky windage" in

FIGURE 1.3. Technology Threatens the Pilot. The automated technologies and their associated "black boxes" that were being crammed into new aircraft in the late 1950s threatened to rob the fighter pilot of his skill and historic professional identity. In contrast to the imagery of the fighter pilot as heroic knight, this cartoon, which appeared in the fighter pilots' professional journal, *Fighter Weapons Newsletter* (September 1959), reduced the historically celebrated warrior to a small, decrepit troglodyte occupying the very black box designed to replace him. MSFRIC.

the air, a skill historically gained only through extensive aerial gunnery practice.[41] A few years later in 1955, engineers working on the US Navy's F-4 Phantom II, which was later incorporated into the Air Force fleet, sought to eliminate the need for the pilot to aim his weapons altogether, electing to remove the aircraft's guns and instead transfer that requirement to the Sparrow radar-guided air-to-air missile.[42] A few more years after that, the engineers tasked with designing the F-15 Eagle invoked additional advanced automation and mechanization in the form of HOTAS (hands on throttle and stick) and a HUD, which in combination would allow the pilot to control most of his aircraft's radar and weapons functions without ever needing to take his hands off the aircraft controls or having to look inside the cockpit.[43] This arrangement, the Eagle engineers proclaimed, "allows the pilot to 'keep his head out of the cockpit' and frees him to press the attack during critical phases of combat," a curious echo of the prediction that greeted the fielding of the A-1C(M) in the F-86E Sabre decades earlier.[44]

Almost all fighter pilots clamored for the opportunity to fly one of these remarkable new machines.[45] Of his Sabre, Evans reflected, "I don't suppose there was ever a fighter that was such a pure delight to escape with into the wild blue as the F-86 Sabre. . . . Gazing back along those first swept-wings, one could imagine all sorts of exciting possibilities."[46] While few Phantom pilots would ever use such glowing terms to describe the F-4's flight characteristics—it was a common joke that "the F-4 represent[ed] a triumph of thrust over aerodynamics"—Phantom pilots nevertheless quickly developed their own special bond with their steed.[47] One Phantom pilot described the F-4 as "the world's greatest tactical toy, filled with all kinds of 'gee charlie whiz bangs' guaranteed to make any fighter pilot go into a state of terminal ecstasy."[48] When the F-15 finally arrived, pilots could once again boast of flying a beautiful airplane, and an unbelievably powerful one, too. Describing his transition from the Phantom to the Eagle, one early F-15 pilot likened it to "going from an F-150 pickup truck to a Corvette."[49] Another exclaimed, "To say that the Eagle's afterburner takeoff was impressive would be an understatement of the highest order. . . . What we have here is truly a Superfighter."[50]

Paradoxically, these improvements in aircraft and weapons functionality often conspired to diminish the historic persona of the fighter pilot. Take, for example, the importance of better-than-perfect vision. In World War II, Chuck Yeager earned a reputation for seeing the enemy long before anyone else, and he excelled in the skies over Europe because of it.[51] But by the time of Desert Storm, in the words of one F-15 aviator, it didn't matter "how thick the guy's glasses were."[52] Now, the measure of a good pilot wasn't so much who could spy distant specks on the horizon but rather who could look at a four-inch-square radar scope and most quickly and accurately transform its abstract symbology into a three-dimensional thirty-mile mental "picture" of the airspace ahead. Similarly, novel aerodynamic designs and new control systems yielded aircraft that possessed better, smoother flight characteristics. Whereas the Phantom was notorious for its sudden, violent, and often deadly spins, even test pilots wringing out the first F-15s had a difficult time forcing them out of control.[53] Consequently, the historic significance once ascribed to a pilot's "stick and rudder" skills also declined.

Despite these changes to the aircraft weaponry and the skills needed to employ them, the universal measure of a fighter pilot's performance remained the same—aerial victory, aka "the kill," often annotated on the side of the victor's fuselage by a small star, enemy flag, or enemy aircraft silhouette. Even in Vietnam, when Phantom aircrews were thrust into combat and repeatedly tasked to drop bombs on targets of questionable military value, the chance to tangle with a MiG represented a rare opportunity to dispense with the "rubbish" of Washington-dictated bombing targets and become once again a fighter pilot in the spirit of the great Red Baron. As Brigadier General Robin Olds, a World War II fighter ace and later a storied Phantom pilot, explained, air combat was "the most exciting. You can get your adrenaline pumping in air-to-air like you would not believe. . . . It is a one-for-one thing, really. . . . You feel very personal about that son of a bitch that is trying to kill you. . . . It is every fighter pilot's dream."[54]

There were practical challenges, too, that accompanied the new equipment. For example, the traditional tools used to document pilot performance in the air were rendered ineffective by the new weapons systems.[55] In World War II and Korea, gun cameras mounted in an aircraft's nose were used to record victory, capturing images of an enemy aircraft disintegrating ahead under a steady stream of straight-flying bullets.[56] During training missions, the same gun camera could be used to document the pilot's performance as he maneuvered into an offensive position and gunned down an aerially towed target banner.[57] With the advent of long-range missiles that could turn and seek out the enemy, though, kills rarely occurred directly out in front of the attacking fighter and in view of the fixed-position gun camera. Moreover, the complex missiles were becoming increasingly expensive, so firing them just for practice became infeasible. And, with the missiles often suffering from their own internal catastrophic failures independent of pilot action, their ability to hit a target successfully was beginning to lose its relevance as a fair measure of pilot performance.

Because of the preeminence of the radar missile and the lack of an internal gun in the F-4, engineers elected to replace the traditional forward-looking gun camera with a camera that instead recorded the Phantom's radar display. To augment the images of the radar screen and capture important radio chatter, Phantom pilots began carrying personal portable cassette recorders in the cockpit.[58] The forward-facing gun camera was reintroduced in the F-15, but rather than mount the camera in the nose of the aircraft, engineers positioned it inside the cockpit and had it take pictures through the pilot's HUD display. This allowed the F-15 pilot to record both the image of a glorious fireball and the weapons parameters, as displayed in the HUD, that yielded the victory. Pilots still needed to carry portable voice recorders, though, to capture the radio coordination that was becoming ever more important in air combat.[59] An additional revolutionary training tool, ACMI (air combat maneuvering instrumentation), was developed to help evaluate pilot performance in advanced aircraft like the F-15. Using ACMI's real-time data stream, which relayed aircraft position and weapons employment parameters, success in air combat was no longer measured with grainy images of a blip on a radar screen but with schematic coffins appearing during a computer-generated recording of an aerial engagement and an accompanying godlike voice announcing over the radio, "Blue 4 [or the appropriate call sign], you're dead."[60]

As the human-machine relationships evolved, so did the fighter pilots' interpersonal relationships. This transformation was perhaps most dramatic in the F-4 Phantom. When the engineers designing the Phantom elected to transition to all-missile ordnance, they decided it would be unwise to saddle an individual pilot with the cumbersome tasks required to operate the aircraft's complex radar and avionics.[61] Hence, the Phantom would forever be a two-seat fighter: the pilot in the front seat would worry about flying the plane and launching the missiles; the back-seat occupant, commonly referred to in Air Force circles as the "GIB" or "guy in back," would be responsible for operating the radar. Although some prior Air Force fighter planes had used a two-person crew, the mere presence of the second crewmember in the Phantom's cockpit was a serious affront to the iconic fighter pilot image. To fly a single-seater was for many fighter pilots a badge of honor that could be traced back to the legendary aces of World War I, and a few fighter pilots in the 1960s opted to fly outdated F-100 Super Sabres and F-105 Thunderchiefs rather than subject themselves to the ignominy of sharing their cockpit with another man.[62]

There were other interpersonal challenges as well. Although often portrayed as lone eagles patrolling the skies, fighter pilots since the middle of the First World War understood the advantages of flying together in a formation. Reflecting the pilots' experiences, a US Army Air Corps report from 1922 declared the lone "'ace' . . . as passé a phenomenon as the champions of Homer."[63] Indeed, before the end of the war, flight-training programs had already begun to emphasize the importance of sticking together in the air, and emerging tactical doctrine assigned specific responsibilities to the wingmen and the leaders flying together in a formation.[64] The more experienced flight leader was supposed to shoot down the enemy. The younger, less experienced wingman was there to watch and learn from the leader, to occasionally serve as bait, and to always defend the leader from attack. Survival as a wingman proved that one possessed the requisite fighter pilot skills needed to become a successful flight leader.

When Evans took to the skies as a wingman in his Sabre, his responsibilities and the position he flew were remarkably similar to those developed decades earlier. Although now rarely used as bait (the lower- and slower-flying bombers served that role), the wingmen in Korea were still expected to place themselves between their leader and an attacking enemy. However, relegating a wingman to this strictly defensive, subservient role in the Phantom or Eagle ignored the incredible offensive potential afforded by the new fire control systems and longer-range weapons. Recognizing that the historic formation roles were becoming passé as well, a handful of fighter pilots proposed eliminating the wingman's close-in "fighting wing" position and the corresponding rigid flight lead-shooter and wingman-looker assignments, but they encountered stiff resistance. Opponents argued that fighter formations would be too difficult to manage if a young, inexperienced wingman was granted greater offensive autonomy.[65] Left unsaid was that the new formations would also demand a greater level of mutual dependence between the flight lead and his wingman, potentially unraveling fighter aviation's forty-year-old social structure.[66] However, a pair of congressionally mandated joint Air Force and Navy flight tests of air combat weapons and tactics—the Air Intercept Missile Evaluation (AIMVAL) and the Air Combat Evaluation (ACEVAL)—poignantly illustrated the absolute necessity of accommodating the younger wingmen's offensive potential, and the fighter pilot community acceded. Aerial victories, and the glory that accompanied them, could no longer purposefully

FIGURE 1.4. The Brains behind the Technology. Confronted with ever-increasing technology, fighter pilots worried they were at risk of being consumed by esoteric and pedantic topics like gigahertz radar frequencies, and they routinely mocked the individual scientists and engineers who worked on these unexciting concepts. Still, the pilots occasionally devoted a small section of their professional journal to the technical discussions. This cartoon appeared in an article in the January 1954 edition of *Fighter Gunnery Newsletter* that discussed maintenance procedures for the APG-30 radar. MSFRIC.

be reserved for the more senior flight lead. The formation and the "sorting" plan used by Sveden and Prather when they targeted the pair of Iraqi fighter jets was one manifestation of this eventual shift.

Finally, there were individual, personal challenges. Each new technology that was thrust into the cockpit, along with its accompanying systems knowledge requirements, threatened to transform the fighter pilots into the "long-haired" scientists and systems managers that they had routinely mocked (figure 1.4). Evans complained during the Korean War that "a pilot today is expected to know all the complexities of every system in his bird: electric, hydraulic, fuel, oil pressurization, *ad infini-tum* [*sic*]."[67] Colonel Francis "Gabby" Gabreski, a Sabre wing commander in Korea and the top American ace from World War II's European theater, similarly grew so disenchanted with the complexity of his Sabre that he reportedly once remarked that he preferred to use "a piece of chewing gum in the windshield" as an aiming reference rather than rely on the Sabre's A-1C(M) gunsight.[68] But the technical knowledge requirements of an F-86 pilot paled in comparison to what was later demanded of F-4 and F-15 pilots. Whereas Evans's flight manual for his F-86E weighed in at about 250 total pages, of which about 13 pages were devoted specifically to the Sabre's armament equipment, to fly the F-4, pilots had to slog through a 518-page flight manual, plus an extra 406-page second volume devoted solely to the Phantom's complex armament systems.[69] Although some Phantom pilots may have tried at first, they quickly realized that they couldn't bluff their way through their manuals or checklists as their predecessors had. Going forward, a fighter pilot would have to complement his flying skills with book-smarts and intensive study.

In sum, the new air combat technologies, always fielded with the intent of aiding the fighter pilot, instead threatened not only his historically celebrated tasks and skills but also the social relationships and hierarchy that governed his life as a fighter pilot. In February 1952, *Life* alluded to the looming clash between the historic fighter pilot profession and the new and increasingly automated technologies of modern air combat. Amid this technological transformation, would the Air Force fighter pilots in the 1950s, '60s, and '70s continue to enjoy their celebrated status as innately gifted and heroic aviators, akin to the "Flying Knight" Richthofen, or would they be transformed into the scientists *Life* predicted? Moreover, if fighter pilots were now relegated to controlling an aircraft that could almost fly itself, and tasked with manipulating electronics that relied on the physics of the electromagnetic spectrum to destroy a foe, would the Sabre pilots' Korean War exaltation "Every man a tiger!" ring hollow?

Before we can begin to address these questions, however, we first need to understand and appreciate the myth of the fighter pilot that emerged from the cauldron of the Great War. The myth's portrayal of the relationship between a fighter pilot and his technology would exert a powerful influence over all successive generations of American fighter pilots.

CHAPTER 2

The Myth of the Fighter Pilot

The word flying has always been synonymous with fighting, first against prejudice, then against wind, weather and the malice of inanimate matter, then for the laurels of records of height, speed and duration and finally against a host of enemies in the air.

Pursuit Text, US Army Air Corps Training Regulation 29-229, 1929

In 1955 Major Frederick "Boots" Blesse was commander of the 3569th Combat Crew Training Squadron. Based out of Nellis Air Force Base (AFB), the self-declared "Home of the Fighter Pilot" located about fifteen miles northeast of Las Vegas, Nevada, Blesse's squadron took newly winged pilots fresh from training and in a few weeks transformed them into *fighter* pilots. With the Air Force tripling in size, the pace was hectic and the training programs jammed pack. Blesse had to educate the young pilots on the more advanced systems installed in the Air Force's front-line fighters, train them how to fly the faster and more maneuverable machines in simulated air combat, and show them how to use the aircraft's weapons to hit a slew of practice targets. Despite the pressure, Blesse excelled, both as a unit commander and as an individual fighter pilot.[1] That year, he took first place in all of the individual events at the USAF Worldwide Gunnery Meet, an unprecedented and never repeated feat.[2] He also authored what became the definitive tactics manual for jet combat. For its title, Blesse selected the mantra he coined while battling MiGs in his F-86 Sabre en route to becoming one of America's top-scoring aces of Korea: "No Guts, No Glory."[3]

Blesse was born on August 22, 1921. The only son of a successful Army doctor, he earned his "Boots" moniker for stomping around the house in the polished shoes his father wore with his uniform. When he was nine and his family stationed at Fort Leavenworth, Kansas, the young Blesse befriended a neighbor who was an Army Air Corps pilot and thereafter spent many afternoons mesmerized by tales of the pilot's aerial exploits. Blesse applied to West Point several times before he was finally accepted and ordered to report to the campus on July 1, 1942. He was elated. In his memoir, he remarked that his standard prayer as a teenager was, "Please God, let me go to West Point and let me become a great fighter pilot." Now he was finally on his way to fulfilling his "lifelong ambition." As he later acknowledged, "I had dreams of becoming an ace as long as I could remember. . . . I grew up reading *G-8 and His Battle Aces* and *Tailspin Tommy* in the funny papers, dreaming of the fighter pilot's life. 'Captain Eddie' [Rickenbacker] was my hero, now joined by World War II aces like Gabreski, Mahurin, Bong, McGuire, Thyng, and others. The picture was always in the back of my mind—fighter pilots diving, climbing, turning, finally destroying the enemy aircraft; bringing honor and glory to themselves and their country."[4]

Blesse's attachment to the legendary aces of old and the classic imagery of air combat was not unique to him. Robin Olds, an ace from World War II who would achieve renown as a fighter pilot and wing commander during Vietnam, remembered meeting Eddie Rickenbacker as a child but being "too awed to say anything."[5] Rickenbacker, the American Ace of Aces in World War I, was also an acknowledged "hero" to Walker "Bud" Mahurin, one of Blesse's influential aces from the Second World War.[6] Too young to have a personal attachment to the Great War aviators, Mike McCarthy, a Phantom pilot in Vietnam, revered "the fighter pilots of the Battle of Britain—the legendary 'Few.'"[7] Whether walking out to a P-51 during World War II, an F-86 in Korea, or an F-4 in Vietnam, these aviators, like most, carried within them a powerful personal attachment to the past warrior-aces, imagining themselves as members of the privileged "few" who had the opportunity to climb into an airplane to battle "wind, weather and the malice of inanimate matter . . . against a host of enemies in the air," hoping like Blesse to capture "honor and glory" for "themselves and their country."[8]

The powerful rhetoric and the captivating imagery associated with the earliest military aviators were, however, elements of a myth. Whereas myths often carry connotations of elaborately constructed falsities, philosopher Roland Barthes's study of myths and mythology demonstrates that today's pejorative conceptions are usually unwarranted. According to Barthes, myths are "a language" or "a type of speech." They are defined not by their type of message or even by their intended purpose but only by their construction. In a myth, an object with understood significance is presented but in an abstract or purposefully incomplete manner that implicitly separates the object from its detailed historical "richness." For Barthes, "it is this constant game of hide-and-seek between the meaning and the form [or message] which defines myth." It "is neither a lie nor a confession: it is an inflexion" that "transforms history" into something that instead seems natural.[9]

Invoking Barthes's definition, the fighter pilot myth was not an outright lie but a purposeful opaquing of historical events.[10] It consisted of three interrelated messages: that air warfare represented honorable combat fought by honorable individual warriors; that it was the unique flying skill of the pilot that dictated success; and that the tally of aerial victories recorded by the pilot and reported in hometown newspapers was an appropriate measure of his success. During the interwar years through World War II, the myth was ritualized as the US Army Air Corps (after 1941, the US Army Air Forces) studied and formalized its processes for selecting and training its fighter pilots. When successive generations of pilots took to the skies over Korea, Vietnam, and Nellis, it was this myth of the fighter pilot that provided the basis for their expectations, not the actual earlier experiences of air warfare. Stripped of its historical "richness" and "contingency," the myth of the fighter pilot remained remarkably consistent despite the pilots' ever-changing relationships with their machines and their fellow aviators.[11]

Crafting the Mythical Ace

Belying the timeless significance it would eventually assume, the *ace* distinction had surprisingly inauspicious origins. The term was first associated with a skilled pilot when a Paris newspaper in 1915 declared the recently felled Adolphe Pégoud "l'as de notre aviation" (the ace of our aviation). Two years earlier, Pégoud had been wowing crowds of two hundred thousand with his masterful displays of aerobatics; he then enlisted his aviation prowess in service to his nation and downed six enemy aircraft before his death.[12] After its initial appearance in print, the ace appellation quickly cemented itself into the French lexicon. The French Air Service thereafter decided to regulate the term's usage, reserving the ace distinction for those pilots who had scored five victories over enemy aircraft. The other warring nations followed suit.[13]

It is perhaps fitting that the ace term was manufactured by neither the military nor the pilots, but by literary publicists. As Pégoud's earlier aerial exhibitions revealed, aviation already solicited an enthusiastic popular response before the Great War, especially within France. But as historian Robert Wohl argued, it was the propagandizing efforts of the belligerents' national media that played an "indispensable role" in transforming skilled aviator-showmen like Pégoud into the war's mythical aviator-aces.[14] Peter Fritzsche and Peter Hart likewise showed how the "image of the war ace" and his aerial exploits bore little resemblance to the reality of air combat in the latter half of the First World War.[15] In the hands of skilled publicists, the aces were custom-crafted to fulfill desperate demands for worthy and inspiring national heroes. As Europe watched its sons being consumed in the trenches by the technology once considered "the surest proof of their civilization's superiority," the tales of gallant aces who seemed to abide by the time-honored principles of "honor, duty, effort, courage" provided a powerful analgesic.[16]

Pégoud may have been the first anointed with the title "l'as," but it was another French aviator's ingenuity that would establish the enduring image of the solitary, self-reliant, self-sufficient ace-pilot. Roland Garros, like Pégoud, was a sportsman-aviator before the Great War. When war erupted, Garros too enlisted in the French Air Service, but he quickly grew dissatisfied. He was assigned to fly a two-seat reconnaissance aircraft, and it was his observer, not him, who was responsible for shooting down enemy aircraft. To make matters worse, his observer wasn't hitting anything. Following one unsuccessful mission, Garros griped of the arrangement, "Our armament has shown itself to be insufficient." Thereafter, Garros, with the help of his mechanic Jules Hue, elected to get rid of his human baggage and instead mount a Hotchkiss machine gun directly in front of his seat, aligned with the aircraft fuselage. To avoid shooting off his own propeller, Garros and Hue attached a set of steel deflectors to the blades. Garros took his new contraption to the front in mid-March 1915, and within three weeks, beginning April 1, he had downed three German aircraft.[17] When news of Garros's accomplishments leaked to the French press, the nation exploded in adulation. *L'Illustration* proclaimed, "He piloted the plane he himself created, a Morane monoplane, ingeniously, lovingly adapted and perfected by him, a light and beautiful machine, similar to a lark, delicate and tiny in comparison to the hawk it had to attack, but formidably armed with a machine-gun. And Roland Garros mounted alone this machine, his child."[18]

As Garros's innovation spread, fighter pilots welcomed the new single-seat aircraft and its promise of unprecedented aerial autonomy.[19] No longer would a pilot have to weigh down his "lark" with another body; now he could rise up into the sky and snatch glory for himself. "I believe in the saying that 'the strong man is mightiest alone,'" German ace Oswald Boelcke exuberantly wrote his parents on July 10, 1915, after learning he would begin flying the new single-seat Fokker E. "I have attained my ideal with this single-seater; now I can be pilot, observer and fighter all in one." A few days later, Boelcke was still gushing: "My little single-seater possesses the advantage of giving me complete independence; I can fly when, where, how long and how I will."[20]

Unlike the masses slated for slaughter in failed assaults on opposing trenches, individualism became the hallmark of the aviators' conquests in the skies. Their aerial triumphs were readily quantifiable, individually assignable, and, with some artistic license, always captivating. The pilots' own serialized descriptions of their battles stoked the image of aerial combat as a "game—an amazing game," in the words of one French aviator. "A game of adventure, of countless thrills, of soul-stirring excitement, a game in which courage, daring, resource, determination, skill, and intelligence achieve honor in life, or if the fates so decree, glory in death."[21] German ace Ernst Udet remembered fondly of an engagement in which he and his foe paused long enough to wave at each other, "Suddenly, I had the strange feeling that I wasn't confronting an opponent but practicing turns with a comrade."[22] Additional tales of squadrons on both sides of the front raising glasses and dropping wreaths to honor fallen foes contributed to the notion that, despite the barbarity transpiring on the ground, "an honorable war among men" was being fought in the sky. Aerial combat was, in the words of German flyer Erwin Böhme, a contest that pitted "man against man, with equal weapons and equal chances."[23]

Soon the names of the young lieutenant flyers began appearing besides, if not above, those of the great generals.[24] In France, in addition to Garros, Georges Guynemer became a prime beneficiary of the journalists' exaltations. Despite being born into a "comfortable, country-squire life" and destined for studies at the prestigious Ecole-Polytéchnique, the young nineteen-year-old Guynemer tried on several occasions to enlist in the French army, but each time he was turned away due to his "feeble constitution." Through dogged persistence, and perhaps some help from his influential father, Guynemer eventually secured an opportunity to begin flight training. Even then, he was initially ridiculed as "un petit jeune homme" and a "filette." Once turned loose over the front, however, Guynemer earned a reputation as a fearless aviator, often flying several missions each day and typically scoring his victories from near-point-blank range. He was lauded as a "pilot of the highest quality, a model of dedication and courage."[25] When Guynemer mysteriously disappeared during a mission in September 1917—heavy artillery barrages in the area delayed discovery of the fallen pilot and his wrecked plane for several weeks—French journalists constructed a tale in which Guynemer was plucked from his aircraft by angels. "*The ace of aces* one day flew so high in combat that he never came back to earth," wrote one in *L'Illustration*.[26]

Across the lines in Germany, newspapers eagerly kept tally as Boelcke and Max Immelmann jockeyed for position as their nation's leading scorer, and the country's High Command began recounting the aces' individual victories in their wartime communiques. Showered with honors and decorations, the two early aviators were portrayed in the German media as "idealist[s], pure in soul and body, fearless and disciplined," each ready to "sacrifice his life willingly, gladly, even lightheartedly to the cause of the *Vaterland*."[27] When Richthofen rose to replace the two fallen aviators, he too was molded to reflect national virtue, transformed by the state's publicists into a model, humble servant of the state.[28]

Britain and America were initially more circumspect in their adulation of their ace flyers. According to Hart, the British "preferred to maintain the illusion" that their combatants "were all part of a team with no 'star turns.'"[29] The Americans harbored a similar sentiment, made explicit in a circular distributed after the war: "The US Air Service does not use the title 'Ace.' . . . It is not the policy of the Air Service to glorify one particular branch of aeronautics, aviation, or aerostation at the expense of another."[30] But, try as the two nations might, there was no avoiding the ace frenzy. Captain Albert Ball, Britain's leading flyer at the time, was mauled by the press when he returned to England in October 1916. Declared by his town's mayor to be "the greatest living expert in aerial warfare," Ball was reportedly a model of modesty at the ceremony; "Conceit was not in his nature," Ball's biographers noted in their 1918 text.[31]

For the Americans, it was Captain Eddie Rickenbacker who stole the nation's attention. A rabble-rousing youth from Columbus, Ohio, Rickenbacker exhibited from an early age an uncanny ability to court and then skirt death. When he was eight, he along with the other members of the local Horsehead Gang went to the nearby gravel mine to race steel mine cars down the slope. When a car tipped, Rickenbacker was nearly crushed, and his leg sliced open to the bone. Following his father's untimely death a few years later, Rickenbacker was forced to temper his reckless behavior with honest work. After a smattering of odd jobs, he became a skilled automobile mechanic, but his thirst for adventure never disappeared. Within a few years, he was racing speedsters at Daytona and in the Indianapolis 500.[32]

When America entered the war in 1917, Rickenbacker had already fallen victim to the captivating ace narrative, and he hatched a plan to recruit several of his fellow race car drivers for "aviation work." He would later comment, "The excitement of automobile racing did not compare with what I knew must come with aeroplane fighting in France." Technically too old to become a pilot—he was then twenty-seven—Rickenbacker set sail with the American doughboys in May 1917 destined to be a chauffeur for General "Black Jack" Pershing. Later assigned to drive for the American airpower advocate General William "Billy" Mitchell, Rickenbacker eventually scored his coveted aviation assignment. In March 1918, he was assigned to the newly formed 94th Hat-in-the-Ring Squadron under the leadership of Major Raoul Lufbery. Rickenbacker would go on to succeed Lufbery as commander of the 94th and eventually became America's Ace of Aces with twenty-six victories.[33]

In his immediate postwar memoir, *Fighting the Flying Circus*, ghost-written by aviation enthusiast Laurence La Tourette Driggs, Rickenbacker's sense of derring-do was paired with a strong devotion to the principle of "fair play" and a keen sense of responsibility for his men's well-being, honorable traits appropriately reflective of American ideals. In the book, for example, Rickenbacker boasts of his resolution to "never shoot at a Hun who is at a disadvantage, regardless of what he would do if he were in my position." Moreover, Rickenbacker asserted that, "with American flyers the war has always been more or less a sporting proposition and the desire for fair play."[34] Driggs's foreword to the text called attention to the ace's "feeling of responsibility that every true leader must know. He is frankly scared over the additional risks that leadership must force upon him. Yet he carries on, assuming these additional risks, which in truth not only limit his private successes but constantly threaten to sacrifice his life for a companion."[35] These were powerful lessons for the American republic and its citizens, and Rickenbacker's book, as described by his biographer W. David Lewis, was hailed as "an inspiration to red-blooded Americans" and snatched up by "aficionados of aerial warfare" who took Driggs's embellishments of the ace's carefully crafted stories of air combat as "gospel."[36] The fact that Rickenbacker was one of the few great aces to survive combat, thereby granting him the unique opportunity to continue accumulating accolades and successes postwar, contributed to his enduring magnetism in American culture.[37]

The iconic fighter pilot persona that emerged from the Great War combined elements from each of these legendary aces: Garros's ingenuity and the status afforded to the self-reliant, single-seat fighter pilot; despite his timid appearance, the dogged persistence and fierce tenacity of Guynemer; the sportsman heritage and selflessness of the Germans Immelmann, Boelcke, and Richthofen; Ball's modesty; and Rickenbacker's daring adventurism paired with a sense of teamwork and fair play. In the hands of the publicists that celebrated the flyers' victories and shrouded their defeats in mystery, the aces were transformed into an honorable "brotherhood of knights" that exerted an irresistible influence over those in the trenches and on the home front. The mere sight of an ace's aircraft was said to produce among the soldiers "a magnificent surge of enthusiasm," while at home, the aces' "adventures thrill the bosoms of countless thousands whose duties, if not their inclinations, shackle them to *terra firma*."[38] There was simply something captivating about "the dizzy height, the deadly plunge into space, and the plain simplicity of plane against plane, of aerial squadron against aerial squadron."[39] In an age of mechanistic warfare that was relentlessly consuming the individual, the ace flyer was hailed as living proof that man could indeed reassert control over his machinery and achieve glory and honor in the process.[40]

Revisiting the History

The myth of the fighter pilot proved inspiring, but as Barthes observed of mythologies, it purposefully stripped away much of the historical richness in favor of its message. The pilots of the First World War recognized this. Guynemer divulged the inconsistency in a letter to his father, penned while the ace lay recuperating from a recent wounding in battle, "I've read while in bed that the crowd gave me an ovation in Paris. It's a result of ubiquity. Modern science really does wonderful things, the journalists also."[41] Despite what the popular narrative claimed, aerial combat was no mere "sport," and it was rarely conducted according to some honorable, agreed-upon aviator code. Moreover, the aces' tallies that the newspapers regularly fawned over were not appropriate measures of individual pilot skill: new formation tactics, advancements in aircraft technology, and the inherent inaccuracy in the pilots' reporting of aerial combat all misrepresented the individual ace-flyer's significance.

An Unfair and Deadly Game

Every pilot knew that despite what was said about Guynemer, there was no angel ready to pluck them from a tumbling aircraft or spare them from being consumed in a fiery inferno. For the average British fighter pilot on the Western Front, life expectancy ran only about ten weeks. They may not have been toiling in the trenches, but the pilots' lives were nevertheless infused with macabre reminders of ever-present danger and death.[42] Moreover, they quickly learned that skill alone could not ensure their survival; chance played a large role, too. The pilots were haunted not only by the prospect of having their wings clipped by some lurking enemy fighter plane but also by the rash of accidents, mechanical failures, and ground fire that likewise extracted a terrible toll.[43] On the morning of July 9, 1918, Britain's leading ace, Captain James McCudden, took off in his brand-new SE5a, circled the aerodrome, and then proceeded to plummet nose first into the ground from two hundred feet, the victim of an apparently faulty carburetor. The ground crews and off-duty pilots that ran to the wreckage were greeted with a typically grotesque scene: the ace's shattered body lay crumpled on the ground next to a broken wing. At least he hadn't burned to death, the onlookers told themselves.[44] Less than a month later, another of Britain's top scorers, Captain Edward "Mick" Mannock, was downed by bullets from a German machine gun post, crashing to the ground "in a burst of flame."[45] The fact that legendary pilots like McCudden or Mannock could be felled not in aerial combat but through hidden flaws in their aircraft or lucky fools firing up from the ground had a disquieting effect on all pilots. Even the great Richthofen became increasingly distraught as he witnessed more and more of his friends succumb to the enemy and to chance.[46]

Moreover, if aerial combat was a sport, it was most definitely a blood sport characterized not by munificence but by ruthlessness. In the words of Cecil Lewis, a British pilot during the war, "The Angel of Death is less callous, aloof, and implacable than a fighting pilot when he dives [on a foe]."[47] Udet may have fondly remembered exchanging waves with an adversary during battle, but such "displays of charity," Fritzsche wrote, "were exceedingly rare."[48] Even Rickenbacker, America's champion of fair play, later acceded, "Fighting in the air is not a sport. It is scientific murder."[49] The goal was to down enemy airplanes, and it didn't matter what type of craft or the skill level of the opposing pilot. In this "survival of the fittest" contest, as described in a retrospective 1929 Army Air Corps text, the most prolific aces preyed on lumbering aircraft piloted by the distracted and the inexperienced.[50]

Pilots learned that the best results were obtained when they successfully snuck-up behind an unsuspecting pilot and unleashed machine gun fury from close range. Boelcke famously explained his success to his young protégé, Richthofen: "I fly close to my man, aim well, and then, of course, down he falls."[51] A June 1917 account of air combat by French ace Albert Deullin explained that "all the most successful French and English pilots" preferred attacking "at under one hundred and fifty feet." Explaining how to get to that close position, Deullin in a later report called attention to the single-seater's unique ability to "play tricks, hide himself more easily in the sun, the mist or the clouds, and take advantage of the blind spots of his opponent to carry out crushing attacks."[52] Rickenbacker even shut down his own engine on one occasion to avoid alerting his prey as he dove down before "pull[ing] both triggers for a long burst. [The enemy pilot] made a sudden attempt to pull away, but my bullets were already ripping through his fusilage [*sic*]."[53] Pilots on both sides of the lines quickly learned the value of the adage, "attack ahead while looking to the rear."[54]

Within the popular fighter pilot narrative, these victories were cast as the destruction of hostile aircraft, not the killing of enemy pilots.[55] The pilots in the cockpits, however, had a unique opportunity to see both sides. Sometimes they subscribed to the semisterile perspective. One British aviator explained that he considered "the German pilot and his aeroplane" an "impersonal thing to be destroyed before it destroyed you. . . . I did not continue to think of the enemy pilot as a man."[56] Boelcke, trying to allay his mother's fears that he was becoming too calloused in his reports from the front, assured her in a letter dated September 16, 1917: "We have nothing against the individual [pilot]; we only fight to prevent him flying against us." That was why, Boelcke reasoned, "it only counts as one victory when two inmates are killed."[57]

But other times, the Janus-faced nature of aerial victory was divulged by the pilots. Mannock earned the admiration of the young pilots in 85 Squadron because "he wasn't interested in just killing [the Huns] himself. He wanted a lot of them killed, and he trained us how to do it."[58] Curiously contrasting his praise for Americans' fair play in air combat, Rickenbacker offered a gushing description of his friend Lieutenant Doug Campbell's fifth victory in May 1918: "Lieutenant Campbell first tried a diving attack, from above and behind. . . . He had an excellent chance of killing the observer with the first burst. . . . But no such easy victory awaited him." The battle continued for fifteen minutes before Campbell "saw a curious sight. The observer was standing proudly upright and his arms were folded! From the edge of his cockpit the empty ammunition belt floated overboard and flapped in the wind. He had indeed exhausted his ammunition and now stood awaiting his doom. . . . As Doug said later, he was so impressed with the bravery of the action that he felt he could not continue the combat against an unarmed enemy. . . . Upon second thought Lieutenant Campbell realized this was not a game in which he was engaged. It was war." Campbell proceeded to shoot the enemy crew out of the sky. They were both killed, and, Rickenbacker recalled, "Douglas Campbell that night received the heartiest congratulations from all the boys in the squadron as the first American ace."[59] Even shifting attention from the machine to the opponent pilot—an individual who likely shared a love of aviation and who, under different circumstances, might have been a lasting friend—could not derail the fighter pilot's lust for victory.[60]

And what of the "fighter pilots diving, climbing, turning" that so enchanted the dreams of later youths like Blesse? They may have been there early in the war, but as more and more aviators were thrust into battle, and as tactics premised on surprise proved successful, the importance of aerobatics for anything other than simply gaining familiarity and confidence in a new aircraft quickly waned.[61] Richthofen professed only disdain for those who chose to engage in a swirling dogfight, writing, "We need no aerial acrobats; rather, daredevils."[62] Although large swirling maelstroms of fighters might still occur, they, like Udet's observed acts of charity, were becoming a rarity. Another British pilot reported, "The days of dogfights where machines rush at one another and circle around each other, firing all the while, seem to be gone."[63]

FIGURE 2.1. Fokker Aircraft Advertisement, World War I. Appearing in the German periodical *Motor* in mid-1917, this advertisement for Fokker aircraft uniquely portrays the human-machine codependence that emerged during World War I. The Blue Max medal in the top left corner is itself ambiguous: it could signal the glory and fame due the human pilot for his pluck in battle or for his courage mastering the machine. Library of Congress.

Thus, despite the popular portrayals, most air combat over the Western Front was not an aerobatic sport arranged between friendly competitors. It was rather "the exact antithesis of the 'sporting attitude,'" in the words of one American pilot: "an attempt to figure out a system whereby the enemy has no chance."[64] And that *system* needed more than just a courageous and gifted pilot; it relied on his marriage to a set of tactics and a powerful machine adequately armed. One image in particular captured this reality, a wartime advertisement for Fokker aircraft in which the man and the machine merge at the gunsight, becoming almost indistinguishable, while a Blue Max medal beckons from above (figure 2.1).

Man and Machine (and Tactics)

No doubt individual skills like superior airmanship could prove valuable in combat. Powerful eyesight and the ability to detect "just a twinkle of the sun on a bit of metal" were also cited by pilots as skills "of paramount importance" that would extend their longevity and help bolster their tallies of kills.[65] Nevertheless, while a pilot possessing the aerial sense of Richthofen or McCudden or the visual acuity of Mannock might do well patrolling alone, his chances of survival and success were much higher if he instead elected to roam with others.[66]

Early experiments at coordinated action among fighter aircraft focused simply on setting traps: an inexperienced pilot was told to fly slightly below and in front of his leader, who promised to dive down and "make short work" of any "unsuspecting enemy" that took the "bait."[67] However, with the emergence of the Germans' offensively oriented Jagdstaffeln under the leadership of Boelcke, more advanced fighter tactics emerged. By the time of Richthofen, McCudden, and Mannock, the lone patrolling fighter on both sides of the lines had become extinct. All, including the legendary aces, now prowled in "triangular" formations of varying size.[68] Upon finding an enemy aircraft, the group of planes would descend on the unlucky foe, as one Englishman analogized, "like a pack of bloodthirsty foxhounds who had caught up with their prey."[69] The American pilots likened the experience not to a duel but to "gang fighting."[70]

It wasn't pure bedlam, though. Rules and roles were quickly established to maximize the formation's efficacy and efficiency. New pilots were paired with those more experienced, which provided the new recruits with an opportunity to learn firsthand "certain essentials" of air combat.[71] The arrangement was predicated on maintaining strict discipline within the formation, however, a quality antithetical to the classic portrayal of the daring flyer. Boelcke addressed the paradox explicitly in his *Dicta*, declaring: "Foolish acts of bravery only bring death. The Jasta must fight as a unit with close teamwork between all pilots."[72] A standing order within British major Sholto Douglas's unit declared, "No pilot was on any account to leave the formation, even to take on what looked like an easy opportunity to shoot down an enemy aircraft."[73] The Americans adopted a similarly stern approach, explaining, "It is the patrol leader only who should have the thinking part until contact with the enemy is gained. Before this time the pilot's place is to follow."[74]

The initial attack was highly scripted. The leader signaled his decision to pounce by wiggling his wings. Then, adopting language usually reserved for descriptions of warfare in the trenches below, each accompanying pilot immediately assumed "his part as cog in the prearranged scheme."[75] As the leader dove to attack, the accompanying pilots all followed him, protecting his rear.[76] Douglas later described the rationale: "Being the most experienced of all the pilots, the leader was the most capable of shooting down the enemy quickly and effectively; and with the squadron behind him and acting as a buffer against any attack from the rear he could afford to concentrate all his powers on the destruction of the enemy without having to peer all the time over his shoulder."[77] Thus, for Douglas, the leader provided the offensive capability; the accompanying pilots were there for defensive purposes. This particular model of formation air attack would survive for decades.

Although all adhered to the principle, not all enforced Douglas's division of labor as strictly. The Frenchman Deullin offered his pilots slightly more leeway, alerting them to "some splendid opportunities" that may arise for the "protector" who doesn't follow his leader "too blindly or too closely."[78] The Americans also offered more autonomy to their younger pilots once engaged in the battle: "The leader opened the attack," one pilot recalled, "but from there on, the individual acted as he thought best." Still, the message was clear: "A pilot is an asset just so long as he remains part of [the formation] unit."[79]

These well-crafted tactics that linked otherwise isolated pilots to one another facilitated greater effectiveness during both attack and defense. But the tactics, as well as any individual's unique flying skill, were of little use if the pilots were not flying capable aircraft. As one American report noted after the war, "No matter how stout of heart a pilot may be, he is at the mercy of the plane he flies; all his skill and courage avails nothing should his motor stop."[80] At the outset of war, the best fighter planes (produced by the French) were powered by a 70- to 85-hp engine, reached speeds of seventy miles per hour, and possessed only the armament that the pilot could carry aboard: carbines, revolvers, grenades, and occasionally bricks (figure 2.2). By early 1918, a 185-hp BMW 3a engine pushed the German Fokker D.VII to speeds in excess of 120 mph, and the pilot had two fast-firing machine guns mounted directly in front of him (figure 2.3). In a span of just over three years, aircraft performance had nearly doubled.[81]

For the individual pilot, the quickening pace of aviation development translated into a wartime experience spent trying to master several distinct aircraft, all possessing different flight characteristics and unique quirks. For example, those who took to the skies in the famed British Sopwith Camel learned quickly that it "was a very tricky aeroplane to fly."[82] Still, given the choice between continuing to fly a plane one was already comfortable with and wringing out a new, unproven, and potentially dangerous aircraft, most pilots opted for the uncertain promise of the latter.[83]

Even then, the improvements in engine horsepower and aircraft performance that made it easier for a pilot to position himself behind his foe were for naught unless that pilot could then bring down his prey with a stream of well-placed bullets. After all, "the pursuit airplane," one American postwar report noted poignantly, "is a flying machine gun" with one mission—"to destroy any type of hostile aircraft encountered."[84] For Richthofen, a "well-firing machine gun" was "better than a smooth-running engine," and becoming a "master" of the gun was the quintessential fighter pilot respon-

FIGURE 2.2. French Morane-Saulnier Type L, 1915. One of the first aerial combatants, the French-produced Morane-Saulnier was described by the press as a "beautiful machine, similar to a Lark." Roland Garros modified this type of aircraft for single-seat aerial combat by attaching steel plates to the propeller blades to deflect bullets fired from his fixed, forward-facing machine gun. The aircraft depicted here was reportedly captured by the Germans. Bain Collection, Library of Congress.

sibility.[85] It could be a challenge. The guns frequently misfired and jammed due to the jarring flight conditions and the pilots' short, frequent firing bursts. Pilots were therefore strongly encouraged to learn how to repair their machine guns quickly, even while still airborne. Those that "neglected this essential principle," Deullin explained, "always regretted it."[86]

For most of the war, the pilots' preferred armament was a pair of .30-caliber machine guns with a high rate of fire. There was some variation, though. For example, on the SE5 the British elected to mount a lighter-weight Lewis machine gun on the upper wing and a single Vickers machine gun on the top of the fuselage. A trigger on the control stick allowed the pilot to fire both guns.[87] Other aircraft designs experimented with incorporating more machine guns or larger-caliber weapons, but the added weight of the extra or larger guns severely degraded aircraft performance with no appreciable firepower benefit. The Americans eventually concluded that it was not the size of the bullet but the "maximum number of effective rounds per second" that proved of "ultimate" significance.[88]

As the designers toyed with varying gun arrangements, they also searched for new tools to improve the pilots' "appalling" aim.[89] One was a telescopic glass sight, which

FIGURE 2.3. German Fokker D.VII. By early 1918, the "delicate and tiny" monoplanes like Garros's Morane-Saulnier had been replaced with larger, more stout machines, such as Germany's Fokker D.VII. This fast, high-powered plane could reportedly "make a good pilot out of mediocre material." This photograph was taken after the war in 1919; hence the presence of the American airman. US National Air and Space Museum.

Deullin labeled a "Chretien collinateur." Approximately twenty-eight inches long and one-and-a-half inches in diameter, stuffed with "a series of lenses and prisms," the sight promised to help pilots estimate the amount of lead fire required to score a hit against a target moving at various angles and airspeeds. However, on the basis of Deullin's nearly page-long, tortured attempt at describing the sight's operation, it is doubtful that it was ever used to its full potential. Deullin's assessment of the sight, buried within his description, was telling: "He who spends too much time trying to work out the proper correction will find the chance gone. . . . Most pilots rely on their tracer bullets."[90] An American report likewise dismissed the utility of the telescopic sight, noting that it frequently fogged over due to moisture within the tube and that its use demanded "too much attention" and "encouraged many pilots to open fire at too great a range."[91]

Far easier to understand and use in the heat of battle was the classic open ring-and-bead (or ring-and-post) sight, seen in the popular Fokker advertisement (figure 2.1). Akin to aiming a rifle, the pilot steered his plane first to align the target under the far bead and then to center both target and far-aiming reference in the center of the nearer ring. There were challenges associated with using the simple sights: any head movement by the pilot would yield different aim points, and the pilot also had to constantly shift his eye focus from the distant target to the near sight and back again.[92] But despite its inaccuracies, the sight had the indisputable advantages, according to one Army Air Corps text, of being "simple to construct, install and adjust, [and] fairly sturdy and reliable." Moreover, with most battles taking place inside of one hundred yards, and with pilots instructed to move closer if possible, the sight didn't need to be that precise.[93] To help them judge when to open fire, the same text noted that many pilots carved "a series of notches across the top of the fuselage at right angles to the direction of flight" to serve as a "crude but effective range finder"; the spacing between the notches corresponded to "the wing spread of the type of hostile airplane most frequently encountered . . . at varying distances."[94]

Once in range and after commencing firing, pilots were instructed to adjust their aim by steering their aircraft to achieve "a good sprinkling" of bullets over the target in hopes that one might prove lethal.[95] Some aviators were obviously better than others. One pilot commented, "I have seen pilots 'sitting on the tail' of enemy machines, and only a few feet off, fire away all their ammunition, and still the Hun tootles along unhurt. Then again, I have seen a careful pilot fire only about 10 shots—and down goes his opponent."[96]

Questionable Glory

The combination of "gang fighting" tactics and the preferred method of spraying bullets at a target hoping for a lucky shot produced a situation ripe for confusion when it came to awarding credit for a victory. As a consequence, the perfect objectivity of the aerial victory, a key element of the fighter pilot myth that so entranced the populace, was illusory. On the surface, each nation attempted to protect the sanctity of the aerial kill by requiring confirmation by a source other than the claiming pilot, but Hart concluded that the chaotic nature of mass air combat meant that all pilots, "one way or another, over-claimed."[97] In truth, there was little incentive to guard against inflated totals. In an environment where numbers of confirmed kills became the measure of the fighter pilot, and the fighter pilot was in turn seen as a measure of the nation, the immense pressure to post superior numbers wasn't limited to just the individual aviator. For example, when Mannock was killed in July 1918, he had amassed only sixty-one victories before Royal Air Force leaders elected to posthumously award him credit for an additional twelve kills. The opportune tweaking pushed Mannock's total to one more than the Canadian major "Billy" Bishop, thereby ensuring that an Englishman would go down in history as Great Britain's Ace of Aces.[98]

The attention devoted to recording victories and achieving ace status significantly distorted the other important contributions of fighter pilots, as well as those of their brethren who were flying reconnaissance and bomber aircraft.[99] There was more to the air war then battling other planes. For example, in April 1918, as the Germans were launching their latest spring offensive, Allied fighter pilots were told by their supreme commander, Ferdinand Foch, that their "first duty . . . is to assist the troops on the ground, by incessant attacks, with bombs and machine guns, on columns, concentrations or bivouacs. Air fighting is not to be sought except so far as is necessary for the fulfillment of this duty."[100] Even before Foch's message, though, fighter pilots had already begun flying low over the trenches to drop small bombs and strafe enemy troops and supplies.

They did it, but they did not like it. Many pilots considered the ground attack mission more dangerous, partly because they had less control over their own survival. During "these low-level missions," a pilot revealed, "one became a mere target to be shot at all the time and it was a matter of sheer luck whether one got hit or not." Consequently, he further explained, several pilots simply tried to empty their ammunition belts "as quickly as possible," preferably on a single diving pass, so they could then return home.[101] There was an additional reason why the pilots despised the mission. Another aviator confided: "If a pilot can shoot down several enemy aircraft observed to crash, he gets personal credit and possibly a decoration. If he does equally or more valuable work for a month by shooting up transport vehicles, troops or aeroplane sheds, his feats tend to be included in the general results of the flight or squadron. . . . The individual has no confirmed and positive result to show. This 'individualism' is not talked about; it may be petty, unmilitary and regrettable, but it exists."[102] For the fighter pilots, to fly and fight in the air brought fame, but to gun down an enemy soldier in a trench brought only increased risk. Although the aerial victory credit may have been ambiguous in practice, the reward system that accentuated it was not, and the consequent effects could be pernicious.

In the end, the World War I aces did account for more than their fair share of aerial victories, but they never matched the uplifting rhetoric that was purportedly based on their exploits.[103] Their victories, while quantifiable, were neither objective nor necessarily accurate. The pilots' flying skills, their vision, and their marksmanship mattered, but only when paired with a capable aircraft. They may have flown alone in their aircraft, but they roamed the skies in gangs. And although they may have lived a life of privilege at ten thousand feet, they were still within easy reach of death's grasp, which often arrived via a bullet in the back.[104] A symbiotic relationship was being forged between the fighter pilot and his machine, but it often went unnoticed. In combat, the pilots occasionally recognized the man but usually focused only on the machine. The popular narrative, in contrast, often acknowledged only the man; his machine typically resided in the hazy background. Yet it was only in combination that either could function. Without his aircraft, the pilot was doomed to an ignominious life as a youth too feeble for the army (Guynemer) or as a chauffeur transporting generals between command stations (Rickenbacker). And without the pilot, the new, more powerful airplanes simply sat on the grass.

The myth of the self-reliant fighter pilot who could capture glory in an aerial duel against an evenly matched foe proved powerful. Germany's General Erich Ludendorff is said to have remarked that the Red Baron "was worth as much as two divisions of German Infantry."[105] It was not Richthofen the pilot per se, but Richthofen the carefully crafted symbol that Ludendorff was referencing. The myth, not the pilot, inspired pride in his fellow troops and countrymen. In the ensuing decades, the symbolism of the great fighter pilot aces like the Red Baron would continue to exert a powerful influence over future flyers, especially within the United States.

Ritualizing the Myth

Following the First World War, the United States experienced an explosion in popular enthusiasm for all things aviation. Hollywood epics re-created on the silver screen the fabled aerial engagements of the war's ace-flyers.[106] Newspapers trumpeted pilots' record-setting jaunts to higher altitudes and faster speeds while traversing longer distances, especially Charles Lindbergh's transatlantic crossing of May 20–21, 1927.[107] Freshly published memoirs and pulp fiction stories, including Blesse's and Olds's favorite series *G8 and His Battle Aces*, likewise fueled the nation's appetite for exciting aerial adventure.[108] Meanwhile, flyers fresh from France barnstormed out across the nation in army-surplus JN-4 Jennys evangelizing what historian Joseph Corn later termed the "winged gospel."[109]

Accompanying this national surge, leaders of the fledgling aviation industries strove to extend aviation's appeal beyond that of spectator sport, enticing would-be travelers and hobbyists with seemingly endless possibilities. But to do so, the image of the pilot as daredevil, "intrepid birdman," or "modern superman" had to be updated. The push to democratize aviation was reflected in one encouragement from the time, "You and I and cousin Libby can all be Lindberghs."[110] While there were some spectacular individual successes, the effort as a whole faltered. Mass demand for personal aircraft with simple and intuitive controls like Fred Weick's Ercoupe, which promised to propel "uncounted thousands" of "groundlings" into the air, never materialized.[111] Similarly, the emerging airline pilot profession reinforced the perception that only white males possessed the necessary physical traits and aptitude demanded of professional aviators.[112]

There was no complementary attempt to recraft the pilot's persona in the military. If anything, the image of the daring and uniquely skilled "birdman" pilot became more institutionalized and ritualized within the military during the interwar years. A 1929 Army Air Corps text commented on the war's important lesson of assigning "only the best" and "most skillful pilots" to the "faster, lighter and more maneuverable" fighter aircraft. "The successful pursuit pilot," the text noted, "must be an individual whose qualifications peculiarly fit him for the type of work to be performed."[113] The text also reasserted the "knight of old" analogy, referring to the aerial war as being "more . . . a sporting proposition" where "wholesome respect was felt for worthy opponents" and a "spirit of chivalry prevailed."[114] Although the aviators likely knew better, there was little incentive to rewrite the popular narrative that had been so carefully crafted years earlier; the pilots were enjoying the esteem of the nation, and their fledgling institution was locked in a struggle for scarce dollars within a torpid Army bureaucracy.[115]

Military aviation was different from civilian flying, military flyers contended, and consequently they rebuffed the prospect of mollifying the military aviation experience to match the democratizing efforts within the civilian sector. "You can teach anybody to fly but you can't make *pilots* out of them," became a common rejoinder of military flight instructors.[116] As one aviator realized midway during his training, to be a military pilot demanded something that couldn't be taught: "You weren't even being tested on something you studied, really, but on what you were. If you were a flier, you passed; if you weren't, you washed out—fell out of the air, and became a lower order of being. . . . Some people were natural fliers, and some weren't."[117]

Determining the physical, emotional, and intellectual attributes that predestined an individual for greatness as a "flier" in fighter aircraft became a major topic of study for the military. Late in the First World War, the prestigious British medical journal the *Lancet* published a profile of the successful fighter pilot: "usually the enthusiastic youngster, keen on flying, full of what one might call the 'joy of life,' possessing an average intelligence, but knowing little or nothing of the details of his machine or engine." The fighter pilot was "endowed to a high degree with the afore-mentioned quality, 'hands' "—which the authors described as a "congenital" trait, similar to that of a "horse-rider," that permitted the pilot to sense "unconsciously the various movements of the aeroplane" and correct "any unusual or abnormal evolutions almost before they occur."[118] Almost a quarter-century later, the military's assessment of fighter pilots had changed little. As a 1942 article published by the Army Air Force's air surgeon's office noted, "The *pilot* must have the ability to learn complex motor skills involving gross muscular coordination. In addition to this, it is desired that the pilot exhibit superior speed of reaction and the ability to make quick and accurate observations and judgments. While he should be confident and aggressive to be a fighter pilot, he should also be mature and have the ability to command."[119] In short, fighter pilots, according to the military and its doctors, had to be uniquely endowed male specimens.

Selecting individuals for military flight training on the basis of their aggressiveness and self-confidence proved problematic. Already inculcated into the myth of the fighter pilot as celebrated in the popular movies, books, and comics of the day, many of the young and pugnacious aviation cadets experienced a rude awakening when they were inserted into a flight training regimen predicated on strict discipline. "Training had to rein in the young hotspurs," one historian noted, but it also had to do so without squelching the cadets' aggressive spirit that the service prized.[120] Those cadets that struck the proper balance and continued to demonstrate supreme self-confidence were rewarded with the service's most cherished aircraft—single-seat fighters.

It wasn't always the most objective process. One instructor recalled of the assignment process: "We'd say, 'Well this fellow is aggressive and he ought to go to fighters.'"[121] Robin Olds experienced the subjectivity firsthand. His "hopes for a fighter pilot career dashed" by an impulsive senior officer who declared him unfit for fighters, Olds later earned the respect of his flight instructor and was rewarded with a fighter assignment after he flew "under every bridge on the Hudson from Albany to New York City," a bold but fortunately successful gamble.[122] Many other budding fighter pilots, not as skilled as Olds or perhaps not as lucky, paid the ultimate price when they tried to prove their fighter worthiness. Going through training in the 1930s, one flyer remembered that of the ninety-six other graduates in his flight-training course, "within a year, fifteen were dead. Crashes—mostly pilot error, and most of it was from high-spirited behavior. . . . Doing stunts, flying under things, flying low, especially, and pulling up."[123]

By the late 1930s, the ascendancy of the strategic bomber in airpower doctrine began to threaten the princely status of the fighter pilot. Offensive action, the airpower advocates at the Air Corps Tactical School at Maxwell AFB in Alabama determined, was now to be the purview of the team of men crowded into the fuselage of multi-engine bombers. Rather than attack plane-for-plane with fighters, it was far more efficient, the doctrinaires preached, to knock enemy fighters out of commission while they were on the ground at their airfields; better still would be to attack the aircraft when they were still parts or raw materials waiting to be assembled or manufactured. In this model of future air combat, small single-seat fighters would play at most a defensive role.[124]

Nevertheless, the fighter pilot myth persisted, and many pilots still longed to fly them. Describing the popular sentiment, one pilot commented, "To choose any course except single-engine planes . . . would have seemed cautious, unromantic, almost middle-aged, like wearing your rubbers or voting Republican."[125] The allure of self-reliance symbolized by the single-seat cockpit was a significant motivator as it had been decades earlier for Boelcke. So too was the nagging worry of being trapped in a lumbering bomber, unable to maneuver, about to be gunned down by an enemy fighter.

War's Next Test

Walker "Bud" Mahurin was, like Olds, originally slated for bombers when he completed initial flight training in early 1942. Horrific losses in the Americans' first daylight bombing raids over Europe had resulted in a sudden and severe shortage of bomber pilots. For trainees like Mahurin, that meant "it didn't make much difference to the authorities what the individual wanted, he was needed in bombers and to bombers he would go." The young twenty-three-year-old from Fort Wayne, Indiana, was not thrilled about the assignment. "Nothing could make up for my disappointment. . . . I simply didn't like being a bomber pilot. It seemed that all we did was fly around the sky in a big gaggle of aircraft." Then fate intervened. The day of graduation, Mahurin's class was told that anyone shorter than five-feet, ten-inches was being reassigned. It wasn't the typical method for assigning aircraft, but Mahurin was ecstatic: "I was 5′9½″. Fighters after all!"[126]

After some additional training, Mahurin found himself a member of the 56th Fighter Group, the first P-47 Thunderbolt unit sent to the European theater. Arriving in England in early 1943 and piloting, in his words, the "nation's newest and most powerful single-engine flying machine," Mahurin quickly distinguished himself as the first American pilot to down ten enemy aircraft in the skies over Germany and France.[127] But once in Europe, Mahurin soon realized that the reality of air combat didn't quite match the popular rhetoric and imagery he had grown up with. As one of the leading aces at the end of 1943, Mahurin was asked to contribute some words of wisdom for a new manual on air combat. He chose to address the discrepancy explicitly:

> In my opinion aerial combat isn't half of what is shown to be in the movies. . . . We like to think that the battle will assume proportions equal to those of the movies. You know how it is—one pilot sees the other, they both grit their teeth to beat hell, and finally the deadly combat begins with violent maneuvering by both parties. This field of thought is entirely erroneous. The combat usually takes place at a hell of a speed; the enemy plane is only seen for a few seconds. In nine cases out of ten the victor never sees his victim crash. As a result of the wrong idea, the new pilot who first sees a Jerry [German] ship, goes in to attack hell bent for election, and winds up feeling futile as the dickens because he didn't even succeed in frightening the Hun. I know, because I've done it myself many times.[128]

In truth, many of Mahurin's experiences of air combat over Europe resembled those of the ace-flyers from the First World War. For one, although now flying an aircraft that routinely reached speeds six times that of Garros's Morane monoplane, and which now packed the striking power of eight .50-caliber machine guns, Mahurin along with most of his squadron mates still needed to get excruciatingly close to his target to ensure a kill. On a few occasions, he was able to score hits from five hundred yards, but more frequently, he teased, "I have to get close enough to the Hun to reach out and club him before I can hit him. Usually, even that won't work."[129] When it did, Mahurin was close enough to see the aftereffects of the engagement, which reminded him that he was shooting down people, not "objects." On one occasion, he saw a German aerial gunner try to jump from his flaming ME-110 only to have his foot get pinned in the cockpit: "The rest of his body [was] sprawled back along the top of the fuselage.

His arms were flapping wildly in the airstream." Mahurin admitted of the sight, "It shook me badly."[130]

The practical instruction on fighter gunnery also mimicked that from 1916, despite the newer aircraft being equipped with more sophisticated gunsights (covered in the next chapter). Major "Tommy" McGuire, who would become for a period the second leading American ace in the Second World War before being killed over the Philippines in January 1945, advised his pilots, "Go in close, and then when you think you are too close, go on in closer."[131] Echoing Deullin's decades-earlier advice to deliver "a good sprinkling" of machine-gun fire over the enemy, Captain Reade Tilley in 1943 recommended that pilots "close in to very short range, center the aircraft's wing span in your windshield, then commence firing and pump the stick gently back and forth."[132] As before, pilots were instructed that it was best to achieve the close-range position through surprise attack. Missing that one critical element of a successful "bounce," Mahurin said, and "Jerry is about ten times as difficult to bring down."[133]

Surrendering a measure of individual autonomy to fly in formation was another similarity, as were the formations themselves. The pilots, like their predecessors, were expected to remain in position in the "triangular" formations at all times.[134] The formations were alternately called "close V" or "four-ship V," or occasionally "fingertip" because the four aircraft in the flight aligned themselves in positions roughly corresponding to the fingertips of an extended hand. Within the formation, American fighter pilots usually flew about three hundred feet apart from each other.[135] Responsibilities within the formations likewise mimicked those established decades earlier, although some of the leniency shown by Deullin during attacks had by now disappeared. Younger wingmen were responsible for protecting their more experienced leader's tail while he attacked the enemy. "Stick close to the man you fly with," Mahurin told his pilots. "Watch behind and let him look out in front."[136]

The collective emphasis on surprise "bounces" against an unaware foe, the aerial gunnery tactics, the strict requirement to stay in a V-shaped formation, and even the individual responsibilities within that formation, all echoed themes encountered during the First World War. But there was one significant difference between the two wars: in the First World War the massive swarming and swirling dogfights that everyone imagined were a rarity; in the Second, and especially after 1943, they were more frequent. For a typical mission, American fighter squadrons in Europe would launch between sixteen and twenty-four fighters, arranged into flights of four. When joined by additional aircraft from the other squadrons in the fighter group, massive formations of forty-eight or more fighters would take to the sky, sidling up alongside hundreds of American heavy bombers en route to their continental targets.[137] The Luftwaffe fighters below faced a choice: either come up and challenge the fighters in the air or risk getting pummeled on the ground by the bombers. Germany chose the former.[138] Mahurin remembered on one mission spotting "50 to 60 enemy aircraft" approaching the B-17 formations; on another, he encountered "150 enemy fighters."[139]

Huge aerial battles ensued. The "gang fighting" of the First World War was alive and well, but now the gangs were veritable armies duking it out in the skies over Europe.

To help the pilots make sense of the increasingly large melees, American fighters came equipped with two new technological tools. One was the aircraft radio. The inability to reliably communicate between aircraft was a frequent complaint of the early flyers. A 1929 text on pursuit flying declared the "lack of adequate communications . . . probably the greatest stumbling block in the development of air tactics." While the manual noted that "neither the radio telephone nor the telegraph can yet be considered entirely satisfactory," by America's entry into the Second World War the radio was standard equipment in America's newer fighter planes.[140] In theory, the radios allowed the flight leader to coordinate action within his massive formation, but in practice pilots had not yet learned how to use the radio in combat. It was more than just pushing a button and talking into a microphone. With all the aircraft crowded onto a single radio channel and with each pilot thinking he had something important to say, once the battle began, the radio became an incessant chatterbox. The importance of "good R/T [radio-transmitter] discipline" was a recurring theme in World War II tactics reports. One scolded the pilots: "Don't say anything unless it is of importance; don't shout over the R/T, take it calm and easy with proper wording and identification. Many lives will be saved and Huns shot down if the 'babbler' shuts up."[141]

The other tool that the pilots took with them into combat was a gun camera. Built into the front of the aircraft and focused in line with the aircraft's bullet stream, the gun camera automatically snapped images anytime the pilot squeezed the trigger. In theory, the images would provide incontrovertible evidence of enemy downings and eliminate any potential controversy over aerial victory credits. The grainy images could also help intelligence officers infer performance data on enemy fighters. And, when they found their way onto newspaper pages or into newsreels, the exciting images would provide powerful visual testaments to the superiority of American fighter pilots and their contribution to the war effort.[142] Regrettably, the gun cameras' performance was often suboptimal. It wasn't just that the camera might not run or that the film might not turn out, although both were recurring issues. More significantly, the camera could record images of battle only straight ahead. If the pilot maneuvered his aircraft before the enemy fighter started spewing smoke or before its pilot bailed out, there might be little visible evidence to support the pilot's victory claim and enthrall the nation. Mahurin complained that he lost credit for at least one kill because he "had to pull away" before the camera could "record the last of the bullets striking the enemy aircraft."[143]

The introduction of the gun camera to help substantiate claims was a tangible manifestation of the continued importance ascribed to aerial victories and ace status. As before, these metrics overshadowed all others. Fighter pilots of the Second World War knew that glory could be earned only by downing enemy aircraft, and consequently most preferred air combat to ground attack. As Olds later commented of his World War II experience, "Strafing trains and supply convoys was fine. Bombing the occasional bridge or supply area was necessary. But it was air combat that we wanted. Bombers drop bombs. Fighter pilots fight. It was simply the way it was meant to be."[144] And, like before, when they encountered enemy aircraft, "the business of claims" remained, in Mahurin's opinion, "a funny one. Although our methods of establishing claims were rather stringent, pilots would often imagine that they had seen things that had actually not happened at all. . . . Heat of combat, excitement, fear and anxiety, all help the imagination along."[145]

The pilots' imaginations were also fueled by the persistent, popular myth of the fighter pilot—that of lone individuals battling worthy foes in a test of flying skill. Indeed, the air combat experience during the two world wars remained largely comparable despite the new technological tools outfitting the newer aircraft. Both generations of pilots needed to visually detect their enemy. Both still needed to possess airmanship so that they could deftly maneuver their speeding aircraft into position to surprise their unsuspecting foe. Both generations of pilots needed to be relatively close to their target when they opened fire with their machine guns. And both were cognizant of the inconsistencies that separated their reality from that of the idealized myth. Given the choice, they preferred the myth's representation.

The myth of the fighter pilot continued to stoke the nation's imagination, too. As before, American fighter pilots during the Second World War became celebrities all out of proportion to their age or military rank. Emblematic of the public's continued fascination with their fighter pilot heroes, when America's leading ace, Major Richard I. Bong, was killed test flying America's newest jet fighter, the *New York Times* considered the story front-page, above-the-fold, two-column-width worthy news. The date was August 7, 1945. The only headline that received larger typeface that day was the banner proclaiming, "First Atomic Bomb Dropped on Japan."[146]

Conclusion

Writing of the Union Navy's ironclad *Monitor* of the American Civil War, historian David Mindell observed that the myth surrounding the machine and its operators far outweighed the combat significance of either. Seen as a great technological advancement, the *Monitor* was in reality a poorly designed floating "iron coffin" in which the sailors toiled. Hailed as uniquely skilled, without their one-of-a-kind ship, the men of the *Monitor* were likely destined for anonymity. However, when the men and their machine came together in one thunderous but inconclusive battle on March 9, 1862, in the waters of Hampton Roads, they collectively achieved mythical status. Neither could have done so without the other.[147]

So too with the aviators of the world wars. Carefully crafted during the First World War, the image of the fighter pilot assumed mythological significance, inspiring not only flyers but also nations and entire segments of Western civilization. By the end of the Second World War, the myth of the timeless, naturally gifted fighter pilot who engaged in chivalric battle reminiscent of the knights of old was firmly entrenched in the American psyche. A few years later, in the midst of the Korean War, two Air Force recruiting advertisements consciously evoked the image of the iconic fighter pilot mystique. The first, "America's Knights of the Sky," was depicted in figure 1.2. The second invoked the spirit of Rickenbacker's rabble-rousing youth and Guynemer's determination and tenacity. It read: "In the US Air Force, it's not the size of the man in the fight—it's the size of the FIGHT IN THE MAN! . . . If you're good enough . . . tough enough . . . smart enough . . . if you can take it while you're learning to dish it out. . . . You'll be equipped to fly the latest, hottest planes."[148] In 1951 those "latest, hottest" planes came equipped with a revolutionary new gunsight that promised to dramatically and automatically increase the pilot's lethality.

CHAPTER 3

Sabres over Korea

"Every Man a Tiger"

It was like money: it did not matter how it had been acquired, but only that it had. That was the final judgment. MIGs were everything. If you had MIGs you were a standard of excellence. The sun shone upon you. The crew chiefs were happy to have you fly their ships. The touring actresses wanted to meet you. You were the center of everything—the praise, the excitement, the enviers. If you did not—although nothing was shameful about it, and there were reasons, allegedly valid, for any man, no matter how capable and courageous, to have failed to get victories—still you were only one of the loose group in the foreground of which the [aces] gleamed. If you did not have MIGs, you were nothing.

JAMES SALTER, *The Hunters*, 1956

Lieutenant Doug Evans stumbled out the door of his squadron operations building the morning of November 2, 1951. An icy blast greeted him, sucking the air from his lungs. Water welled in his eyes and ran down his nose as he and his flight mates trekked across the tarmac at Kimpo Air Base in South Korea toward their individual aircraft. They moved "like mummies," he said, stiff from the cold and awkward under the weight of their flying gear. Evans wore a pair of long johns, wool shirt and trousers, a knit sweater, a wool flying suit, a cashmere scarf, and a leather flying jacket. Somewhere under those layers he also wore his G suit, which would help keep him from passing out as his blood began to pool in his lower extremities during aggressive, high-G maneuvers.[1] On his feet, Evans wore layers of cotton and wool, flying boots, and rubber galoshes. He also had on his safety and survival gear: the escape and survival vest, complete with "hunting knife, signal mirror, extra gloves and socks, first-aid kit, chocolate bars, and an extra signal flare;" an orange Mae West inflatable life preserver; his parachute; and a .45 pistol sidearm. Evans opted to wear an earflap hat and carry his flight helmet—his "brain bucket"—along with his oxygen mask and mission map in one wool-lined-leather-gloved hand; the other lugged an additional heavy survival kit. By the time they got to their assigned aircraft, "we felt like pack mules," Evans recalled, "and it was a relief to get settled down in the cockpit."[2]

But this particular aircraft was different from the others Evans had flown earlier in combat. Trickling into the theater since the summer, this newest model of the Air

Force's high-speed, day fighter-interceptor, the E-model of the F-86 Sabre, boasted a revolutionary flight control system.[3] Evans had flown an F-86E a few weeks earlier as part of a familiarization sortie around the airfield. Although he later remarked that he enjoyed the new Sabre's handling characteristics, he remained skeptical of the innovative but "pretty complicated" flight control system, and he worried about putting his "fate in the hands of hydraulics without a manual backup."[4]

Evans's assigned F-86E for his November morning mission incorporated another improvement over the more prevalent F-86A models that populated the flight line. Unsure how to operate the new cockpit device, Evans "called one of the pilots over for a little info on it and he said, 'Just turn on the sight and gun switch. When you get something, the diamonds will snap onto radar ranging to span, a yellow light will come on to show radar lock-on, and you just have to keep the pip on target—nothing to it.' "[5] Thus concluded Evans's introduction to the most sophisticated gunsight yet developed for fighter aircraft.

Evans almost scored a victory during the morning's mission using his new A-1C(M) radar-ranged gunsight. He and the other three members of Fox flight arrived in MiG Alley just as a pack of enemy fighters showed up. Among the "MiG formations scattered all over below us," Evans spied a lone pair of MiGs off to the side. Fox flight dove to attack. "Down the chute we went on those two MiGs—what a beautiful bounce. . . . Everything was going great. I put the pip on MiG number two and the diamonds opened and closed and then spanned perfectly; that lock-on light sure looked pretty." Just then, the four Sabres came under attack by another formation of MiGs, "diving down like bombs falling right on top of us." The members of Fox flight immediately broke off their attack and turned to defend themselves. After the mission, Evans remarked that they were so close to victory that "we were ready to slap red stars" on our planes. But, he kidded, the MiGs "who jumped us were all ready to stick on white stars. Everybody went to bed hungry, yuk, yuk."[6]

Evans's reference to the red stars that adorned the fuselages of the top American fighter pilots, each star representing an aerial victory, was but one reflection of the continued allure of individual glory that was a critical component of the fighter pilot myth. As the thirty-two-year-old Air Force Sabre pilot James Horowitz, writing under the nom de plume James Salter in his acclaimed novel *The Hunters*, observed, "MIGs were everything" to the Sabre pilots. "If you had MIGs you were a standard of excellence." Moreover, "the fifth was more than just another; it was beatification, the step across the gulf."[7] If they wanted to get a kill, and especially if they wanted to become an ace with five, Sabre pilots were told they needed to be aggressive and tenacious like the aces of old. The pilots that fit the bill were referred to as "tigers"; those that fell short, "pussycats." The pilots' motto during the war became "Every man a tiger."[8]

The new radar-ranging A-1C(M) gunsight fielded in the F-86E was designed to help the fighter pilots in their quest to prove their worth as "tigers." Touted as a revolutionary improvement in fighter aircraft technology, it would simplify the multitude of tasks required of a pilot in the heat of battle. It also promised to extend the lethality of the Sabre's existing armament, allowing pilots to score hits "with extreme accuracy at hitherto impossible ranges."[9] Reports boasted that, when used properly, the new generation of gunsight doubled the pilot's chances at downing a MiG, instantly turning any junior, mediocre pilot into an expert ace.[10] Even six months before the Korean armistice, individuals were already identifying the new gunsight as the critical advantage that allowed the American Sabre pilots to rack up their impressive tally over their MiG-15 foes.[11]

For most pilots seated in the Sabres' cockpits, however, the new gunsight was at best a modest improvement. Many aviators considered the new sight too complex, too unreliable, and too impractical for use in the skies over Korea. So unhappy with the new sight, a group of leading Korean aces approached the chief of staff of the Air Force (CSAF) in the summer of 1952 and beseeched him to terminate further development of radar-ranged gunsights. Several pilots argued that the new technology addressed only the final "10%" of a mission. "Approximately 90% of your air-to-air combat is positioning," one Korean ace preached. "If your [*sic*] not in position your [*sic*] not going to get a kill."[12] Moreover, the ace pilots argued that the new gunsight's extra weight significantly degraded the overall performance of their aircraft, limiting their ability to capitalize on their flying acumen as they tried to maneuver into the proper firing position. Capturing the pilots' popular sentiments, the president of the Air Force Association concluded after the war, "In our enthusiasm for the electronic gadgets which take the thrill out of aerial gunnery and grab the controls from the pilot, we are inclined to forget that science has yet to produce leadership by remote control."[13]

Thus, in the story of the F-86E, we see clearly two narratives emerging: one of a "great machine" that incorporates cosmic technologies to simplify pilots' tasks; the other of a "great pilot" who somehow triumphs in spite of the new and poorly designed machinery. Amid it all, the myth of the fighter pilot exerted considerable influence over both the engineers' and the pilots' understanding of what air combat was in the past and what it should be in the future. When the engineers tried to codify the public essence of the fighter pilot—his ability to flame an enemy aircraft in glorious battle—they quickly realized that aerial gunnery was not readily reducible to mechanical computations. A certain element of artistry and luck remained. Correspondingly, the pilots in summer 1952 also realized that they might indeed benefit from the new technologies, but only if they spurned the image of being naturally gifted aviators and instead got their heads into their flight manuals and learned the details of the new systems. It was this collaboration between the pilots and their gunsight machinery, not the performance of the technology itself, that led to the Air Force's 1952 decision to continue development of its advanced gunsights. In the end, though, despite the novelty of zooming into battle at closing speeds approaching twelve hundred miles per hour, the fighter pilot ace emerged from the Korean War more akin to his goggle-clad, knight-of-the-air predecessor than different, but also armed with a new appreciation of the importance of advanced fire control technologies in an increasingly complex air combat environment.

A New Solution to an Old Gunnery Problem

The state of aerial gunnery in the Second World War was, like that of its predecessor, dismal. Ace-pilot Walker "Bud" Mahurin writing in late 1943 told his fellow pilots, "If I knew how to practice shooting, I would spend all my waking hours at it." Unfortunately, he didn't, so he primarily concentrated on just getting close enough to his target where he essentially couldn't miss. "If we, and I speak of the Air Forces as a whole, could only shoot perfectly," Mahurin predicted, "we would double our score with no effort at all."[14]

Unbeknownst to Mahurin, Leighton Davis, then a lieutenant colonel recently assigned to the Army Air Forces' Armament Laboratory at Wright Field in Dayton, Ohio, was working on a potential solution. A 1935 graduate of West Point and an Army pilot, Davis was well aware of the deficiencies in aerial gunnery. During his first assignment in Hawaii flying P-12s and P-26s, Davis witnessed stray aircraft bullets injure the ground personnel who changed out the gunnery targets between the fighters' passes. After two stints teaching engineering at West Point, with an intervening year spent at MIT earning a master's degree in aeronautical engineering under Professor Stark Draper, Davis was ordered to combat in early 1943. When he arrived at his new unit, Davis was "amazed" to discover that the aircraft he was being assigned to fly in the Second World War came equipped with the same simple ring-and-bead sight he had used in the P-12 almost seven years earlier. Davis immediately dashed off a letter to his superiors bemoaning the laggard state of gunsight technology in the Army Air Forces. In it, he explained that he himself had worked on a more advanced lead-computing gunsight for the Navy years earlier while at MIT. Within a few weeks, Davis's flying orders were rescinded, and he was told to report instead to Wright Field to begin work on a new gunsight.[15]

When he arrived in Ohio in 1943, Davis immediately understood why development of the Air Forces' gunsights had stalled. "There were two chaps there in the fire control branch," Davis recalled, "and they were optics people," working on "making better lenses for these fixed sights. On the other hand, . . . there were 150 people working [on] making the .50 caliber [machine gun] more reliable." The arrangement didn't make much sense to Davis: reliable guns were still worthless guns if the pilot couldn't hit anything. Davis wanted the shop to concentrate its efforts instead on improving the pilot's aim.[16] For his efforts, Davis would eventually be awarded the Legion of Merit.[17]

The Aerial Gunnery Problem

Shooting down an aircraft with another is difficult. Rarely can the attacking pilot simply point directly at his target; that works only when he is exactly behind the target and at extremely close range. Rather, the attacker must first adjust his aim to account for gravity's effect on the bullets as they exit the muzzle and travel toward the target, thereby applying a gravity drop correction. Second, if the enemy aircraft appears to be moving across the horizon relative to the attacker, then the attacker must aim not where the target is but instead where the target will be when the bullets arrive; that angular difference is known as the lead angle. A third correction, unique to aerial gunnery, is required when the attacking aircraft is maneuvering at conditions higher than 1-G, such as during a level turn. In these instances, a bullet's flight trajectory diverges from the aircraft's longitudinal axis. This velocity jump effect causes the bullets to appear to droop away from the gun as they exit the muzzle, similar to that due to gravity, except that this droop is not necessarily in the pure vertical. The bullet's flight path may be altered further due to a crosswind, or it may be distorted due to vibrations when the gun fires. However, the corrections due to gravity drop, velocity jump, and lead angle tend to be the most significant (and predictable) components when solving the aerial gunnery problem, and of the three, compensation for the lead angle due to target motion tends to dominate. Accounting for these three corrections yields the prediction angle, which is the total angular difference between where the target is and where the guns need to be pointed to score a hit (figure 3.1).[18]

All three primary corrections are significantly affected by the amount of time it takes for the bullet to travel from the attacker to the target. The longer the bullet is in the air (its time of flight), the more pronounced the effects on the bullet's trajectory due to gravity and velocity jump. Additionally, more time allows the target to travel farther across the horizon, affecting the lead angle. Hence, the bullet's expected time of flight to the target must be known to produce a valid gunnery solution. The bullet's travel time is itself a function of several factors. Most significant is the range to the target, with longer ranges obviously requiring longer bullet flight times. The density of the air also plays an important role, with denser air at lower altitudes or higher temperatures slowing the bullet and increasing its time of flight. Conversely, bullets tend to travel faster when fired at higher altitudes or colder temperatures. The bullet's speed as it exits the muzzle has an understandably significant effect, and although this muzzle velocity is largely fixed by the design of the gun and the round, it too can vary due to erosion within the gun barrel, slight variations in the amount of powder packed into each casing, and changes in ambient temperature. The attacking aircraft's speed also influences the bullet's time of flight because the bullet's velocity through the air is a resultant vector of the muzzle velocity and the aircraft's velocity. Similarly, if the range to the target is increasing because the target is outpacing the attacker, then the bullet will take longer to reach the target.[19]

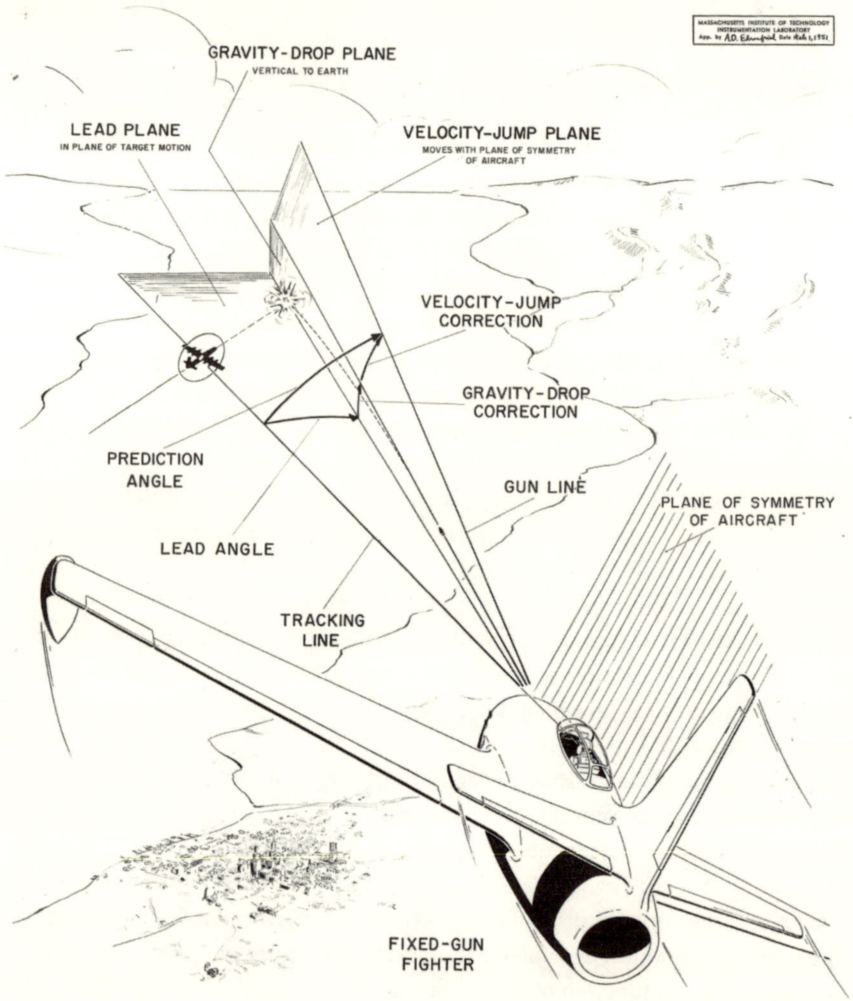

FIGURE 3.1. Three Critical Components of an Aerial Gunnery Solution. As depicted in an MIT Instrumentation Laboratory report on the *General Operating Principles of the A-1*, gunsight aiming corrections are required in three separate planes of motion. Of the three, the lead angle for the target's motion across the horizon (lead plane) dominates the final solution. All three corrections are also dependent on the bullet's time of flight, itself primarily a function of the range to the target.

The net result of all these trajectory and timing corrections and variations is that, even if the attacking aircraft's gun could be aimed perfectly along a computed prediction angle and fired under the most carefully controlled conditions, it is highly unlikely that two rounds would ever hit the exact same point in space. Instead of firing a straight line of bullets, the gun produces a cone of fire whose diameter expands with increasing range. This bullet dispersion, however, can work in the pilot's favor, since it is equally unlikely that any of the bullets he fired would hit the precise location he was aiming at. As one gunnery manual noted, "Although dispersion may make 100-percent hits impossible, it may raise the actual hits from zero to a useful amount."[20] Thus, while a combination of pilot skill and instincts might get the gun aimed in the correct general vicinity, luck (or the laws of probability) always played a role in scoring a hit.

Even then, finding the appropriate general vicinity to aim could be extremely challenging. It was virtually impossible for a pilot to methodically account for all the variables, determine the appropriate aiming corrections, position the aircraft properly, and then fire at the correct time. That is why Mahurin and so many others were befuddled by long-range gunnery. It remained part instinct and part luck, neither of which could be codified to be taught in a classroom or committed to a manual. Mahurin once complained that one of the top gunners in World War II, Jerry Johnson, wouldn't explain the secret of his success. It was more likely that Johnson simply couldn't.[21] Long-range aerial gunnery was an art, and for the not-so-gifted majority, the accepted solution was simply to "fill the windscreen with target and let 'er rip."[22]

Now Davis's team was going to try to transform every fighter pilot into a long-range gunnery expert. Identifying the relevant variables affecting the gunnery problem was only the first step. The engineers still needed to develop a way to measure the variables, determine the appropriate aiming correction, and then display that solution to the pilot. And, because the variables could all change quickly and unexpectedly, the gun solution had to be continuously updated if it was going to be of any use. It was a daunting task. Fortunately for Davis and the others working on the new fighter gunsight, two recent technological developments would aid them: the gyroscopic lead-computing gunsight (LCG) and airborne radar.

New Tools at Hand

The first of the new tools was the gyroscopic lead-computing gunsight. While studying at MIT, Davis worked with Draper on an antiaircraft version of an LCG for the Navy. The Navy's gunsight was simpler than that required for aerial gunnery, primarily because the sailor shooting from the deck of a ship was not moving and rolling in three axes to the same extent that a pilot would be in an airplane, but both ship-mounted and aircraft-mounted LCGs were based on the same fundamental principle.

If a target continues to travel in a straight path at a constant speed, the lead angle component of the gunnery problem is simply a product of the target's angular rate (relative to the gunner) and the time of flight of the bullet.[23] For example, a distant aircraft traveling across the horizon would have a lower angular rate than a nearer aircraft traveling on a parallel flight path at the same airspeed. But because the bullet would take longer to travel to the more distant aircraft, the required lead angle may be similar to or greater than that of the nearer aircraft with a higher apparent angular rate. Knowing that the correction for lead angle dominated the gunnery solution, Draper's team focused on measuring these two variables—angular rate and range to the target—and transforming them into a reliable and intuitive method of directing a sailor's aim.

Their ensuing Mark 14 gunsight, produced by the Sperry Gyroscope Company, found widespread use during the Second World War. The gunsight could be affixed to any 20-mm gun emplacement on a Navy ship. Standing behind the guns and looking through the gunsight's window, the sailor-gunner would see an illuminated reticle appear to hover within the viewing frame. Inside the gunsight were two single-axis gyroscopes, one mounted in the horizontal plane, the other mounted ninety degrees perpendicular in the vertical plane. When the sailor-gunner swung his antiaircraft guns left or right and up or down, the spinning gyroscopes detected the movement. If the sailor-gunner were to smoothly follow a target with his guns, adjusting them in both elevation and horizontal deflection to keep the reticle in the gunsight's window superimposed over the target (a process known as tracking), the total angular rate of the target could be measured. Once supplied with an estimate of the target's range, the gunsight could then mechanically compute the necessary lead angle. Draper's team decided that the best way to communicate this solution to the sailor-gunner would be to adjust the position of the reticle in the gunsight's window. Hence, the sailor-gunner would simply need to keep the moving reticle superimposed over his moving target.[24]

To understand how the system worked in practice, imagine a gunner looking through the window of a gunsight that is mounted on his guns. He sees a target moving level across the horizon, left to right. The gunner initially turns the guns to place the reticle over the target. Because the target is moving left to right, the gunner must then rotate his guns to the right to keep the gunsight's reticle superimposed over the target. Inside the gunsight, the gyroscopes measure how fast the guns are rotating. Using a provided range estimate, the gunsight computes the necessary lead angle. The gunner looking through the gunsight then sees the reticle start to drift to the left, behind the target. He adjusts the rate at which he is rotating the guns to re-center the reticle on the target, after which the reticle appears to stabilize in position within the gunsight's window.[25] With the reticle now closer to the left side of the gunsight window but once again superimposed over the target, the gunner resumes rotating the guns at a rate commensurate with the target's motion over the horizon.

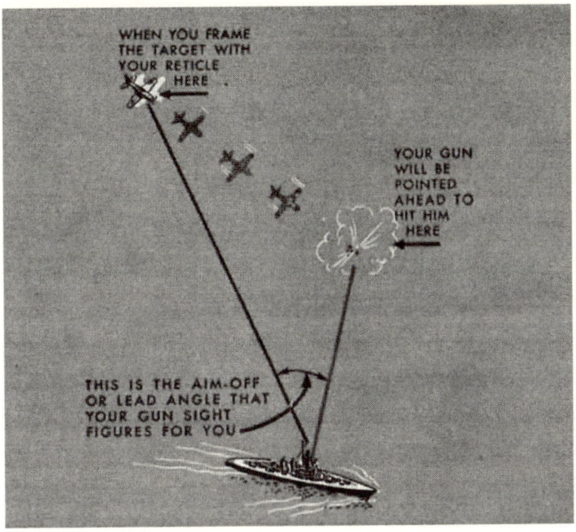

FIGURE 3.2. Mark 14 Antiaircraft Gunsight. The Navy's Mark 14 gyroscopic gunsight, designed at MIT, was used by sailors to aim their ship's deck-mounted antiaircraft guns. The gunsight automatically adjusted the position of the sailor's aiming reference to provide appropriate lead fire against the target. This depiction of the sight's operation was included in a Navy operating bulletin. The design and development of the Mark 14 would have a significant influence on the Air Force's later A-1 aircraft gunsight. DTIC.

The net effect is that the guns have now been aimed in front of the target at the necessary lead angle, but for the sailor-gunner looking through the gunsight, he simply continues to see the reticle superimposed over the target (figure 3.2). As one Navy manual explained the process to its gunners, "Remember, you're working with a LEAD-COMPUTING GUN SIGHT. This means that when you have your target framed in the reticle and are tracking smoothly, your gun is being aimed ahead of the target AUTOMATICALLY. . . . Don't lead with the reticle. Do keep the reticle on the target. Let your sight work FOR you!"[26]

While the internal gyroscopes in the Navy's Mark 14 could automatically measure angular rate as the gunner tracked the target, the gunsight required an additional operator to manually input the target's range. Normally, a sailor-gunner was paired with a "range setter" to accomplish this extra task.[27] The range setter would estimate the range to the target and then adjust the Mark 14's range knob, located on its right side, accordingly. However, with the range setter visually estimating the distance to the target, and with the Mark 14's range knob only adjustable in four-hundred-yard increments, there was little chance the ranging would be accurate. It didn't necessarily matter, though, because the range setter could tweak the setting according to the observed path of the tracer bullets. For example, if the tracers were tracking well in front of the target, and if the gunner had announced that he had the target under the reticle, then the range setter would know to dial back the range knob.[28]

While not perfect, the Navy's Mark 14 possessed the "right combination of precision, ease of use, and simplicity" needed to outfit the thousands of guns adorning the decks of Navy ships.[29] Boasting of the Mark 14's capability, Sperry president Tom Morgan wrote in 1945, "The new sight broadens the mental powers of the gunner, frees him from tasks requiring judgment, and enables him to devote his entire attention to accurate 'tracking' of enemy aircraft." One Navy officer who oversaw the Mark 14 project proclaimed that the gunsight offered "fire control for the masses."[30] The reputation of the remarkable new sight was sealed when Mark 14–equipped gunners aboard the USS *South Dakota* reportedly downed thirty-two Japanese aircraft during the Battle of Santa Cruz on October 26, 1942.[31]

The immediate success of Draper's Mark 14 gunsight aboard Navy ships renewed the Army's interest in LCGs for use in its aircraft. Earlier while he was still at MIT, Davis had been unable to convince the air ordnance officers overseeing the Army's gunsight development of the feasibility of Draper's solution.[32] In March 1942, however, just as the Mark 14s started to emerge from Sperry's workshops, the director of the Office of Scientific Research and Development, Vannevar Bush, sent a memo to General Carl Spaatz, then serving as chief of the Army Air Force Combat Command in Washington, DC, alerting him to the recent developments in "the gyro-type of lead computing sight." Bush was enthusiastic in the memo, telling Spaatz, "I think, personally, that it has much application, and I hope that our people will have the best thing possible when they are called on to use it."[33]

Unfortunately, the Army's subsequent attempts to develop an LCG for aircraft use floundered. So too did the Navy's for its aircraft. Consequently, both services elected to copy a British design: the Army called its version the K-14; the Navy designated its the Mark (Mk-) 18.[34] Both were not fielded until late in the war, and both would continue to see service in the immediate postwar years. In fact, Air Force officials elected in 1949 to use the Navy's Mark 18 gunsight in their new F-86A Sabres then rolling off assembly lines.[35]

Despite using only a single gyroscope, the K-14 and Mark 18 gunsights were functionally similar to Draper's Mark 14 two-gyroscope design. Because their single gyroscope was allowed to precess in both elevation and deflection, the K-14 and Mark 18 could measure the total angular rate of the target relative to the attacking aircraft. Then, with an estimate of target range, the gunsights could compute the appropriate prediction angle mechanically and adjust the optical aiming reticle to show the attacker where he should point his aircraft in order to aim his guns and score a hit.[36] Like the sailors using Draper's Mark 14, pilots using the K-14 or Mark 18 simply needed to place the aiming reticle over the target, track it for a second, and then shoot.

The pilots, however, had to complete one additional task. Whereas the sailor using the Mark 14 usually had an assistant nearby to visually estimate the target's range

FIGURE 3.3. Mark 18 Gunsight. The Navy-designed Mark 18 was installed on the initial batch of F-86A Sabres. For the gunsight to function properly, the pilot first had to set the span scale selector lever (located on the front of the sight unit) to the anticipated enemy target's wingspan. The Mark 18 used this information to determine the target's range, which was needed to calculate the final gun solution. NMUSAF.

and then feed that information into the gunsight, the pilot in a single-seat fighter needed to perform that task himself. Moreover, his gunsight demanded a more refined range estimate to accurately compute the gunnery solution. The method the pilot used to determine the target's distance, known as *stadiametric ranging*, was not new. The same method was employed decades earlier when World War I pilots carved notches onto the top of their fuselage to serve as a "crude but effective range finder."[37] In the K-14 and the Mark 18, however, the stadiametric ranging process was mechanized to automatically feed the target range information into the gunsight for use in its calculations.

With the Mark 18, which equipped the first F-86As used in Korea, the pilot initially set the "span scale selector lever," located on the front of the sight unit, to the wingspan dimension of his anticipated foe (figure 3.3). Then, once the pilot found his adversary and maneuvered into a position near its six o'clock, he adjusted the size of the "gyro reticle circle," outlined by six illuminated diamonds and projected in front of him, to visibly span the target's wingspan (the critical target dimension entered using the span scale selector lever) (figure 3.4). To make the gyro reticle circle larger, corresponding to a shorter range to the target, the pilot twisted his throttle control clockwise; conversely, twisting the throttle control counterclockwise reduced the reticle size. The Mark 18 compared the size of the pilot-spanned gyro reticle circle with the dimension entered using the span scale selector lever, mechanically computed the target range, and then used that information to update the commanded, displayed prediction angle.[38]

Hence, operating the Mark 18 sight was simple in principle, reduced to only three steps in the F-86A pilot's manual:

1. Identify target; then set span scale selector lever to correspond with span of target airplane.

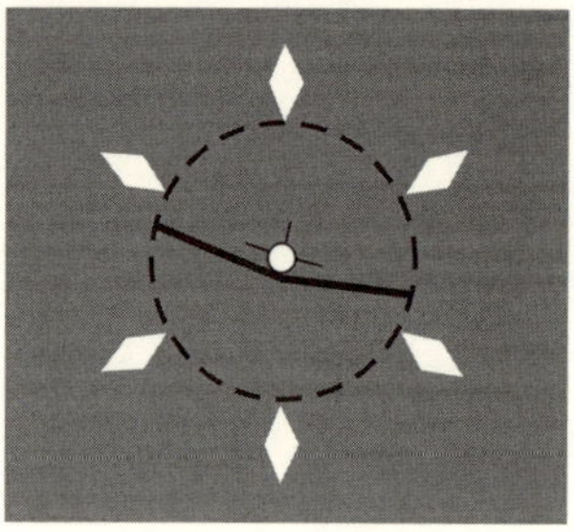

FIGURE 3.4. Mark 18 Aiming Reticle. The illuminated reticle for the Mark 18 gunsight consisted of the central aiming pipper (or pip) and a set of six diamonds arranged in a circle around it (the dashed circle was not displayed). The pilot would rotate his throttle control to size the diamonds around the critical target dimension that he had entered using the span scale selector lever (located on the sight unit). The gunsight's mechanical computers would then use these two settings to compute the range to the target and adjust the aiming solution accordingly. Although simple in practice, this stadiametric ranging process proved especially cumbersome in combat. NMUSAF.

2. Fly airplane so that target appears within gyro reticle circle, and rotate range control until the diameter of the gyro reticle circle corresponds to size of target.
3. Keep target with [*sic*] gyro reticle circle by rotating range control as range changes. Track target for one second; then fire.[39]

The third step in the Sabre's flight manual, however, hinted at the difficulties a pilot might experience when trying to use the Mark 18 in practice. Air combat was not a static exercise. Rarely would the attacking aircraft be able to assume a perfectly stabilized position at a constant range behind the target. This meant that the attacking pilot needed to constantly twist the throttle control lever, adjusting the reticle size to account for changing distance to the target. Moreover, rarely would an attacking pilot get to the direct six o'clock position behind his target, which distorted his viewing perspective of the critical target dimension that he was supposed to span with the reticle (usually wingspan), a problem known as *foreshortening*.[40] Finally, the target aircraft would usually react defensively when it came under attack. These aggressive defensive maneuvers, in addition to altering the closure rates and further distorting the attacker's perspective of the critical target dimension, required the attacking pilot to focus on following the target while trying to keep it under the shifting reticle as the Mark 18 computed new prediction angles.[41]

In sum, to use the Mark 18 gunsight, the pilot needed to maneuver his aircraft in three dimensions to position a target that was also moving in three dimensions under a reticle that was moving in two dimensions, which had to be constantly resized to a dimension that was not always visible. Evans later admitted of his time using the Mark 18 that the process of "trying to do everything at once" could quickly become overwhelming.[42] One Air Force study noted that imprecision in the stadiametric ranging process itself routinely produced average errors of 22 percent of the distance to the target. Under these less-than-ideal circumstances, the effective range of the LCG with .50-caliber ammunition was no more than one thousand feet.[43] Still, despite the difficulties associated with the manual ranging process, the new aircraft LCGs introduced late in the Second World War reportedly yielded a fivefold increase in hits and a threefold increase in kills.[44]

One MIT engineer would later describe the problem of "ranging" as "the bugaboo of aerial gunnery."[45] To address the obvious shortcoming of current aircraft LCGs, Davis's and Draper's teams turned to another innovation coming to life on MIT's campus—airborne radar. Ever since the delegates of the British Tizard mission unveiled the resonant cavity magnetron to a group of astonished American military and scientific leaders in late summer 1940, the American scientific establishment, and particularly those at MIT's Rad Lab, had been furiously developing a plethora of short-wavelength radar-based gadgets for the US military. Initially, the Rad Lab's top wartime priority—Project I—was to develop a ten-centimeter wavelength radar that could detect an aircraft but was also compact enough to be carried aloft in a medium-size bomber. The Rad Lab scientists and engineers hoped to have their airborne interceptor (AI) radar project completed by early March 1941. They came close, outfitting a B-18 with an AI radar for a first flight on March 10. Although the first mission produced "poor results," performance steadily improved.[46]

These first AI radar sets were heavy, bulky contraptions that could be carried only in multi-crew, multi-engine aircraft. For example, although much-improved over its predecessors, the SCR-720 radar in the P-61 Black Widow, fielded in early 1944, weighed more than four hundred pounds and required a dedicated crewmember to operate it. Its operation reflected the basic radar principles that would guide all future radar designs. The radar transmitter housed inside the aircraft generated short, intense bursts of energy and fed them to the radar antenna. The antenna, mounted on powered gimbals at the front of the aircraft, focused the radar energy into a narrow beam and transmitted it out into the atmosphere. A timer or synchronizer in the radar equipment regularly interspersed the brief radar transmissions with long periods dedicated to receiving any reflected radar energy. Measuring the time between when a radar pulse was transmitted and when any reflected radar energy was received yielded the range to the target, which was displayed to the operator on a cathode ray tube (CRT). To determine the relative bearing to the radar target, the radar set would note the position of the antenna when the reflected energy was received. That relative position was also displayed to the operator on a second CRT.[47]

While useful in the P-61, the arrangement was obviously ill-suited for use in a small, single-seat fighter, and the military continued to search for alternative designs.[48] The Army Air Forces' landmark, multivolume, postwar study *Toward New Horizons*, published iteratively between 1945 and 1946, made mention of the lingering requirement: "It has been found that the most successful day-fighter pilots are those who can judge the range to the enemy most accurately and who hold fire until the range has closed to an effective firing value. This range data can be supplied for the pilot simply and automatically by a lightweight radar, which can be used to light a green light when it is time to fire. New, fast planes, such as the P-80, will particularly need this facility, for the firing time in an encounter may be short indeed."[49]

The Army Air Forces' newest gunsight would capitalize on these late-war advancements in LCGs and radar technology. Describing the resulting system, one manual proclaimed, "Further development and improvement of the K-14 gunsight and far reaching advances in the field of radar resulted in a gunsight almost completely automatic in its operation." Because it finally offered a viable solution to the air gunnery problem, the new A-1 gunsight was hailed as "the most versatile fighter type sight in use today."[50]

The A-1C(M) Gunsight

It took Davis nearly two years, but on March 26, 1945, his efforts in the Armament Laboratory finally culminated in an Army Air Forces contract with MIT and Draper's Instrumentation Laboratory for a "Computing Combination Bomb Sight and Gun Sight."[51] Work on the new sight had already begun earlier in November 1943 as part of Army Project MX-402, but it would still take another three years after the contract was signed before the sights were ready for installation in the newest service's newest fighters (the Air Force became an independent branch of the military in September 1947). As work on the sight was progressing, Draper in 1946 offered a rather understated appraisal of the project: "The A-1 Sight offers a form of fire control equipment that should produce operationally worth-while results in fixed-gun fighter installations."[52]

The A-1C(M) installed in the F-86Es that began arriving in Korea in summer 1951 represented the third iteration of Draper's new aircraft gunsight (figure 3.5). The previous A-1B version of the sight had been installed in some F-84 Thunderjets and a handful of early F-86 Sabres beginning in March 1948. However, flight tests later that year revealed significant problems with the new sight and its tendency to display "reticle jitter" whenever the aircraft guns were fired, transforming the precise, illuminated aiming sight into just "an orange blur." Sperry Gyroscope and the AC Spark Plug Company, which had received the production contract for the A-1 sights, tried to correct the jitter problem by adding new "stainless steel stiffeners" to the sight, yielding the A-1C in April 1949. However, the Air Force was unsatisfied and demanded that the two firms continue to work on improving the sight's usability in the field. The resulting, newly modified sight, now referred to as the A-1C(M), boasted "a more efficient sight head and computer heating system, a brighter reticle, and special stiffeners to reduce reticle vibrations."[53] While all three versions of the A-1 possessed the capability to direct gunfire, rocket fire, and bomb delivery, most F-86E pilots during the Korean War used only the gunfire mode.[54]

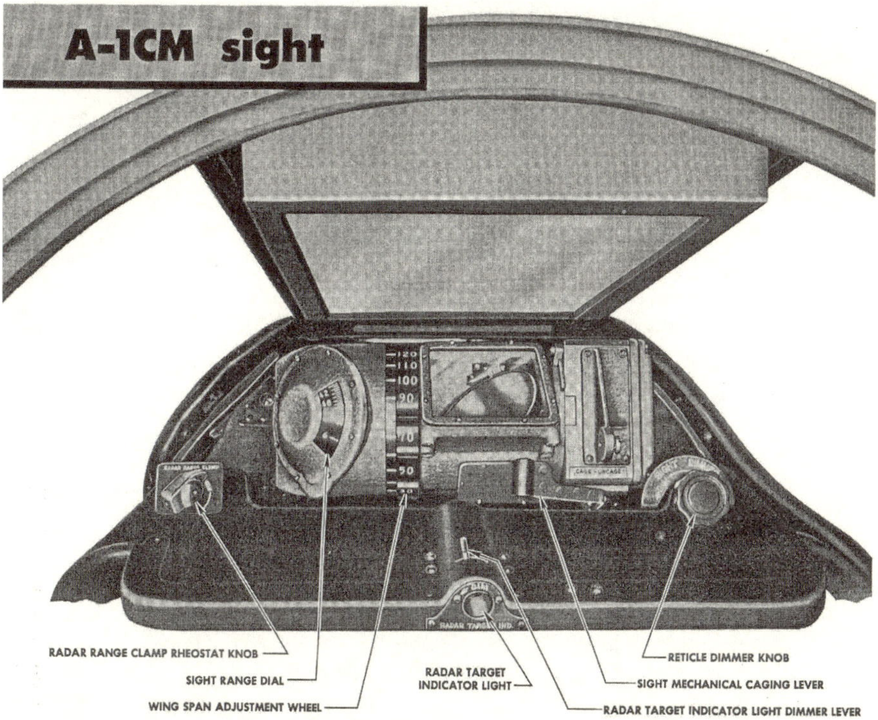

FIGURE 3.5. A-1C(M) Sight Unit. The A-1C(M) sight unit was mounted atop the instrument panel. It included a wing span adjustment wheel, which was functionally similar to the span scale selector lever on the Mark 18, but the A-1C(M) mechanical computer used that information only if the pilot disabled automatic radar ranging. The sight range dial to the left of the wing span adjustment wheel displayed the A-1C(M)'s computed target range, based on either the radar input or the pilot's stadiametric ranging of the target. The single radar target indicator light, located just below the sight unit, was the pilot's only direct source of information on the radar's operating status. Note: The radar range clamp rheostat depicted to the left of the sight range dial was added as part of the Jenkins range limiter improvement, discussed later in this chapter. NMUSAF.

Earlier with the Mark 18 gunsight, the effectiveness of the Sabre's guns was limited to a range of one thousand feet; with the new A-1C(M), the effective firing range of the same .50-caliber armament was predicted to increase nearly threefold.[55] A good portion of that improvement could be traced to the A-1's new source of range data. Rather than relying on pilot-provided estimates of target range obtained through stadiametric ranging, a radar mounted in the fighter aircraft would now "automatically and continuously" feed precise range data to the gunsight. Although technically not one of the eight components that composed the A-1 gunsight system, the radar was nonetheless critical to the gunsight's operation.[56]

General Electric's AN/APG-30 was the radar of choice for the A-1C(M).[57] It was a relatively simple set compared to the larger AI radar systems used in earlier aircraft like the P-61. The APG-30 had no CRT display for the pilot and provided no information on target location or relative bearing; it simply measured range to a target. Its advantage, however, was that it could fit in the tiny F-86: the radar's small antenna was housed behind the pouting upper lip of the Sabre's intake (figure 3.6).[58] Like its gunsight counterpart, the APG-30 was designed to function with very little input from the pilot. The pilot simply turned the unit on and, after a ten- to fifteen-minute warm-up period, the radar began automatically transmitting its 9.3 GHz waveform at eight hundred pulses per second. Advertised to be effective out to ranges of nine thousand feet, if any object was detected within the antenna's fixed eighteen-degree beam width, the APG-30 would automatically lock on, and a single "tracking indicator" would illuminate in the cockpit. The range dial on the A-1's sight unit would also provide indication of a lock-on as it quickly rotated to display the radar-measured range to the target in hundreds of feet.[59]

The automaticity of the radar posed a practical problem, however. If there were multiple targets in front of the aircraft and within the radar's field of view (including large nonaircraft radar targets such as mountains), the radar would simply lock onto the first target it detected. *What* the radar had decided to track was never indicated directly to the pilot. To verify the radar's selection, the flight manual suggested the pilot "check [the] range dial indication against [the] estimated range of target." If the pilot still remained unsure, or if he determined that the radar was not locked to the desired target, then the manual instructed him to depress the "Range Gate Out" switch (sometimes referred to as the radar target selector button) on his control stick to interrupt the tracking. The radar would then break lock, return to search, and lock the next target it detected. But there was still no guarantee that the new target would be the desired one. The pilot was told to repeat the process as necessary until the correct target was finally locked.[60]

Once the radar was tracking a target, the radar-measured range information was sent to the A-1's range servo to be converted into a compatible electrical signal (figure 3.7). From there, the range signal was carried to both the sight unit (discussed later) and the amplifier. In the amplifier, the range signal was combined with an electrical signal based on the relative air density, as measured by an aneroid bellows located within the amplifier. The combined signal that emerged from the amplifier, referred to as the sensitivity current, then went to the sight's prediction computer.[61]

Like the Navy's ship-based Mark 14, the prediction computer in the A-1 housed two spinning gyroscopes: one to measure changes in elevation, the other to measure changes in horizontal deflection. Draper's team used an unbalanced mass attached to the elevation gyroscope to measure and account for the effects of gravity drop and

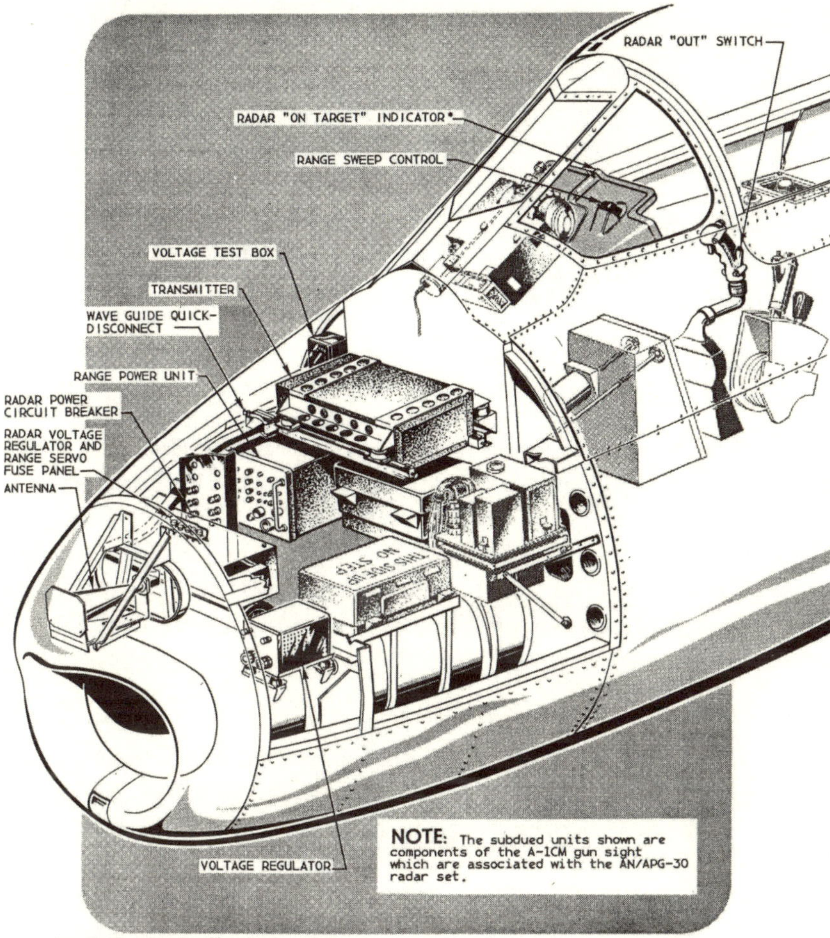

*On Airplanes F-86F-15 and subsequent, the "ON TARGET" indicator is installed in the A-4 sight head.

FIGURE 3.6. AN/APG-30 in the F-86E. Although critical to operation of the A-1C(M), the APG-30 radar set was not originally considered a component of the gunsight system. With no single organizational or manufacturing entity responsible for integrating the different radar and gunsight components in the aircraft, reliability and maintenance of the system in toto proved troublesome. Aircraft technicians working on the Sabres in Korea eventually concluded that with each individual component precisely tuned at the factory for use in a particular radar set, failure of any one radar component required replacing the entire set. Radar maintenance was further hindered due to insufficient documentation, test equipment, and training. NMUSAF.

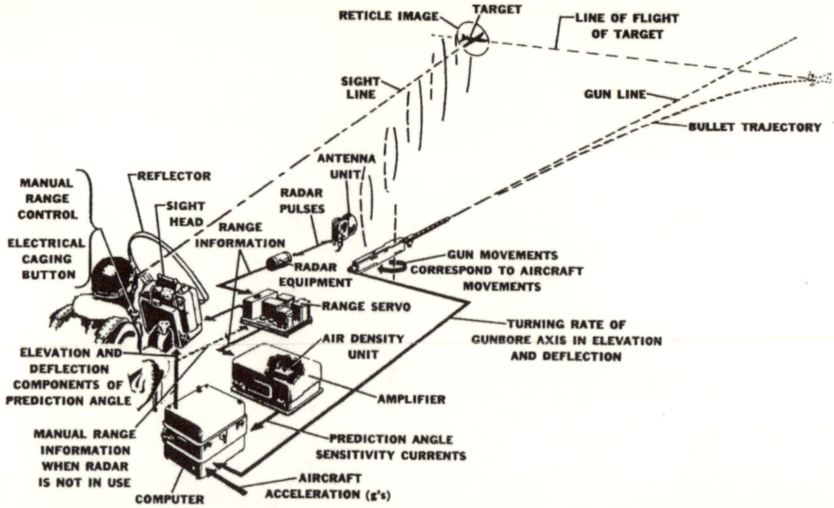

FIGURE 3.7. A-1 Operation, Pilots' Perspective. This schematic, included in an Air Force *Harmonization and Firing Techniques* manual, traces the flow of radar-provided range information through the A-1 gunsight system. Of special interest, the pilot occupies a prominent position in the foreground of the diagram and his controls over the gunsight system are appropriately annotated (such as manual range control and electrical caging button). Contrast this depiction with that used by the A-1 engineers (figure 3.13). NMUSAF.

velocity jump (figure 3.8). To correct for any rolling tendencies of the attacking aircraft, referred to as cross-roll, the deflection gyroscope was canted ten degrees off horizontal. The pilot could electrically cage the gyroscopes to prevent their precession by depressing an electrical caging button located on the aircraft throttle control. He was instructed to do so whenever he was actively maneuvering into position, thereby eliminating spurious inputs to the gunsight computer while also limiting potentially distracting reticle movements in the sight.[62]

Once the pilot was tracking the target aircraft and he released the electrical caging button, the gyroscopes were free to detect any changes in angular movement. Changes in flightpath caused the gyroscopes to apply a torque, proportional to the turning rate of the aircraft, to their respective computer shafts (figure 3.9 shows the elevation computer; the deflection computer was mechanically similar but with the gyroscope oriented in the vertical). The rotation of the computer shaft was electrically restrained, however, to account for the bullet time of flight according to the sensitivity current produced by the amplifier. A viscous damper also restrained movement of the computer shaft in order to "smooth and stabilize the prediction computation" over time.[63]

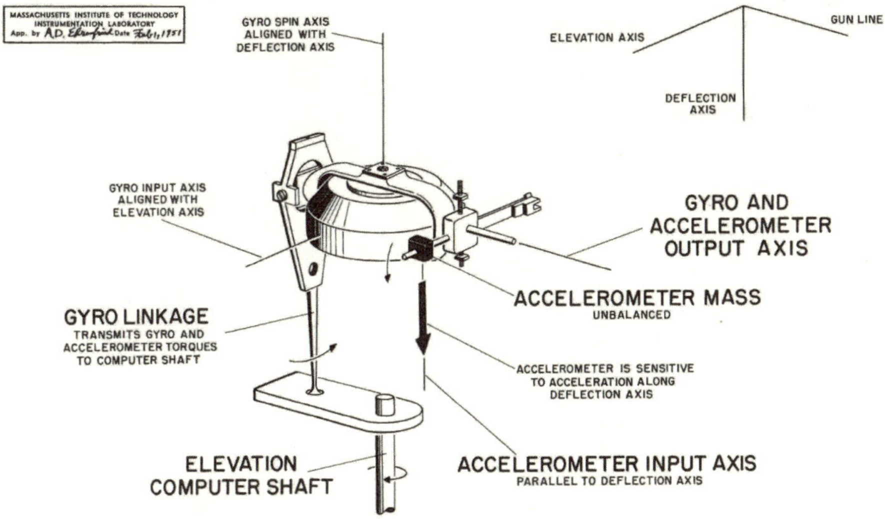

FIGURE 3.8. A-1 Accelerometer and Elevation Gyroscope. The "unbalanced mass" attached to the elevation gyroscope in the A-1 was a simple but effective way to compensate for the effects of gravity drop and velocity jump, provided the aircraft was flown in a coordinated fashion. Copyright Massachusetts Institute of Technology. Used with permission.

As the computer shaft rotated, the torque generated an electrical signal that was sent back to the amplifier. There, the elevation and deflection computers' prediction signals were compared with the electrical signals produced by the sight unit, which were based on the position of the reticle imaging mirror. If the signals from the sight unit were not coincident with the signals from the prediction computer, the amplifier generated a new signal to drive the mirror in the sight unit into correct alignment. Focused light from a pair of light bulbs located in the sight unit was directed through a collimating lens and projected onto the imaging mirror, which in turn projected the aiming pipper and reticle onto the windscreen in front of the pilot at the appropriate prediction angle.[64]

The sight unit also received the range signal directly from the range servo. This allowed the sight unit to display the target's range in the range dial. Additionally, the sight unit used the range input to automatically scale the size of the circular reticle. A distant target had a correspondingly small reticle, and as the range to the target decreased, the size of the reticle increased. The reticle never necessarily spanned the target, however. The variable-range reticle appeared as a solid circle, sized between 20 and 160 mils (milliradians), with a 2-mil pipper located at the center (left image in figure 3.10).[65]

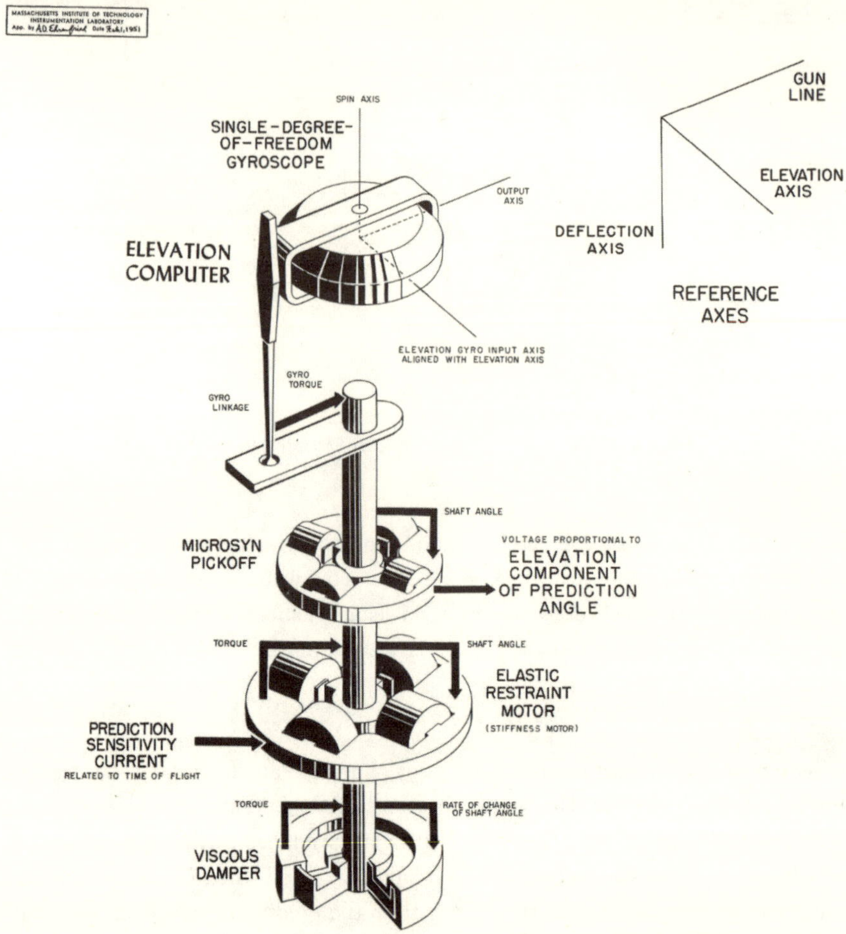

FIGURE 3.9. A-1 Elevation Computer. Changes in aircraft pitch caused the elevation gyroscope in the A-1 to apply a torque to the computer shaft through a mechanical linkage. A microsyn pickoff in turn measured the applied torque and generated an electrical signal for use in the prediction computer. Rotation of the computer shaft was restrained, however, by both a viscous damper and a restraint motor. The viscous damper helped mitigate excessive, momentary deflections of the gyroscope caused by erratic aircraft maneuvering. The restraint or stiffness motor was electrically adjusted according to the bullet's expected time of flight, which itself was primarily a function of the range to the target. Note: Arrows show the information flow through the computer.

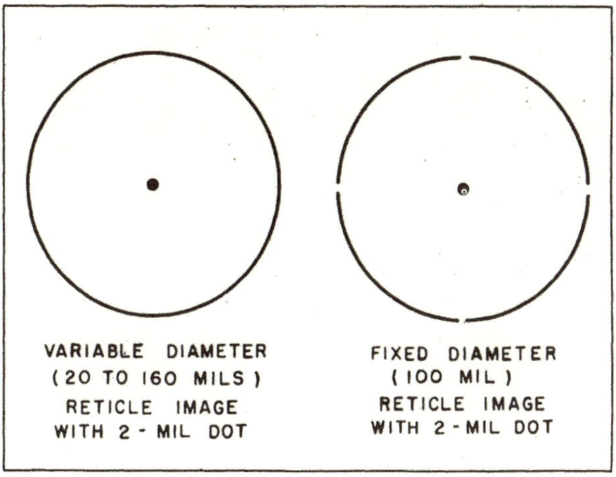

FIGURE 3.10. A-1 Aiming Reticle. The illuminated A-1 reticle had two modes: variable diameter, which was used during both automatic- and manual-ranging operations; and fixed diameter, which was used when the gunsight was mechanically caged (disabled). During automatic ranging, the A-1's reticle auto-adjusted in size, with small reticles corresponding to distant targets, but it did not necessarily span the critical target dimension. When the A-1 was used in manual ranging mode (if the wingspan selector wheel had been set to the enemy target's critical dimension), the pilot needed to size the reticle to span the target dimension just as with the earlier Mark 18. The difference in reticle sizing between the two modes led to occasional confusion in the cockpit. NMUSAF.

Knowing that the radar would likely experience tracking issues at low altitudes and in poor weather conditions, and also because the contract demanded it, the A-1's designers included a pilot-controlled stadiametric ranging capability for use as an "emergency measure."[66] Its mechanization mimicked that of the earlier Mark 18. The pilot first engaged manual ranging by twisting the throttle control clockwise. He then adjusted the size of the reticle by continuing to twist the throttle until the reticle matched the target's size (figure 3.11). If the pilot released the applied torque, the throttle was spring-loaded to return to the counterclockwise position, restoring the radar's automatic ranging input. If the pilot engaged manual ranging after having set the dimension of the target using the wingspan selector wheel located on the sight unit (similar to the target span scale selector lever on the Mark 18), the A-1 sight could estimate the target's range and would feed that information to the range dial on the sight unit and to the amplifier for eventual use in the prediction computer.[67]

Engineers also included a way for the pilot to disable the automatic lead-computing function of the A-1 sight. Sliding a lever on the sight unit from UNCAGE to CAGE

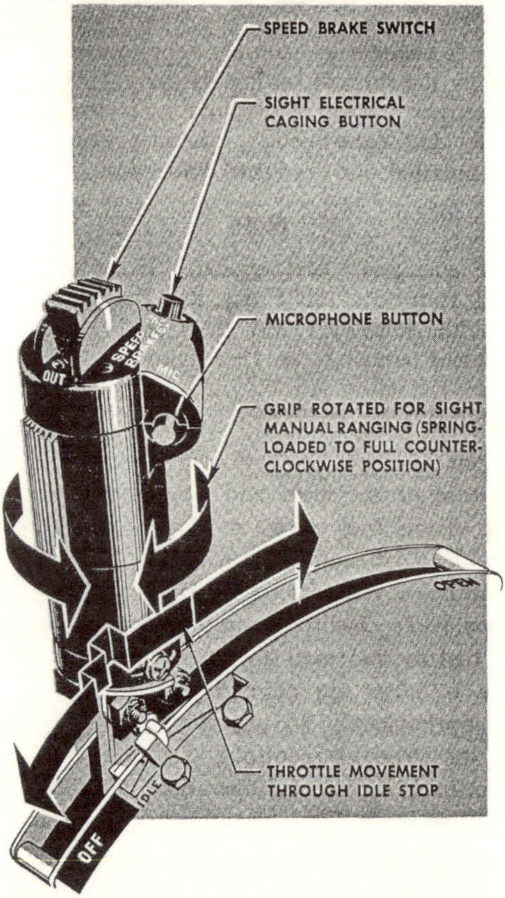

FIGURE 3.11. F-86 Throttle Grip. Mechanization of the manual-ranging mode of the A-1 in the F-86 mimicked that of the earlier Mark 18, with the pilot rotating the throttle grip to size the aiming reticle around the target. The grip was spring-loaded to the full counterclockwise, automatic ranging position. The electrical caging button on the top-forward position of the throttle grip restrained the A-1's computer shafts; pilots were instructed to depress the caging button as they maneuvered into position behind a target to avoid injecting spurious tracking inputs into the A-1's computations. NMUSAF.

turned off the power to the A-1's amplifier and mechanically caged the gyroscopes in the prediction computer and the reflector mirror in the sight unit. The variable-sized reticle was extinguished, and a fixed, four-segmented one-hundred-mil diameter reticle was projected onto the windscreen along with the central two-mil aiming pipper (right image in figure 3.10). No prediction angle was computed, and the reticle remained locked in its fixed position along the gun boreline. Returning the switch to the UNCAGE position reactivated the sight.[68]

Although he always retained the option to engage these "emergency measures," the pilot was expected to use the full, automatic capability of the new sight and its accompanying radar. Paradoxically, however, the additional automaticity of the new gunsight demanded more complex operating procedures. Whereas operation of the Mark 18 required only three steps (and five accompanying notes) in the F-86A flight manual, the A-1C(M) had fourteen steps, plus an additional six if the sight was to be used in manual ranging mode. The first nine helped the pilot prepare the delicate A-1C(M) sight for use. After, among other things, checking the radar power inverter, cycling the cage-uncage lever, and checking the throttle grip in the full counterclockwise position, the pilot was directed to:

10. Set wing span adjustment to wing span of target airplane, so that manual ranging can be set up in a minimum amount of time if radar ranging fails.
11. Adjust sight reticle dimmer control. . . .
12. Depress electrical caging button to stabilize reticle image, and begin tracking, estimating the lead.
13. After radar target indicator light comes on . . . , release electrical caging button.
14. Continue to track target smoothly, without slipping or skidding, for approximately one second after releasing caging button; then fire.[69]

The reference to not "slipping or skidding" the aircraft in the final step of the A-1's firing procedures reminded pilots of the three critical assumptions that engineers invoked when designing the sight (figure 3.12). First, the sight assumed the target flew a "straight and level [constant velocity] course in the same plane as the [attacking] fighter." Second, the attacking fighter was assumed to fly a "lead-pursuit course," as commanded by the sight, "with smooth and accurate tracking of the target." Finally, the attacking fighter needed to fly "in a coordinated maneuver with net acceleration and lift vectors in the aircraft plane of symmetry." While not ideal, these three simplifying assumptions allowed the engineers to exploit the "special interrelationships among the problem variables," reducing the inputs required to compute the gun solution to a few readily measurable quantities. The net result was a "simpler, smaller and more reliable" gunsight that nonetheless still functioned over "an extremely wide range of intercept conditions." However, if any of the assumptions were violated, the "usefulness of the computer," one operating manual warned, would become "somewhat limited" due to the "appreciable error" in the computed solution.[70]

The A-1 gunsight was initially praised as a significant improvement in air combat lethality, and Air Force test pilots enthusiastically declared the A-1 "superior to any other fixed Gun Fire Control system."[71] By using a radar to "automatically and

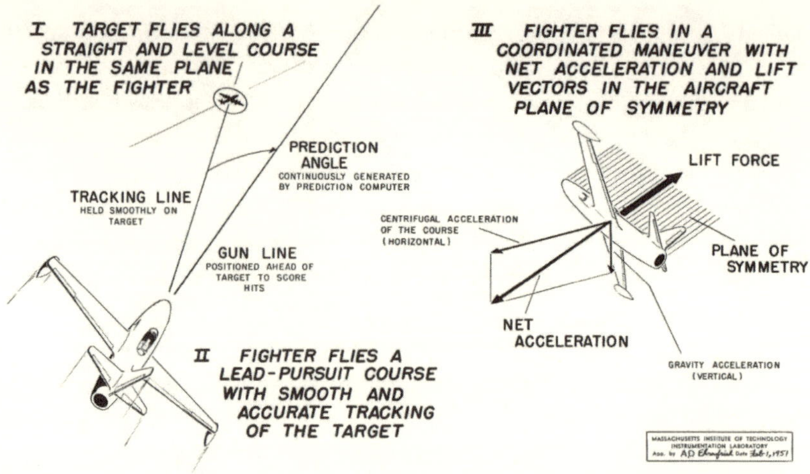

FIGURE 3.12. A-1 Simplifying Assumptions. By invoking three assumptions, the A-1's designers greatly simplified the variables associated with calculating a gun solution, which allowed them to produce a smaller and less complex gunsight. However, if any of the three assumptions were violated, the gunsight's calculated solution would be incorrect.

continuously" feed target range information to the prediction computer, the new gunsight eliminated the pilot's cumbersome requirement to manually span the target with a reticle. Accordingly, several manuals triumphantly declared that the new sight finally allowed "the pilot to direct his full attention to the selected target," a curious but likely unrecognized allusion to Sperry president Tom Morgan's earlier announcement regarding the powers of the Navy's Mark 14.[72] The A-1's accompanying radar also eliminated a major source of error in the calculated gunnery prediction angle—the pilot's notoriously inaccurate range estimates to the target. Consequently, the Air Force predicted that long-range, high-deflection gunshots would soon become the norm, and commanders in the field were told "to encourage long range firing in their units."[73] As the unit pilots began practicing with their new sights, the A-1's designers actively discouraged the pilots from doing anything but simply following the commands of their "precision instrument." In the engineer's model, the pilot was on board simply to aim the guns by steering his aircraft in accordance with the gunsight's direction (figure 3.13); no other intervention or interaction was required or desired.[74]

Davis's efforts had successfully shifted the armament engineers' focus from hitting the target with more reliable guns to more reliably aiming those guns. However, as the pilots in the skies over Korea soon learned, the sight that was touted as being "almost completely automatic" wasn't; it demanded close supervision to ensure it was

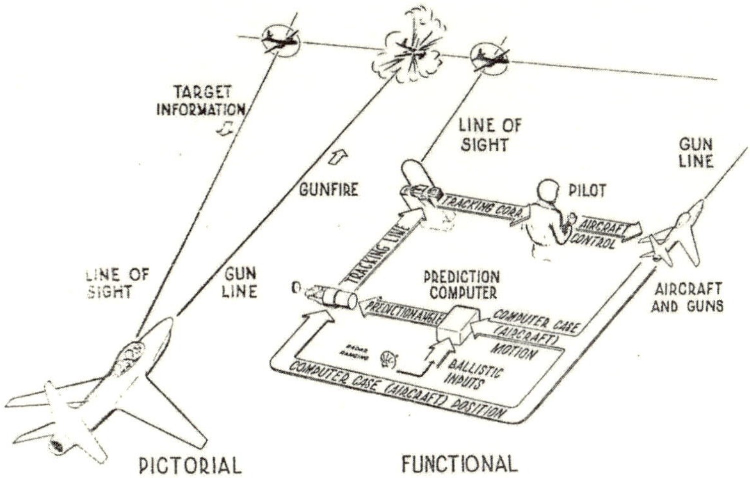

FIGURE 3.13. A-1 Operation, Engineers' Perspective. In contrast to the looming image of the pilot set in the foreground of the Air Force's depiction of the A-1 in figure 3.7, the pilot plays a much smaller role in the engineers' scheme presented at a mid-January 1953 conference at MIT. Here, the pilot is reduced to simply one more functional component of the system, executing aircraft control commensurate with the tracking solution provided to him. He has no direct input into the gunfire solution. DTIC.

operating correctly. A-1 sight operation entailed more checklist steps and, paradoxically, required the pilot to visually estimate the target's range in order to verify the sometimes-suspect radar-supplied range. Moreover, there were still several intermediate steps that needed to occur between the time the pilots trudged out to their aircraft and when they squeezed the trigger. The Sabre pilot still needed to get airborne, travel hundreds of miles to MiG Alley, and then visually detect the small, high-flying Communist fighter planes. He then had to maneuver into a stabilized offensive position behind his target and maintain that position long enough to develop a firing solution before the maneuvering MiG spoiled his tracking. Throughout it all, the Sabre pilot had to maintain constant awareness of his aircraft (lest he run out of fuel) and the sky around him (lest he or his wingman rapidly transform from hunter to hunted). In short, although the A-1C(M) promised unparalleled automatic aiming accuracy, "the electronic wizard" could not guarantee a successful combat outcome in the frenetic skies over MiG Alley.[75] The sight was far more likely, some aviators lampooned, to crush its fighter pilot operator (figure 3.14).

FIGURE 3.14. A Bright Idea Goes Horribly Wrong. A cartoon published in the Air Force fighter pilots' professional journal, *Fighter Gunnery*, shortly after the A-1 started appearing in Korean air combat in 1951 reflected the pilots' growing frustrations with their new gunsight. What had emerged from MIT as a good idea was now figuratively crushing the life out of the pilot and was far too cumbersome to be of practical use in an F-86 Sabre. MSFRIC.

Thrust into War

Few would have predicted that the new A-1C(M) would first see action in the skies over Korea. Yet that's where it found itself in mid-1951 as F-86 pilots battled Communist foes flying Soviet-built MiG-15 fighters. Almost a year earlier, on June 25, 1950, North Korean troops raced south across the thirty-eighth parallel into South Korea. At the time, the North Korean Air Force (NKAF) consisted of a motley collection of World War II "hand-me-downs" bequeathed by the Soviets. Despite being neither numerically nor technologically impressive, the North quickly commandeered control of the sky that first day and scored several blazing victories.[76]

The North Koreans' domination in the air was, however, extremely short-lived. Two days after the initial invasion, American F-82 Twin Mustangs and F-80C Shooting Stars downed a total of eight NKAF fighters.[77] As the Communist troops advanced farther south toward Pusan during the summer, the US Air Force launched a massive effort to eradicate the NKAF. American crews in B-29 Superfortresses assigned to the Far East Air Forces (FEAF) pummeled the North's airfields, while their Fifth Air Force friends in B-26 light bombers and F-51 fighters strafed any remaining planes left on the ground. Pilots flying above in F-80C jet fighters scored easy victories over any NKAF aircraft that managed to get airborne.[78] By early October 1950, following General Douglas MacArthur's daring amphibious landing at Inchon, North Korea's army was in full retreat, and its air force was in shambles.[79] The aerial campaign was so lopsided that many suggested that the Air Force's new, expensive jet fighters were unnecessary extravagances.[80]

All of that changed on November 1 when Communist China intervened on a massive scale. The US and South Korean army positions near the Yalu River were decimated, and the accompanying United Nations (UN) troops commenced a hasty retreat back south.[81] China's foray into the Korean War was also announced that afternoon in the air. Just before two o'clock, a force of six sleek, swept-wing fighters took off from a nearby base in Manchuria and darted south of the Yalu River near Sinuiju, hoping to pounce on a group of unsuspecting British and American aircraft. Miraculously, the Allied pilots in the British Mosquito and American F-51 Mustangs survived the lightning-quick onslaught. The twin-engine Mosquito sped back to its temporary base at Pyongyang to file details of the engagement and spread the astonishing news—"very, very fast" Soviet-built, Chinese MiG-15s were now patrolling the skies over Korea (figure 3.15). The arrival of the new jet-powered MiG instantly "rendered obsolete every American plane" in the theater.[82]

Operating from their bases just inside Manchuria, the MiGs immediately began to challenge the American and UN forces for air superiority. The high-speed jet fighters easily pounced on American B-29 bombers sent north to fell the bridges spanning the Yalu. Even when the B-29s were alerted to the MiGs' attacks, the automated self-defense gun system on the Superfortresses, a technological marvel when it was

FIGURE 3.15. Soviet MiG-15bis. The bis, a later version of the Mikoyan-Gurevich MiG-15, boasted a larger, more-powerful engine, which pushed the lightweight Soviet fighter even higher above the Americans' heavier Sabre. Note the midmounted wing and the horizontal stabilizer mounted high on the vertical tail—common features on all MiG-15s. Lieutenant Ro Kum Suk, a North Korean pilot, defected to the South on September 21, 1953, in this MiG-15, which was quickly moved into a hangar at Kimpo Air Base. NMUSAF.

developed in the latter stages of World War II, could not reliably shoot down the enemy because the MiGs simply outran the mechanical servos that controlled the bombers' machine gun turrets.[83] Nor were the Americans' F-80s much-better equipped to deal with the MiG threat; although both were jet-powered, the MiGs were more than one hundred miles an hour faster than the F-80C, leaving the once-impressive Shooting Star looking "as if it were anchored in the sky."[84]

Seven days after the MiGs were first sighted, the CSAF, General Hoyt Vandenberg, declared his intention to ship his service's premier fighter aircraft to the war. A week later, F-86As from the 4th Fighter-Interceptor Wing (FIW) at Wilmington, Delaware, and F-84Es from the 27th FIW at Bergstrom AFB, Texas, began arriving in California to be loaded onto four tankers and three Navy escort carriers for their trans-Pacific

FIGURE 3.16. North American F-86E Sabre. The E-model of the F-86 started trickling into Korea in mid-1951, initially replacing the earlier F-86A models on a one-for-one basis. In contrast to the MiG-15, note the Sabre's low-mounted wing and horizontal stabilizer. Heavier than the Soviet MiG, some Sabre pilots likened their well-appointed aircraft to a Cadillac. NMUSAF.

journey. The first batch of Sabres arrived in Japan the first week of December; the second batch arrived the following week on the thirteenth (figure 3.16). After some hasty repairs made necessary by the long ocean voyage, the Sabres were flown to Kimpo airfield, located just northwest of Seoul, the closest suitable airfield to the MiGs based in Manchuria.[85] However, Kimpo was already overcrowded with aircraft when the Sabres arrived, so only a portion of the 4th's F-86s were placed on the peninsula. With just seven serviceable aircraft and fourteen pilots, Detachment A of the 336th Fighter-Interceptor Squadron, 4th Fighter-Interceptor Group (FIG), launched the first F-86 missions of the Korean War on December 15, 1950. Their primary assignment, the detachment commander reported, was "to seek out and destroy hostile aircraft in operations on the South side of the Yalu River."[86]

Two days later, the F-86s had their first opportunity. On December 17, 1950, Lieutenant Colonel Bruce Hinton led three of his squadron mates on the second Sabre mission of the day. Spotting a group of MiG-15s below, Hinton quickly positioned his flight to attack. One history of the engagement described the MiG pilots' initial response as "perfunctory and leisurely," the Red pilots having grown accustomed to battling "those type aircraft whose speed and climbing ability was somewhat inferior to their own." A fifteen-minute dogfight ensued before Hinton, using the Sabre's Mark 18 gunsight, successfully fired "fifteen hundred rounds of fifty caliber ammunition into" one of the MiGs. "Multiple hits were observed; debris falling away and finally, the entire rear section of the aircraft was enveloped in flame as it began spinning," the postmission report detailed. The 4th triumphantly declared that its F-86s "had engaged and destroyed an aircraft reputed to be the highest performing aircraft known in the enemy's airforce."[87]

Vying for Air Superiority

A thirty-month-long battle ensued as MiGs and Sabres fought for control of the sky over the Korean peninsula. For most of the war, American military leaders were intent on using aerial bombing to stem the tide of Communist supplies crossing the Yalu River from Manchuria into North Korea. To do that, they needed the sky to be made relatively clear of enemy MiGs. While the Americans would have preferred to destroy the enemy jet aircraft in bulk on the ground by bombing their airfields in Manchuria, political restrictions prohibited that course of action.[88] Instead, the Americans relied on their Sabres to destroy the MiGs in the air. The Communists understandably did not want to cede air superiority, and their MiG pilots fought vigorously to upset the American plans. The opposing fighters typically tussled with each other over the northwestern corner of the peninsula in an area known as MiG Alley (figure 3.17).

The importance of gaining air superiority as a prelude to an aerial bombing campaign, or any campaign on land or sea for that matter, was one of the dramatic lessons of the Second World War. Reflecting on the earlier wartime experience, US Army spokesperson Robert Patterson testified in 1947, "Without command of the air, the launching of a military operation on land or sea was virtually unthinkable."[89] In Korea, the significance of the Sabres to the aerial campaign was demonstrated in dramatic, albeit negative, fashion shortly after their arrival. With Chinese troops advancing on Seoul, the Americans were forced to abandon their base at

FIGURE 3.17 (*opposite*). MiG Alley. Most air combat during the Korean War took place in MiG Alley, an area in the northwest corner of the peninsula between the Chongchon and Yalu Rivers, directly across from one of the Chinese MiGs' main operating bases at Antung.

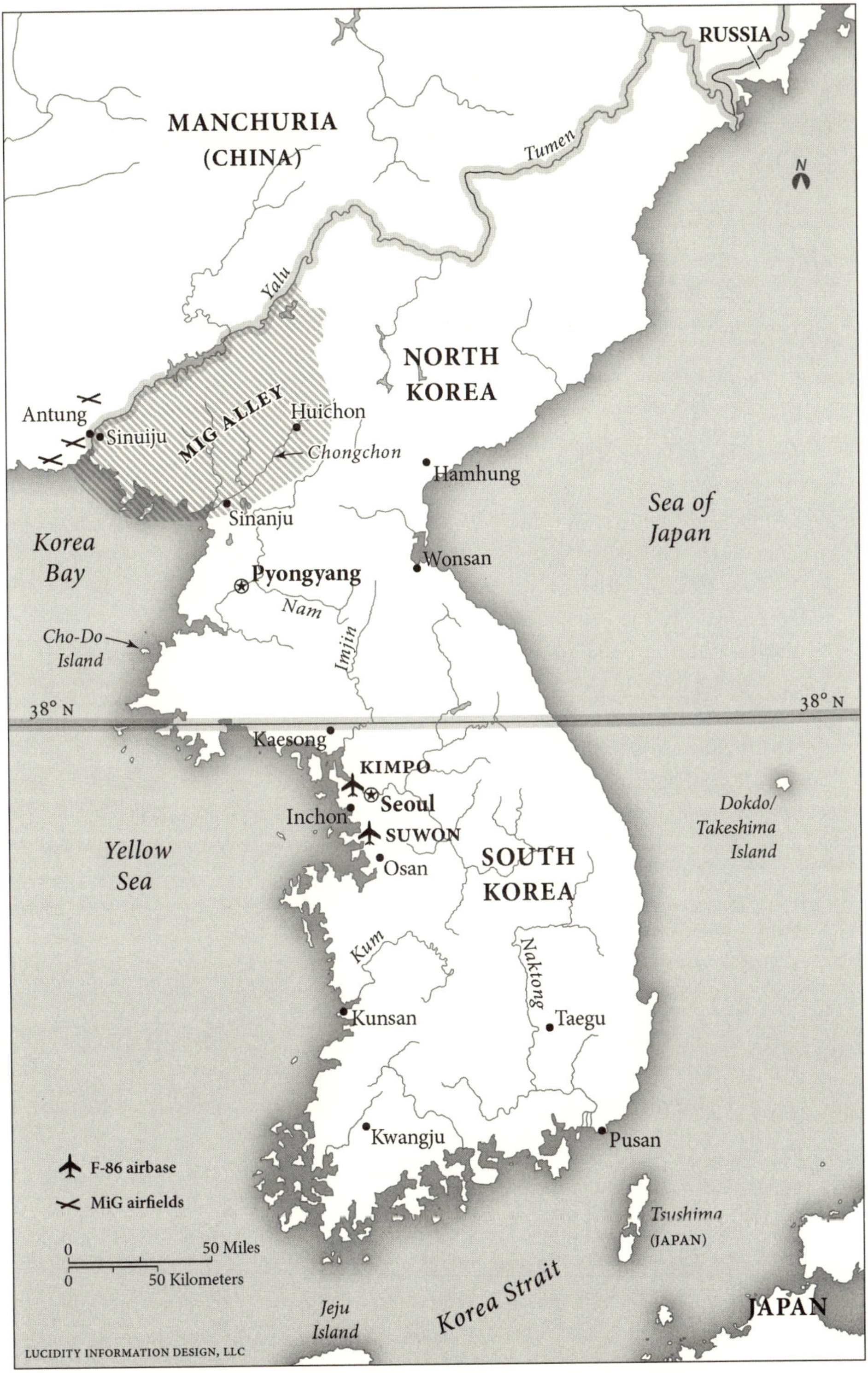

RUSSIA
MANCHURIA
(CHINA)
Tumen
N
Yalu
NORTH
KOREA
MIG ALLEY
Antung
Sinuiju
Huichon
Chongchon
Hamhung
Sea of
Japan
Sinanju
Korea
Bay
Wonsan
Pyongyang
Nam
Imjin
Cho-Do
Island
38° N
38° N
Kaesong
KIMPO
Seoul
Inchon
SUWON
Osan
Dokdo/
Takeshima
Island
Yellow
Sea
SOUTH
KOREA
Kum
Naktong
Kunsan
Taegu
Kwangju
Pusan
F-86 airbase
MiG airfields
Tsushima
(JAPAN)
0
50 Miles
0
50 Kilometers
Jeju
Island
Korea Strait
JAPAN
LUCIDITY INFORMATION DESIGN, LLC

Kimpo on January 2, 1951, and relocate their fighters to Japan. Lacking the air cover provided by the Sabres, FEAF's bombers reluctantly discontinued their attacks on the far regions of North Korea. The Chinese and North Koreans took advantage of the bombing lull to train additional MiG-15 pilots and make extensive repairs to their facilities south of the Yalu River. Only after UN troops recaptured the airfields at Suwon and Kimpo, and the Sabres returned to their bases near the thirty-eighth parallel, could the FEAF bombers be put back into action. That didn't occur until early March.[90]

Even then, despite their best efforts, the Americans couldn't make the aerial bombing campaign in Korea match that of the Second World War. The inability to strike lucrative targets in Manchuria, to include the MiGs' airfields, was only one aberration. Unlike in Germany, North Korea possessed precious few large, industrial targets amenable to massive strategic bombing. Additionally, many of the bomber escort tactics perfected by the fighter pilots during the prior world war were rendered obsolete by the speedy new MiGs.[91] Consequently, the Americans were never able to exploit the synergy between bomber and fighter that years earlier had led to the destruction of the Luftwaffe. Facing a numerically superior Communist foe that often retained the tactical initiative, the F-86 pilots were left trying to defend the bombers and win air superiority one downed MiG at a time.[92]

It was a daunting assignment, marked by frequent, heavy losses of the service's prized B-29 bombers. In October 1951 the Air Force chose to suspend its B-29 daylight precision bombing efforts rather than watch its bomber inventory continue to dwindle. The B-29s were shifted to night duty, leaving the more maneuverable but less destructive F-84 fighter-bombers to fly the daylight bombing sorties.[93] Although an acknowledgment of temporary defeat, the shift in strategy was welcomed by the Sabre pilots, who loathed the difficult bomber-escort missions. Now the Sabre pilots were free to perform "fighter sweep," the "most desired and enjoyed" mission because, they said, it presented "the best possibilities for an individual to demonstrate his ability and his aggressiveness as a fighter pilot."[94]

Still, the air war for the limited number of Sabres in theater remained challenging. The MiG pilots were experts at exploiting their nearby sanctuary in Manchuria to maximize their tactical advantage. One early assessment of Korean air combat noted that "the enemy could choose if he wished to meet the F-86 or not, and, if so, could choose the time, relative number of aircraft and usually the method to begin the encounter."[95] For the first few months of the Sabre-MiG battles, the MiG pilots tended to use "hit-and-run passes" during their attacks. The MiGs would launch from their sanctuary bases in Manchuria in flights of four, climb high above the Sabres while remaining just north of the Yalu River, and then sweep south, splitting up into elements of two, for a one-pass attack against the American aircraft before speeding back north to safety. Often, the MiGs tried to time their passes so that the first MiG element could act as a "decoy," overshooting the

American fighters and luring them into a position so that the second element of MiGs could score an easy kill.[96]

Beginning in mid-1951 and lasting through the following April, American pilots noticed a shift in the complexity of the MiGs' tactics. For example, newly observed "Yo-Yo" tactics involved up to twenty MiGs cycling through coordinated air attacks. The MiGs' "pincer and envelopment tactic" was even more complex, with two massive formations of sixty to eighty MiGs each carefully choreographed to fly south, turn to meet up near Pyongyang, and then proceed north in search of UN fighter-bomber aircraft and low-on-fuel F-86s that were flowing south. During this period, it was not infrequent for the F-86s to be outnumbered two or three to one in a dogfight. Occasionally, the numerical disadvantage even reached fifteen to one. Recalling his October 5, 1951, mission, Evans described seeing a "blob" of eighty MiGs—"an awe-inspiring sight—in fact, I was so awed I figured we'd all had it."[97]

Despite the abundance of MiGs in the sky and their occasional bouts of aggressive, offensive action, most days the MiGs chose not to battle the Sabres. FEAF intelligence officers eventually reasoned that the Communists must have been conducting "mass training," allowing their "new pilots [to] observe the application of classroom theory" as "their more experienced brothers demonstrated combat tactics through occasional engagements."[98] It was an incredibly unsatisfying period for the Americans as they watched numerous MiGs soar high above them. On one November 1951 mission, Sabre pilots reported seeing thirty MiGs "high in contrails headed southward but no contact was made." Some Sabre pilots tried goading the MiGs to fight, but more often than not, the MiGs did not take the bait.[99] Paddy Harbison, then a British-exchange pilot flying F-86s with the 4th at Kimpo, recalled the difficulty he experienced "enticing the MiGs to engage. It was very frustrating to fly to the Yalu, . . . see these enormous formations tracking back and forward in contrails above us, and not being able to engage them."[100]

Upping the Ante

On October 22, 1951, General Vandenberg finally had enough. For more than a year, the Air Force had tried to fight the Korean War with a shoestring force. Following a particularly disastrous week in which several B-29s were lost to MiGs, Vandenberg ordered Air Defense Command to surrender seventy-five of its eighty-one newest F-86Es for immediate transport to Korea. Shortly thereafter, he also initiated contracts to purchase more F-86s then starting to roll off production lines in Canada. Most of the new F-86Es were originally destined for the 51st FIW at Suwon, which had agreed to exchange its F-80Cs for the new Sabres, but a portion eventually found its way to the 4th at Kimpo. By December 1951, there were 165 Sabres—a mixture of A- and E-models—located on the peninsula, but the Americans remained outnumbered by a combat-ready MiG force five to ten times larger located just across the border in Manchuria.[101]

Compounding the numerical inequality, the American Sabre pilots who walked through the Torii gate at Kimpo on their way to their jets had a nearly two-hundred-mile journey back and forth to MiG Alley (figure 3.18). The Sabres began carrying jettisonable 120-gallon external tanks to augment their fuel capacity, but even then their time spent patrolling MiG Alley was limited to between twenty and twenty-five minutes due to the long transits. Worried that they would run out of gas during a heated battle, some pilots took to carrying alarm clocks to remind them when it was time to head home. Several did run out of gas, including Blesse on what would be his final mission in Korea. On other occasions, the external tanks would not jettison, which meant that the encumbered Sabre needed to abandon its flight and turn around for home, further dwindling the number of American fighters available for air patrol.[102]

Additionally, the Mikoyan-Gurevich MiG-15 maintained a slight edge over the North American F-86 Sabre in key performance parameters, notably climb rate, maximum altitude ceiling, and horizontal acceleration, all stemming from the Soviet fighter's lighter-weight design. The MiG could also turn tighter, and it possessed more powerful but slower firing guns, consisting of one 37- and two 23-mm cannons. When fielded in mid-1951, a more powerful VK-1 engine boosted the MiG-15's performance even more, and soon American pilots reported seeing MiG-15 formations cruising at altitudes approaching fifty-five thousand feet, almost a mile above the highest-flying Sabres.[103]

The Americans' heavier Sabre did, however, enjoy an advantage over the MiG in maximum dive speed. The F-86 also had faster firing .50-caliber machine guns, but the guns' range was limited, and the shells' destructive power, or lack thereof, left many aircrews longing for something that packed a bigger punch.[104] Still, with its redundant flight control systems, "heavy armor plate, bulletproof glass, adequate cockpit pressurization and temperature controls," and other extravagances like better gunsights, Mahurin likened the Sabre to a well-appointed Cadillac; the MiGs were more like stripped-down Fords. "There were many times when we felt that we would rather be driving Fords," Mahurin recalled, "but every time one of our pilots brought a shot-up F-86 home for a safe landing we were damn glad we had the Cadillacs." Nonetheless, most young American pilots considered the MiGs superior to their Sabres in the high-altitude dogfights of MiG Alley.[105]

The Americans' first update to their Sabres, the F-86E, began trickling into Korea in mid-1951 to replace the older A-models on a one-for-one basis. Vandenberg's October 1951 decision to reapportion Sabres from stateside bases to the war effort accelerated the E-model's combat fielding. Although the F-86E's arrival was applauded by the airmen, the aircraft retained the same A-model engine and was weighed down with 565 pounds of additional equipment, much of it due to the new A-1C(M) gunsight and its accompanying APG-30 radar. The Sabre pilots' demands for an aircraft that could match the MiGs' high-altitude performance would not be answered until the

FIGURE 3.18. Walking to the Flight Line at Kimpo Air Base. The Torii Gate at Kimpo, which became an iconic symbol of the Korean air war, subtlety reminded Sabre pilots of the tactical disadvantage they faced battling the MiGs. Whereas the Americans had to transit nearly two hundred miles one way to reach the battlespace, their MiG opponents at Antung were based less than ten miles from MiG Alley. NMUSAF.

following summer when the first F-86Fs began arriving, equipped with a higher-thrust engine and, more importantly, a new wing design. Armed with growing numbers of F-86Fs, the Americans in early 1953 finally achieved a significant technological advantage over their MiG-equipped adversary.[106]

But even before the combat debut of the more capable F-86F, the Americans were already routinely schooling the MiGs in the air. Looking back on the war, Air Force officials were astonished at their success despite the seemingly insurmountable odds. Initial claims pegged the Americans' superiority at fourteen to one, later refined in FEAF's final *Report on the Korean War* to ten-to-one based on its count of "792 MiG-15's . . . destroyed at a cost of only 79 F-86's."[107] When asked to identify the source of the American success, officials relied on a common refrain: the superior training, skill, and aggressiveness of American Sabre pilots.[108] That and their gunsights.

Although the assessments are usually caveated with the phrase "when it worked," several histories of Korean air combat and MiG Alley single out the new radar and A-1C(M) gunsight combination. For example, aviation historian Kenneth Werrell claimed that the Americans' "superior gunsight" was the "most significant" advantage "American pilots had over their Communist adversaries." FEAF's final report on the Korean War similarly concluded that the "superior armament and fire-control system" were critical contributors to the Sabres' lopsided victory over the Communist MiGs.[109]

Despite the platitudes, victory in air combat in MiG Alley relied on many things besides just the "superior gunsight" or just the "aggressive" pilot. The Sabre pilot's eyesight needed to be good enough to detect the MiG; the MiG had to give battle by descending to the F-86's altitude; the American pilot needed to be skilled enough to maneuver his aircraft into position near the MiG; the radar had to track the appropriate target; the A-1C(M) needed to compute the correct prediction angle; the Sabre's machine guns had to fire properly without jamming; and there needed to be some way to verify the MiG claim. Only then would the pilot have a chance to adorn his aircraft with a red star. As in wars past, the popular narratives of air combat conveniently overlooked this intricate chain of events that necessarily preceded a pilot celebrating a victory that night in the squadron bar.

Capturing Glory

The day after he disembarked from his trans-Pacific troopship on June 6, 1951, Lieutenant Evans shuffled through his remaining in-processing requirements at the 4th's rear echelon location at Johnson Air Base, Japan. The day's activities included attending a briefing offered by two Sabre pilots, both of whom had recorded a MiG victory. Evans emerged wide-eyed from the "well-organized lectures," noting in his diary, "From the sound of things so far, there are shades of Buck Rogers in this aerial war between the F-86s and MiG-15s."[110]

Powered by its six-thousand-pound-thrust J-47-GE-13 jet engine, the Sabre burst into Korea with a roar. But other than the rumble of a jet engine replacing the whirl of a spinning propeller, and in the F-86E a small radar that automatically fed range information to a lead-computing gunsight, not much else had changed in the fighter world since the later stages of World War II. The machine guns still fired .50-caliber rounds. The radios were equivalent. The formations were similar, including their emphasis on the flight lead-shooter and the wingman-observer roles. And the individual battles bore a remarkable resemblance to those from earlier, for although the A-1 gunsight promised successful attacks from greater distances and with larger deflection angles, pilots still strove to catch their opponent unaware before firing from a close-range, six o'clock position. Combat in the newer jets was simply now conducted at greater speeds, higher altitudes, and with more forceful G-maneuvers. With a few months of combat under his belt, Evans concluded, "Dogfighting hasn't changed, basically—it's still a matter of seeing first and trying to get on the other's tail or shaking him off yours."[111]

In Korea, as in wars past, glory went to the individual pilot who pulled the trigger and sent an enemy aircraft spiraling to the ground. But despite the continued aura of simplicity and objectivity, victories in the air remained subject to interpretation and shrouded in uncertainty. In an ostensibly "pure meritocracy," the chance of earning glory was not equally distributed, but bound instead in formations that were as much rooted in tradition as they were a product of tactical necessity. And air warfare between planes, although still celebrated as honorable and often depicted as relatively bloodless, remained as brutal, ruthless, and perilous as before. Nevertheless, bolstered by the official assessments that attributed the Sabres' success over the more numerous MiGs to the "well-trained, experienced and aggressive [American] pilots," the ace figure thrived in Korea.[112]

The Allure of Acehood

Evans was thrilled to be flying the F-86 Sabre. It was, he thought, "the hottest assignment in Korea," and he joked of the "jealous prop and straight-wing pilots" that only dreamed of mounting the sleek, swept-wing Sabre. Continuing the policy of the interwar years, the Air Force tended to send its best pilots to single-seat fighters, and by 1950 the Sabre was clearly at the top of the pecking order. In the words of Salter, only "those with the least élan would go to bombers and attack aircraft." Evans, validating his status as a Sabre pilot, proudly recalled one occasion when he and a flight mate traveled to Japan to pick up some new Sabres: walking through the operations building, "everybody just gaped in awe at us. After all, we were the only 86 Sabre outfit in the war, the big MiG hunters, and considered very hot stuff wherever we went—especially heading back for combat."[113]

While Evans may have overstated the reception he and his flight mate received, his story was not totally divorced from reality. Those flying the F-86 had by far the best chance at becoming ace celebrities. While there were handfuls of pilots in the F-84 and the F-80C, and even a few B-29 gunners, who successfully downed an enemy plane during the Korean War, only forty Americans downed at least five, earning the timeless distinction of ace, and they all flew the F-86. Leading the charge was Captain Joseph McConnell with his sixteen victories. McConnell and the other top four aces combined for nearly 10 percent of the total claims awarded to the Sabre pilots (72½ downed aircraft). Blesse quickly joined the ace ranks, eventually earning distinction as a double ace with ten kills. Mahurin, having flamed more than twenty enemy aircraft in World War II, came up two short of ace status during the Korean War, his attempts cut short by a sixteen-month stint as a prisoner of war (POW) in China after he was shot down on May 13, 1952. Evans too missed out. Charles "Chick" Cleveland, then a lieutenant, didn't join "that elite group" of aces until fifty-five years later in 2008, when access to Soviet records from the Korean War finally yielded confirmation of an earlier-deemed "probable" victory.[114]

As they had in the two wars prior, the ace pilots captured the nation's attention. *Life* magazine on July 7, 1952, proudly published the photographs of America's then-seventeen Korean jet aces, proclaiming them "a new and deadly addition to the ranks of air warriors."[115] Unlike in World War II, the fighter pilots in Korea did not have to compete for prestige with the service's bomber pilots. North Korea's dearth of suitable industrial targets meant that the lumbering bombers no longer traveled in awe-inspiring packs of hundreds, and their new missions to blow-up irrigation dams or drop napalm on villages as part of the limited "police action" left some feeling uneasy. Consequently, the Air Force elected to popularize the aerial conquests scored by its fighter aces and their sleek, jet-powered aircraft. Both offered a compelling story: the pilots once again in the romantic role of "aerial knight" dueling in the skies; their aircraft, a testament to the vision of the new, future-minded, technologically savvy Air Force.[116] Adding more fuel to the "great pilot" narrative was the pilots' own stories of combat, the symbolism of the single seat in their fighter plane, and the narrative that continued to emphasize the successful pilots' natural flying skills. The myth of the fighter pilot was alive and well.

Many Korean War pilots, like their predecessors, struggled to find an appropriate analogy to convey their experience of air combat. Some invoked sporting analogies, complete with connotations of sportsmanship and fairness. Evans described the squadron operations building after one victorious mission as "a madhouse, . . . like the locker room of the winning team of the game of the year. After that, who needs games?" Others simply relied on superlatives. Blesse for one recalled his MiG battles as "the greatest thrill of my life."[117] Regardless of the chosen analogy, the experience of combat was for many Sabre pilots intoxicating. When faced with the choice of going home with only four kills or extending his tour, Blesse opted to extend, having been "caught up in the Ace syndrome long before I got to Korea." He added, "I wanted that fifth kill so bad I could taste it."[118] Cleveland twice opted to extend his tour hoping to get his confirmed fifth kill. Writing to his wife of his decision, he explained, "This is a lifetime experience—it is the best part of my Air Force life, by far the best—all flying after this will seem lackluster." When he nearly ran out of gas chasing his goal, Cleveland was finally ordered home by his commander, still one confirmed kill short.[119] Salter noted, "It was not duty, it was desire" driving the pilots.[120]

The single seat in the F-86 likewise continued to fuel the image of the individual ace-pilot capturing glory. As Boelcke had earlier, the pilots in the Korean War reveled in the status, responsibility, and independence they enjoyed as their aircraft's lone occupant. Mahurin proudly declared, "A jet-fighter pilot is responsible to himself alone."[121] Evans similarly recalled the "very personal relationship each fighter pilot must establish alone with a single-seater."[122] Salter most poetically captured the feeling of "spiritual isolation" common to single-seat fighter pilots:

> You lived and died alone, especially in fighters. Fighters. Somehow, despite everything, that word had not become sterile. You slipped into the hollow cockpit and strapped and plugged yourself into the machine. The canopy ground shut and sealed you off. Your oxygen, your very breath, you carried with you into the chilled vacuum, in a steel bottle. If you wanted to speak, you used the radio. You were as isolated as a deep-sea diver, only you went up, into nothing, instead of down. You were accompanied. They flew with you in heraldic patterns and fought alongside you, sometimes skillfully, always at least two ships together, but they were really of no help. You were alone. At the end, there was no one you could touch. You could call out to them, as he had heard someone call out one day going down, a pitiful, pleading "Oh, Jesus!" but they could touch you not.[123]

Moreover, echoing the fighter pilot narratives of the past, the success of the ace-pilot in Korea, seated alone in his fighter aircraft and engaged in a contest often likened to a sports match, again appeared dependent upon his naturally endowed flying acumen. Many continued to believe that the necessary fighter pilot skills could not be taught, only refined. The prevalent attitude seemed confirmed when, toward the middle of the war, a shortage of fighter-trained pilots necessitated Sabre units to start accepting "re-called reserve officers who had no fighter background." Colonel Francis Gabreski, then commander of the 51st FIW at Suwon, complained that these new majors and lieutenant colonels "constituted a very serious problem" because they had "no desire to be fighter pilots, showed no enthusiasm for their work, and a few worked up fear of combat to the point that they were dangerous as a member of the combat team."[124] The 4th's pilots were more condescending in their assessment: "It is a waste of the government's time and money to take a 30 or 32 year old man whose service has been in troop transport or multi-engine flying, and attempt to train that man to engage in the highest form of aerial combat that has ever been developed."[125]

Even if properly culled from the ranks of Air Force pilots, the Sabre pilots still required significant practice. In Evans's words, it was "a constant honing of the edge—analyzing style, refining, polishing, dueling, *always* dueling—separating the pros from the amateurs."[126] For many, the honing commenced immediately upon arrival in the theater. After Evans had finished his in-processing at Johnson Field but before he went forward to his unit at Kimpo, he was thrown into "Clobber College," a semiformal training regimen designed to prepare green replacement pilots for combat.[127] There, the new pilots flew as often as they could, gaining greater familiarity with the Sabre's performance in a variety of flight regimes. Combat demanded, they were told, an instinctive ability to fly "at all the edges of an airplane's envelope, as fast as you can go and as slow as you can go."[128] The program included precious little detailed instruction on the skills needed to score a victory in the air, but then neither did much of the Air Force's fighter training of the period. There was some gunnery practice in the various curriculums, mainly shooting at a banner that was being towed by another aircraft along a preplanned course.[129] And there was occasional dogfight training, which emphasized aggressive, violent maneuvering and maintaining the offensive, six o'clock position behind a more experienced Sabre pilot.[130] But there was little intersection between gunnery training and dogfight training.[131] Furthermore, there was minimal formal flight instruction included in either regimen, and any learning was largely gained through individual trial and error or through often-embellished stories told around the squadron bar.

Nor was there any significant instruction during Clobber College on the pilots' aircraft or its new gadgetry, like radar-ranged LCGs. The lack of book study reflected the pilots' predilections. They much preferred to spend their time either talking tactics or perfecting their dogfighting skills, things they thought would be useful in combat.[132] Pilots needed to know how to max-perform their airplanes relative to their adversary, but knowledge of how that performance was attained, via either hydraulically boosted flight controls or a radar-aided gunsight, was considered by many to be mere pedantry.[133] In fact, many pilots of the period boasted of their *lack* of knowledge of the aircraft they flew, reinforcing the long-held belief that good pilots were *natural* pilots, capable of flying any winged machine solely through intuition.[134] Evans rationalized the pilots' nonchalant attitude toward formal instruction and technical knowledge, citing a World War I French Air Force saying, "The pilot should not be overly familiar with the intricacies of his machine—it is apt to destroy his dash." Reflective of the prevalent sentiment among his fellow pilots, Evans opined: "My feeling is, if the pilot can't solve it or affect it from the cockpit, why clutter his mind with all that junk? Like the French saying infers: He may learn so much about what *might* happen that he's just as liable to get too smart to do this job—then what?"[135] Unfortunately, many fell victim to their own bravado before they had the opportunity to regale their friends with tales of their machismo.[136]

The young fighter pilots entering Korea still looked to the historic image and the myth of the fighter pilot—conveniently stripped of its nuances and contingencies—as they sought to understand what an ace was and what it would take to become one. The service's ritualization of the myth and the corresponding "great pilot" narrative during semiformal instructional programs like Clobber College exacerbated the confusion. Fighter combat, it was understood entering Korea, was a young man's sport that demanded quick reflexes, exceptional eyesight, and the brashness of youth, all of which were known to degrade with age.[137] Pairing their finely honed flying skills with a ferocious and aggressive attitude, pilots planned to pounce on their foe and deftly maneuver their aircraft into a position to flame their Communist adversary.

Manufacturing the Ace

The Air Force's first jet ace, Captain (later Major) James Jabara, didn't fit the typical mold of the successful fighter pilot. Born in 1923 and a veteran flyer from the Second World War, Jabara was "short," "had weak eyes," and smoked "a dozen cigars a day." He was aggressive though, a "feisty fellow" and a borderline "discipline problem" in the words of the 4th FIW's commander, Colonel John Meyer, himself the seventh leading American ace from World War II with twenty-four kills. Meyer also described Jabara as a "hot shot Charlie type . . . who sang the loudest in the club and made more noise than the other people and dressed a little bit on the extreme side for the military." But when the Air Force demanded an ace-figure for the current fight, and officials at Fifth Air Force in Japan spurred their subordinate flying commanders to hurry along the process, Jabara had already distinguished himself with two MiG kills. Hence, Meyer identified Jabara as "the most likely candidate" for acehood. From then on, Jabara flew only the best missions that had a high likelihood of encountering MiGs, and he always flew as the flight lead to maximize his chances at downing a MiG. After having earned four kills but while still chasing the elusive fifth, Jabara was granted special permission to remain in Korea flying missions even after his squadron had rotated back to the Wing's rear-echelon base.[138]

Jabara also earned a reputation as a flight lead not particularly interested in his wingman's well-being. When wingmen were lost to the MiGs—a recognized risk of combat—most flight leads expressed anguish. For example, Mahurin wrote that he "was crushed" after one of his wingmen was shot down.[139] Pilots flying on Jabara's wing weren't so sure that the cocky flight lead was looking out for them. One, Bruno Giordano, complained that Jabara cared "less about what happened to his wingman and more than once put them in harm's way."[140] For example, during his celebrated ace mission, Jabara and his wingman became separated. Jabara subsequently explained that it was due to differences in aircraft performance, but others suggested that Jabara's wingman got "jumped by MiGs." In either instance, Jabara elected to press on alone, leaving his wingman to scrap for himself.[141]

Despite the hidden drama, Jabara became an instant celebrity when he downed his fifth and sixth MiGs on May 20, 1952. Most of the nation's leading newspapers featured history's first jet ace, accompanied by his photograph, on the front page. Many readers were treated to Jabara's own description of the impressive dual victory, told "in his usual calm manner," *New York Times* editors noted, "as though he were discussing a duck hunt." Jabara recalled: "I tacked on to three MiG's at 35,000 feet and picked out the last man and bored straight in. . . . The MiG burst into flames. . . . Just as the pilot bailed out the MiG disintegrated. Then I climbed back to 20,000 feet to get back in the battle. I bounced six more MiG's. I closed in and fired two bursts. . . . When my second burst caught him square in the middle he burst into flames and fell into an uncontrolled spin. All I could see of him was a whirl of fire." Although Jabara's account of "bor[ing] straight in" on the "last man" doesn't elicit the image of gallant, fair play usually featured in the standard air combat script, the narrative still reflected the common image of bloodless victory. The *Washington Post* report included with the description of Jabara's victories the standard air combat tally: "No American planes were lost in today's action that brought the Fifth Air Force's score against the Red jets in air-to-air combat to 49 destroyed, 12 probably destroyed and 124 damaged."[142]

After Jabara captured headlines, the nation's papers regularly updated readers on the ensuing race to become America's Ace of Aces in Korea. "13 Red Jets Shot Down in 2 Battles," trumpeted the December 14, 1951, edition of the *Washington Post*, "the greatest all-jet victory of the war." The action, reported the paper, also marked a "new record for United States jet pilots," with Major George Davis accumulating credit for "12 planes shot down in 16 days."[143] When Colonel Royal Baker, the commander of the 4th FIG from mid-1952 through early 1953, surpassed Davis as the top ace, the *New York Times* deemed the news front-page worthy.[144] Manuel Fernandez's and McConnell's tenures as Ace of Aces received similar coverage.[145] Most of the news stories included the thrilling details of the victorious air battles. After devoting precious paragraphs to the enthralling tales of "furious dogfights" in which "Sabre jet pilots blasted . . . MIGs out of the northwest Korean skies," the newspapers had only a few lines remaining at the bottom of the column to report on the ground action. In one *Chicago Daily Tribune* report, having devoted fifteen paragraphs to McConnell and the air battle over Korea, the paper summed up the action on the ground in a single sentence: "Ground fighting raged for the third day near outpost Horseshoe—a bend in the Pukhan river along the east-central front."[146]

Curiously, the pictures of the aces in the papers belied the image of the stereotypical fighter pilot. These celebrated aviators were not the usual young twenty-somethings characteristic of prior wars, but were instead often pushing thirty or beyond. Some called them the "'old pros' of MiG Alley."[147] Of the forty Korean War aces, one shy of half were field-grade officers holding ranks of major, lieutenant colonel, or colonel, and four of the eleven top scorers had also claimed victories in the prior world war. A veritable fighter pilot dinosaur, Lieutenant Colonel Vermont Garrison scored his fifth jet victory after his thirty-seventh birthday—he would go on to score five more and became a double ace in Korea.[148] In contrast, field-graders in World War II accounted for less than a quarter of the confirmed kills.[149] One FEAF study tasked with investigating the demographics of successful fighter pilots in Korea concluded that "some 68 percent of pilots who had destroyed MiG's were over twenty-eight years old, while 67 percent of the pilots who had scored no kills were less than twenty-five years old."[150] Another FEAF report noted the apparent contradiction, concluding that Korea showed that a pilot's "maturity, judgment, and experience," not "youth with its attendant peak physical condition," were "of greater concern in modern jet warfare."[151]

Perhaps a more significant influence on the disproportionate success of the senior aviators, rather than just the introduction of "modern jet warfare," was that for the first time in American history, the US military could take advantage of a large cadre of older, seasoned fighter pilots at war's outset, men who not more than a decade earlier had proved their mettle battling German Messerschmitts and Japanese Zeroes. Harbison noted that Korea "wasn't a young man's war; . . . most of the people were World War II veterans."[152] Another pilot explained that, consequently, these veteran fighter pilots had already learned the importance of aggressiveness and "fly[ing] to the sound of the guns."[153] Unmentioned was the fact that these older and more experienced pilots always flew in the all-important flight lead-shooter position in the fighter formation.[154]

New Tactics, Old Formations

Initially, the fighter pilot veterans tried to carry their tactics from World War II over into the jet age. The fighter formations proved semicompatible but not the bomber-escort tactics. One problem was the Sabre's appetite for jet fuel. Like most early jet aircraft, the Sabre consumed a prodigious amount of fuel at low altitude, which made it impractical to waste time forming up into massive forty or fifty aircraft formations like the pilots had done before in their P-47s and P-51s. The Sabre's high fuel consumption and limited supply also reduced the time it could devote to protecting the slow-moving bombers. The pilots tried to address this limitation by flying at extremely high altitude and slow airspeeds, but that left them at a significant speed disadvantage and potentially far out of the fight if a pack of MiGs suddenly appeared, diving on them or the bombers. Moreover, unable to draw the MiGs to battle, the Sabre pilots found it impossible "to predict where or when the MiG would strike next," which in turn made it "difficult to place our large formations in the right place at the right time."[155] In the end, the 51st FIW's tactics manual concluded that the "World War II fighter sweep practices . . . restricted both the exploitation of the tactical advantages of the Sabres, and the aggressiveness of the personnel operating them."[156]

The Sabre pilots consequently developed a new tactic, which they described as "offense in depth" and labeled the "jet stream." Rather than launch all the Sabre aircraft en masse, individual four-ship flights were launched at short, specified time intervals to provide in aggregate a constant presence over the target area as the bombers and fighter-bombers cycled through. If a Sabre flight spotted any MiGs, then in theory the flight leader could quickly summon the other airborne F-86s to the battle using the radio. Boasting of the new tactic, the pilots in the 51st declared, "With each flight responsible to itself, these highly mobile, maneuverable, self-supporting forces that can slash at formations of the enemy with relative impunity, have increased the normal aggressiveness of the individual Sabre pilot."[157]

Unlike the escort tactics, which did require modification, the Sabres' formations themselves, and the responsibilities specified therein, remained unchanged from those used during the past three decades. All the aircraft in the formation were there primarily to support the single flight lead flying in the number 1 position. As the most experienced pilot, he was responsible for leading the formation to and from the patrol area. If MiGs were detected, he would decide if, when, and how they would be engaged. The 4th's tactics manual noted that the flight leader "must be able to think alone, possessing that essential ability of being able to assess a combat situation quickly and accurately. He must be aggressive or all his other capabilities are wasted."[158]

The second-most experienced aviator flew in the number 3 position as the element lead. Together, the two leaders (flight and element) were designated as the "gunners" or "shooters" in the formation. The element leader's job was to protect the flight leader by maneuvering to thwart any potential enemy attack, "thus allowing the lead element that period of time necessary for the destruction of the aircraft he is attacking" (note the singular form, conflating the element with the element leader). At war's outset, the element leader typically flew 250 feet away from the flight lead, but later during the war he was granted more flexibility and instructed to fly "500 to 1,000 feet above the leader and at any position out and in back of the lead element where he can provide the best possible coverage."[159]

Both leaders had a wingman assigned to them. As "the subordinate member[s] of the team," the wingmen served as "protectors" or "lookers" for their leaders, leaving the more experienced pilots free to concentrate on gunning down MiGs. The wingman's "sole duty is to assure his element leader security and freedom of action for initiating and completing attacks," the 4th's tactics manual explained. If a wingman detected a threat, he was instructed to inform his flight mates but never to leave the formation; he needed "to stick like Scotch tape to his leader."[160] Flying only 150 to 200 feet beside and slightly aft of his leader (later expanded to a spacing of 200 to 300 feet)—close enough to read the stenciled numbering placarded on the leader's fuselage—minimized the chances of the wingman becoming separated when his flight lead initiated an abrupt maneuver. It also placed the wingman in a position where he could immediately transition to the offensive if his leader needed offensive support—for instance, if the leader's guns suddenly jammed or if he ran out of ammunition. Most significantly, though, the tight positioning allowed the wingman to serve as a veritable shield for his leader since any MiG bullets would likely need to pass through him before they reached the leader (figure 3.19).[161] "Watch the leader. Watch out for bogeys. Turn and look. Turn and look. That's the very fabric of survival," the wingmen were told.[162] Acknowledging the wingman's difficult task of constantly searching for enemy aircraft while simultaneously following an aggressively maneuvering leader chasing a wild and unpredictable MiG, the 4th's tactics manual observed, "It takes a cool, determined, reliable, skillful pilot to fly wing on a good flight commander."[163]

One author likened the wingman experience to being "blooded," explaining, "You can't be a good leader . . . until you've flown wing."[164] Lacking sufficient formal combat training, the wingman experience provided the junior pilots with an on-the-job practical education that would prepare them for eventual element and flight lead responsibilities. Through "every day training, precision, and strongly supervised and standardized formations and procedures," the 4th's tactics manual noted, the wingman would learn how to strike the tenuous balance between "sufficient discipline" and "individual initiative" demanded of combat flight leaders. It was trying, but, Mahurin explained, a wingman "takes solace in the fact that one day he will be a leader."[165] The training wasn't confined solely to the air either. William Brown, Jr., who would later retire as a three-star general, recalled being assigned errands around the base at the behest of his Sabre flight lead.[166] Cleveland was tasked with building his squadron's ping-pong table.[167] And whenever a wingman was sent to Japan to retrieve an aircraft that had been undergoing repair, there was a clear expectation that upon his return, his flight suit pockets better be bulging with the fifths of liquor needed for the squadron's rabble-rousing, evening martini-drinking contest.[168]

In retrospect, it is clear that the Sabres' formation was more than just a tactic for use in air combat. It represented a tool for the senior pilots to maintain control over the community, to garner glory for themselves by capitalizing on the flight lead-shooter position, and, most importantly, to imbue in their community's newest members the treasured fighter pilot ethos. Serving as a wingman in a combat unit became a rite of passage that one had to endure before being admitted into the fighter pilot fraternity. Most young lieutenants accepted their place at the bottom of the fighter pilot social order; as Sabre pilots, they were thankfully still well above all other aviators.[169] However, the "re-called reserve officers" weren't so enthusiastic about starting out again as lowly wingmen along with the junior-ranked pilots, and their reluctance to adhere to the fighter community's norms was one reason for the disdain shown them.[170]

The significance of the wingman position as a social element in fighter pilot life was also thrust to the forefront whenever a wingman somehow betrayed the

FIGURE 3.19. F-86 Four-Ship Formation. In the standard four-ship Sabre formation, the flight leader (number 1) and the element leader (number 3) were responsible for shooting down MiGs. Flying close aboard (approximately 150 to 300 feet away), the wingmen (numbers 2 and 4) were responsible for defending their leaders. This illustration of the Sabres' formation was included in a January 1952 report on Korean air combat filed by the Fourth Fighter Interceptor Group. AFHRA.

experience and began garnering kills, and consequently glory, before he had completed the prescribed initiation. The prime protagonist in Salter's *The Hunters* is not a MiG but rather a "pale lieutenant" named Ed Pell who, after successfully downing a MiG, demonstrates a penchant for seeking glory at the expense of protecting his flight lead.[171] Salter based the character on Lieutenant James Low, who, like the fictional Pell, lost his flight lead and then shot down a MiG. Low's flight lead was almost shot down in the ensuing melee, and the situation caused quite a stir among the squadron pilots. With only one month of combat time in Korea, Low explained, "I got lost and I got screwed up. I knocked an airplane down; I guess that vindicated me, but it wasn't my choice." Yet Low continued to rack up victories, becoming an ace only six months after finishing flight school. He also kept mysteriously losing his teammates—his flight lead when flying as a wingman, his wingman when flying as a flight lead. Some in the squadron wanted him grounded. Others reportedly threatened court-martial. But, as Blesse noted, it was "difficult to discipline a guy that comes back with a kill, even though he's broken the rules."[172]

These deeper social implications embedded within the pilots' formations and their influence on the fighter pilot community went unappreciated at the time. Lining up in formation at the end of the runway, ready to blast off toward MiG Alley, the pilots' minds were fixed solely on the task at hand. And for that task, the tactics and the formations were well suited for the ensuing battle. As members of the 51st FIW declared, armed with the "jet stream" tactic and arranged in "fluid four" formation, the flight leader enjoyed "unlimited freedom in maneuvers" and could execute "high speed tactics" with "absolute confidence that his wingman and element will remain with him to support his fire. . . . This formation makes a 'bounce' on an enemy as simple as is possible."[173]

Battling the MiGs

To execute a bounce, the Sabre pilots first had to detect the enemy MiGs using their "Mark I eyeball."[174] Visually searching for the MiGs was strenuous. Pilots had to constantly refocus their eyes out to infinity and slowly, methodically scan each sector of the sky, searching for any tiny black specks that might be far-off MiGs or any glints of sunlight that might be reflecting off a distant enemy's wing.[175] The task was made more challenging by the Sabre's near stratospheric patrol altitudes, which reduced the pilots' visual acuity.[176] Evans told of "the eye strain" the pilots developed "trying to see 100 miles in all directions at once. . . . A tiring exercise as it must be continuous; you can't afford to relax."[177] Many pilots, including ace Harold Fischer, took to carrying binoculars in the cockpit hoping to aid their visual search, but Salter and a few others thought the practice laughable, "like Nelson holding a telescope up to his blind eye."[178] Other pilots relied on different tools to improve their perceptive prowess, like one pilot's "special pair of half-mirrored distance glasses."[179] With the MiGs so difficult to see, and with the crop of older, veteran pilots leading most of the Sabre formations, many of the younger wingmen joked that they were brought along just so they could help their leader find the enemy.[180]

If a formation of MiGs was detected, the flight leader was supposed to calmly alert the other Sabres to its presence. An informative radio call would allow the other nearby Sabres to locate the fight quickly and speed toward it to provide assistance. Mahurin offered an example of one sortie's radio exchange: "The call should have been: 'Eagle Red Leader, this is Eagle Red Four. I think I have something spotted at twelve o'clock high.'" However, what came across the radio that day was: "Hey, I think I see something," followed by, "One of them is on my tail! Get him off," and "You'd better pull in tighter. He's shooting mighty near you." Mahurin, whose flying call sign was "Honest John," repeatedly inquired about the location of the rapidly escalating fight before finally shouting into his microphone, "Listen, you bastards, this is Honest John. If you guys want help, tell me where the fight is and tell me now." Their response, "To hell with you, Colonel. Find your own damn Migs."[181]

Mahurin's radio exchange was atypical, not for its content but because an actual dialog occurred. The AN/ARC-3 very-high-frequency (VHF) radio installed in the Sabre wasn't especially reliable, and it had only eight channels that couldn't be retuned in the air. Of the eight channels, five were needed for operations around the air base, leaving only three available for communications over MiG Alley.[182] With dozens of F-86 pilots all clamoring on the radio, along with the radar controllers at Kimpo or Cho-Do Island, once a fight erupted pilots often couldn't get a word in edgewise. "Look out for those two!" was a commonly heard call that relayed no useful information but instantly sent everyone's pulse skyrocketing.[183] To encourage better radio discipline, the senior pilots in the 51st borrowed the advice of earlier World War II aviators and told their fellow pilots, "If that bogey ain't close enough to shoot—or get shot at—keep quiet."[184] But in the heat of battle, few followed the advice, which left others grumbling about the incessant "yak on the radio." Evans once recalled a "maddening moment" when he saw "a stream . . . of MiGs" fly "right beneath our formation. I . . . was literally hopping with excitement trying to call out" the MiGs, "but the radio had erupted into such a babble I couldn't get in a word." On another sortie, he complained that the radio was "so boggled with fight chatter" that "I simply couldn't coordinate or communicate with my wingman, and this really torqued my jaws." It also almost cost him his life, since Evans's wingman was unable to inform him that he was lost and that Evans was on his own.[185]

Most engagements in MiG Alley were short, frenetic affairs as both sides sought to exploit the element of surprise. Sometimes, they lasted mere seconds. It was "over like that. Snap!" one Sabre pilot explained. "Somebody comes in and makes a pass and gets shot down."[186] The abruptness of an aerial engagement was central to Salter's vivid description of battle:

> Suddenly a plane flashes by beneath: huge tail, red stars, incredibly close. I turn after it, glance quickly behind, my heart pounding. It's clear, but Kasler [Salter's wingman] cries, "Check your right! Look right!"
>
> Not two hundred feet away, plain, foreign-looking, is the wingman. I turn hard towards him and then partway back. He seems fixed, frozen there, like a hare in the headlights. I'm nearly behind him. It will be point-blank. Before I can fire there are four of them almost on top of us, coming in from the other side. "Break left!" Kasler is calling. They turn with us, like cars on a speedway, and we are going down, I can't see if they are firing. Then we are alone; they've broken off when we didn't see it. It's over. Above us the contrails are already fading.[187]

Quick strikes weren't exclusive to the MiGs that magically appeared out of the blue, though. Blesse recalled his fifth kill "happen[ing] so fast it took me a moment or two to realize I had just gotten Number Five."[188] Cleveland's third kill also "wasn't much of a dogfight, just one diving turn and catching up to him."[189]

Occasionally, the Sabres were able to detect the MiGs at long distance as the enemy fighters streaked across the sky at high altitude, their massive formations drawing out "spider webs" with their wispy, white contrails. Upon observing the MiGs above, the American pilots would try to gingerly coax their aircraft up into an offensive position, but they usually fell short due to the better altitude performance of the Soviet interceptor. When that happened, the 4th's pilots planned to "place ourselves in a position very close along side [*sic*] and below the enemy formation while moving in the same direction as they." There was one caveat: "If, however, we are reasonably sure we have not been seen by the enemy, we often try to stay directly under him, or under and slightly behind him and wait for him to descend" directly into the path of "three pairs of .50 caliber guns." In either case, the Sabre pilots had "no choice but to keep [their] Mach high" and "wait until [the MiGs] made the first move."[190]

When the MiGs finally did pounce, they would slice down into the Americans' formations with guns blazing before streaking northward toward their Manchurian sanctuary. Occasionally the aerial contests would devolve into swirling dervishes with repeated high-speed passes and fleeting, slashing gun attacks. Sabre pilots did their best to maneuver into the good tracking position that the A-1C(M) demanded during these "Korean merry-go-rounds," but achieving the stabilized position was a rarity as the MiGs rolled, dived, and zoomed trying to get away.[191] Almost always outnumbered, the Americans also had to be constantly alert to new MiGs seeking a quick kill. As Salter learned, often just as a leader was moving into an offensive position against one MiG, his wingman would call "Break!" on the radio and the flight lead would have to immediately shift his focus from glorious victory to mere survival. That usually entailed a "steep, high Mach, high 'G' descending spiral," an exhausting and physically demanding maneuver designed to lose the MiG.[192]

The American pilots had to watch out for other Sabres jumping into their fight too, ostensibly to assist, but often seeking to claim a MiG for themselves. Fischer remembered almost colliding with another flight of Sabres as both groups chased down some MiGs "since neither [Sabre] formation wanted to give way and lose the advantage."[193] At least Fischer and the accompanying Sabres didn't mistakenly open fire on each other. With multiple fighters swirling around, and with the MiG and Sabre similarly defined by their cigar-shaped fuselage and swept wings, it was sometimes difficult for pilots on both sides to distinguish friend from foe. On one occasion, Mahurin and his wingman became separated after they turned in opposite directions; Mahurin said "Break right and up," but his wingman heard "Break left." After turning 270 degrees, the two Sabres were "face-to-face," and Mahurin, "taking no chance at all" that the aircraft in front of him might be a MiG, "started firing immediately." He finally took his finger off the trigger when he heard his wingman screaming over the radio, "Watch it, it's me."[194]

Like in air combat before, the best results were obtained when pilots could sneak up close to their prey and wail away from between fifteen hundred to five hundred feet back. Extolled in their tactics manuals to "*Shoot to Kill!*," Sabre pilots longed for the unaware MiG pilot who would let them achieve such an optimum attack position. But, more typically, with at most only a few seconds available to line up behind a MiG and without having received extensive training in high-deflection "snap" shooting, many an F-86 pilot resorted to just spraying some bullets out in front hoping for a lucky shot.[195]

Unfortunately for the Sabre pilots, even if they successfully got to close range, hits were not always guaranteed.[196] And even if hits were observed, the Sabre's six .50-caliber M3 Browning machine guns, essentially a faster-firing version of the .50-caliber machine gun from World War II, were not always capable of crippling the MiG target. In Salter's opinion, the Sabre's machine guns were "almost feminine in comparison" to the MiG's "cannon—the maw of a MiG seemed swollen and menacing."[197] Exacerbating the plight of the Sabres, the MiG-15's stern quadrant proved particularly resilient to the Americans' smallish and low-velocity .50-caliber shells. Some estimates suggested that up to 95 percent of bullet strikes against the MiG's rear would be ineffective at catastrophically disabling the aircraft.[198] On one occasion, a disheartened F-86 pilot reported getting "good 'strikes' along the right wing and fuselage" of his MiG-15 target, only to then watch the presumably damaged Soviet plane initiate "a 60 degree climb" to "at least 6,000 feet above the point where the F-86 had to level off to keep from stalling" and then fly to safety "across the Yalu River."[199]

Frustrated by a political border that they couldn't cross, flying aircraft that couldn't climb to reach their foe, and using bullets that couldn't bring their enemy down, but still enchanted with the allure of acehood and aerial victories, many Sabre pilots began taking matters into their own hands. When one of their external fuel tanks refused to jettison at the start of an engagement, both Blesse and Jabara defied higher guidance mandating that they return to base and chose instead to battle with the MiGs; their decisions nearly cost them their lives.[200] Evans and Mahurin likewise recalled their fellow pilots getting so MiG-hungry that they began "taking extremely hazardous risks" like "purposely forget[ting] to call 'Bingo,'" the code-word for low fuel, in the hopes that a few extra minutes might bring along an easy MiG target.[201] Blesse actually ran out of fuel shortly after getting his tenth kill and nearly fell into North Korean hands, ejecting just a half-mile off the enemy coast.[202] The leading Sabre scorer at the time, Major George Davis, elected on one mission to abandon his squadron and took his wingman to go hunt MiGs alone—Davis was shot down and killed that afternoon.[203]

Perhaps the most egregious example of pilots' reckless pursuit of glory occurred when Sabre pilots started going north of the Yalu River searching for MiGs. Cleveland recalled the first time he became aware that pilots were routinely violating the Manchurian sanctuary. He was standing outside waiting for a bus in the pouring rain when a car stopped, the window rolled down, and the occupants shouted, "Get in, Lieutenant." Inside, Harrison Thyng and Mahurin, the 4th's Wing and Group commanders respectively, were bantering about their recent missions. Then Thyng proudly announced, "You know my latest one was just after he took off. When you get 'em just after take-off and they're all full of fuel, they burn." Hearing the two colonel-leaders, fighter aces, and "role models" describe their trips north was sufficient justification for Cleveland, and he explained that it soon "found its way into my psyche that's where the action was. I remember Colonel Mahurin saying, 'Well, wherever you get them, north of the river, over the river, south of the river, every time you shoot one down, that's one less that's going to be up the next day hunting you and your buddies.'"[204]

Mahurin had started streaking past the Yalu chasing MiGs months earlier when he was flying with Gabreski in the 51st. The two colonels came out of a meeting one day and concluded that the Communist sanctuary, which allowed the MiGs to "lurk on [their] side of the border, dart over to our side, shoot at an F-86 and return . . . at will, . . . was a distinctly unfair advantage and it was costing lives." Even if it was strictly against the rules, they both agreed that it was time the American F-86s finally caught "a break." Besides, they reasoned, better to be alive seated at a court-martial than to be shot dead in the skies over North Korea.[205] The 51st's first mission north of the border was a resounding success. The "Maple Special" flight claimed three MiGs, the enemy pilots surprised no doubt by the Sabres' presence in Manchuria. Other Sabre pilots learned of the action, and soon whole squadrons assigned to the 51st were hunting for MiGs on the wrong side of the Yalu. When Mahurin transferred out of the 51st to become the 4th's Group commander, the pilots in that unit quickly followed suit. Although he continued to justify the flights on the basis of "forc[ing] the Communists to come to terms more rapidly," Mahurin also proudly observed that "with the border restrictions off, our pilots began to gain in victories over the 51st."[206]

For the Sabre pilots, killing MiGs north of the Yalu River might have eventually helped thin the ranks of Red fighters appearing over MiG Alley, and it could have also generated a significant psychological advantage over the Communist pilots, but it would do little to further their individual quests for glory unless they could figure out a way to claim the victory. The American pilots couldn't very well report the facts of an engagement conducted in clear violation of higher guidance, and they certainly wouldn't forsake credit for the kill. Consequently, they concocted a standard cover story: "We had followed the enemy aircraft as they climbed for altitude, but kept well south of the border. When they came down into our territory we attacked, shooting down three while the rest escaped us by flying north. The three we destroyed fell to the ground on the Manchurian side of the border." Mahurin proudly noted that the ruse "satisfied our intelligence man," but that represented only the first step in eventually earning credit for the victory.[207]

Getting the Credit

Initially in the Korean War, there were three criteria that could earn a pilot credit for an aerial victory: if an enemy aircraft was observed to "strike the earth" by another pilot or someone on the ground; if "gun camera film showed" or another pilot "witnessed aircraft disintegration, shedding of a major component necessary to sustain flight, a persistent fire in the engine or tail section, or pilot bail-out"; or if the enemy aircraft was seen by another pilot "in such a position that known limitations or circumstances would preclude possible recovery." According to the guidance, if two pilots could prove that they both contributed to the kill, then the Air Force would recognize both, but only with a half credit apiece.[208] In the turbulent skies over Korea, with pilots flying only a set number of missions before rotating back to the States, and with the war frequently on the verge of ending at the negotiating table, gathering only half-kills rendered the chances of a pilot accumulating the sought-after *five* that much more remote.[209] Luckily for the flight leads, the Sabres' formation practices ensured that they would usually earn full credit for any downed aircraft. Collectively, the policy and the practice ensured that the glory of air combat remained tightly focused on the individual heroic pilot.[210]

Cleveland had relied on newly released Russian documents to verify his claim and earn credit for his all-important fifth kill fifty-five years after his victorious sortie. Other Sabre pilots likewise relied on intelligence reports to corroborate their claims, although the reports were usually more timely.[211] Most of the pilots, however, used more conventional sources to validate their aerial triumphs. One was the small AN-N6 gun camera mounted in the Sabre's air intake. Unfortunately, despite occasionally capturing some spectacular imagery (figure 3.20), the World War II–era camera was ill-suited for jet-age air combat. The high-speed, slashing attacks that characterized Korean dogfights made it difficult for the camera's fixed twelve-degree by

FIGURE 3.20. Validating a Kill. When the nose-mounted gun camera in the F-86 functioned correctly, it could capture some spectacular imagery, used to both validate a pilot's victory claim and thrill the public at home. However, proper operation of the Sabre's camera was never a surety. Fortunately, if gun camera footage was not available, the leader usually had a wingman alongside him who was eager to validate the claim. NMUSAF.

seventeen-degree field of view to capture damage to a MiG target, especially when it appeared on only a few frames of film.[212] Additionally, on occasions when a close-range, stern attack was made, the Sabre pilot's attempt to film his kill was frequently thwarted by the diminutive striking power of his aircraft's .50-caliber ammunition: hits might be observed, but the film evidence would invariably fail to reveal significant damage. On one sortie, Fischer "fired several long bursts" into a MiG from close range, but the enemy aircraft kept flying. As Fischer was maneuvering back for another attack, the MiG pilot ejected, and the enemy plane went into a steep dive toward the ground. Knowing "that there was no positive verification [of the kill] on the gun camera film," Fischer followed the empty plane down and "strafed the wreckage for confirmation purposes."[213]

Other times, the camera simply failed to work. The film wouldn't run, it would run out at an inopportune time, or it would be too blurry or otherwise distorted to reveal the necessary images. With the Kimpo base photo lab processing up to twenty-three thousand feet of typically lackluster combat film a month, a pair of 1952 Air Force inspections declared the Gun Camera Program "not functioning."[214] The pilots were understandably frustrated. Several Sabre pilots estimated that their cameras worked between only 10 and 20 percent of the time, and they were incensed that their victory credit while "flying a quarter-million-dollar airplane" depended on "a two-bit sight and camera." The unreliable camera had already cost Evans one kill, and after landing from another exciting mission, Evans fumed, "my film didn't come out again. It makes a man feel like a jackass to have no film to show after a good scrap."[215] Of course, there was always the chance that the camera did work, and when the Sabres started prowling in Manchuria, the pilots had to take great care ensuring that their violations were not documented on film. Mahurin recalled, "We had to kill a lot of combat-film footage."[216]

Fortunately, whether the camera worked or not, there was usually a wingman, properly indoctrinated on the importance of deferring to his flight leader, ready and eager to validate a victory claim. Snuggled close to the lead aircraft, the wingman was usually in prime position to witness and verify the aftermath of his leader's attack. Wingmen quickly realized that validating their leaders' claims was one of their key responsibilities, just as important as shielding their leaders' six o'clock. Cleveland recalled, for example, one postmission discussion that ended with an unsolicited "Yeah, I saw what you saw and whatever you claim" from his junior wingman.[217] Sometimes, a distracted or momentarily absent wingman resulted in the loss of a confirmed kill, which helped explain why flight leads were continually harping on their junior wingman to pay attention and maintain formation.[218] Ace George Davis often made a point of alerting his wingman to an impending kill by calling out "Look now, Baker Two" just before flaming a MiG; sadly, those were also his last words.[219]

There were occasional squabbles between a wingman and his flight leader over a victory credit. Mahurin remembered one clash between Thyng and his wingman, Paddy Harbison. Thyng, already a World War II ace, chased down two MiGs and quickly flamed them both, marking his fourth and fifth kills and earning him ace status for the Korean War. "[Paddy] was really bitter about Harry's being so greedy," Mahurin remembered. "Paddy claimed that Harry should have shot the first Mig down and turned the other one over to him, sharing the wealth, so to speak. Harry couldn't see it, because those two victories made him a jet ace."[220] Flying on Jabara's wing (after Jabara had already become an ace), Giordano recalled one occasion when he returned to base ecstatic at having earned a kill, only to have Jabara sharply rebuke him and instead claim the kill on the grounds that he as the flight lead had initiated the attack.[221]

As the Sabre pilots traipsed north of the Korean border and padded their victory totals with MiGs they caught during takeoff or landing, and with some leaders pressuring their wingmen to validate otherwise questionable claims, the Korean War pilots experienced a revelation shared by their predecessors in the First World

War: despite the popular narrative, the glory associated with individual aerial victories was not without blemish. Mahurin remarked that his boys "considered it slightly dirty pool to shoot [the MiGs] down just as they were trying to land."[222] Salter likened himself and his fellow Sabre pilots to "overcoated gangsters in limousines, high above North Korea . . . we too were sleek for murder."[223] Still the pilots continued, knowing that "each victory added up" both for the war effort and for them as individuals.[224] Moreover, the swashbuckling "bravado" with which the American pilots went about their tasks was, in Mahurin's estimation, a "veneer to cover up the fact that they were well aware of the dangers of combat flying."[225] One pilot quipped, "It sounds great to be a fighter pilot, but when you see those 30-millimeter cannon balls coming at you (and it only takes one to blow the tail of an airplane off), then people say wait a minute, this is a little too sporty for me, and then they back off."[226]

On the surface, many pilots continued to popularize the "clean, impersonal" imagery of aerial combat, with victory signaled from a distance by a gracefully descending white parachute, the occupant no doubt shaking his first and cursing at the victor.[227] However, the relatively close-quarters fighting and the overarching desire to confirm any kills produced haunting sounds and searing images during the aerial contests. Mahurin recalled seeing his wingman get hit by a lucky MiG cannon shell, then hearing frantic, "labored breathing" over the radio and calls, "Am I on fire? I've been hit personally. . . . Tell me, am I on fire? Am I on fire?" The wingman ejected and Mahurin "never saw him again."[228] Another American remembered seeing a stricken F-86 pilot "squatting in his seat trying to shove the canopy off the F-86 when his airplane went straight into the ground."[229] Fischer had the misfortune of witnessing one of the more disturbing images of the air war as he passed a flaming MiG and saw the "pilot beating on the canopy, trying to escape. The heat must have been intolerable for the canopy was changing color and the smoke was intense." The sight had a powerful effect on Fischer: "Up to that moment the enemy had been impersonal, each aircraft a target that had little meaning and not associated with flesh and blood. But the sight of another man trapped in the cockpit of a burning aircraft with no power and no place to land was impossible for me to forget."[230]

As in the past, this sordid "richness" of air combat was rarely acknowledged. Instead, the public and the pilots were treated to stories and images that celebrated the glory of air combat and its gallant aces, warfare conducted at a comfortable distance from the enemy over beautiful landscapes and against a vibrant blue sky (figure 3.21). But, as before, the pilots knew that air combat remained a brutal and gruesome business of killing. Korean War aviators, like their predecessors, sought victory wherever they could find it and however they could win it. Blesse later remarked that he "was surprised at how cold and calculating I had become" when he selected his prey and initiated his attack.[231] The 4th implored its pilots, "When in doubt, attack."[232] Mahurin remembered doing just that: "We shot them down in the landing pattern, we shot

FIGURE 3.21. Jabara becomes History's First Jet Ace. Captain James Jabara became America's first jet ace when he downed his fifth and sixth MiGs on May 20, 1952. The historic ace victory was not without complications: one of Jabara's external fuel tanks failed to jettison, and he and his wingman had become separated during the battle. Stephen Doyle's painting of the engagement reflects the dominant themes within the fighter pilot myth, with air combat conducted by isolated individuals at a close-but-still-comfortable distance away, set against a beautiful landscape and a vibrant blue sky. *Jabby's 6th Kill*, painting by Stephen Doyle. Used with permission.

them down on local test hops, we shot them down on training flights, and we shot them down anywhere we could find them."[233]

With the emphasis on claiming glory through aerial victory, one would imagine that the Sabre pilots would have welcomed their new gunsights, which promised to increase their chances at earning another all-important red star for their fuselage. Alas, most shunned the new technology.

Using the New Gunsights

When the newer Sabres with their A-1C(M) gunsight and its accompanying APG-30 radar started arriving in theater in mid-1951, the Sabre pilots had already started encountering the massive MiG formations circling high above. A year later, the maddening experience of helplessly watching scores of potential MiG kills flying just out of reach above was fresh on the minds of the fourteen F-86 aces who gathered before the CSAF for a late-summer meeting. There, the aces "vigorously" argued their case to General Vandenberg: the new A-1C(M) radar gunsight in the F-86E wasn't working. Moreover, the once-touted system was hindering the American pilots' struggle for aerial supremacy. They would be better off, they told the chief, if they could rip out the new sights and replace them with the older, more reliable, and lighter-weight K-14 or Mark 18 gunsights. Swayed by their arguments, Vandenberg on September 4, 1952, ordered the Air Proving Ground at Eglin AFB, Florida, to conduct an immediate, "accelerated comparison" of the new, radar-ranging sights to the older, manually ranged LCGs. The outcome of the test would decide the future of radar and advanced gunsights in Air Force fighters.[234]

Some histories attribute the controversy over the new radar-ranged gunsight to the older combat pilots' reluctance to adapt their practices and their brazen, self-serving attempts to preserve their place at the top of the fighter pilot hierarchy. For example, aviation historian Thomas Wildenberg claimed that "older, more veteran air aces" like Gabreski "had the most to lose" when the A-1C(M) arrived in theater: "As an elite fraternity within the Air Force, their status depended upon their skill in downing enemy aircraft—skills that took many missions and combat encounters with the enemy to perfect." With its promise of being "almost completely automatic in its operation," the new gunsight, according to Wildenberg, threatened to dilute this guild of aces with crops of younger pilots who possessed significantly "less combat experience."[235] When Lieutenant Low became the youngest ace, it was, according to Wildenberg, his success that riled the established aces, not his propensity to break formation. Furthermore, the success of youngsters like Low, according to the argument, demonstrated the simplicity and utility of the new sights if one was not hidebound by obsolescent practices.[236]

Even after almost fifty years, the Korean aviators' opinions of the new gunsights remained sharply polarized, with one group claiming they "were not bad in the F-86E" and others declaring that "the improved radar ranging system that came with the E model was not much better" than the earlier sights.[237] It is through this controversy that we most clearly see the competition between the "great machine" and "great pilot" narratives, both colored by the myth of the fighter pilot, and the consequent coevolution of the human-machine system as pilots struggled to learn the intricacies of their new equipment while those same technologies were simultaneously adjusting to the reality of Korean air combat.

Shooting (and Aiming) to Kill

As in wars past, the Korean War fighter pilots preferred to get as close as possible to their targets before opening fire. Jabara for one preferred to do his work from within seven hundred feet.[238] Blesse too had a tendency to get right behind his adversary. On one occasion, he steered his formation so close to his target that when it began to disintegrate from his bullets, "a big hunk of . . . stabilizer came off" and got sucked into his "wingman's airscoop"; the wingman ended up having to eject over the water. Another time, Blesse opened fire on an LA-9 from only two hundred feet away; as the North Korean plane started coming apart, it sprayed aviation oil all over Blesse's Sabre, leaving his glass canopy looking "like someone had pulled a shade over it."[239]

The Sabre pilots knew that firing from close range simplified the gunnery problem and substantially increased their chances of scoring a hit. It also helped compensate for the lackluster performance of the F-86A's Mark 18 gunsight. Designed for propeller-driven, World War II aircraft, the sight was not properly configured to deal with the high angular rates and excessive G-forces that characterized air combat in the jet age. One combat report on the Sabre's Mark 18 complained that the gunsight's "ranging control [was] much too stiff and erratic" for the rapid closure rates in modern dogfights, which made it nearly impossible for the pilot to properly span the enemy target. Pilots also complained of the "excessive movement of the sight pipper" during tracking and "excessive movement of the reticle . . . when [the] guns were fired." Furthermore, the Mark 18 was optimized for use at an airspeed of 357 knots, but the Sabre routinely traveled more than 100 knots faster than that. In the end, many Sabre pilots deemed the Mark 18 "worthless." Rather than rely on the sight's lead-computing functionality, several resorted to using the gunsight in its fixed, iron-sight mode.[240]

Without the benefit of the LCG, the Sabre pilots were not much different from the pilots of the First World War, endeavoring to get so close to their targets that their bullets essentially couldn't miss. But sneaking up behind a target to destroy it at close range remained difficult in practice, especially given the MiGs' preference for quick, slashing attacks initiated from high altitude. Additionally, if a Sabre pilot did find a suitable MiG target, his enthusiasm for the looming kill and his worry that his MiG prey might suddenly maneuver to spoil the shot often led to less-than-ideal gunshots. During his first air-to-air engagement in Korea, Blesse went chasing after a MiG proudly declaring over the radio, "Colonel, I'm going to get me a MiG." However, Blesse was too far out of range when he opened fire, and his target sped away unscathed.[241]

Consequently, while many strove for the perfect close-range, directly astern position, and while some pilots may have thought that they had achieved it, few ever did. Echoing earlier assessments of air combat from World War II, a review of Korean War gun film identified a "wide-spread" problem of pilots "grossly" underestimating their "ranges and, to a lesser extent, angles-off" from their targets.[242] The Sabre pilots might have been able to close to within fifteen hundred feet of their prey, but they were often more than ten degrees offset from the target's tail.[243] Harkening back to Mahurin's warning to his fellow World War II aviators not to believe the movie portrayals of air combat, officials in 1952 likewise concluded that the Korean War pilots' exposure to "films released for public consumption, showing close-in, highly effective attacks on enemy aircraft" was having a pernicious effect on their performance in actual air combat.[244] So too was the pilots' exposure to nonrealistic gunnery training.

These were the exact issues Davis's team at Wright Field had hoped to address when they set out working with MIT to develop the A-1C(M)—a gunsight that would help pilots take more effective, longer-range, higher-deflection gunshots with minimal training. But there was a catch. As noted earlier, the accuracy of the gunsight's prediction angle was based on three assumptions, and in Korea those assumptions rarely held true. For example, rarely did the Sabres have the opportunity and the time to match the target's plane of motion, the critical step that allowed the A-1's gyroscopes to measure the rate of change needed to calculate the proper prediction angle. The 51st's tactics manual noted the irony, reminding pilots, "Proper tracking is mandatory for the [A-1C(M)] sight to compute properly"; it then declared, "Tracking is not possible in high 'G' maneuvers or acrobatics flights and reliance on the fixed sight, manual ranging procedure, and tracers becomes the general rule in these cases."[245] The A-1C(M) sight might have done well shooting down large bomber aircraft under more docile, controlled conditions, but pilots were discovering that the gunsight was not especially well designed to deal with the small, nimble, maneuvering MiG fighters over Korea.

There were other A-1 design limitations that became apparent in MiG Alley. For example, when the Sabre pilot pulled more than 3Gs, a second reticle image would appear above the primary—the result of the G-distorted optics on the windscreen. "Pilots have been known to fire using the false image," one Sabre pilot reported. Echoing others' concerns, the same pilot also complained that the "range drum is too difficult to read and requires too much concentration" to interpret; in addition, the reticle image and pipper needed to be a different color, like "lime green," so they were easier to see when the target was in line with the sun. Another pilot noted that the variable diameter reticle was insufficiently scaled, so that at long ranges it did not properly span the target, "which sometimes give[s] pilots the impression that the sight is not ranging properly." As was common with the Mark 18, pilots frequently griped about the A-1's "nervous pipper" when they fired their guns.[246]

Another major critique of the A-1's design was the pilot's inability to control the radar, which, when coupled with the radar's ability to lock targets far out of range of the guns, produced wildly erratic and distracting reticle movements at long range. It was, the engineers acknowledged, "a first-class headache to the inexperienced pilot." Draper's team had anticipated this situation, which is why the pilot was instructed to depress the electrical caging switch on the throttle grip until he was ready to begin tracking the target. But everyone soon learned that it was impractical to fly around while constantly holding the caging button down, so another solution was introduced in mid-1952.[247]

The Jenkins range limiter, so named after its designer, Jack Jenkins, incorporated a variable rheostat mounted in the cockpit that allowed the pilot to adjust, for practical purposes, the "maximum operating range" of the sight. As Jenkins described it, the range limiter was "based on the concept that there is no fundamental reason for flying a perfect lead pursuit course [as determined by the gunsight] at ranges appreciably in excess of the effective range of the weapons carried. The use of a range limiter allows the pilot to sweat out the initial transients and get on an approximately correct lead pursuit course while avoiding some of the undesirable sight dynamics which are prevalent at the longer ranges." With the limiter installed, if the radar reported a range in excess of the selected maximum range, the prediction computer in the A-1 ignored the radar data and instead used the preselected range to compute the bullet's time of flight, thereby reducing the commanded prediction angle and the sight's sensitivity. Once the radar range collapsed to less than the preselected value, the actual radar range to the target was used and the correct prediction angle displayed. The Jenkins-modified sight also provided a visual cue to the pilot whenever the radar-computed range to the target was less than the selected maximum range, obviating the need for the pilot to continually cross-check the radar range drum located on the side of the A-1C's sight unit. Despite being "a rather crude affair thrown together on a dark night," Jenkins's range limiter "proved popular immediately." With its introduction, the Sabre pilots at Kimpo noted that "the fire control system . . . approached more closely the ultimate of complete automatic function."[248]

Sabre pilots' attempts at scoring a longer-range kill were stymied not just by the early limitations of their gunsight but also by the quality of their ammunition. The ineffective striking power of the .50-caliber bullet meant that, even against a stable target, the longer-range, high-deflection aiming capabilities of the A-1C(M) could not be fully exploited. Many Sabre pilots began complaining about their armament almost immediately after they started tangling with the MiGs. One of the first combat intelligence reports detailing a MiG-15 battle noted that it took "an unusual amount of .50 caliber ammunition" to achieve "lethal results."[249] A subsequent report noted that by December 1950 "the consensus [among Sabre pilots] is that fire power of the F-86 is not sufficiently destructive, and should be modified with a caliber heavy enough to insure [*sic*] structural damage with a minimum number of hits."[250] Later analysis concluded that it took on average almost eight and a half seconds of concentrated .50-caliber fire, equating to 1,024 rounds, to bring down a MiG, an extraordinarily long time for a "pilot concentrating on a gunnery problem" to be in a position "most vulnerable to enemy attack." In early 1951, members of the 4th Wing submitted an "Unsatisfactory Report" on their M3s, recommending that the Sabre's six .50-caliber machine guns be replaced with four 20-mm cannons.[251]

The dissatisfaction with the M3 was not, however, universal. The question, "Will four 20mm cannon do a better job than six .50 caliber machine guns?" became, according to one September 1951 report, "probably one of the most widely debated points among the combat crews of the F-86 in Korea." Gabreski considered the Sabre's existing "fire power . . . adequate for fighter vs fighter operations." Thyng reported that "if you are within range and in position, the 50 caliber machine gun is more than adequate." Baker defined that range as two thousand feet. Others claimed that the .50-caliber was suitable out to ranges of only fifteen hundred feet. The 4th's report on F-86E operations was more pessimistic, declaring that the maximum combat effective range of the M3 was only a thousand feet. The 4th also highlighted the prime concern of the pilots—gun dependability—noting that most pilots would prefer a 20-mm cannon "providing that it is *dependable*, and will deliver a minimum cyclic rate of 1,500 rpm. If it has a high percentage of stoppages, pilots here would prefer to keep the M3 . . . and keep pressure on for [getting a more powerful engine] to put the airplane within range where the .50s are effective."[252]

In January 1953, the Air Force initiated Project Gun Val to investigate replacing the Sabre's armament with the sought-after 20-mm cannons. Eight specially modified F-86s armed with the T-160 20-mm cannon subsequently flew 284 combat sorties with the 4th Wing. The combat statistics for the 20 mm-equipped Sabres—hits on twenty-one MiG-15s, destroying six, probably destroying three, and damaging twelve—were slightly but not significantly better than the standard .50-caliber armed F-86s. Of greater interest were the twenty heart-stopping compressor stalls, with their accompanying loud bang and immediate loss of engine thrust due to the disruption in engine airflow, that the Gun Val Sabres suffered when firing their 20-mm cannons. The Air Force concluded that, although the 20-mm gun system showed some promise, between its reduced ammunition load and the troubling compressor stalls, it was not ready for worldwide deployment. Sabre pilots would have to continue relying on their .50-caliber machine guns for the duration of the Korean War.[253]

Maintenance Headaches and Training Issues

Aside from the ineffective ordnance and the new sight's operating limitations, the pilots' number one compliant about their armament system in late 1951 and early 1952 was that it simply didn't work. Before the F-86E's arrival, maintenance troops were already struggling to keep the Mark 18s in the F-86As properly serviced. The introduction of the significantly more complex A-1C(M) sight and its APG-30 radar aggravated the problem, and the reliability of the gunsights plummeted. On one of his first missions in Korea, Mahurin was zeroing in on a MiG when "all of a sudden the light in my gun sight went out. . . . In desperation," he recalled, "I kept blazing away, blindly hoping that one stray bullet might hit the Mig"; none did.[254] On several occasions,

Evans lined up behind a MiG target and fired away with his guns only to later discover that his sight was inoperative.[255]

The problems the Air Force experienced in Korea with its new equipment were not unforeseen. When the A-1C(M) first began to appear in Air Force aircraft months before the Korean War, several commanders complained that their squadrons were woefully unprepared. Of note, after one training exercise in January 1950, a participating squadron concluded that its "present personnel are definitely inadequately trained and equipped to maintain the [A-1C(M)] sight."[256] MIT Instrumentation Lab representatives were on hand to watch the exercise, and they were appalled to see that the unit had "no written operating instructions" for the sight, no personnel trained to maintain the sight, and no test equipment for the sight. Months later, after observing the USAF Gunnery Meet in March and April 1950, the MIT engineers again warned Air Force officials of the pressing "need for additional devices for ground and flight checking the A-1C sight systems"; otherwise the officials risked squandering the finite time they had to instill in the pilots "a high degree of confidence in [the sight's] ability."[257]

Inexperienced personnel, shortages of critical components, and inadequate testing equipment would dog the A-1C(M) system throughout its time in the Korean War. Nearly twelve months after the engineers' first prescient warnings, the same problems remained, now exacerbated by wartime pressures and long-distance supply routes from the States. Reflecting on the situation "from mid-1951 to mid-1952," FEAF concluded that F-86 "sight maintenance was extremely poor."[258] The early, trickling replacement of F-86Es for F-86As on a one-for-one basis only exacerbated the chronic shortages of A-1C(M)-trained personnel, parts, and testing equipment at the Sabre bases. For example, Meyer complained in May 1951 that "the arrival of six aircraft with A1C gunsights created an immediate problem since" he had "only three trained maintenance personnel," was missing "approximately 40% of the testing equipment," and had "no stock level of spare parts."[259] The sudden influx of new F-86Es after Vandenberg's October 1951 decision added more aircraft and some additional parts but did little to address the supply and maintenance backlog.

When skilled maintainers, parts, and testing equipment were found, inadequate design of the A-1C(M) for rough wartime operating conditions and poor modular integration of its APG-30 radar hampered timely repair. MIT had designed the sight to be maintained and calibrated in a temperature-controlled building, yet none of the Korean air bases had a suitable structure. Nor were the gunsight's "fragile" mechanical linkages designed to be bounced along a rough, steel-planked Korean runway several times a day. Additionally, when one electronic component in the radar failed, ground technicians were required to replace the entire radar system, "not just the faulty module."[260] In February 1952, Gabreski reported that more than 60 percent of the 51st Wing's A-1C(M) gunsights were unserviceable.[261]

Many of the mechanical deficiencies of the A-1C(M) were eventually corrected in its A-4 successor, which was incorporated into the F-86F that began arriving in Korea in late 1952.[262] However, for the F-86E pilots and maintainers stuck in Korea in early 1952, the A-1C(M) proved a constant headache. Gabreski lambasted the A-1C(M) for its burdensome financial and personnel requirements, observing that "the gross value of equipment needed to maintain an A-1CM gunsight is approximately $256,595.52. This is to support 58 aircraft, and only includes the cost of test equipment. In addition to the prohibitive price of test equipment, it is estimated that it takes approximately 10,000 man hours per month to maintain the A-1CM sight systems in 58 aircraft."[263] After Gabreski's February 1952 bombshell, the director of operations at Fifth Air Force declared that the A-1C(M) sight was "too complicated to be maintained." A month later, another officer in Fifth Air Force reported that "the majority of F-86E pilots have little faith in the sight."[264]

With the uproar over the poor condition of the new sights reaching a crescendo, officials at Air Materiel Command, spurred by the commander of Fifth Air Force, finally launched a comprehensive program to rehabilitate the A-1C(M) and its accompanying APG-30 radar. The program was known as Project Jaybird. Beginning in April 1952, a group of Air Training Command instructors and military and civilian technical specialists spent nearly three months traveling throughout Japan and Korea repairing the faulty gunsight and radar equipment. Simultaneously, they offered in-depth training on the system to both ground maintenance personnel and pilots. It took "a considerable amount of work . . . to put the sights in proper condition," the Jaybird technicians reported after the program. "About half of the sights required some attention," and, because of the system's poor modularity, any malfunction, "however minor, was sufficient to make the sight inoperative."[265] It was during Project Jaybird that the A-1C(M) sights were modified to accommodate Jenkins's range limiter.

While the Jaybird team was working to restore and upgrade the sights in theater, officials at Fifth Air Force vowed to tackle the supply issues that had contributed to the dramatic decline in the system's combat functionality. They revamped the command's supply processes to guarantee acquisition and retention of adequate on-hand bench stock and quick replenishment of hard-to-find sight components and testing equipment.[266] In the individual units, commanders elected to consolidate all of their gunsight technicians and maintenance actions into a single group-level organization, achieving a synergy not achievable when each squadron performed its own repair work on their gunsights and radars. Although not immediate, the performance of the system steadily improved under the increased scrutiny.[267]

The pilots in particular benefited from the additional, remedial training provided by the Jaybird team. After the war, FEAF officials concluded that most Sabre pilots "arriving in the theater had no prior training on the sight," which suggested to them that "a significant proportion of the difficulties [with the A-1 sight] must be attributed to improper operation."[268] It wasn't just that the pilots didn't understand the complex physics, mathematical relationships, or mechanical linkages that generated and positioned the aiming reticle on their windscreen. Rather, as a 1952 Operational Survey of the A-1C(M) discovered, only a few pilots could even anticipate how the reticle would work in a dogfight. Moreover, the pilots' poor understanding of the system exacerbated the maintenance difficulties because the pilots couldn't adequately distinguish between proper and faulty operation of the sight.[269]

The pilots' lackadaisical attitude toward the very equipment that was designed to help them garner glory in the skies over MiG Alley reflected their general tendency to shun technical knowledge—an attitude that, as noted earlier, could be traced back to the first assessments of air combat and its protagonists. Still, the tactics manuals for the two Sabre wings in Korea both stressed that "complete familiarity with the aircraft is imperative." The 51st's manual even drew special attention to the importance of a pilot's "operating knowledge of the A1-CM gunsight" as a "prerequisite for accurate gunnery."[270] But, recalling Evans's impromptu instruction on the A-1C(M), for most pilots adequate operating knowledge of the gunsight could (and, by extension, should) be conveyed in a quick, two-minute pitch.[271] Pilots didn't consider it a useful allocation of their time to study in detail a system that rarely worked and that was installed in only a handful of airplanes then parked on the flight line.

After Jaybird helped restore the sights' functionality, and with their chances of using the sight increasing as more A-1-equipped Sabres populated the flight line, the pilots began to devote more energy to understanding the gunsight. Reflecting its new priority (and the availability of spare parts), the crews at Kimpo in July 1952 constructed a simulator mock-up of the gunsight. The mock-up used the "exact aircraft wiring and cabling" and sight components, and was mechanized such that "stick movement would cause the same reticle movement in the mock-up as in the aircraft in flight." The 4th's officers concluded that the new apparatus would "be a valuable training aid in maintenance instruction as well as pilot instruction."[272]

Project Jaybird in mid-1952 thus marked a turning point in the pilots' relationship with their gunsight equipment. Although the sight continued to suffer from its inherent design limitations, Project Jaybird helped restore the pilots' trust in the system. With greater reliability and enjoying the practical benefits of the range limiter, the pilots who remained in combat after summer 1952 began to write more approvingly of their A-1C(M) gunsight and expressed optimism about its A-4 follow-on. For example, Low remarked in a July 1952 report that his "experience with the A1CM sight has given me complete satisfaction. . . . The ease of operation and accuracy are two of the best points of the A1CM."[273] Affirming the sight's original promise, Baker remarked that he especially liked the way the "sight takes all the work off of the pilot and allows him time to concentrate on positioning."[274] Even Cleveland remembered the A-1 working "to perfection" during his first MiG encounter on July 11, 1952.[275]

But for the pilots who flew with the A-1C(M) and finished their tours before mid-1952, before Project Jaybird, there was little appreciation for the sight. They railed against the gunsight's weight, complexity, and general unreliability.[276] Colonel Albert Schinz, Gabreski's deputy in the 51st, concluded in his postwar report, dated July 10,

1952, "It is very doubtful that this sighting system, as difficult to maintain as it is, will be of much value for world-wide application."[277] Gabreski, who finished his postwar report just days before Schinz, was one of the A-1's most vocal critics. As the Air Force's premier two-war ace with 34½ kills, 6½ of those earned in the Sabre over Korea, it was Gabreski who led the charge later that summer to convince the Air Force chief of staff of the need to abandon the radar-ranged gunsight.[278]

Deciding the Future of Advanced Fire Control Systems

The fourteen aces who confronted the CSAF in late summer 1952 were determined to rid the Air Force of the troubling A-1C(M) gunsight, its APG-30 radar, and the A-4 follow-on. Fresh from their Korean assignments, where they repeatedly had to wait for a high-flying MiG to pounce before they could attack, only to be then thwarted by a faulty, unreliable, and "too sophisticated" gunsight, the aces thought they had an obvious solution to their frustrations. Eliminating the complex and, in their experience, rarely functioning radar-ranged gunsight would yield a simpler and therefore more maintainable armament system. Its removal would also lighten the F-86, boosting the Sabre's acceleration performance and maximum ceiling, which the pilots could then use to outduel the MiGs at altitude. Following the meeting, General Vandenberg summarized the aces' position: because "an extremely high percentage of kills will be at short range . . . during a stern chase, . . . it is most unwise to burden our fighters with 205 extra pounds of gunsight complexity and unreliability to take care of the rather remote chance of a long range, high deflection kill."[279]

On September 4, 1952, the CSAF ordered his weapons testers to conduct an immediate and accelerated comparison of the service's radar-ranged and non-radar-ranged gunsights. Since the A-1C(M) was already in the process of being replaced, the newer A-4 gunsight with the APG-30 radar (now collectively referred to as the J-2) was used during the test. The Air Force selected its venerable K-14 as the non-radar-ranged gunsight. Scarcely three weeks later, Major General Patrick Timberlake, commander of the Air Proving Ground at Eglin, concluded that, along with the A-4, "the APG-30, even in its present state, should be retained in existing air superiority fighters" and that "efforts to improve the reliability of the APG-30 radar should be intensified."[280] In recommending retention of the radar, Timberlake was effectively vetoing the recommendation of six of the Korean aces invited to fly in the test.

Colonels Gabreski and Meyer, Majors Whisner and Jabara, and First Lieutenants Kincheloe and Kasler quickly reached two conclusions during their test missions: first, the old K-14 gunsight was wholly inadequate for use in the F-86; and, second, the A-4 represented a significant improvement over the A-1C(M). However, these six aces thought the APG-30 was still too problematic, so they set out to determine if the A-4 should be fielded with or without its accompanying radar. They concluded in favor of the latter, recommending "the APG-30 Radar be deleted as a requirement for air superiority fighters and a gyro sighting system such as the A-4 be utilized with stadiametric ranging." While acknowledging the "JAYBIRD fixes," the test pilots concluded that the radar "is still not suitable for use in the F-86E due to its unreliability, limited performance at low altitudes, inadequacy against jet fighter targets, erratic performance in the presence of clouds, excessive maintenance and personnel requirements, etc."[281]

The pilots' assessments of the radar reflected their earlier distasteful combat experience with the A-1C(M). In the test report, for example, Gabreski focused on his ordeals generating sufficient numbers of F-86s while commanding the 51st FIW at Suwon, warning that the "intricate, highly complicated electronic equipment" in "the radar section of the J-2 system . . . will create a terrific maintenance problem in the field."[282] Lieutenants Kincheloe and Kasler based their recommendations *solely* on their combat experience; the report notes that the two "did not arrive" at Eglin "in time to participate in the flying portion of the test."[283] To justify their recommendation, the six aces claimed that the manual-ranging and radar-ranging sights were "comparable at short ranges" where they believed the majority of combat firing took place.[284]

Only one test participant, the 4th's former wing commander, Meyer, seemed to display a concern for future armament requirements. In his comments on the test, Meyer suggested that on the basis of "future programming, . . . with the increased speeds of aircraft in the air superiority field, and with the direction toward increased caliber of weapons and shorter time of fire, a very real requirement exists for automatic ranging features in future air superiority aircraft." Although he signed the report indicating that he concurred with its "evaluation, conclusions, and recommendations," Meyer's individual comments offered a dissenting recommendation: "that automatic ranging features of an improved design to the APG-30 be included in the requirements for all air superiority fighters. Further, that those fighters presently equipped with APG-30 continue to use them" while "imposing maximum effort" on enhancing "reliability and maintainability as well as improved performance of automatic ranging features."[285]

Timberlake and his subordinates at the Air Proving Ground sided with Meyer. Testing officials noted that the aces' hopes of quickly lightening the Sabre by removing the radar set were misplaced, because removing the equipment would then require adding ballast to the nose of the aircraft to maintain its proper center of gravity. More significantly, Timberlake declined to base his recommendation on the pessimistic recollections of a handful of Korean aces and the test's orchestrated firing passes against a slow banner target, which he noted "is not realistic or comparable to fighter vs fighter combat."[286] Using a collection of Korean air combat gun camera footage, researchers working on contract at the University of Chicago and the Museum of Science and Industry concluded that "the bulk of the firing in combat is conducted at ranges and angles off in excess of 1500 *feet* and 10°. . . . At these conditions," Timberlake's cover memorandum read, "the kill probability using the A-4 sight with radar ranging is *twice* that attained with either the A-4 or K-14 sights with manual ranging."[287]

Timberlake was also not without support from the field when he chose to dismiss the aces' arguments. When Vandenberg ordered his testers at Eglin to initiate the test, he also asked for a new round of input from the current Korean aviators, men who had more experience with the sights post-Jaybird. Forwarding the chief's request, a FEAF–to–Fifth Air Force message noted, "it is probable the 14 aces Gen Vandenberg referred to were not well trained on the A-1C sight prior to arriving in Korea and were possibly serving in Korea during the period of time that maintenance difficulties prevented the full effectiveness of the A-1C sight. Perhaps the newest crop of jet aces might have different views on sight effectiveness."[288] They did.

In a report dated September 17, nine days before Timberlake penned his cover letter for the final test report coming from Eglin, the "'present crop' of jet aces . . . well versed in the use of the A1CM gunsight through all phases of operation including its present improved operation as against its period of unreliability earlier this year" argued that "all of the automatic features of the A1CM gunsight should be retained." The pilots reasoned that while an experienced combat pilot might achieve good results with "a Mark 18 type sight, the superior pilots are the exception rather than the rule." As Meyer did, the current FEAF pilot-aces looked toward the future and concluded that soon younger and more inexperienced pilots would be flying fighters and that they would "need, to be effective in the air superiority role, all of the automatic features afforded by the A1CM sight." Moreover, the FEAF pilots concluded that "with the improved armament expected in the near future" and a more aggressive enemy, "high deflection shooting" and an "automatic computing, automatic ranging sight" would "become much more important." Hence, they argued, "removing the A1CM gunsight" would be a step "made backward instead of forward," irreversibly hindering the "development of tactics, training and employment of an automatic ranging and computing gunsight for use in any future conflict."[289] The Air Force in the end elected to continue fielding the J-2 (the A-4/APG-30 combination) in its F-86 Sabres. Later, the J-2 would be redesignated the MA-3, and it would go on to equip the follow-on to the Sabre, North American's F-100 Super Sabre, which entered service in fall 1954.[290]

The radar-ranging gunsight survived not because of its performance in the rough-and-tumble air combat environment of Korea but in spite of it. After the war, the Air Force's Materiel Command "conceded that the decision to produce the A-1C sight had been a little premature, considering its state of development."[291] Even by 1956, issues with the complex APG-30 radar remained. That year, Jenkins authored an article for *Fighter Weapons Newsletter* in which he acknowledged that "airborne range-only radars . . . are heavy, bulky, costly, difficult to maintain, unreliable and erratic in operation." But, he countered, "the only subject . . . that's more revolting than radar ranging is *manual ranging*."[292] The decision to continue developing the radar-ranged gunsight represented a choice for the Air Force between the lesser of two evils, but it also depended on the shift in pilots' attitudes that occurred in mid-1952. The pilots' acknowledgment that they needed technology's aid and their new willingness to learn how to use that technology, combined with the new emphasis on properly maintaining the sight and its adaptation to better serve the conditions of Korean air combat, allowed the gunsight and the pilot to evolve into a more effective human-machine system.

Conclusion

After having battled the MiGs for months, the 4th's Group commander, Lieutenant Colonel Glenn Eagleston, in July 1951 declared to his fellow Sabre pilots:

> The prevalent, but extremely uneducated, "school of thought" that the fantastic speeds, fuel, and altitude problems would eliminate the possibility of aerial combat, jet vs jet[,] has been summarily squelched by the scorching, kerosene consuming, maneuvering, 50 plane mass dog fights over the Yalu River. When two or more aggressive fighters meet in the sky and sufficient determination is maintained by the pilots concerned to desire a positive decision as to who is the victor, a maneuvering "dog fight" will ensue and continue until the decision is made self-evident by the withdrawal, damage to[,] or destruction of one or more of the participants.[293]

Because the contest required the Sabre pilots to joust in topsy-turvy, slashing, high-speed aerial battle, to persevere though frequently outnumbered, to constantly shift back and forth from hunter to hunted, and to fly a comparable but not superior aircraft outfitted with a finicky and questionable radar gunsight, Korean air combat, the men were told, demanded pilots who were aggressive "tigers," aviators who possessed the requisite "guts" needed to brave unenviable odds for a shot at "glory."

The Sabre pilots sought to prove their aggressiveness and machismo at every opportunity. Though flying unprecedentedly complex aircraft, they boasted about their ability to fly without checklists and other aids, and they occasionally perished because of it. Pushed to perfect their flying skills and join in alcohol-fueled camaraderie, they had little incentive to spend time studying the physics that both propelled their jet aircraft to lofty heights and yielded an electromagnetically determined range for their gunsight. Motivated by the glory of aerial victory and the allure of acehood, they resorted to hunting MiGs in off-limits Manchuria, at times even gunning down the enemy aircraft as they landed. And although the pilots were told that they were part of a formation team—that there was no place for individuals—they quickly realized that glory always went to the leader.

Aerial warfare remained a ruthless business. The advent of jet combat had not changed that element. But now dogfights became even shorter affairs characterized by blazing fast speeds, high altitudes, quick turns, and high-Gs. In a matter of seconds, a speck on the distant horizon could transform into a massive, red-nosed MiG spewing bright orange balls of fire before disappearing back into the azure sky a moment later. Although the A-1C(M) radar-ranged gunsight was designed to aid Americans pilots in their quest to become an ace, these types of battles rendered it difficult for the Sabre pilots to use their gunsight as the engineers had anticipated.

The fact that the A-1C(M) and its accompanying APG-30 radar rarely worked compounded these problems. Because the sights were victims of a complex design ill-suited for wartime field repair, and with their critical parts and test equipment stretched across a transoceanic supply line, reliability plummeted. With it, so did the pilots' trust in the new armament system. The advertised automaticity of the new A-1C(M) gunsight and its APG-30 radar was never realized, and when the sight failed to perform correctly, the pilot was left to intervene, relying on the same skill sets that were supposedly obviated by the new sight. Then, in mid-1952 with the future of Air Force radar-ranged gunsights hanging in the balance, something changed and a more effective human-machine system emerged.

On one end, the gunsight technology began to evolve. New efforts to rehabilitate the sight during Project Jaybird increased the gunsights' reliability, which helped restore the pilots' trust in the system. Concurrently, the gunsight engineers concluded that the pilot played a much greater role in scoring a kill than simply placing a mechanically aimed pipper over the target and squeezing the trigger. Still, these changes would have been for naught unless the pilots' attitudes toward the new technology had not also evolved. Pilots had to move beyond the decades-old stereotype that had defined their profession. Detailed knowledge of their aircraft and its weapons systems wouldn't necessarily dilute their tiger spirit, the fighter pilots realized. It might actually render their tiger-like instincts that much more lethal for their opponent.

The fighter pilots' experience in the Korean War revealed to them that the new, advanced technologies might augment but would never supplant a pilot's fighting instinct. Consequently, the senior flight leads continued to work to inculcate that aggressive spirit in their younger apprentices during their tenure as a wingman. They made sure that the junior pilots received a proper schooling in the fundamentals of being a fighter pilot, not just tactics but also the roles and responsibilities that separated flight leaders from wingmen. In Korea, the wingman formation position, once borne of tactical necessity, began to resemble a rite of passage that one needed to complete before successfully gaining entry into the fighter pilot fraternity.

In the end, the new equipment might make it easier to score a hit, but as Jabara liked to remind others, "After all a fighter is simply an airborne gun platform. The pilot must turn, dive and climb to get into position to fire."[294] The 4th's tactical manual was explicit on this point. It was the fighter pilot who had to possess "complete mastery of his aircraft" so that *he* could "get maximum effective performance out of his ship in order to defeat the enemy in maneuvering battles." *He* had to be able to "direct his fire expertly into the enemy aircraft." To declare "mission accomplished," *he* had to perform each function perfectly and without hesitation, from "the initial sighting of the enemy," through "the maneuvering phase [and] positioning." Only then could *he* commence "firing" and send the MiG smoking toward the ground.[295] The A-1C(M), even if it worked perfectly, could only offer hope for the final "firing" phase. But, the pilots eventually reasoned midwar, why not accept help in that final phase if it might widen the potential pool of future MiG-killers?

Thus, in spite of the new automated tools in their cockpits, it was easy for the Sabre pilots of the Korean War to sustain the still-cherished, still-popular fighter pilot myth that emphasized the glory of aerial victory won by an individual pilot's supreme skill and daring. In this light, the battles fought in the skies over Korea didn't seem that different from those fought by the hard-charging Mustangs over Germany scarcely a half-decade earlier, or those fought by the fabric-covered biplanes three decades before that. Similarly, the jet fighter pilots of this new generation appeared little-changed from their brave and noble forebears, still specially endowed men with eagle-like eyesight and superb flying skills. These men were not scientists; they were clearly aerial knights still looking to capture glory in the air. And they were tigers.

A little more than a decade later, the fighter pilots' treasured tiger-like spirit would be challenged by even greater automation. If new, advanced weapons could be launched without relying on a pilot's steady aim and sharp eyesight, and if the projectiles themselves could then maneuver in midair to chase down a foe, then what of the fighter pilot's flying and fighting skills that were so-prized over MiG Alley? Ostensibly built to improve a pilot's lethality in air combat, the new guided missile armament wasn't the only challenge thrust upon the fighter pilot when the McDonnell F-4C Phantom II started appearing on Air Force flight lines. The F-4 also required the fighter pilot to forsake the historic allure of the single-seat aircraft and open his cockpit to another person, a GIB—a "guy in back." The next generation of fighter pilots would be forced to confront these existential issues as they tangled with ever more wily MiGs in the deadly skies over Vietnam.

CHAPTER 4

Phantoms over Vietnam

The "GIB"

> We felt like young tigers and now we were going to be put in the back seat with another guy. We were told, "You're in the back seat. Learn something. The guy in the front has got all the experience." So we accepted it. And the guys in the front treated us pretty nice. . . . The front seaters even had a pet nickname for us. They called us "GIB." They said it stood for "Guy in Back." . . . We were proud to be a GIB. However, I must admit that some years later I was flipping through a real thick Websters, and I happened to chance upon the word gib. G-I-B, it was in there, and the definition they had in there was a "castrated tomcat."
>
> COLONEL HOWARD J. HILL, USAF (Ret.), 1991

It was late December 1966, and First Lieutenant Charles "Charlie" Clifton had just returned from his ninetieth mission flying over North Vietnam (NVN). That evening, he joined his squadron mates at the bar to celebrate the milestone. Considering the standard practice of not sending a pilot with ninety missions under his belt back to Route Pack 6, the most heavily defended area around Hanoi, Clifton expected his last ten missions to be a little less harrowing. A few close calls earlier had left the twenty-four-year-old's aircraft riddled with holes and metal shrapnel, earning him the nickname "Magnet Ass" in the process. The evening's revelry, however, was abruptly interrupted when two military policemen escorted the lieutenant out of the bar and told him to report to the wing commander immediately. Clifton raced off. He found the colonel in a large auditorium, giving a mission brief to a packed audience. Clifton sheepishly entered the room at the back, but the colonel saw him and halted the presentation. He walked over to Clifton, put his arm around the lieutenant's shoulder, and said in a hushed tone, "I know you just completed your ninetieth, but I'd like you to come with me on this one."[1]

Born in Toledo, Ohio, in May 1942 but raised in Fort Wayne, Indiana, Clifton graduated from the US Air Force Academy in Colorado Springs, Colorado, in June 1964. Not an especially stellar student, Clifton once remarked that he "ultimately majored in graduation." After the academy, he went straight to pilot training at Laughlin Air Force Base in Del Rio, Texas. There, he said, he learned the secret of being a fighter pilot: "It's not just learning to fly an airplane, it's being able to climb in the cockpit and feel like you're getting dressed, like you just got out of the shower and put on a 'fresh skin.'" Clifton performed well enough to earn an assignment in the F-4C Phantom, and after training at MacDill AFB, Florida, he was off to Vietnam in July 1966 as a member of the "Triple Nickel," the 555th Tactical Fighter Squadron (TFS). Now ninety combat missions later, Clifton was looking forward to whisking his "very pregnant" wife off to his next assignment in beautiful Spangdahlem, Germany.[2]

Upon hearing the colonel's request, Clifton didn't hesitate. "You got it, Boss," he replied. The mission was going deep into NVN, within twenty miles of Hanoi and directly over the large MiG airbase at Phuc Yen. A few days later on the morning of January 2, 1967, after the wing commander issued a last rousing, "OK, you Wolfpack, let's go get 'em!," Clifton climbed the boarding steps up into his McDonnell F-4C Phantom II and strapped himself into the pilot's seat. The aircraft was configured solely with air-to-air missiles for the mission. There were four twelve-foot-long, 380-pound, radar-guided AIM-7E Sparrow missiles snuggled into the recesses carved into the belly of the Phantom. Another four smaller, infrared (IR)-guided AIM-9B Sidewinder missiles were hung under the wings near the fuselage, two on each side, attached to the aircraft's inboard pylons. Farther away from the fuselage on the left outboard pylon hung a large, 370-gallon external fuel tank. A larger 600-gallon fuel tank was suspended under the aircraft's fuselage along its centerline. Normally there would have been another 370-gallon fuel tank on the right outboard pylon, but for the first time during the war it was replaced by a small, cigar-shaped QRC-160 electronic countermeasures (ECM) pod.[3] Without the extra 2,500 pounds of fuel on the right side to counterbalance the external fuel tank on the left side, takeoff that morning was described by one aviator as "a thrill. . . . You can imagine with this asymmetric load what happened to you the minute you broke ground. You just had to come in with full rudder and hope."[4]

The goal of the afternoon mission was, like Evans's years earlier, to find enemy MiGs and shoot them down.[5] But unlike Evans in his Sabre, Clifton in his Phantom didn't need to get behind the enemy and within gun range to get a kill. First off, Clifton's F-4 had no guns. Secondly, his AIM-7 missiles could technically be launched from any angle relative to the target. Plus, with the Sparrow's ability to home in on the radar energy emanating from the Phantom and reflecting off the enemy target, the radar-guided missiles could be launched from more than five miles away. It had taken another decade after the A-1C(M), but the fighter pilots in the Phantom finally possessed a system that allowed them to kill their targets at long range and with high-deflection angles, or so they thought.

Approaching the enemy airfield from the northwest at an altitude of about fifteen thousand feet, Clifton's Phantom was in the lead, joined by three other Phantoms as part of Olds flight. They were followed by more Phantom flights, spaced at five-minute intervals; Ford flight was next in line behind them. Hoping to lure the MiGs to battle, the Phantom pilots that morning had disguised themselves as a pack of less maneuverable F-105 Thunderchief or "Thud" fighter-bombers. The jamming pods slung under the Phantoms' wings were part of the ruse, since only the Thud had yet carried the QRC-160 into NVN. Adding to the deception, the Phantoms also adopted Thud formations, communications procedures, and typical routing into enemy airspace.

The Americans weren't sure their plan had worked, though. As Olds flight passed over Phuc Yen, the pilots still had not detected any MiGs with their Phantoms' radars, and a solid deck of clouds below them at seven thousand feet obscured their view of the airfield. The Phantoms continued southwest "for a full minute" before starting an "erratic turn" back northwest. Rolling out of the turn, Olds 3 reported a possible radar contact seventeen miles ahead at twelve o'clock, but then the "low, very fast" contact disappeared. The Phantoms passed Phuc Yen again; still no MiGs. Ford flight was nearing the enemy airfield, and Olds flight turned to the southeast once more.

According to one postmission report, "Everything then happened at once." Several enemy MiG-21s suddenly streaked up through the clouds. The first emerged right behind Clifton in Olds 1. Olds 2, flying in fighting wing formation about fifteen hundred feet to the left of the lead Phantom, saw the MiG. So too did the members of Ford flight, one of which beat Olds 2 to the radio, calling, "Olds lead, you've got two MiG-21's right on your ass." After all the careful mission planning and orchestration, it was, one pilot remembered, "a hell of a way to start a fight."[6]

Olds flight stated an immediate defensive break to the left. Midway through the turn, another MiG appeared at eleven o'clock. Clifton already had the radar set up for a close-in engagement and locked on immediately. With a squeeze of the trigger followed by another, two Sparrow missiles dropped from the Phantom's belly and rocketed off. But then the radar abruptly broke lock, and the two missiles careened off course. A third missile, a Sidewinder, was fired at the MiG ahead, but it too missed when the enemy plane dove into the clouds below.

There were now MiGs everywhere—fourteen to be exact—and the members of Olds flight, Ford flight, and the approaching Rambler flight were soon ensnared in a raging, fifteen-minute-long aerial battle around Phuc Yen. During the melee, Olds 1 spied another MiG one mile away off to the left, in a left-hand turn. Clifton's Phantom zoomed high above the horizon, executed a barrel roll to the right toward the MiG's tail, and pulled down into perfect position behind and below the unaware foe. From forty-five hundred feet away and about fifteen degrees off the bandit's tail, Olds 1 fired two Sidewinders. The postmission report described the result: "Suddenly the MIG-21 erupted in a brilliant flash of orange flame. A complete wing separated and flew back in the airstream, together with a mass of smaller debris. The MIG swapped ends immediately, and tumbled forward for a few instants. It then fell, twisting, corkscrewing, tumbling, lazily toward the top of the clouds."[7]

Clifton later earned credit for using a Sidewinder missile to destroy the MiG that day, but he did not pull the trigger. The trigger for the missiles was located on the control stick in the front cockpit of the F-4C. Clifton was sitting in the "pilot" position in the rear cockpit of the Phantom—he was a GIB, a guy-in-back. Colonel Robin Olds, the successful World War II ace and now commander of the 8th Tactical Fighter Wing (TFW) at Ubon Royal Thai Air Force Base (RTAFB), was sitting in the Phantom's front, "aircraft commander" seat during the famed Operation Bolo mission.[8]

Operation Bolo would go down in history as the Air Force's greatest aerial triumph of the Vietnam War. Masterminded by Olds, the afternoon mission claimed seven MiG-21s, more than 40 percent of the North's top-of-the-line, delta-winged, Soviet fighter inventory. Following the loss of two more MiG-21s in a Bolo-like ruse four days later, NVN effectively grounded its MiG-21 force from combat operations for nearly three months.[9] For Olds, Operation Bolo cemented his status as "larger than life: a legend" and earned him recognition as "probably the greatest aerial warrior America ever produced."[10] In contrast, Clifton's name was essentially lost to history; he retired in relative obscurity, eventually opening his own civilian flight school in California.[11]

Olds's success during Bolo would not have been possible without Clifton, just as his later aerial victories similarly depended on Lieutenants William Lafever and Stephen Crocker—his GIBs on his other victorious missions of May 4 and May 20, 1967.[12] But the GIBs' contributions to aerial victory often went unacknowledged by those occupying the front seat. For example, describing one engagement in which he "found two more MIGs way off at nine o'clock," Major Bill Kirk, who would eventually finish his combat tour with two MiG kills, stated: "*I* got behind them, *I* got all armed up, got a radar lock-on and *I* fired off two Sparrow missiles at them."[13] Similarly, Major Sam Bakke noted that on his victorious mission of May 14, 1967, "*I* must have attacked about six MIGs that day firing a missile at most of them until *I* finally did get one."[14] Reading Kirk's and Bakke's narratives, one might never realize that there was someone else seated in the cockpit with them. Nor was the oversight unique to those two particular front seaters. When recalling his second MiG kill, Olds remembered to acknowledge his wingmen, "Dick Pascoe, Kirk, and Allen," but he made no mention of his or his wingmen's GIBs.[15] It may have been unintentional, but the denotation that accompanied the pet name for the back seaters that former-GIB Howard Hill happened upon one day while flipping through *Webster's* was not far off. Although the GIBs through 1969 were all trained pilots, they remained in the minds of many front seaters only pseudopilots, veritable second-class citizens within the ranks of the fighter pilot fraternity. Not quite tigers, the GIBs were indeed more like "castrated tomcat[s]."[16]

The two-seat cockpit in the Air Force's F-4C Phantom represented a profound divergence from the historic representation of the fighter pilot. Fighter pilots since the time of Garros and Boelcke had reveled in the autonomy associated with flying a single-seater. The presence of the GIB sullied the historic persona of the fighter pilot as lone eagle—or, in the case of Korea, lone tiger. Salter's eloquent description of the fighter pilot "as isolated as a deep-sea diver" now rang hollow, and Evans's experiences of being "sealed off to the peculiar whines and whispers of my own aircraft" were soon drowned out by the incessant and annoying sound of another man's breathing over the intercom.[17] The Phantom flight manual's reference to the radar-manipulating, rear-seat occupant of the F-4C as the "pilot" added to the ignominy.[18]

The use of missiles likewise clashed with the historic fighter pilot image. Since the First World War, aerial victories were recognized as the cardinal measure of a fighter pilot's flying prowess, chronicling both his skill maneuvering his aircraft and his keen sense of when to squeeze the trigger to flame his foe. Olds and Clifton earned their victory during Bolo with a single trigger squeeze from more than forty-five hundred feet away. Scarcely a decade earlier, Sabre pilots in Korea were extremely lucky to score a hit at even one-third that distance. Plus Olds's and Clifton's short-range Sidewinder missile was guided by an internal lead sulfide (PbS) detector that homed in on the heat plume produced by the MiG-21's jet engine before the missile automatically detonated itself once it determined it was close enough to do sufficient damage. Where was the skill in that?

Even the iconic images of aerial battle over Vietnam reflected the divergence. For example, contrast the image that celebrated Jabara's ace-victory in the Korean War (figure 3.21) with the image of Olds's and Clifton's victory in Operation Bolo (figure 4.1). The former exudes the romantic imagery of dueling dogfighters, but in the latter the enemy is barely visible, a mere speck in the distance. Rather than focusing on the aerial contest or even the F-4's two occupants, the viewer's attention in Keith Ferris's iconic *MiG Sweep* is instead drawn to the brutish power of the Phantom's thundering afterburners and its gleaming white missile armament—the technological tools for a new era of air warfare.

All told, the shift from the .50-caliber machine guns and radar-ranged A-1C(M) gunsight of the F-86E Sabre to the rocket-propelled guided missiles and intricate fire control system of the two-seat Phantom threatened to upend the treasured, decades-old myth of the fighter pilot. The Phantom's new armament represented a carefully calculated effort to democratize fighter aviation by automating out the pilot's historic skill set, transferring it to a guided missile. It was a shift encouraged by the Korean aces still flying combat in late 1952, and it would have profound implications for not only the individual fighter pilot and his relationship with his aircraft machinery but also the pilot's social relationships with others in the aircraft, within the formation, and across the Air Force. With Phantom aircrews now worried about seemingly esoteric details like PbS IR detectors, gigahertz radar frequencies, and antenna train angles and intercept geometries, had the fabled silk-scarf-wearing aviators finally morphed into the long-haired scientists (figure 1.4) that *Life* predicted in 1952, or the small, decrepit troglodytes crowded into black boxes (figure 1.3) that the fighter pilots imagined themselves as at the turn of the decade?[19]

FIGURE 4.1. Maneuvering for the Kill during Operation Bolo. Keith Ferris's *MiG Sweep,* which has become one of the iconic images of Vietnam air combat, commemorates the Air Force's greatest aerial victory of the war, Operation Bolo. In sharp contrast to imagery of air combat from wars prior (see, e.g., figure 3.21), Ferris's representation emphasizes not the individual pilot or the "fair contest" of air combat but rather the Phantom's extraordinary technological hardware. *MiG Sweep* © 1975 Keith Ferris, U.S. Air Force Art Collection.

A New Approach to the Gunnery Problem

When the fourteen ace pilots approached the chief of staff of the Air Force in summer 1952, they argued the best way to gain the upper hand against the troublesome MiG-15 was to rip the too-complicated A-1C(M) gunsight and its accompanying APG-30 radar out of the F-86E Sabre, lightening the aircraft and instantly boosting its performance.[20] There was, however, another way. What if the Sabres' bullets could be improved? What if the bullets could be loaded with extra propellant so that they could fly higher and farther? Moreover, what if they could be made to steer while in flight, even updating their trajectory in response to changes in the target's flightpath? Such a projectile might wholly eliminate the requirement to precisely solve the troublesome gunnery problem. It was an intriguing approach to the decades-old problem of aerial combat, now rendered possible through continuing advances in rocketry and radar.

As the early versions of the A-1B gunsight were first being mated (rather unsuccessfully) to Air Force fighters in 1948, Air Force officers attending the Air Tactical School (ATS) at Tyndall AFB, Florida, were already receiving a one-hour lecture on the potential benefits of this alternative approach to aerial combat. The lesson plan opened with an unnerving assertion: "It would appear that our faster aircraft, especially ones still on the drawing board . . . may be fine to carry a pilot from one point to another in a great hurry but may be of little or no use in air to air combat." The instructor's notes identified two components to this emerging "dilemma." The first was "the effect of the high speed on the pilot," specifically his ability to solve the gunnery problem in a more dynamic air combat engagement. The students were told that the Air Force had already taken commendable steps toward addressing this deficiency by developing a new gunsight for their fighters—the A-1. Regrettably, the lesson plan explained, there was less tangible progress toward resolving the second issue, that of the fighter's armament. Recognizing that the standard .50-caliber machine guns installed in fighter aircraft were too limited in range and their striking power wholly inadequate, the lesson plan suggested that

> it would appear that a weapon of some kind is required which will enable a pilot to stand off at least 10,000 feet away and fire at a target with fatal results to that target. One such weapon is the air-to-air guided missile. As presently visualized, the missile has the following advantages over armament now mounted in our aircraft:
> 1. Much longer effective range
> 2. Controllable all the way to the target
> 3. Powerful enough to insure [*sic*] a kill.[21]

The quest to realize the promise of the air-to-air guided missile would consume aircraft and weapons designers for decades. Several different armament variants were developed over the ensuing years leading up to Vietnam, ranging from the unguided rockets that equipped the D-model interceptor version of the Sabre, to the Falcon missile family developed for the Air Force's "1954 Interceptor," to the Navy-developed Sparrow and Sidewinder missiles that outfitted the two-seat Phantom, which eventually found its way into Air Force inventories and then into the skies over Vietnam. As the various missiles worked their way through the development process, the fighter pilot was slowly and subtly remolded from a brash, barnstorming aviator into a technically minded systems manager, at least in the minds of the engineers developing the new fire control systems.

Rockets, Missiles, and Radars

Three years before Air Force students sat in a windowless room at Tyndall AFB learning about the benefits of air-to-air missiles, around the time that the Army Air

FIGURE 4.2. Hughes JB-3 Tiamat. The Army Air Forces' first air-to-air guided missile, the Tiamat, was test fired on August 6, 1945. The massive missile, which looked more like a "bomb with wings," was canceled after only ten test launches. AFHRA.

Forces was finalizing its contract with MIT to develop the A-1 gunsight, engineers at Hughes Aircraft were already hard at work applying the emerging lessons from German rocketry and radar-homing technology to air combat. Their efforts culminated in a test firing of the Army Air Forces' first high-speed, air-to-air missile on August 6, 1945. Ironically, a single bomb dropped by a B-29 Superfortress half a world away that day presaged an age in which these faster-, farther-, and higher-flying air-to-air missiles would assume increased importance defending the nation from fleets of high-flying bombers ready to unleash atomic Armageddon.

Alas, the Hughes JB-3 Tiamat missile tested that late-summer morning was not a viable solution (figure 4.2). Although featuring an internal FM radar homing device, a top speed of six hundred miles per hour, a range of five to nine miles, a massive 100-pound warhead, and the ability to attack aircraft flying at altitudes of up to fifty thousand feet, the "very cumbersome" 625-pound missile—"essentially a 100-pound bomb with wings"—was relegated to the ash-heap of failed military prototypes after only ten test launches. Following the Tiamat's demise, the service rededicated and accelerated its efforts toward acquiring a more "'practical' air-to-air missile that could be developed within two years."[22] A flurry of missile development activity ensued. At one time, the Air Force was pursuing two separate contracts each for a fighter-launched missile (to attack bombers) and a bomber-launched missile (to attack fighters). However, the funding streams soon ran dry, and by April 1949 only a single guided missile program remained—Hughes's Falcon missile. It would take another six years before the first Falcons reached the field.[23]

The Air Force also initiated work on a new generation of fighter aircraft that would carry the sleek air-to-air missiles. About the same time that it contracted with North American for the P-86 (later F-86) Sabre in summer 1946, the Air Force asked Northrop to develop the experimental P-89 Scorpion, a two-seat, twin-jet, all-weather fighter. Whereas the Sabre entered production a scant two years later, development of the F-89 languished. When the anticipated arrival of the Scorpion slid into the next decade, along with its preferred Falcon guided missile armament, the Air Force in late 1948 turned to Lockheed and requested it build a two-seat, radar-equipped, gun-armed interceptor version of its F-80 Shooting Star, which became known as the F-94 Starfire. Later, it contracted with North American to design an interceptor version of the F-86. Lockheed's F-94 emerged first in May 1950. A gun-only version of the Northrop Scorpion (F-89B) appeared a year later in June 1951. The F-86 interceptor, designated the North American F-86D Sabre Dog (and for a brief time before that, the F-95), didn't arrive until April 1953.[24]

The Air Force's sudden thrash to field a radar-equipped, all-weather interceptor aircraft, even if only armed with guns, was a reflection of the uneasiness that gripped the nation following the Soviets' successful detonation of their own atomic bomb in August 1949. At that time, the Air Force's fighter aircraft were constrained to patrolling only during daylight and in relatively cloud-free skies—conditions that allowed the pilot to visually detect and then attack his enemy. The new interceptors, on the other hand, could theoretically use their powerful on-board radars to hunt down enemy nuclear bombers lurking in bad weather or trying to exploit the cover of darkness. Eventually, a new multibillion-dollar computerized air defense system known as SAGE (semi-automatic ground environment) would complement the Air Force interceptors. From 1958 through 1962, twenty-one separate SAGE sector sites were constructed along the US coastline and northern border, their uniformed controllers standing ready to vector America's interceptor pilots toward any menacing Soviet bombers.[25]

At the turn of the decade, though, before SAGE and before the March 1956 arrival of the first F-89Hs armed with the Falcon radar-guided air-to-air missile, Air Force interceptor pilots took to the skies armed with a mixture of .50-caliber and 20-mm guns aimed by an A-1C(M) gunsight, now paired with an APG-33 search-and-track radar and collectively referred to as the E-1 fire control system. The pilots' chances at success, however, were not good. Sturdier and better-armed enemy bombers meant that an interceptor pilot equipped solely with guns had only a one-in-ten chance of scoring a kill, according to one fire control system manual, "largely because of the

FIGURE 4.3. North American F-86D Sabre Dog Interceptor. One of a trio of early 1950's Air Force interceptor aircraft armed with unguided air-to-air rockets, the F-86D Sabre Dog was the only one to have a single-seat cockpit. To aid the lone pilot, additional automation was built into the aircraft's E-4 fire control system, which incorporated a more powerful and pilot-steerable radar antenna housed under the new Sabre's larger radome. NMUSAF.

interceptor's poor chance of surviving" the necessary short-range gun attack.[26] With the service's favored guided missiles still not ready, but with the deficiencies of its aircraft gun armament becoming more severe with each passing year, the Air Force decided to transition its interceptors from guns to unguided rockets. The test bed was the new F-86D (figure 4.3), which was armed exclusively with twenty-four unguided 2.75-inch "Mighty Mouse" folding-fin aerial rockets carried in an extendable tray in the belly of the aircraft just aft of the nose gear. Subsequent versions of the F-89 (F-89D) and F-94 (F-94C) quickly followed suit.[27]

The interceptor aircraft's shift from guns to air-to-air rockets necessitated a new fire control system. In the F-86D, unlike the two-seat F-89D and F-94C, the system would have to be highly automated and optimized for use by only a single pilot. Air Force officials eventually selected Hughes's E-4 fire control system for their Sabre Dogs.[28] Emphasizing its remarkable sophistication, one contemporary author likened the E-4 to a "flying IBM machine . . . built to calculate everything but an enemy pilot's pulse and blood pressure."[29] As to be expected, Gabreski, now serving in an administrative role at Norton AFB in California, objected to the complex system, but his concerns were summarily dismissed.[30]

The E-4's operating instructions declared that the system was made "as simple and straightforward as possible."[31] For the pilots trying to fly the F-86D while simultaneously operating the radar, it was neither. One report complained years later that "the pilot actually needed about three arms and four eyes to perform all the functions required of him during an interception."[32] Staring at a small, circular radar scope mounted between his knees, the F-86D pilot looked for a "blip" that illuminated after the *radar range trace*, a vertical line that swept left and right across the display, passed over a target. Once the pilot noticed the blip, he pressed a button on the radar's hand control, a separate lever located near the aircraft throttle control, to take command of the antenna and slew the range trace line back to the azimuth of the target blip. The pilot then adjusted the radar's elevation using the same hand control until the blip achieved its maximum brightness. Releasing the button on the hand control commanded the radar to lock on to the target. Staring heads down at the radar scope looking for momentarily illuminated blips and then manipulating the radar controls to lock on to one, all while flying an extremely maneuverable aircraft, could easily disorient even the most veteran pilot, so the engineers included a separate horizon line on the radar screen. Defeating its purpose, though, the engineers warned the pilots that the attitude indication wasn't always accurate.[33]

If the radar successfully locked on to the target, the E-4's computer then calculated the necessary steering angle for the intercept and displayed a steering dot on the radar scope. The pilot would subsequently maneuver his aircraft to position the steering dot in the center of the scope to complete the intercept. Unfortunately, the range trace line on the radar screen tended to obscure the target blip itself, leaving the pilot with scant information about the actual range to the target during the course of the intercept. Consequently, the pilot was forced to estimate the remaining range according to the target's range rate or closing velocity (Vc), which was displayed on the outer ring of the radar scope. When the computer determined that there were about twenty seconds left in the intercept, that range rate circle would suddenly start to shrink in size. On this cue, the pilot gently leveled his wings and depressed and held the trigger. The E-4 computer would then automatically lower the tray of rockets and fire them at the appropriate time. A large X appearing on the radar display indicated to the pilot that the intercept was over and that he should initiate an aggressive maneuver away from the target, an important step in the procedure because, as the manual noted, "the pilot cannot rely upon there always being ample clearance between the interceptor and its target" at the end of the intercept.[34]

"Flying the E-4 system is like flying instruments," one Hughes manual declared.[35] Although meant to reassure the F-86D pilots, it was an unfortunate analogy. Flying instruments remained one of fighter pilots' more loathsome tasks. When instrument flying was first introduced in the 1920s and '30s, many rebuffed the thought of subordinating their own intuition and sensory stimuli to the instruments' round dials and wavering needles. It was, after all, a serious affront to the myth of the fighter pilot as an innately gifted "birdman." After thirty years of experience, fighter pilots still only reluctantly trusted their instruments, and had it not been for mandated training requirements, most would have opted to never fly heads down.[36] The engineers at Hughes failed to appreciate that their new system represented as acute a challenge to the myth of the fighter pilot as that posed by instruments decades earlier. Moreover, the new fire control systems of the day had not yet earned a comparable level of pilot trust, despite what the manuals suggested.

For the interceptor crews in the F-94C and F-89D, or the lone pilot in the F-86D, following the fire control system's directions explicitly was only part of the challenge; they still had to contend with the troubling performance of the new armament. The unguided rockets had a much longer range of nearly forty-five hundred feet, but they remained subject to many of the trajectory perturbations that had made the original gunnery problem so challenging. A proximity fuse that detonated the rocket when it approached the target helped, but target destruction was still far from certain. The rockets' limitations became all too clear on August 16, 1956, when a pair of F-89D Scorpions, each outfitted with the improved E-6 fire control system and 104 rockets, tried to shoot down a wayward Navy F6F Hellcat target drone that had veered off course toward downtown Los Angeles, California. The Scorpions fired all 208 of their rockets on three firing passes, and they missed the pilotless plane every time. The drone eventually ran out of fuel and crashed into a remote field east of Palmdale. The flurry of rockets from the Scorpions, however, started "three good-sized fires and numerous smaller blazes," sent shrapnel through one woman's front window and into her pantry, narrowly missing her six-year-old son, and blew up the pavement in front of a car, leaving the vehicle riddled with holes. Fortunately, all survived, and the comical incident later became known as the "Battle of Palmdale."[37]

The Air Force's family of Falcon air-to-air guided missiles theoretically promised better results and no more embarrassing gaffes over the California desert. Hughes produced several individual variants of the Falcon missile, but there were two primary types: a radar-guided missile and a heat-seeking IR-guided missile.[38] Both were approximately the same size (6½ feet long), shape (fixed canards with delta-finned control surfaces at the tail), and weight (about 135 pounds), designed such so that they could be carried in the internal weapons bays of the Air Force's future high-speed "1954 Interceptor" aircraft (figure 4.4). However, after accounting for the necessary rocket propellant and guidance sections, the smallish design left little space in the missile body for the warhead and no room for a proximity fuse. Hence, for all the Falcons, the fuse was attached to the missile's rear fins, which would physically have to contact the enemy target in order to detonate the diminutive 2.7 pounds (in later versions, 5 pounds) of explosives in the missile's warhead. With the missile enjoying the

FIGURE 4.4. Hughes Radar-Guided Falcon Missile. There were two primary types of Hughes Falcon guided missiles: a radar-guided version and an IR-guided (heat-seeking) version. Both were similarly sized so that either could be carried in the internal weapons bays of the Air Force's newest interceptors under development. Regrettably, the missiles' relatively small size left scant room for the explosive warhead and no room for a proximity fuse. Here, an early prototype of the radar-guided GAR-1 is displayed.
USAF.

advantage of guidance throughout its trajectory, and with the target assumed to be a large Soviet bomber with plenty of surface area, the design tradeoffs were deemed acceptable.[39]

Development of the radar-guided Falcon preceded that of the IR-guided version, and the resulting GAR-1 Falcon entered testing in mid-1951. Touted as having "electronic 'eyes' and 'brains,'" the missile used semi-active radar homing (SARH) to find its prey.[40] In this guidance scheme, the interceptor aircraft used its radar to illuminate the target aircraft. Once the missile was launched, a small radar antenna in the nose of the missile sensed the radar energy reflected off the target, measured the relative change in line-of-sight between the missile body and the radar reflections, and then steered its control fins to zero out any relative changes, thereby effecting a collision intercept. To provide the necessary target illumination, the interceptor aircraft needed to stay pointed near the target with its radar locked on throughout the entire missile time of flight. Unlike the earlier gun and unguided rocket armament, the new missiles therefore demanded flawless performance of the pilot, his radar, and his weapon from before the trigger squeeze until the moment the missile impacted the target.[41] Any hiccups in the interceptor radar's performance, any aggressive maneuvering by either the interceptor or the target, or any faults lurking within the weapon itself would cause the missile to miss.

With the missile alone stuffed with seventy-two finicky vacuum tubes, and the interceptor's radar and fire control system relying on countless more, the system's reliability left much to be desired. Persistent design problems resulted in numerous production delays. The first GAR-1 Falcon-equipped squadron of F-89H Scorpion aircraft was not declared operational until March 1956, almost two years after the originally scheduled delivery date.[42] Hughes attempted to correct some of the missile's deficiencies in later variants, notably the GAR-1D, which was fielded in 1956 (about the same time the GAR-1 finally became operational), and the later GAR-3, released in 1958. Simultaneously, engineers worked to enhance the missile's capabilities. For example, reporting on the improved GAR-3, a *New York Times* article boasted that it had "a longer, higher, and deadlier reach than that of any other air-to-air missile," allowing it to "climb far beyond the altitude capabilities of the interceptor and destroy an enemy H-bomber in any kind of weather." Nevertheless, problems persisted.[43]

Hughes's heat-seeking variant of the Falcon, initially based on the GAR-1 and designated the GAR-1B but later modeled after the GAR-1D and renamed the GAR-2, suffered from an equally tumultuous development process. Because it relied solely on the heat plume emanating from the target's engine for guidance information, the IR-guided Falcon was in principle simpler than its radar-guided sibling. One 1956 Air Force evaluation report summarized the anticipated advantages of the heat-seeking missile design, noting that the "GAR-1B can be used at lower levels (no ground clutter); against multiple targets (it will select a target); and it has greater accuracy since the missile homes on a point source of heat rather than seeing the entire target. Additional advantages are that it is a passive seeker, it is immune to electronic countermeasures and it can be launched with less specialized fire control equipment."[44] But, because the Falcon's small IR-detector, which was mounted under a glass dome in the nose of the missile, needed to be close enough and in a position where it could detect the target's exhaust, the missile had a much shorter range and a more restricted firing envelope than the radar-guided Falcon. For the GAR-2, that translated into attacks of less than a mile and only when launched from within a small thirty-degree cone extending from the target's tail. Consequently, even from the program's initiation in November 1951, the IR-guided Falcon was never intended to supplant the radar-guided Falcon; only to complement it.[45]

Beyond the known restrictions in the IR-guided missile's firing envelope, engineers encountered challenges getting the guidance system to work properly outside of the lab. Atmospheric conditions, especially sun-lit clouds and haze, tended to confuse the IR-tracker and cause the missile to careen off course. After a series of dismal test performances in 1959, the engineers at Hughes finally developed an improved IR-guidance unit that used a filter to better distinguish the target's heat signature from that of shorter-wavelength IR-energy in the atmosphere. Later improvements to the GAR-2 eventually culminated in the GAR-4 and -4A, fielded near the turn of the decade.[46]

As with the earlier rockets, the introduction of Hughes's guided missiles demanded new electronic devices that could effectively link the aircraft's radar, pilot, and weapon together to form a coherent fire control system. The first Falcon-compatible fire control system was the E-9, developed for use in the F-89H. In the two-seat Scorpion, the engineers at Hughes chose to saddle the rear occupant with the responsibility for working the E-9's radar and fire control computers, leaving the front seater to concentrate on flying the aircraft. However, with Falcon missiles destined for use in the coming single-seat "1954 Interceptor," Hughes needed to develop a more automated system like that in the F-86D, one that would allow a single pilot to seamlessly fly the aircraft, work the radar, and launch the weapons.[47]

The resulting fire control system for the "1954 Interceptor" proved nearly as ambitious as the aircraft. The plane itself would not emerge until April 1956, and then only in an interim form as the F-102 Dagger Dart; it would take another three years before the ultimate ambition of the "1954 Interceptor" was realized in the F-106 Delta Dart.[48] Work on the interceptor's fire control system began well before that in mid-August 1950. The Hughes MX-1179 was to be highly integrated, capable of controlling the aircraft's radar, communication, and navigation systems. It would even interface with the aircraft's autopilot. During an intercept, the pilot would essentially relinquish control of the aircraft to the MX-1179 immediately after takeoff. The MX-1179 would then use the aircraft's autopilot to fly an optimum intercept against a specified target that was passed to it via a computer datalink from a SAGE command center. At the appropriate range, the system would automatically activate the radar, lock the prescribed target, and fire the missiles. After the target was destroyed, the MX-1179 would fly the interceptor back to its base and execute an automatic ground-controlled approach to landing. During the mission, the pilot would "merely" need "to maintain surveillance over the maze of gadgetry which surrounded him." But even that, one Air Force report reasoned, would still require a "highly capable" individual possessing "intense concentration and mental dexterity."[49]

Like most components of the new interceptor, the MX-1179 fell behind schedule, and an interim solution was devised for use in the compromise aircraft that became the F-102. Hughes's resulting MG-10 system still achieved a high level of sophistication and semiautomation, but it did not reduce the pilot to "an equipment monitor and safety factor" as envisioned with the MX-1179. Rather, the pilot in an MG-10-equipped F-102, as in the E-4-equipped F-86D, needed to follow the displayed steering commands and fly his aircraft along a commanded intercept course. However, once on course, the MG-3 radar (part of the MG-10 system) would, like the planned MX-1179, automatically lock the appropriate target and fire the missiles at the proper time.[50]

Despite the rapid pace of development, most fighter pilots in the Air Force's Tactical Air Command (TAC) remained blissfully unaware of the new Falcon guided missiles and their associated advanced fire control systems. Those were the tools of the bright-orange-flight-suit-wearing *interceptor* pilots of Air Defense Command (ADC), not the olive-drab-flight-suit-wearing *fighter* pilots of TAC. At the time, the distinction made sense. Interceptor pilots were trained to shoot down menacing enemy bombers approaching the United States, and their guided-missile ordnance was optimized for that particular type of target. Fighter pilots, on the other hand, were trained to secure air superiority over enemy territory by either destroying enemy fighter aircraft in the air or bombing them on the ground.[51]

The fighter pilots' professional journal, *Fighter Weapons Newsletter*, reflected TAC's near-total ambivalence to the changes taking place within ADC during the latter half of the 1950s. For example, as the F-86D pilots in ADC were learning in 1955 how to use their E-4s to execute rocket attacks according to "thunderbolt geometry"—attacking at high speed from a ninety-degree angle—the non-interceptor-Sabre pilots in TAC were reading Blesse's "No Guts, No Glory" and learning techniques to achieve an optimum close-range, low-angle gunshot.[52] When Falcon missiles started appearing on the wingtip pods of the service's F-89H Scorpions in 1957, fighter pilots in TAC were concentrating on learning how to use the A-4 gunsight in their F-100s to execute supersonic gun attacks.[53] Except for one very technical guest article on the "Application of Infrared Radiation" published in November 1956, there was virtually no discussion in the fighter pilots' journal on potentially adopting or adapting ADC's guided missiles or its advanced fire control systems for use in TAC fighters.[54]

The journal did, however, reflect the fighter pilots' growing appreciation for the increasingly technical aspects of their aircraft and air combat. In mid-1956, Lieutenant Bruno Giordano, the pilot who a few years earlier had been frustrated flying alongside Jabara in Korea, published nearly four pages detailing how to check the proper operation of the A-4 gunsight on the ground before flight.[55] A few years later, readers were treated to a lengthy ten-page treatise on the A-4 and APG-30 (MA-3) system in the F-100, complete with equations.[56] Captain John Boyd's attempt at mathematically describing the perfect "pursuit curve attack," the progenitor to his famous theory of energy-maneuverability (E-M), also graced the pages of *Fighter Weapons Newsletter* during this period.[57]

TAC fighter pilots finally took notice of the evolving weaponry outfitting ADC aircraft in early 1958 when they began testing an IR-guided missile to augment the cannon armament on their F-100D Super Sabres. Since the Air Force's heat-seeking Falcons were still mired in developmental delays, the Air Force elected to use the Navy's new Sidewinder IR-guided missile, which the Air Force labeled the GAR-8.[58] The new missile elicited an immediate, enthusiastic response from the F-100 test pilots:

> We feel that GAR-8 (Sidewinder) is just the beginning of a whole new family of Fighter Weapons which will greatly increase our offensive potential. Don't misunderstand us, we do not advocate removing the guns nor do we intend to stop gunnery training. We feel there will be a definite need for air-to-air gunnery training even after the missile has become our primary weapon. It is always possible to find yourself in a position where you are unable to fire the missile, and must make a gunnery attack. . . .

> [But] the elusive MiG flying 10,000 to 15,000 feet above us in Korea would be real meat for GAR-8. We might have to revise our standards for qualifying as an ACE in future conflicts.[59]

The pilots' assessment would prove prescient three years later when the Air Force decided to acquire the Navy's exclusively missile-armed F4H Phantom II fighter.

The Phantom Appears

In 1961 the Air Force was desperate to update its self-described "obsolescent" tactical fighter and reconnaissance force with new aircraft possessing superior speed, altitude, range, and all-weather attack capabilities. The Air Force's F-100 Super Sabre design was a decade old. Its newer F-105 Thunderchief, built to replace the service's F-84 of Korean War–vintage, had clearly sacrificed any fighter-like performance in favor of the *bomber* portion of its fighter-bomber mission. And, although TAC had begun work in 1958 outlining requirements for a Mach 2, sixty-thousand-foot, all-weather fighter, that project had since been transformed into the joint Air Force–Navy TFX (Tactical Fighter Experimental) program and was languishing in development by joint committee.[60]

Coincidentally, the Navy had just begun taking deliveries of its new F4H Phantom II, produced by McDonnell Aircraft Corporation. Aware of the Air Force's dilemma, executives at McDonnell initiated an intense marketing campaign that targeted Air Force senior leaders and the budget-minded civilian "whiz kids" in the Pentagon. The lobbying ultimately proved successful when in December 1961, the Air Force elected to procure virtually unmodified the Navy's F4H as an "interim fighter" that would sustain TAC until the arrival of the TFX. The Air Force planned to rename the Navy's F4H the F-110A Spectre, but the F-110A designation was abandoned following an August 3, 1962, decree by Secretary of Defense Robert McNamara demanding the services standardize their aircraft nomenclature. Hence, the Navy's preproduction F4Hs became F-4As, its production model F4Hs became F-4Bs, and the Air Force's F-110 became the F-4C; all versions retained the "Phantom II" moniker.[61]

The F4H was designed in the immediate wake of the Korean War to serve as the US Navy's newest ground-attack aircraft, or reconnaissance aircraft, or electronic warfare aircraft, or even air-to-air interceptor. McDonnell engineers planned to achieve this remarkable versatility by pairing a standard aircraft fuselage and wing section, overpowered by two massive J-79 afterburning engines, with interchangeable nose assemblies stuffed with the electronic equipment optimized for each unique mission. Responding to the Navy's initial wishes, McDonnell expected the ground-attack configuration to be primary, and it therefore planned to arm the aircraft with four 20-mm cannons or fifty-six unguided rockets. However, in mid-1955 the Navy balked at the radical design and instead dictated that a single nose assembly be permanently affixed to the aircraft fuselage. A few months later, the Navy announced that it wanted its new aircraft optimized not for ground attack but for intercepting enemy aircraft and defending the carrier fleet. Following the announced shift in the primary mission, the engineers at McDonnell quickly reappraised their aircraft's proposed cannon and rocket armament.[62]

Like the Air Force, the Navy had begun experimenting with guided air-to-air missiles immediately following World War II, and in 1947 it contracted with Sperry Gyroscope to build the Sparrow missile. Whereas the Air Force's radar-guided Falcons used SARH guidance, the Navy's first Sparrows, which entered the fleet in mid-1954, were beam-riding missiles, a simpler design but one that also proved to be more aerodynamically inefficient when attacking maneuvering targets.[63] Douglas Aircraft Company hoped to enhance the radar-guided missile's performance with its new Sparrow II design, which would use an autonomous active radar seeker mounted in the missile's nose to guide it independently and directly to the target. To ensure compatibility, Douglas planned to retain the same size and shape of Sperry's original Sparrow design. Never intended to be carried internally, the Sparrow measured twelve feet in length, almost twice that of the Air Force's Falcon, and it capitalized on its size to pack a considerably larger wallop than the Falcon, too—a sixty-five-pound warhead along with a proximity fuse to detonate it. Despite the Sparrow's large size, however, the Douglas engineers could not find a compatible radar that was compact enough to fit inside the missile's nose, and Sparrow II production faltered.[64] Raytheon threw its hat into the air-to-air missile ring in mid-1951 when it began working on its Sparrow III design. Like Douglas, Raytheon also retained the original size and shape of the Sperry Sparrow, but it chose to adopt the SARH guidance scheme used by the Air Force's Falcons. The Sparrow III performed reasonably well during testing, and in 1957, Sperry's Sparrow I was discontinued in favor of the Sparrow III. The Navy set sail with the Sparrow III the following July (figure 4.5).[65]

The competition between the three different versions of the Sparrow occurred while the engineers at McDonnell and their Navy counterparts were wrestling with how best to arm their new F4H fighter. In August 1955, following a successful round of missile testing, Captain Frank Timmes, the Navy officer in charge of the F4H project, decided to adopt Douglas's Sparrow II for use on the Phantom II. Echoing the Air Force's rationale, Timmes and other Navy officials reasoned that the new guided missiles would provide their aircraft with more offensive firepower. Although the abrupt switch challenged the aircraft engineers to integrate a more complex fire control system and a larger, more powerful radar into their Phantom, most appreciated the shift because it ultimately provided them with greater design flexibility. With missiles, as historian Glenn Bugos explained, the engineers had freedom to "decide which tasks—how much guidance or speed—should be built into the missile and which built into the aircraft, and how these tasks should be shifted between the aircraft and the missile as the technologies changed."[66]

FIGURE 4.5. Navy F-4B Launching a Sparrow III. After several design iterations, the McDonnell F-4B Phantom II catapulted into the air in the early 1960s to serve as the Navy's next fleet defense fighter. Its primary punch would come from four radar-guided Sparrow missiles. Unlike the Air Force's Falcons, the Sparrow was never intended to be carried internally, which freed the designers to build a larger, more powerful missile. Worried that a single pilot would be unable to simultaneously fly the aircraft and operate the complex radar needed to guide their aircraft's missiles, McDonnell engineers successfully lobbied to install a second seat in the Phantom's cockpit. Robert L. Lawson Collection, Naval Aviation Museum.

The Navy at the time was also working on its own IR-guided missiles. Similar in concept to the already-discussed heat-seeking Falcons built for the Air Force, the Navy's Sidewinder used an IR-sensor mounted under a glass dome in the missile's nose to detect a target's hot jet exhaust and then steer itself toward the target. Like the Falcon, the Sidewinder also had a correspondingly limited effective range and restricted firing envelope. But, at 9½ feet long, the Sidewinder had room for a larger twenty-five-pound warhead and a proximity fuse to detonate it, dramatically enhancing its lethality over the Falcon's. The results spoke for themselves. During one set of joint missile tests conducted at Holloman AFB in June 1955, the Navy's Sidewinders posted a perfect score; the Air Force's Falcons, not so much. Observing the tests was the commander of the Air Force's Missile Development Center, General Leighton Davis,

FIGURE 4.6. Navy F9F-8 Carrying Sidewinders. In contrast to the industry-built Air Force Falcon and Navy Sparrow missiles, the Navy's IR-guided missile—the Sidewinder—was designed in-house by Naval Weapons Laboratory personnel at China Lake, California. Their resulting missile posted impressive results during its testing, often outperforming the Sparrow at half the cost per missile. With its own IR-guided Falcons faltering, the Air Force elected to adopt the Navy's Sidewinder for use on its fighters. Here, an early version of the Sidewinder is carried under the wings of a Navy F9F-8 Cougar. Robert L. Lawson Collection, Naval Aviation Museum.

the same MIT graduate who had once spurred the Air Force to develop the A-1 gunsight. Suitably impressed with the Sidewinder's performance, he immediately pledged Air Force support for the Navy missile. The first Sidewinders (AIM-9As) became operational in the Navy in July 1956 (figure 4.6). A year later, an improved version was fielded in both the Navy (as the AIM-9B) and the Air Force (as the GAR-8).[67]

By June 1956, the Navy had decided to also incorporate its Sidewinder missile into the F4H Phantom design. Then, with Sparrow II development proving more challenging than Douglas and Timmes first suspected, Navy officials opted to switch to Raytheon's Sparrow III missile version. Hence, the Sparrow III radar-guided missile would serve as the Phantom's primary armament, the IR-seeking Sidewinder secondary, and the original 20-mm cannon tertiary. Ten months later, the guns were purged from the Phantom design altogether, a move intended to improve the aircraft's aerodynamic profile, reduce weight, and free up additional space for the advanced fire control system that the missiles required. The "archaic" aircraft guns of yesteryear had finally given way to the futuristic guided missiles of tomorrow.[68]

As the engineers at McDonnell wrangled over the Phantom's armament, they became increasingly convinced that only a two-person crew would be able to manage the sophisticated technologies they planned to pack into the Phantom. The idea of a two-person fighter plane, however, encountered stiff resistance among naval aviators. Like their Air Force brethren, the Navy's pilots had no desire to relinquish their single-seat autonomy, and they insisted that they could adequately manage any complex radar or armament system then envisioned. But as the F4H design slowly morphed into an all-weather interceptor solely dependent on guided missiles, the Navy's position began to soften. Bugos summarized McDonnell's argument for the dual cockpit: "A second crew member was mandatory in some roles . . . but superfluous in none."[69]

In April 1957, the same month the gun was finally stripped from the F4H design, the Navy finally acceded, and the Phantom permanently became a two-seater. Reflecting the conventional distribution of tasks, the back seater would be responsible for operating the radar, while the front seater focused on flying the plane. The Navy also decided that it would create a new naval flight officer, specially trained in radar theory and radar operation, to man the Phantom's back seat. Because the new officer, colloquially referred to as a RIO (radar intercept officer), was not going to be a trained pilot, the Navy decided not to install any flight controls in the rear cockpit of the Phantom II.[70] After a series of successful F4H flight trials in September 1958, one Navy official declared triumphantly, "The single seat fighter era is dead, and no more should be initiated." Another agreed, proclaiming, "The greater effectiveness of the two seat airplane in adverse conditions is decisive."[71] By October 1961, Navy Phantoms were flying off carrier decks in the Mediterranean and the Western Pacific manned by a pilot in the front seat and a RIO in the rear seat and armed with four radar-guided Sparrows, four heat-seeking Sidewinders, and no gun.[72]

When Air Force officials decided in late 1961 to procure the Navy's Phantom as a "multi-purpose fighter bomber," they left the Navy's design virtually unchanged. The decision is often attributed to McNamara's demand for defense commonality and efficiency, but the Air Force's actions were also a reflection of the urgency it attached to getting its new fighter into the air as rapidly as possible. In an internal memo dated November 7, 1961, Air Force officials disclosed the "sufficiently firm" possibility that their service would acquire more than 500 Navy F4Hs. The memo also noted that initial consultations with McDonnell representatives had identified only five changes required to adapt the Navy's Phantom for Air Force use. Some of the necessary fixes included replacing the Navy's probe-and-drogue in-flight refueling system with one compatible with the Air Force's boom-equipped tankers; equipping the Air Force version with a self-contained engine-starting capability; and incorporating provisions for the aircrews to employ the Air Force's GAM-83 Bullpup radio-guided air-to-surface missile.[73] The list of aircraft changes would grow over the next few months as the Air Force sought to bolster its Phantom's air-to-ground capability. A self-contained inertial navigation system (INS), a radar altimeter, and a new APQ-100 radar that could better accommodate ground mapping were all later deemed critical enhancements to the Navy's design.[74]

From the beginning, though, the Air Force was adamant that its version of the Phantom include a full set of flight controls and instrumentation for the rear cockpit. The November 7, 1961, memo declared that the service wanted the ability to "fly and land" its Phantoms from the back seat, but it did not specify whom the service planned to put in the fighter's rear cockpit.[75] Any uncertainty was removed by the time the Air Force published its Specific Operational Requirement (SOR-200) for the F-4C on August 29, 1962. The document made clear that there would be no un-winged RIOs sitting in the back of the Air Force's Phantom; the service planned to put a fully trained pilot in both the front and the back seat of its newest fighter. Officially, the front seater would be the "aircraft commander" and the back seater the "pilot systems operator," but the crews usually referred to the front seater informally as the *pilot* and the back seater as the *GIB*.[76] While later implementation of the manning decision suggests that the service intended to use the Phantom's back seat as an opportunity to groom its fledgling fighter pilots, some contend that the Air Force's two-pilot policy was simply an attempt to increase the rolls of its favored aircrew.[77]

Regardless of the motives, Air Force officials, after completing a series of flights in the back seat of a Navy F4H in late November 1961, determined that the Phantom's rear cockpit required thirty-six "provisions" before it could support full flight operations. These included, among other things, a full suite of flight instruments; engine performance indicators; controls for the landing gear, flaps, and brakes; and a control stick and rudder actuators. The original Navy F4H included some of these items in the rear cockpit, but many would need to be added or upgraded to support the Air Force's desire for a two-pilot crew.[78] Facing a potential delayed delivery schedule and increasing modification costs, the Air Force shortly thereafter tempered its expectations. By the time SOR-200 was released, the Air Force had decided that the rear cockpit required only "sufficient" equipment to "land the aircraft safely . . . *in emergencies*." On the basis of this subtle change, the engineers reasoned that normal actuators for the landing gear and flaps were no longer required in the rear cockpit; a onetime emergency extension capability would suffice. Similarly, the engineers couldn't foresee a requirement for the back seater to light the engines' afterburners, so they removed that capability.[79] The GIB did receive most of the requested flight instrumentation, but it was cobbled together and not arranged in the commonly approved layout (figure 4.7). Decades later, one GIB griped at the injustice of having to fly from the Phantom's back seat using "this little rinky-dink attitude indicator and a little rinky-dink heading indicator" while the front seater got to use his giant, "cosmic" flight instruments.[80]

The Air Force's first YF-4C prototype was delivered on May 27, 1963, sixty-five days ahead of schedule. Six months later, on November 20, 1963, the Air Force's F-4C version of the Navy fighter entered operational service at MacDill AFB, Florida, armed with four Navy Sparrow III radar-guided missiles, four Navy Sidewinder IR-guided missiles, no gun, and two pilots.[81] Now the fighter pilots in TAC just needed to learn how to operate it.

Figuring Out the F-4C

It later became popular for Air Force fighter pilots to deride their service's decision to acquire the *Navy*-designed, *two*-seat, *missile-only* F-4. Both the dual-seat configuration of the Phantom and its lack of a gun were anathema to the myth of the fighter pilot, threatening to transform the celebrated protagonists into mere missile managers.[82] The pilots' criticisms, however, were levied with the benefit of hindsight. At the time in 1961, many pilots were still haunted by the memory of MiGs overflying them in Korea, and consequently any technological advancements that promised to reduce the likelihood of similar, future frustrations were generally greeted with enthusiasm.

Hence, the new, beastly looking Phantoms, like the earlier GAR-8s installed on the F-100Ds, initially elicited a rhapsodic response from the Air Force's fighter pilots. The Navy's F-4 had already established numerous aeronautic records, including several for time-to-climb and top straightaway speed. One aviation publication described the Phantom as "one of the hottest machines that fifty-plus years of airplane building have been able to produce."[83] A June 1962 article in *Fighter Weapons Newsletter* similarly gushed over the Phantom and its "amazingly responsive" engines and smooth, "comfortable" flight characteristics. With such features coupled to a "Mach 2 plus max [speed]" and "a deadly fire control system (including outstanding radar detection and lock-on ranges)," one Air Force pilot declared of the new airplane, "I'm sold on the F-110A."[84] The World War II veteran Olds later remarked that he "loved the Phantom. Everything about it felt right. It was light on the controls and quickly responsive to power changes, and it gave me a feeling of eagerness not normal in an object weighing more than 17 tons."[85]

The first Air Force Phantom aircraft commanders (front seaters) were culled from the ranks of the service's top F-100 fighter pilots and sent to MacDill AFB for training in borrowed Navy F-4Bs.[86] Within a few years, whole scores of new aircraft commanders were taught how to fly the Phantom as TAC units began trading in their older aircraft for newer F-4Cs.[87] One of those transitioning pilots was Larry Welch. Born in Oklahoma but raised in Kansas, Welch originally enlisted in his hometown's artillery

CAUTION
RUD PED ADJUST

MASTER CAUTION
ILS ONLY
UHF-17
GROUND SPEED
TRUE AIR SPEED
CAUTION
RUD PED ADJUST

unit of the Kansas National Guard. Two years later in November 1953, he elected to join the Air Force through the Aviation Cadet program. After earning his commission and his wings, Welch ended up flying F-84s at a variety of locations before converting to the F-4 with the other members of the 366th TFW. He would eventually become the twelfth chief of staff of the Air Force.[88]

Despite having already proved their mettle flying one of TAC's single-seat fighters, Welch and the others still required more than two months of instruction to learn the intricacies of the F-4C, spending more than a hundred hours in ground academics and another thirty hours in the air.[89] Some of that time in the air was devoted to learning the peculiar quirks of the Phantom, especially how to avoid the "adverse yaw demon" that lurked at high angles of attack (AOA) and could cause the aircraft to suddenly snap out of control and enter a spin.[90] Other portions of the Phantom course familiarized the new front seaters with typical fighter pilot tasks, such as how to take off and land, drop bombs (both conventional and nuclear), and execute air combat maneuvers (ACM), the new term for dogfighting. None of these tasks were radically different from those the pilots had practiced countless times before in their prior fighters. For example, despite being almost a decade old and written for pilots who flew aircraft armed with guns and not missiles, Blesse's "No Guts, No Glory" was still distributed to the F-4 student pilots as an approved manual on ACM.[91] There was however one new task that few of the experienced fighter pilots had ever practiced before, and it consequently consumed a good chunk of training time—over twenty hours of classroom and simulator instruction, almost as much time as that devoted to reviewing the mechanics of the aircraft and its electrical, hydraulic, and fuel systems. The new unfamiliar topic was the Phantom's APQ-100/APA-128 radar and fire control system and, more specifically, how to use them in air combat.[92]

Those twenty-plus hours, however, paled in comparison to the training burden placed on the pilots destined for the Phantom's rear cockpit seat. The future GIBs, usually young officers who had just graduated from pilot training, began their F-4 immersion with an intense, eight-week radar fundamentals course that included over

FIGURE 4.7 *(opposite)*. F-4C Cockpit Layout. In the Phantom's front cockpit (*top image*), the flight instruments were arranged in a common pattern, with the large primary attitude indicator located in the center, airspeed to the left, altimeter to the right, and heading indicator below. Indicative of its anticipated importance, a radar display was mounted prominently in the front cockpit, just above the attitude indicator; however, the front seater had no control over the radar's operation. In the rear cockpit (*bottom*), the radar display was mounted on the center pedestal between the knees of the occupant. The flight instruments in the rear cockpit were smaller and arranged in a less-than-ideal horizontal line (altimeter, attitude, airspeed, and heading positioned left to right), which made flying instruments from the back seat particularly challenging. NMUSAF.

208 hours of classroom and simulator instruction on radar theory, operation, navigation, and basic intercept geometry. After completing the specialized course, the GIBs then transitioned to the F-4 conversion course that the aircraft commanders attended. There they sat through eighteen more hours of academics discussing intercept tactics alongside their front seaters. The program could be mind-numbing for the young lieutenants, but it was extremely important.[93] Flight regulations dictated that it was the sole responsibility of the "pilot" (rear-seat occupant) to "operate and monitor the radar," thereby freeing the "aircraft commander" (front-seat occupant) to "fly the aircraft."[94] The task assignments specified in the regulation were merely a formality, though. Reflecting the common convention, McDonnell engineers had vested all control of the radar functions and operation in the hands of the Phantom's back seater. Although the aircraft commander had a radar scope prominently mounted directly in front of him (figure 4.7), it was simply a repeater display; he had no control over the radar. It became quickly apparent to all that without a properly trained GIB in the back, the powerful radar installed in the front of the Phantom was nothing more than ballast.

The F-4 radar in air-to-air mode used a B-scope display (figure 4.8), similar to that used in the earlier F-86D with its E-4. During search, a vertical B-sweep line slid left and right across the scope as the radar antenna slewed back and forth in azimuth, with target blips appearing as the line swept over them. The vertical position of the target blip on the scope indicated the range to the target, with the bottom of the scope representing zero miles. The maximum display range at the top of the scope was selected by the GIB using the radar control panel. The options were 10 miles, 25 or 50 miles (the GIB toggled between 25 and 50 miles using a button on the radar antenna control stick), 100 miles, or 200 miles, but most GIBs regarded the two longest scope ranges "of no value in air-[to-]air combat." To convey the target's elevation information, the radar used an elevation strobe located on the right side of the scope set against tick marks that represented ten-degree increments above or below level (the center horizontal scribe). It was left to the GIB to mentally compute the trigonometric relationship between the target's elevation angle and its range to determine its altitude.[95]

An individual target blip told the GIB only where the target was at the present moment; it did not necessarily tell the GIB where the Phantom needed to aim to effect an appropriate cutoff. Also, all target blips didn't necessarily correspond to actual aircraft targets; many were spurious radar signals. To distinguish between actual and false targets and to gain an initial understanding of the intercept geometry, the GIB needed to watch the target blip for a few radar sweeps, assessing its relative movement on the scope. A consistent target blip on each radar sweep generally indicated an actual target. If the radar blip was also steadily tracking straight down the scope (target 1 in figure 4.9), then the GIB knew that the two aircraft were on a collision course. In contrast, if the target drifted to one side or the other (e.g., targets 2 through

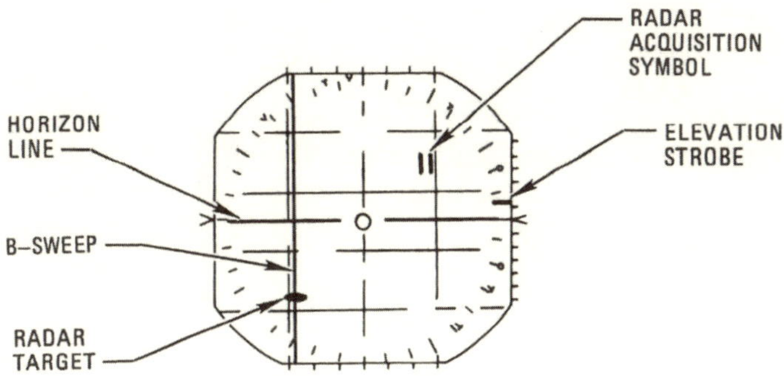

FIGURE 4.8. F-4C Radar Scope in Search. In air-to-air search mode, the F-4C radar used a B-scope display. A "blip" would appear as the B-sweep line passed over a potential target. The left-right position of the "blip" from the center scribe indicated the relative azimuth of the target, measured from the nose of the aircraft. The vertical position of the "blip" indicated the target's range. Relative elevation of the target (above or below the horizon) was indicated by the elevation strobe located on the right side of the display, but the 8th TFW tactics manuals warned aircrews that this indication was often inaccurate. In this example, the target is approximately thirty degrees left of the nose, about twelve degrees high, and, with a twenty-five-mile scope range, about six miles away. NMUSAF.

5 in figure 4.9), then the GIB knew that the two aircraft were not on an approximate collision course. By assessing the amount and direction of the blip's movement on the scope, a well-trained GIB could quickly develop a mental image of the intercept.[96]

When the GIB wanted to lock on to the radar target, he used the antenna control stick (figure 4.10) mounted on the right side of the rear cockpit to maneuver the radar acquisition symbol, a pair of quarter-inch vertical lines, over the target blip on the scope.[97] Once the acquisition symbol was placed over the target, the GIB then rotated the elevation wheel located on top of the antenna control stick until he obtained maximum target illumination, just like in the earlier E-4. To command a lock and automatic tracking, the GIB depressed the paddle switch at the base of the antenna control stick to its second detent (the first detent enabled manual tracking, but it was rarely used).[98]

With the radar in track, the symbology on the scope changed slightly (figure 4.11). The acquisition symbols were removed, replaced by two small dashes positioned along the B-sweep line that indicated the Sparrow missile's estimated maximum aerodynamic range (Ra) and its estimated minimum firing range (Rmin, computed as 0.2 of Ra). An allowable steering error (ASE) circle and an aim dot also appeared. The ASE circle and the aim dot were functionally similar to the steering dot and circle used in the F-86D's E-4: if the aircraft was maneuvered to position the aim dot inside the ASE steering circle, then the aircraft would approach an optimum computed intercept

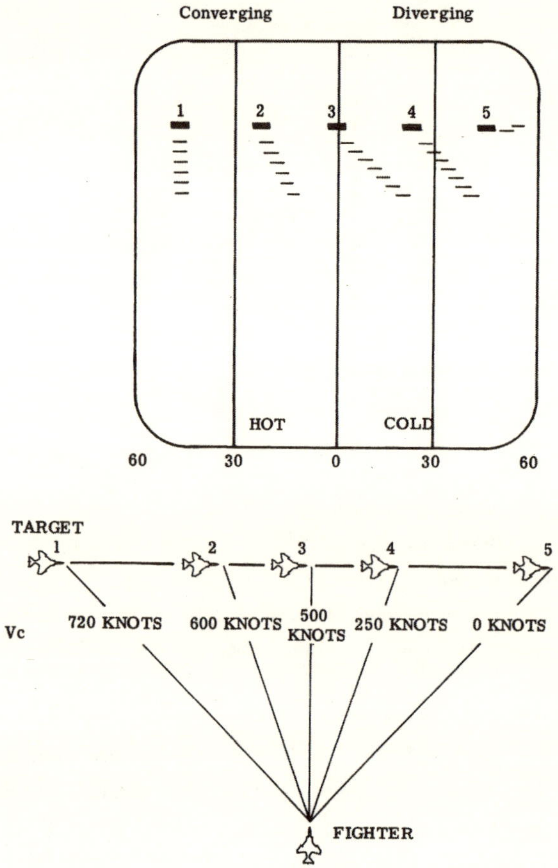

FIGURE 4.9. Assessing Target Drift on the Radar Scope. The initial radar "blip" indicated only where the target was at that moment, not its direction of flight. To determine the intercept geometry, the GIB could either lock the target or, if he was proficient, wait and observe the target's relative drift on the radar scope for a few more radar sweeps. Waiting and assessing reduced the probability of locking false returns. Assessing target drift was part of the intercept techniques instruction provided to new Phantom crews. As one instruction manual explained, placing the target farther on the "hot" side of the scope would cause the two aircraft to converge faster; conversely, placing the target on the "cold" side of the scope would slow down the intercept. AFHRA.

profile. Also borrowing from the E-4, the F-4 used a circle with a gap in it to indicate the target's Vc. But whereas the gap in the E-4's range rate circle corresponded to a graduated scale marked on the scope's face, the F-4's display used relative clock position. Hence, if the gap in the F-4's range rate circle appeared in the nine o'clock position,

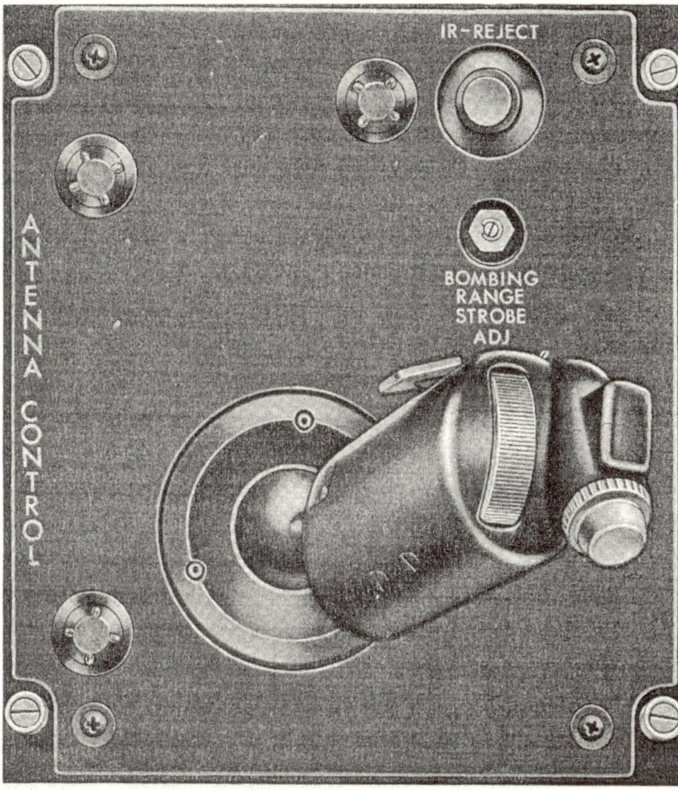

FIGURE 4.10. F-4C Antenna Control Stick. The Phantom's antenna control stick was located in the rear cockpit. Movement of the stick adjusted the radar in range and azimuth; the wheel on the top of the stick adjusted the radar in elevation. The paddle switch at the base of the stick was used to command a radar lock. NMUSAF.

then the target's Vc was 900 knots; a gap at three o'clock corresponded to 300 knots of Vc; and the gap in figure 4.11 represents about 150 knots closure. It wasn't quite that easy, though. Counterclockwise rotation from the twelve to nine o'clock position corresponded to negative Vc. And there was a switch in the rear cockpit that would change the clock position scaling, such that a gap at the nine o'clock position could also indicate 2,700 knots of closure. It was up to the GIB to keep all this straight.[99]

The additional information displayed during radar track made it possible for the GIB to monitor the target and update the intercept geometry. Because the GIB was not flying the plane, however, he could not actually maneuver the aircraft to refine the intercept. That was the job of the aircraft commander sitting in the front seat. To effect a successful radar intercept then, the two trained pilots, individuals whom Air

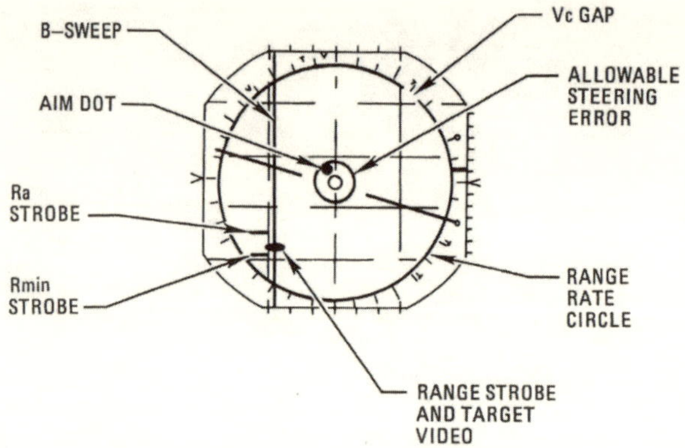

FIGURE 4.11. F-4C Radar Scope in Track. After the GIB commanded a radar lock, the B-sweep locked in position on the azimuth of the target. Two tick marks (strobes) appeared on the B-sweep line indicating the Ra and Rmin ranges for the Sparrow missile. To optimize the intercept, the aircrew was instructed to center the AIM dot inside the ASE circle. The gap on the outer Vc circle indicated the target's closing velocity according to clock position. NMUSAF.

Force instructors had previously designated fighter-worthy after demonstrating appropriate self-sufficiency during flight training, needed to learn to work together in the cockpit. As one of the Phantom's training manuals noted, "Any discussion of the F-4C's fighter versus fighter capability must assume that the aircraft commander and pilot work as a team, continually anticipating the problems and needs of the other in performing the assigned combat missions." That included the GIB "keeping the aircraft commander informed of everything that he observes, either visually or on radar," and the aircraft commander being gentle on the flight controls since "maneuvering under high 'G' forces" made it difficult for the GIB to keep "his head in the optimum position for viewing the radar scope."[100]

The crews also had to learn a new vocabulary for communicating with each other over the Phantom's intercom. Already well versed in the fighter pilot vernacular and recently introduced to the cryptic language of radar and radar intercepts, which included terms like B-sweep, ATA (antenna train angle), and TCA (track crossing angle), the crews now also had to practice using clear, concise "descriptive commentary." The GIB was supposed to report radar targets to the aircraft commander using a standardized A–E–R–O format: *A*zimuth in degrees from the nose, *E*levation in degrees above or below level, *R*ange in miles, and *O*vertake in knots. Reflecting the Phantom's naval origins, "port" and "starboard" were used rather than "left" and "right." Hence, an appropriate radar point-out from the GIB would be "30 port, 2 above, 5 miles, overtake 500." The GIB had a slew of other standardized commands available to him as well, such as "easy port" (a request for a fifteen-degree banked turn to the left), "harder" (increase the current bank angle by fifteen degrees), and "down more" (increase the current descent rate by one thousand feet per minute).[101]

All told, the Phantom with its advanced armament systems was supposed to simplify the fighter pilot's traditional air combat tasks, automating many of them in the aircraft or otherwise allocating them either to the new high-speed missiles or the second crewmember seated in the back seat. One mid-1962 McDonnell advertisement for the Phantom boasted of the "hundreds of test missions" flown in the simulator that pitted the "manual and mental capabilities" of the "two man crew" against an "electronic competitor." These simulations, the company explained, provided the engineers with invaluable opportunities to test and optimize the distribution of critical combat tasks between the human crew and the aircraft's "advanced radar, fire control, instrumentation, and communication systems."[102] Despite the rehearsals though, the Air Force and its fighter pilots quickly learned that the resulting arrangement was still exceedingly complex. It took more than four hundred hours of ground training to qualify just one new Phantom aircrew for basic operations, a fivefold increase over the training requirements of a mid-1952 Sabre pilot and an apt indicator of the dramatic shift in pilot tasks and responsibilities that accompanied the new fighter.[103]

It may have been a thinly veiled jab at the new two-seat, twin-engine Phantom and its future crews, or it may have just been serendipitous, but the editors at *Fighter Weapons Newsletter* decided to include a poem in the journal's June 1962 edition, which they placed immediately after a glowing, first evaluation of the new Phantom. Titled "Ode to a Fighter Pilot" and accompanied by an image of a single fighter pilot looking up at his monstrous, single-engine, single-seat F-105 Thud, the amateur poem reflected the myth of the fighter pilot as an isolated warrior soaring among the clouds.

> Oh, you knights of modern times with armor strapped so close around;
> You fly your noble craft to awesome heights and speeds
> Forever knowing the danger lying close beneath.
> Of *single-powered flight* where lonely screw or loosened blade
> Can halt your rapid flight and bring you quickly down,
> You know not where.

The poem's last sentence proved foreboding:

> One day the nation and the world will be aware of these fine men;
> Until that time the lonely vigilance will last,
> And communism will stand deterred from conflict, large or small.[104]

Within three years, the United States would find itself embroiled in another conflict intended to stop the spread of Communism and, as the poem's author predicted, its citizens would once again have the opportunity to be made "aware" of its "fine" fighter pilots and their exploits.

Thrust into War, Again

The first Air Force Phantoms to journey across the Pacific Ocean departed MacDill AFB, located near Tampa, Florida, on December 8, 1964. Although the eighteen F-4Cs from the 12th TFW weren't then destined for war—their destination was instead Okinawa, Japan—they demonstrated that the service's barely one-year old fighters, along with their aircrews and accompanying maintenance and support personnel, were ready to leave the comforts of home and journey halfway around the world when called upon.[105] That call to arms came four months later when on April 4, 1965, aircrews from the 45th TFS, 15th TFW, taxied their Phantoms into position at the end of MacDill's runway, lit their afterburners, and rocketed off to war. After overnight stops in Hawaii and Guam, the eighteen F-4Cs finally touched down at Ubon RTAFB on the seventh. Having just spent twenty-one of the last sixty hours in the cramped Phantom cockpit, the aircrews were told that they were needed back in the air as soon as possible. So, sixteen hours later, the 45th's Phantoms thundered down the runway at Ubon and turned North, ready to battle MiGs.[106]

By the time of Operation Bolo in January 1967, there were ten times the number of Phantoms supporting the Vietnam War than what the 45th brought with it in April 1965. The Phantom became the Air Force's workhorse during the war, remaining in continuous combat over Southeast Asia (SEA) until the final ceasefire was signed on January 27, 1973.[107] The 45th's tenure in combat, though, was brief. Just four months after its arrival, the squadron returned home to MacDill, having been replaced at Ubon by the 47th TFS in late July. Following the 47th's four-month deployment, which ended in December 1965, the Air Force realized that its plans to rotate intact squadrons through SEA on four-month stints wasn't sustainable, and it thereafter decided to semi-permanently base its aircraft in the region and instead rotate personnel and aircrews through on a yearlong or hundred-combat mission tour, whichever milestone occurred sooner.[108] Countless Air Force aircrews would subsequently have an opportunity to experience air combat firsthand in three distinct versions of the Phantom (F-4C, D, and E) while operating from a handful of bases in Vietnam and Thailand.

The war above Vietnam had already begun by the time the 45th TFS arrived at Ubon. The Gulf of Tonkin incident, in which North Vietnamese patrol boats allegedly attacked US Navy ships on August 2 and 4, 1964, led to the first American air attack against NVN, conducted on August 5 by carrier-based Navy fighters and fighter-bombers.[109] Following that action, the Air Force quickly reoriented its tactical forces within the region, but its first direct attack against North Vietnamese infiltration routes into northern Laos didn't come until four months later when it launched a total of fourteen F-105s and F-100s to participate in Operation Barrel Roll.[110] In early February 1965, North Vietnamese insurgents launched another round of attacks against American Army facilities near Pleiku and Qui Nhon in South Vietnam (SVN). The Americans responded over the course of the following week with Air Force– and

Navy-led attacks—Flaming Dart I (February 7–8) and Flaming Dart II (February 11)—against a variety of North Vietnamese military targets. Little more than a week later, President Lyndon Johnson authorized a program of "sustained US bombing of North Vietnam." The resulting operation, labeled Rolling Thunder, would roll on for nearly three years.[111]

At the outset of Rolling Thunder, the Air Force had approximately two hundred tactical aircraft in the theater, the bulk of jets being F-100 Super Sabres and F-105 Thunderchiefs. For the first Rolling Thunder strikes conducted on March 2, the Air Force used nearly half of its force in the region to attack an ammunition dump: twenty-five F-105s and twenty B-57s provided the bombing effects, while another fifty-eight total F-105s, F-100s, and RF-101s were mustered to provide flak suppression, air cover, and tactical reconnaissance support. That same day, the Air Force sent another quarter of its tactical force to support a South Vietnamese–led attack against a different target.[112]

The Air Force lost five fighters during those first attacks—three F-105s and two F-100s. They were brought down not by MiGs but by ground-fire.[113] Trained to approach their targets low and fast, equipped with weapons designed for the same, and constrained by the narrow geography of the region, Air Force fighter pilots throughout the war proved particularly susceptible to the "unsophisticated" ground-fire that often congregated around the limited number of suitable bombing targets in NVN. Small-arms fire would eventually account for almost half of the Air Force's more than sixteen hundred aircraft lost in combat over SEA from January 1962 to June 1973. Larger-caliber antiaircraft artillery (AAA) claimed another 26 percent.[114] After the war, one Air Force commanding general and World War II veteran likened the intensity of AAA around Hanoi and Haiphong to that of the Reich's most stoutly defended targets.[115]

Another set of ground-based weapons would also menace American aircrews for the duration of the war. These new weapons, one Phantom pilot noted, "were faster than enemy planes and carried a much larger bang than antiaircraft fire. They were guided and homed in on you. . . . They chased you down to kill you."[116] They were SAMs—surface-to-air missiles—and their lethality had been signaled five years earlier in 1960 when the Soviets used an SA-2 SAM to shoot Francis Gary Powers's U-2 spy plane out of the sky. American intelligence officials started observing North Vietnamese troops constructing their own SA-2 SAM sites around Hanoi beginning in April 1965 (figure 4.12). However, the Johnson administration placed the new SAM sites off-limits to attack because of concerns about collateral damage and war escalation. Three months later, after the SA-2 sites had been completed and ringed with defensive AAA, the North Vietnamese opened fire. The first missiles, launched on July 24, 1965, brought down one F-4C and damaged three others in the formation, one severely. The attack occurred just as the 45th TFS was beginning its swap-out with the 47th TFS.[117]

FIGURE 4.12. North Vietnamese SA-2. The Soviet SA-2 surface-to-air missile (SAM), deployed across North Vietnam beginning in April 1965, proved a formidable threat to American aircraft throughout the war. Eventually, the Phantom aircrews would develop individual aircraft maneuvers, new formations, and advanced electronic attack technologies to enhance their survival against the Mach 3+ radar-guided missile. NMUSAF.

At first, the new SAMs were extremely lethal, on average scoring a victory for every eighteen missiles fired. Commonly described as a "flying telephone pole," the SA-2 missile was nearly thirty-five-feet long, reached a top speed of Mach 3.5, and packed a nearly three-hundred-pound warhead. Riding the beam provided by its accompanying ground-based Fan Song radar (like the first Sperry Sparrows), the SA-2 missile could attack targets anywhere between three thousand feet and fifty thousand feet altitude, out to ranges in excess of seventeen miles. Improved American countertactics and electronic countermeasures introduced by 1968 significantly reduced the effectiveness of the SAMs, and eventually it took on average more than a hundred SA-2 missiles to bring down one American aircraft. Nonetheless, with nearly six thousand SA-2s launched between July 1965 and March 1968, the SAMs were a constant and deadly threat.[118] Describing their impact on the Air Force's tactical air operations, the commander of the 47th TFS declared: "The advent of the Surface to Air Missile has changed things considerably. It denys [*sic*] the free use of airspace above 3000 feet AGL [above ground level] in a SAM defended area. To complicate matters further there has been a significant improvement in the employment of AAA and AW [automatic weapons]. In short you can't fly high enough to avoid SAM's and you can't fly low enough or fast enough to avoid ground fire."[119]

In addition to the myriad of AAA and SAM defenses dispersed throughout NVN, the American aircrews also had to contend with the enemy's fighter planes. At the start of Rolling Thunder, the North Vietnamese Air Force (NVAF) possessed thirty-four Soviet-built MiG-15s and MiG-17s, all based at its new airfield at Phuc Yen. The NVAF MiG-15s were nearly identical to those used in the Korean War more than a decade earlier, and the MiG-17s were only slightly faster than their predecessors. But both MiG versions were small and nimble, and the Air Force eventually realized that its advanced supersonic fighters were not well equipped, nor its pilots well trained, to deal with the threat.[120] According to Olds, despite the Phantom's technological superiority, the MiG-17 "could give you a tussle the likes of which you never had before in your life."[121] Moreover, the NVAF pilots astutely observed that they didn't need to shoot down any aircraft to disrupt the American bombing attacks; merely arriving on-scene and threatening the Americans would often cause the bomb-laden fighter-bombers to jettison their heavy ordnance as they turned to defend themselves. Enjoying the benefit of its ground-based radar facilities that could detect the American aircraft approaching from a distance, and seasoned radar controllers who could precisely vector the MiG pilots toward the Americans, the once-discounted NVAF repeatedly demonstrated its ability to hinder the Americans' bombing efforts.[122]

Conventional military doctrine decreed that air superiority needed to be obtained quickly, before initiating an intense strategic bombing campaign. Consequently, military planners had sought immediate approval to bomb the NVAF airfields, aircraft, and surrounding air defenses in order to establish the requisite superiority over the enemy's territory. However, the military encountered stiff resistance from those within the Johnson administration. As in the Korean War, political considerations trumped military doctrine as many of the North's radar sites and all of its MiG airfields were stricken from the list of authorized Rolling Thunder targets. To placate the military's concerns, Secretary of Defense McNamara approved on April 1, 1965, the deployment of a single squadron of Air Force Phantoms to combat the MiG threat. Three days later, the Phantoms from the 45th TFS departed MacDill for Ubon.[123]

The MiGs struck twice before the Phantoms arrived. On April 3, 1965, three MiGs swooped in and attacked a group of Navy fighters that were bombing a North Vietnamese railway bridge, damaging one F-8 Crusader. The next day, as a flight of four heavily laden Air Force F-105s orbited ten-miles south of Thanh Hoa Bridge waiting for its turn to attack, a pack of four MiG-17s dove "out of the haze" from the formation's high six o'clock. As they had the day before, the MiG pilots tore through the

American formation on a high-speed, single-pass attack, with the two lead MiGs closing to within fifteen hundred feet before they opened fire on the two lead Thuds. The MiGs' bullets found their mark, and they downed the two aircraft, killing both pilots. The MiGs' sudden aggressiveness startled the Americans, which is partly why the 45th's pilots were ordered back into the air almost immediately after landing at Ubon.[124]

Loaded with four Sparrow and four Sidewinder missiles each, the 45th's Phantoms started flying MiG CAP (combat air patrol) for the larger Rolling Thunder missions, escorting their Thud brethren to and from the target areas. Perhaps wary of the Phantoms' presence, the NVAF MiGs stopped giving battle.[125] Although two Phantoms sighted a pack of eight MiGs loitering above them at the end of a mission on May 31, and four F-105s were harassed by a pair of MiG-15s three weeks later on June 24, air activity remained relatively calm. The next engagement between Air Force fighters and MiGs would not take place until July 10, 1965.[126] In the interim, with the MiG threat apparently in abeyance, the members of the 45th TFS turned their talents elsewhere, showcasing both their own and their aircraft's remarkable versatility.

Jack of All Trades

Despite the Phantom's origins as a Navy interceptor optimized for carrier defense, Air Force officials from the outset planned to take full advantage of the Phantom's massive engines and the numerous hardpoints on its wings to deliver all sorts of air-to-ground munitions. The Air Force had already swapped out the Navy's radar with a system better suited for air-to-ground work, and its first aircrews had received nearly as many training sorties devoted to practicing ground attack as they had for learning the new air intercept tactics.[127] With no MiGs willing to give battle, Air Force officials wondered if their Phantoms might be better used to support the immediate task at hand. From then on, Phantom aircrews would split their time between hunting MiGs and dropping bombs, with the latter often consuming the bulk of their efforts.

MiG CAP missions soon gave way to hybrid strike CAP missions. During MiG CAP, the 45th's Phantoms' sole mission was to defend the less maneuverable fighter-bombers from air attack. The mission usually required twelve Phantoms, arranged into three flights of four. In addition to their eight air-to-air missiles, the Phantoms also usually carried three external fuel tanks to maximize their loiter time in the target area protecting the fighter-bombers.[128] Strike CAP saddled the Phantom aircrews with an additional task. They would still fly MiG CAP, but if no MiGs were encountered, the Phantoms would then execute a bombing mission alongside the fighter-bombers. The strike CAP-configured Phantoms had to trade out their centerline 600-gallon external fuel tank to make room for three 750-pound general purpose bombs.[129]

In principle, the strike CAP tasking made sense, but the aircrews in the 45th TFS despised it. Their squadron commander called it "probably the most degrading mission given to the F-4C." By attempting to be both an escort fighter and a bomber simultaneously, the Phantom was rendered incapable of doing either well. First off, the "light bomb load" was judged by many pilots to be tactically ineffective. Then, because the bombs replaced much-needed fuel and added significant drag, Phantom aircrews on strike CAP missions routinely ran low on gas and often had to jettison their bombs simply to continue their primary escort duties. Concluding that 25 percent of the Phantoms' bombs carried during strike CAP missions never made it to the target area, the commander of the 45th TFS declared the strike CAP mission "a waste of aircraft and munitions."[130]

The hybrid strike CAP mission was thereafter discarded in favor of straight "air to ground bombing missions." For these, the aircrews traded in their underwing-mounted Sidewinders for bombs, but they were allowed to keep their Sparrows—those missiles didn't take up bomb space and theoretically provided the aircrews with an air-to-air self-defense capability. After some experimentation with air-to-ground tactics, the aircrews in the 45th TFS settled on a forty-five-degree dive-bomb attack with the GIB serving as an "oral altimeter," calling out the Phantom's altitude as it descended "down the chute"—"10-5, 9-5, 8-5, 7-5, ready, pickle"—while the front seater focused his attention on lining-up on the correct target.[131] The Phantoms also began flying various other air-to-ground missions that summer, all of which demanded different types of ordnance, including unguided air-to-ground LAU-3A rockets and SUU-16 external podded 20-mm Gatling guns. The variety of assigned missions and corresponding ordnance, each requiring different tactics and employment procedures, added significant complexity to Phantom operations.[132]

The pace of American bombing over NVN steadily quickened, and by the end of 1965 most Phantom operations were devoted to the air-to-ground task. Of the 1,743 combat sorties flown by the 47th TFS while deployed to Ubon from August to December 1965, less than a third were for the original MiG CAP mission.[133] To simplify interservice coordination of the escalating bombing effort, military planners during this period implemented the Route Pack (RP) structure, dividing NVN into seven geographic areas (figure 4.13). According to the agreement, the Navy's aircraft, operating from their carriers positioned off the coast at "Yankee Station" in the Gulf of Tonkin, were generally assigned targets in RP II, III, IV, and VIB. The Air Force was primarily responsible for targets located in RP I, V, and VIA, which subjected the Air Force fighters to long transits over hostile regions. A massive Air Force aerial refueling operation along the Thai-Laotian border consequently took shape, itself becoming a staple of Vietnam air operations.[134]

Over the next year, the Air Force consolidated its growing force of F-4Cs in the region into three wings: the 8th TFW at Ubon RTAFB, Thailand; the 12th TFW at Cam Ranh AB, SVN; and the 366th TFW at DaNang AB, SVN.[135] The Phantom aircrews also continued to add to their air-to-ground repertoire, especially after the new F-4D variant began arriving in theater. Physically similar to the F-4C, the latest

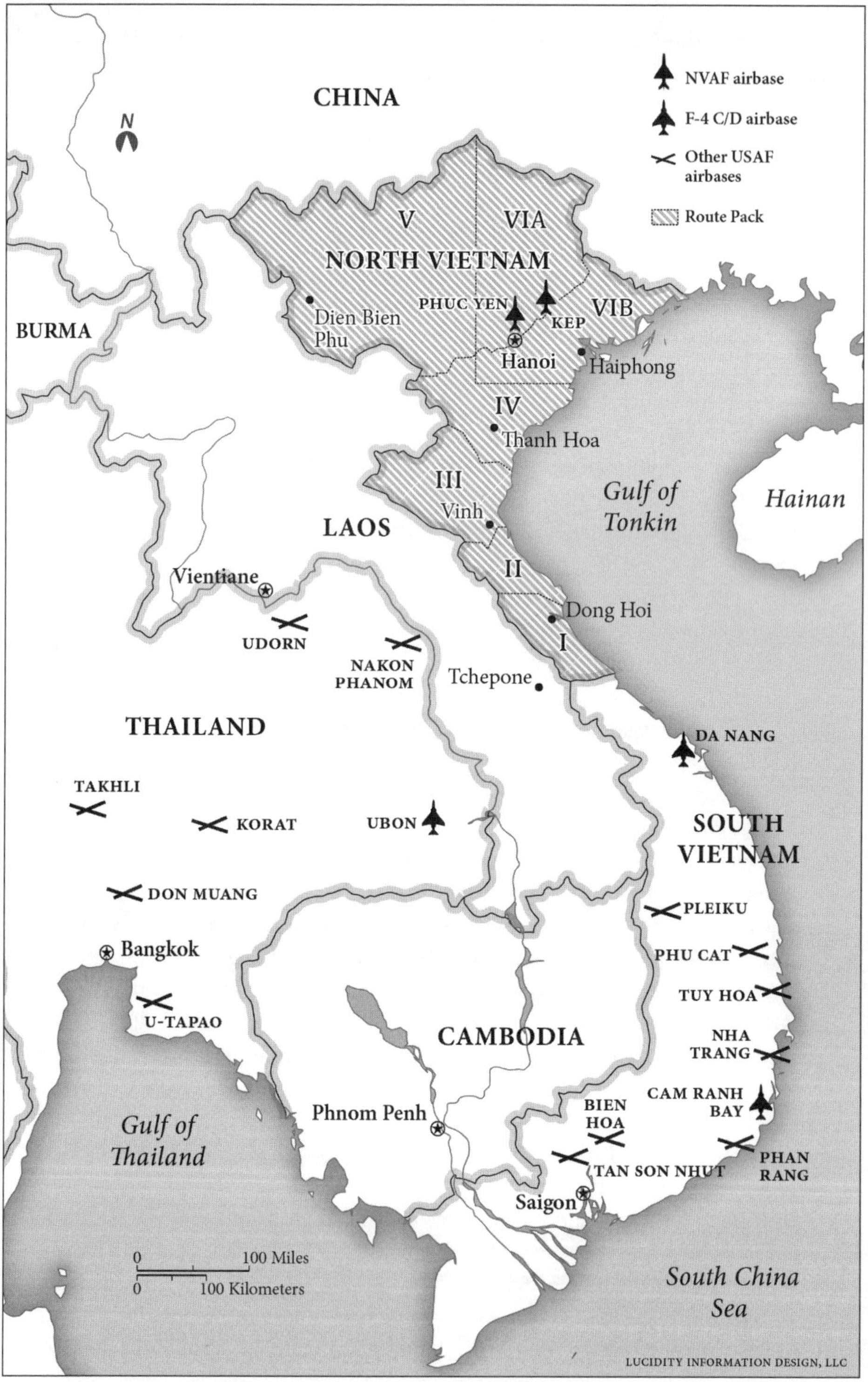
NVAF airbase
F-4 C/D airbase
Other USAF airbases
Route Pack
CHINA
N
BURMA
V
VIA
NORTH VIETNAM
VIB
Dien Bien Phu
PHUC YEN
KEP
Hanoi
Haiphong
IV
Thanh Hoa
III
Vinh
Gulf of Tonkin
Hainan
LAOS
II
Vientiane
Dong Hoi
I
UDORN
NAKON PHANOM
Tchepone
THAILAND
DA NANG
TAKHLI
KORAT
UBON
SOUTH VIETNAM
DON MUANG
PLEIKU
Bangkok
PHU CAT
TUY HOA
U-TAPAO
CAMBODIA
NHA TRANG
CAM RANH BAY
BIEN HOA
Phnom Penh
Gulf of Thailand
PHAN RANG
TAN SON NHUT
Saigon
0
100 Miles
0
100 Kilometers
South China Sea
LUCIDITY INFORMATION DESIGN, LLC

Phantom included an updated, semiautomatic bombing computer, the AN/ASQ-91 Weapons Release Computer System (WRCS), which allowed the aircrews to execute a new bombing tactic known as "dive-toss."[136] The aircrews by now were regularly flying a mix of radar bombing missions, armed night reconnaissance missions, and, as the number of US Army troops in the region steadily grew, close air support missions. The 8th TFW's tactics manual reflected both the growing diversity of the Phantom's air-to-ground missions and their relative weight of effort. Updated in December 1967 shortly after the F-4D arrived at Ubon, the revised tactics manual weighed in at over three hundred pages, almost ninety of which were devoted to air-to-air combat tactics and MiG CAP or strike escort operations.[137]

The variety of air-to-ground missions performed by the Phantom aircrews, as well as their proficiency executing them, is an important reminder that the F-4 combat experience in Vietnam was not focused solely on downing enemy MiGs. Indeed, with the limited number of enemy aircraft and the American pilots' relatively short tours of duty, few Phantom aircrews ever had an opportunity to engage a MiG. While their Rolling Thunder missions might have at times seemed a lost cause as they repeatedly bombed targets they assessed had little political or military significance, fighter pilot and military historian William Andrews observed that the aircrews nevertheless felt "duty-bound to use the immense power [of the aircraft] under their control to help fellow combatants, especially American soldiers."[138] Supporting Andrews's assessment, one Phantom pilot, Jerry Cook, remembered the surge of adrenaline, pounding heart, and "eagerness to help" whenever he heard the call, "ground troops in trouble."[139] Jere Wallace, an F-100 pilot who would go on to fly F-4s and later F-15s, expressed a similar sentiment, "The most rewarding mission for me was getting some [Army] grunts out of trouble."[140]

Still, the allure of a potential MiG kill remained an intense attraction for many Phantom aircrews. The F-4 pilots "all wanted to get a MiG," one Air Force commanding general observed.[141] Cook wrote that he left Vietnam in October 1966 feeling "somewhat unfulfilled . . . I flew every air-to-air mission that I could to the north, but [the opportunity to fight a MiG] just never happened."[142] When some suggested that

FIGURE 4.13 (*opposite*). Southeast Asia. As Rolling Thunder operations intensified, coordinating Air Force and Navy air attacks against North Vietnamese targets became more challenging. To simplify the process, military planners divided North Vietnam into seven geographic regions: targets located in RP II, III, IV, and VIB were generally assigned to Navy aircraft, while those in RP I, V, and VIA went to the Air Force. The Air Force's aircraft operated from a slew of bases spread across Southeast Asia; its F-4C/D Phantoms were primarily stationed at Ubon in Thailand (8th TFW) and Cam Ranh Bay (12th TFW) and DaNang (366th TFW) in South Vietnam. Map by Lucidity Information Design, LLC aircrews should become specialized and fly only select mission types rather than the entire mix, the men howled in anguish, saying that it would be unfair to those of them not assigned the MiG CAP role.[143] The 8th TFW's tactics manual likewise stoked the pilots' enthusiasm with its declaration, "Air-to-air combat is the utmost test of a fighter pilot's skill."[144] In short, like their Korean forebears, Phantom pilots longed for the opportunity to land from a mission, be greeted on the tarmac by a cheering throng, and then paint a red star on the side of their jet.[145] A few members of the 45th TFS had the opportunity to experience the phenomenon firsthand during their deployment. In the process, they also revealed many of the issues that would plague the Phantom in air-to-air combat for the duration of the war.

Air-to-Air Frustrations

On July 10, 1965, a pair of F-4Cs from the 45th TFS scored the Air Force's first MiG kills of the war. The American aircrews had for weeks watched the enemy's fighters practice chasing them down as they left the target area. Assigned to fly MiG CAP that day for a group of F-105s attacking an ammunition depot in Yen Bai (about thirty miles northwest of Hanoi), the Phantom aircrews of Mink flight hoped to catch the MiGs in their usual routine. The four Phantoms deliberately took off fifteen minutes late and snuck into the target area. When the MiGs turned to follow the strike package, the Phantoms raced north to intercept them. Four-minutes later, two MiG-17s had been shot down by Sidewinder missiles.[146] In the *New York Times* article that followed two days later, one of the front seaters, Major Richard Hall, described the engagement as "almost a schoolbook exercise."[147]

In reality, it wasn't. Hall and his GIB, Lieutenant George Larson, were flying in the lead Phantom, call sign Mink 1. They first detected the MiGs at a distance of thirty-three miles, but they couldn't shoot. The Americans' rules of engagement (ROE) required a visual identification (VID) of the target before a missile could be launched at it, a consequence of the Phantom radar's inability to discriminate between friendly and hostile aircraft. To satisfy the restriction, the aircrews in the 45th TFS had planned to send their two lead Phantoms racing ahead eight to ten miles in full afterburner toward any targets once detected. If the targets were in fact hostile MiGs, the lead element would then radio the trailing Phantoms who would still be in an ideal position to fire their Sparrow missiles.[148] That day, though, the Phantom elements achieved only about three-miles' separation, and by the time the visual identification was made, the trailing element was already too close to the MiGs to launch its missiles.

Even had there been sufficient range, the trailing Phantoms would still have had difficulties. Mink 4's radar had spouted a leak in its wave guide tube, rendering it incapable of automatic target tracking. Additionally, the radio intercom that allowed the GIB and the front seater to communicate with one another had stopped working. The crew in Mink 3 was not much better off. Their aircraft's radar also failed, which led to significant confusion in the cockpit when the front seater wanted to shoot a Sparrow but the GIB kept telling him to use a Sidewinder instead.

Nonetheless, Mink 3 and 4 were able to exploit their thrust advantage to power high above the MiGs before diving down into position to fire their Sidewinders. The pair of Phantoms fired a total of eight Sidewinders at the two MiGs. Four of the Sidewinders veered off course from the start, two because the front seaters had failed to first check that their missiles had detected the MiG target before firing them. Of the remaining four missiles, two appeared to detonate near the targets but failed to bring the MiGs down. The remaining two, one against each MiG, finally did the job. It took less than ninety seconds from the time of the identification pass to the shootdown.[149] In the process, the single engagement disclosed several of the challenges the Phantom aircrews would confront during the war: the onerous requirement to perform a VID of a target before launching a missile; unreliable radars, and consequently unreliable radar-guided missiles; finicky heat-seeking missiles, and front seaters who had a tendency to fire them out of parameters; and miscommunications and tension within the Phantom cockpit.

The requirement to positively identify a radar target as hostile before firing a missile at it was not unforeseen. When the Air Force first began procuring the missile-only Phantom, it was expected that a target's identification would be determined and relayed to the aircrews by ground or airborne radar controllers. The F-4C's requirements documents made explicit reference to the assumption, warning that if the F-4s were deployed to an austere region lacking sufficient radar coverage, the aircrews would be "required to use visual means of identifying targets before missile launch. In this event, the advantages of the Sparrow are nullified."[150] In Vietnam the Phantom aircrews faced exactly that hypothetical scenario, which was exacerbated by the constrained geography of the region and the limited coordination among the different American aircrews. For example, Air Force Phantoms approaching targets from the west would frequently see on their radar scopes other Air Force or Navy aircraft approaching nearby targets from the east.

Years later, General William Momyer, who served as the commander of Seventh Air Force during Rolling Thunder and was therefore responsible for most tactical air operations in SEA, complained that because of the VID requirement the Air Force had "forfeited our initial advantage of being able to detect a MIG at thirty to thirty-five mile range. . . . Many kills were lost because of this restriction."[151] Momyer might have been able to change the air-to-air ROE, but he never did. He deemed fratricide more likely and more costly than the chance of losing an American aircraft to the limited number of enemy MiGs. And, despite their grousing about the VID requirement, many aircrews admitted that "they preferred [it] to shooting down one of their own aircraft by mistake."[152] On one occasion, a late VID saved Cook the ignominy of shooting down a pair of Navy F-8s.[153]

Throughout the Vietnam War, the Air Force worked to improve its network of ground and airborne radar coverage, hoping one day to shift some of the fighters' identification and ROE burden to the radar controllers as originally envisioned. Indeed, the morning that the members of the 45th TFS left MacDill for Ubon, Air Force EC-121Ds, specially converted Lockheed Super Constellations with bulbous antennas mounted above and below the fuselage, departed McClellan AFB in California for SVN. The EC-121s specialized in aerial surveillance, but their radar system was optimized for target detection over water, not the mountains and valleys of Vietnam. Using the call signs "Big Eye," "College Eye," and "Disco" during the course of the war, the EC-121s were never able to reliably detect targets flying below eight thousand feet. Consequently, although they could broadcast alerts of general MiG activity—"yellow" provided a warning that MiGs were in the air, while "red" indicated an imminent danger to American aircraft—the EC-121s were never able to relieve the Phantoms of the stringent VID requirement.[154]

Tactics and new technologies were also enlisted to help the Phantom aircrews satisfy the ROE. As an example of the former, Olds had carefully orchestrated Operation Bolo to provide the first three flights of F-4s with clear lanes and temporary "missile free zones," which would allow the Phantoms to immediately fire on any radar target they detected. However, when the MiGs failed to climb up to meet the Phantoms and Olds flight turned back toward Ford flight, Colonel Olds had to abruptly cancel the free-fire zones, "holler[ing] out loud and clear, 'Don't shoot. Come off the missiles free.'"[155] The Phantom aircrews would eventually capitalize on the dual technological developments of Combat Tree, which was installed on some late-model F-4Ds allowing the Phantoms to interrogate the MiG's radar transponder, and TISEO (Target Identification System, Electro-optical), essentially a television camera with a telephoto lens mounted in the F-4E's left-wing root, but neither system arrived in time for use during Rolling Thunder.[156]

The VID requirement also exacerbated known limitations with the Phantom's missile ordnance. As noted earlier, both the Sparrow and Sidewinder incorporated proximity fuses that increased the missiles' chances of destroying a target. However, their inclusion in the design necessitated a larger, safe-minimum missile employment range—three thousand feet—in order to preclude the possibility of the missile fusing and detonating near the launching aircraft. Against the anticipated enemy bomber threat, the missiles' minimum range was not expected to be an issue. But when facing a nimble MiG that could quickly turn to meet an approaching Phantom head on, and that was also so small that the Phantom crew could usually obtain a positive VID only once inside of one mile, the missiles' minimum range proved to be a "definite handicap." Analogizing their predicament, Air Force fighter pilots frequently griped that they were handed a "spear" or "rifle" before being locked in the cramped confines of a "phone booth" or "closet" to battle "knife"- or "pistol"-wielding MiGs.[157]

Complicating matters, the performance of the Phantom's longer-range "spears" was at best lackluster.[158] Entering Vietnam, the Sparrow was expected to hit its target 71 percent of the time. It never came close to that number.[159] For example, at the end of April 1966, one Phantom front seater reported that "confidence in Sparrow was low at this point; there had been 13 firings with no hits in the previous week."[160] The following month, a team of Phantom and Sparrow specialists was sent to SEA to investigate the situation and "recommend the required actions necessary to enhance success of future Sparrow/Sidewinder firings." They reached an unsettling conclusion, reporting that, even with "proper maintenance of both aircraft and missiles, the probability of kill with the Sparrow can be expected to be low."[161]

In response, the Air Force and Navy collaborated to race two new versions of the Sparrow missile into the fight. Regrettably, neither the new AIM-7E nor the AIM-7E-2 matched expectations. The performance of the AIM-7E-2 was especially disheartening. Hailed as the "dogfight Sparrow" because it promised a shorter minimum range and improved capability against small, maneuvering fighters like the NVAF MiGs, the AIM-7E-2 scored only 34 hits out of 281 firings during the course of the war, a dismal 12 percent success rate.[162]

The Phantom's complex radar shared some of the blame for the Sparrow's poor combat tally. The corrosive heat and humidity of the region wreaked havoc on the Phantom's delicate electronic components that were scattered in hard-to-reach places. It was later said that on average the F-4 required thirty-seven man-hours of maintenance for every single hour of flight.[163] While Olds's team at Ubon had been able to execute Operation Bolo with a minimum of aircraft electronics failures, they had also spent weeks preparing and "peaking" their aircraft in preparation for the important mission, often asking maintenance personnel to work sixteen- to eighteen-hour days.[164] The pace was unsustainable. Consequently, the uninspiring performance of the 45th TFS Phantoms' radars in the earlier July 10, 1965, battle was more often the norm throughout the Rolling Thunder campaign. Plus, even when "peaked," fundamental issues with the Phantom's radar remained; its pulsed-radar design rendered it physically incapable of reliably detecting targets flying at low altitude. Whenever a MiG dove toward the ground, the Phantom's radar would inevitably break lock and any Sparrow missiles in the air would be lost.[165]

The Sidewinder performed markedly better than the larger, more complex Sparrow. The IR-guided missile's 28 percent hit rate in April and May 1966, however, still fell well short of its originally advertised success rate of 65 percent. Unlike the technical afflictions that plagued the Sparrow, the poor performance of the Sidewinder could often be traced to improper employment. In the rough-and-tumble dogfights, the Phantom front seaters often had a difficult time achieving the missile's restricted firing envelope, which resulted in frequent shots well outside of specifications.[166] One front seater declared the AIM-9B Sidewinder "totally hopeless in the air-combat environment."[167] Another was only slightly more optimistic, reporting, "It's a damn fine little missile *if* you can get the thing launched under the right parameters."[168]

The Air Force's solution to its Sidewinder woes was to forsake the Navy missile altogether. When the service rolled out its F-4D Phantom, the aircraft came armed with the Air Force's own highly touted IR-guided AIM-4D Falcon, offspring of the Hughes GAR-2B. Regrettably, despite claims of dramatic improvements in missile capability, the new AIM-4Ds performed miserably in combat. The new missile retained the Falcon's diminutive warhead and contact-only fusing, which significantly impaired its ability to down a small, maneuvering target like a MiG. Complicating matters, aircrews found the missile's employment procedures and restrictions even more cumbersome than that of the earlier Sidewinder. By the end of the war, almost 55 percent of the AIM-4Ds had been fired outside of parameters.[169] After experiencing numerous AIM-4D failures firsthand, Olds declared the new Falcons "no good" and ordered them removed from his Wing's Phantoms.[170] Thereafter, the Air Force elected to design its own AIM-9E Sidewinder, but the newer IR-guided missile did not reach the theater until after Rolling Thunder had concluded. Even then, the performance of the Air Force's AIM-9E never matched that of the Navy's own AIM-9D update to the Sidewinder.[171]

The F-4D—AIM-4D combination was an unequivocal failure in the air-to-air arena. But in its defense, the F-4D was never intended to rectify the air-to-air combat deficiencies of the Phantom. That was the job of the Air Force's F-4E, development of which was announced in October 1966. Among other notable air-to-air improvements, the new F-4E would sport an internal 20-mm Gatling gun, which the *New York Times* reported "will complement the plane's missile capability and should give it superiority in both long-range action and close combat" (figure 4.14).[172] The new Phantom was also supposed to incorporate a radical new radar system that would offer an unparalleled ability to detect low-flying targets, but Hughes's coherent-on-receive Doppler system failed to materialize in time. Consequently, when it reached SEA in November 1968, the F-4E came equipped with the smaller and only slightly improved APQ-120 radar set.[173] By that time, though, the Americans had already begun curtailing their Rolling Thunder effort, all but eliminating any chance of an American pilot claiming another MiG kill until hostilities flared again in 1972 during the Linebacker operations.

The Phantom aircrews witnessed several dramatic shifts in NVAF tactics and fortunes during Rolling Thunder. At first, the North's MiG-15s and -17s tended to execute isolated, single-pass slashing attacks reminiscent of Korea. However, in the latter half of 1965, the small, supersonic, delta-winged and IR-missile-equipped Soviet MiG-21 fighter began appearing in the skies over Vietnam (figure 4.15), and more aggressive, coordinated NVAF tactics soon emerged. Success during Operation Bolo temporarily restored American air superiority, but a new round of American bombing in mid-1967 stirred the hornet's nest. By then, the NVAF had significantly improved

FIGURE 4.14. F-4E Phantom II. The F-4E reflected many of the lessons learned from air combat over Vietnam, one of the foremost being the need for a viable close-range weapon. Consequently, the F-4E included an internal 20-mm, six-barrel Gatling gun mounted under the aircraft's nose. NMUSAF.

its tactics. MiG-17s would frequently take to the sky in large numbers and establish a circling "wagon wheel" formation at low altitude, ambushing unsuspecting American fighter-bombers en route to their targets. Meanwhile, MiG-21s exploited their advantage by remaining at high altitude before being vectored by increasingly proficient North Vietnamese radar controllers into missile range behind an unsuspecting American fighter.[174]

Through the summer of 1967, American and North Vietnamese pilots waged an intense battle for air superiority. By August, it appeared that the Americans had finally gained the upper hand, leading General Momyer to proclaim proudly to a Senate subcommittee, "We have driven the MiGs out of the sky for all practical purposes."[175] But Momyer's assessment proved overly optimistic, and the MiGs returned with a vengeance. During the last quarter of 1967, North Vietnamese MiG-21s successfully downed twelve American aircraft, five of them Phantoms, while suffering only one loss. The MiGs' successes and political turmoil stateside had, by March 1968,

FIGURE 4.15. Soviet MiG-21. Small and highly maneuverable, the supersonic, delta-winged Soviet MiG-21 was the American aircrews' first adversary armed with guided air-to-air missiles. Vectored into position behind the American fighters by ground-based radar controllers, the North Vietnamese Air Force pilots could fire an R-3S missile (referred to as an AA-2 Atoll by the American crews) from nearly one mile away. The IR-guided Atoll was functionally similar to the Navy's first Sidewinders. István Toperczer.

effectively driven the American Phantoms from the skies over North Vietnam.[176] Even after hostilities resumed in 1972, the air combat tallies in Vietnam never matched the impressive numbers of the Korean War.[177]

The F-4 aircrews struggled to keep pace not only with their increasingly-proficient enemy but also with the rapidly changing technological requirements of their own Phantoms. New human-machine relationships in both the front and rear of the cockpit had to be formed. Interpersonal relationships that linked front seaters with GIBs and Phantom aircrews with other Phantom aircrews had to be renegotiated. Collectively, these changes were subtlety transforming what *flying* a Phantom in combat meant. The picture that the *New York Times* editors selected to accompany their account of the 45th TFS's victories on July 10, 1965, alluded to one possible outcome. Whereas in the past pilots were usually photographed proudly standing near their aircraft's cockpit, in this case the crew of the victorious Phantom was photographed crouching under the wing of their aircraft, the front seater and GIB each pointing to a Sidewinder missile. The title of the picture made no mention of fighter pilots but instead labeled the two flight-suit-wearing aviators "successful missilemen."[178]

Tension in the Air

In November 1966, shortly after Olds arrived at Ubon and took command of the 8th TFW, the Air Force filed a news release announcing the rare occurrence of having two ace pilots in charge of a fighter wing in Vietnam. Olds had earned his acehood with his 24½ kills in World War II. His vice commander, Colonel Vermont "Pappy" Garrison, now a "wise old sage" at age fifty-one, had claimed eleven victories in the Second World War and then another ten in Korea.[179] Asked to compare their prior combat experience with that in the Phantom over Vietnam, Olds remarked, "These young lieutenants today have a far more difficult task than we had with the less sophisticated aircraft of World War II. It's difficult to compare the F-4C Phantom with the 'seat of the pants' flying in the old P-51 Mustang."[180]

Olds's brief remark made explicit reference to one of the challenges the Phantom aircrews confronted over Vietnam—figuring out their "sophisticated" aircraft. The task was complicated by the fact that, although the aircrews had gone through months of schooling learning how to use their new aircraft and its systems, they soon discovered that neither their aircraft nor their training was especially well suited for the air war they encountered over Vietnam. The 8th TFW tactics manual made plain the disconnect: "The F-4C/APQ-100/APA-157 weapons control system and associated armaments, the AIM-9B and the AIM-7E, are designed to be employed in a non-maneuvering environment using close control. This close control coupled with the long ranges of the armament provide an element of surprise and thus a high probability that the target will be in a non-maneuvering state. Further, the system was designed more as a defensive rather than an offensive system. *The chances of employing the system in this manner in SEA are very remote.*"[181]

Olds's remark, however, also alluded to more than just the pure technological differences that separated the Phantom from the earlier Mustang. Of course, the Phantom was a more complicated piece of technology. Nonetheless, it still was "difficult" for Olds to draw a meaningful comparison. The Phantom, despite its complexity, was in some ways easier to fly and operate. In other ways, it was infinitely more challenging. Each Phantom front seater and GIB would have to navigate the unique challenges associated with operating the F-4C in combat. In doing so, the aviators were forced to confront the uncomfortable question: Were they still "flying knights"? Had their technology finally transformed them into "flying scientists"? Or perhaps they were now something slightly different, something akin to the "successful missilemen" referenced in the *New York Times* caption.

Tensions in the Cockpit

By the end of the Vietnam War in 1973, the Phantom, according to one Air Force historian, was recognized as "the big dog on the porch" within TAC circles.[182] Even before the single-seat F-100s and F-105s had faded into history, however, the Phantom had lost some of the initial luster that accompanied its arrival in the Air Force. Not every fighter pilot was thrilled about sharing his space in the cockpit with another man, and the Air Force's preferred policy before 1969 of assigning junior, newly winged lieutenants to the Phantom's back seat fueled the tension. Confronting the sometimes-acrimonious interpersonal dynamics within the cockpit became as critical to the Phantom's overall combat performance as each individual's efforts sorting through the technical challenges associated with his specific position in the cockpit.

For the young lieutenants emerging fresh from pilot training before 1969, the F-4C Phantom seldom rose to the top of their wish lists.[183] Clifton, Olds's GIB during the Bolo mission, admitted that his first choice of aircraft would have been the single-seat F-105, but the lone Thud offered to his pilot training class was snatched up by the top graduate.[184] Lieutenant Howard Hill was also disappointed when he found out that he was going to be thrown into the back seat of the Phantom. An El Paso, Texas, native, Hill graduated from the US Air Force Academy in 1965, a year after Clifton. Like Clifton, Hill got to experience the thrill of flying the supersonic, needle-nosed T-38 Talon in the skies around Del Rio, Texas, during pilot training. "We had soloed. . . . We had flown formation; we had done all sorts of things," Hill remembered. "A lot of us were thinking we would be single-engine jocks." But when the assignments were revealed, three-quarters of the "young tigers" in Hill's class found out that they "were going to be put in the back seat with another guy." Resigned to his temporary fate, Hill did his best to adopt the "fighter pilot's bravado and the swagger" as he rolled into his F-4 radar training course.[185] But he and the other future GIBs couldn't fool Colonel Frank Everest, the commander of the first Phantom training unit at MacDill. The new GIBs "hated" sitting in the back seat; "it was demoralizing" for them, Everest observed.[186]

The single-engine F-105's appeal to the newly winged lieutenants is telling. It was a big and fast jet, but old, certainly not agile, and, as Olds poignantly noted, it had "a bombay [*sic*] in it, for Christ's sake."[187] The Thud was also suffering grievous losses over Vietnam. A pair of analyses conducted in mid-1966 concluded that an F-105 pilot had only a 70 percent chance of surviving his hundred-combat mission tour. In contrast, the crews in the F-4C enjoyed a 96 percent chance of completing their assignment.[188] The F-105's loss rates weren't a secret either. Many pilots, including Hill, joked that an "optimist was an F-105 driver that smoked two packs of cigarettes a day and thought he was going to die of lung cancer."[189] Facing challenging missions and long odds on a daily basis, the Thud pilots earned the almost-universal respect of the other fighter pilots.[190]

Hence, with the Thud pilots' "valor and performance . . . fast becoming legend," it's not difficult to appreciate the young fighter pilots' enchantment with the thought of *flying* a Thud on a bombing mission as opposed to *riding* in a Phantom staring at blips on a radar screen.[191] Some Phantom instructors tried to entice the new pilots, advertising that "the discriminating young fighter pilot recognizes that the F-4 is not only the most versatile plane in combat today, but that it will be with us for many years to come. Sooner or later, those who want to stick with supersonic fighters will have to transition into it."[192] The sales pitch rarely worked though, and the older single-seat fighters continued to draw the top graduates from pilot training. Still, for the remainder who continued to dream of one day being a fighter pilot, at least the back seat of a Phantom was far better than being relegated to the right, copilot seat of a transport, bomber, or tanker. Plus, Phantom GIBs like Thomas Sokol, who grew up near an airport and spent his time "star[ing] at airplanes all day," could entertain the thought of "hopefully find[ing] a way to get in the [Phantom's] front seat as soon as possible."[193]

It wasn't just those destined to ride in the back of the Phantom who were put off by its two-seat configuration. Many front seaters worried about a new, snot-nosed, young lieutenant pilot-GIB trying to tell them where and how to fly. Even by 1972, Phantom GIBs were still encountering front seaters who had previously flown single-seat fighters, individuals whose attitude was "they had done it all before" and therefore "didn't need anybody back there."[194] In the Phantom, though, they had to have a GIB because many of the critical aircraft systems could be controlled only from the rear seat. One possible solution was to simply use the GIB as an automaton executing the front seater's commands, and several GIBs reported receiving rather simple instructions from their front seater at the start of a mission: "Shut up, go cold mike [turn off the radio intercom], and when I need you I'll tell you."[195] Welch suggested that front seaters like these saw the GIB as a threat to their "manhood."[196] Future Lieutenant General John Burns observed of his time in Vietnam that "an awful lot of the guys in the front seat" showed no interest in learning about the radar or the GIBs' tasks, and similarly exhibited no desire to share their piloting experience with the junior aviators. Thus, although the young pilots were supposedly put in the Phantom's back seat to receive "apprentice training" and learn how to be better fighter pilots, "in practice," Burns admitted, "it didn't work out that way."[197]

The tension within the cockpit was further exacerbated by the Air Force's decision to implement a "one-tour policy" for its combat aircrews. The policy reflected the service's desire to spread combat experience among its aircrews and knock down the castes that had developed within TAC, Strategic Air Command (SAC), and Air Training Command (ATC). However, as the war wore on, the supply of trained and experienced fighter pilots available to fill the Phantoms' front seats steadily dwindled. As it had once before in Korea, the Air Force began sending its nonfighter pilots, individuals who had spent years flying larger transports, bombers, or other training aircraft, to a brief training course and then off to war in the service's premier fighter.[198]

The effects were, in the words of one fighter pilot, "devastating," and the retrained pilots were "universally disrespected. . . . They were good guys . . . but they could never catch up."[199] Another Phantom crew member expressed similar sentiments of the retrained pilots, explaining that an "80-hour training course . . . , if they have no previous fighter time, fighter background, fighter tactics, is just not quite enough to bring them up to par."[200] According to one postwar Air Force analysis, the net effect of the manning policy was to force the service into "fighting seven one-year wars" using a stream of replacement pilots of steadily decreasing skill "instead of one seven-year war" with increasingly combat-seasoned pilots.[201]

As the backgrounds and experience level of the Phantom front seaters became increasingly diversified during Rolling Thunder, the incoming GIBs continued to be pulled fresh from pilot training. Moreover, the future GIBs were still being skimmed from the top of their pilot training classes on the understanding that they would one day become fighter pilots after their initial grooming assignment in the Phantom's back seat. But now the GIBs' apprenticeship was being overseen by "temporary," cross-trained fighter pilots who had not made the initial fighter pilot cut. Riding in the back seat of a Phantom flown by a pilot whose flying skills were often judged to be less than his own, it was easy for a GIB to grow depressed, especially when, in the words of one, he "gets his everliving tail shot off continually and can do nothing about it."[202]

The situation was even more trying for those experienced GIBs who were nearing the end of their combat tour. They were usually paired with a "neophyte" front seater who, although technically in charge of the aircraft, possessed little practical knowledge of combat flying in Vietnam. For example, on Hill's seventieth combat sortie on December 16, 1967, he was paired with a front seater who had never flown a mission into North Vietnam. It ended up being Hill's last combat sortie; their Phantom was shot down by a MiG-21 that had snuck in behind them.[203] Sokol noted that toward the end of his combat tour, he had no intention of dying or becoming a prisoner of war because of his front seater's buffoonery, and he "wasn't afraid to take the airplane away from the guy [in the front seat]" to prevent that from occurring. Sokol joked that in the process he earned a reputation as "an asshole back seater."[204] This was not exactly what the GIBs had signed up for when they were told they could be a fighter pilot in the US Air Force.

The GIBs' reward for their inglorious introduction to fighter aviation was supposed to be a quick upgrade into the Phantom's front seat, preferably while the junior pilots were still serving their first combat assignment. "This was the bill of goods . . . that [the Air Force] sold" them, one front seater dryly remarked.[205] In November 1967, Momyer encouraged his F-4 unit commanders in SEA to honor the unwritten pact, directing them to establish an in-house program "to upgrade highly qualified and experienced F-4 pilots to aircraft commander status as rapidly as combat commitments and the local environment permit."[206] Indeed, as some commanders in the

12th TFW had already learned, the lure of an in-theater aircraft commander upgrade provided a valuable carrot to maintain the "morale and motivation of the squadron pilots."[207]

However, there were practical difficulties with the proposed upgrade program. Although the upgrading GIBs all had extensive combat experience and had spent plenty of time riding in the Phantom's back seat, those experiences and skills did not automatically transfer to the front seat. Hearing their once-GIBs cuss as they fumbled with front-seat tasks and annoyingly misplaced switches, Welch and others snickered that the GIBs were finally learning that flying the Phantom "wasn't quite as easy as it looked" from the back seat.[208] In one case, an upgrading front seater flew a bit too close to his target while trying to drop a bomb and physically "knocked off the roof and part of the second floor [of the building]," much to the chagrin of his instructor.[209] It would take time for an upgrading GIB to learn the necessary combat skills demanded of a front-seat aircraft commander. In the case of the 8th TFW, the upgrade training lasted more than ten weeks and included at least twenty-eight sorties. Unfortunately for the GIBs, as the pace of combat quickened, few units had the extra time or resources to allocate to the upgrade program.[210]

If a Phantom GIB was unable to upgrade to aircraft commander while still in SEA, he was supposed to be first in line for the necessary training when he returned to the States. Again, however, practical difficulties emerged. Ironically, even though the Air Force was scrounging for pilots to fill the front seats of its combat-flying Phantoms, there weren't enough front-seat opportunities for the newly "graduated" GIBs. The GIBs were excluded from another combat assignment because of the one-tour policy. But of the remaining Phantom front-seat assignments located in the States, most were tasked with training the new front seaters for combat, which required an already-experienced aircraft commander. Having dreamed of becoming a fighter pilot and having been promised a front-seat cockpit, many GIBs were crestfallen when they opened their assignment letters at the end of their combat tour and learned that they were once again relegated to riding in the Phantom's back seat. Of course, the GIBs could always volunteer to fill the numerous SAC or ATC aircraft cockpits left vacant as those pilots were shuffled into the Phantoms' front seats for combat, but then the GIBs would no longer be flying fighters.[211] Still, one front seater remembered his GIB saying he'd do just that, "go to training command—anyplace—to get where he was going to fly the airplane rather than ride in the back seat."[212]

The only other way for a GIB to get around the shortage of front seats was to immediately volunteer for a second combat tour, sidestepping the one-tour restriction. If a GIB chose that option, he was guaranteed to receive one of the few front seater training slots stateside. After approximately six months of training, the previous-GIB-turned-front seater would then return to Vietnam for another full combat tour in a new Phantom unit. Many GIBs, stoked by "youthful feelings of invulnerability" and the overriding desire to *fly* a fighter, chose that option. Sokol was one.[213] Clarence

"Dick" Anderegg, a small-town kid from northwestern Pennsylvania who wanted to fly KC-135s until his first aerobatic ride in the T-37 during pilot training at Laredo AFB, Texas, was another.[214]

As former GIBs started populating the front seats of the Phantoms, the situation for all GIBs improved slightly. Anderegg in particular remembered that "it was great when you flew with a former GIB" because they were often more forthcoming and willing to share the piloting tasks.[215] Sokol likewise noted that "ex-back seat pilots treated back seat pilots different than ex-F-100 pilots treated back seaters." After he moved to the front seat, Sokol took pride in the fact that, as a former GIB, he could help "make the back seater a better back seater."[216]

In the end, though, the Air Force's policy of using newly winged pilots as Phantom GIBs was almost universally detested. Front seaters complained that their GIBs weren't interested in becoming proficient radar operators. "The capability of the F-4 is being wasted by having a pilot in the back seat," one front seater concluded in a postmission report. "[We] need a radar expert in the back seat. The pilot back seater's primary goal is to be upgraded to the front seat (aircraft commander) rather than master the radar."[217] Most back-seat pilots agreed. "A radar officer would be more interested in the back seat operations than a pilot would be," one GIB admitted.[218] Anderegg too acknowledged that, as a GIB, he "certainly was interested in getting into the front seat" of the Phantom, and he recalled his front seater's occasional scolding, "You're spending too much time wanting to fly the airplane." But Anderegg rationalized his clamoring for stick time, "That's what my paycheck gets written on. It doesn't get written on whether or not I can do an offset radar bombing run." His argument alluded to a significant institutional oversight: oblivious to the nontraditional tasks required of the GIB in the back seat of the Phantom, Air Force regulations required that the GIBs pass *pilot* performance checkrides each year, not *navigator* or *bombardier* or even specialized-*GIB* checkrides. Anderegg and the other GIBs therefore had good cause to fight "to keep some kind of pilot proficiency."[219]

In late 1968 Air Force senior leaders finally implemented a new manning policy to help address some of these issues. Junior pilots fresh from pilot training would no longer be required to cut their teeth in the Phantom's back seat; they could move straight into the front cockpit. To fill its newly vacated back seats, the service elected to adopt elements of the Navy's RIO program. Going forward, the Air Force's GIBs would be culled from the ranks of trained navigators, and they would now be called weapons systems operators (WSOs). The Air Force stopped short of adopting the Navy's preferred division of cockpit tasks, however. In the Navy's model, the RIO maintained primary control over the aircraft (exercised through the pilot in the front seat) during the long-range intercept portion of the mission; the front-seat pilot took control only when he was close enough to see the target visually. In contrast, the Air Force planned for its front-seat pilot to retain control throughout the entire mission. The change in manning policy was generally well-received by those flying the Phantom.[220] One front seater wrote approvingly, "The WSO formed a team with the pilot, supporting rather than challenging. You got synergy rather than erosion. . . . It was a good solution. Not perfect, but much better than before."[221]

Before the shift, and probably to a certain extent afterward, the GIBs' common joke was that flying in the two-seat Phantom was "like being married."[222] Although the analogy was more often used to refer to the pattern of communication between the two crew members—knowing when to speak and when to be quiet—it's probably a fitting descriptor for the broader context. Each front seater and his GIB brought a different set of experiences and expectations to the cockpit. While in some instances they were compatible, many times they were not. In the case of the two-*pilot* cockpit, it was also a forced marriage that neither participant was too keen on entering and that inevitably "caused friction."[223] Plus, the thought of "being married" to another man didn't exactly conform to the fighter pilot myth, steeped as it was with themes of masculinity and self-sufficiency. Nonetheless, the two crewmembers had to work together if the Phantom was going to function properly. The engineers at McDonnell had hardwired the requirement for intracockpit collaboration into the Phantom's design. The Phantom crews' stateside training was supposed to have reinforced this imperative, but it obviously didn't always produce the desired behavior. That was unfortunate because once the Phantom encountered hostile MiGs over Vietnam, the front seater and the GIB would have their own human-machine issues to attend to in their individual cockpits.

Managing Modes in the Back Seat

An F-4C instructor once counted "fifty-seven different combinations of weaponry and flying modes" available in the Phantom.[224] When designing the Phantom and divvying up the crew tasks, the engineers vested most of the controls for these different modes in the hands of the GIB. The arrangement made sense since it theoretically allowed the front seater to focus on flying the aircraft while the GIB managed the radar and weapons systems. The engineers included some provisions to help automate the information flow between the two Phantom crewmembers, most notably the repeater radar scope installed in the front cockpit. Otherwise, though, most information still needed to be shared between front seater and GIB verbally using the Phantom's radio intercom. The arrangement wasn't expected to be a problem during a long-range intercept because the information exchange would follow a logical sequence: the GIB would use his radar and "descriptive commentary" to guide the front seater into a position where he could then look outside and visually detect the target, launching missiles along the way as required. However, when the Phantoms started battling enemy MiGs over North Vietnam in short-range, visual engagements, the flow of information was suddenly reversed. Now if the Phantom's four Sparrow missiles were to be of any use, the front seater had to talk the GIB's radar onto the enemy target. The F-4C was not optimized for this type of engagement or the pattern of communication it required.

The first *Tactics and Techniques Bulletin* released by Pacific Air Forces (PACAF), dated May 4, 1965, offered some initial guidance for the Phantom's GIBs during close-range air combat. For example, the bulletin instructed that if an enemy target started to maneuver "out of the gimbal limits" of the radar during a dogfight, the GIB should "come right out of the scope and pick up the bogey visually," using "every available handhold" to contort his body as necessary to keep sight of the enemy. If the GIB could see the bandit, which was not at all a surety due to the restricted visibility out of the Phantom's rear cockpit, he then needed to anticipate when and where the enemy aircraft might reenter the radar scan pattern so that he could quickly get his head back into the scope and get a radar lock. To make things a little easier, the bulletin suggested the GIB reconfigure the Phantom's radar before his front seater engaged in a high-G, maneuvering dogfight.[225]

The 8th TFW's tactics manuals identified "gyro out" as one of two preferred radar modes for close-range battles. Normally, the GIB configured the Phantom's radar for "gyro nor[mal]" operations. In this mode, the radar was space-stabilized, meaning that regardless of the attitude of the aircraft, the radar swept back and forth across the horizon. In contrast, when in "gyro out" mode, the radar used the aircraft's attitude as the frame of reference, and the radar antenna swept back and forth along with the aircraft's wingline. The GIB toggled between the two modes using a small switch on the auxiliary radar set control panel located on the left side of the rear cockpit (figure 4.16). If the GIB was going to select "gyro out" mode, the tactics manuals also recommended he select a "10-mile" scope range, "wide" gate, clutter "override," and "1 bar" scan, all located on the primary radar set control, which was installed just aft of the auxiliary panel. The 10-mile scope range made it easier for the GIB to spy a nearby target that would otherwise be crowded in the bottom half of the B-scope radar display. "Wide" gate told the Sparrow missiles to anticipate significant changes in the target's closing velocity due to abrupt maneuvering. Clutter "override" was selected to minimize the chance of the Phantom's radar losing lock due to the target's maneuvers and its changing radar "reflectivity." Finally, switching from "3-bar" to "1 bar" scan locked the radar antenna to a fixed-elevation coverage, the position of which was adjustable using the antenna elevation control wheel on the antenna control stick (figure 4.10). (When in "3-bar" mode, the radar automatically shifted up or down in elevation at the end of each horizontal sweep to increase its effective vertical coverage.)[226]

After reconfiguring the radar, the GIB then needed to adjust the antenna so that it was level with the aircraft. Theoretically, the GIB should have been able to use the antenna elevation control wheel to simply slew the antenna elevation strobe, displayed on the right side of the radar scope, to the level, center-scribe position. However, the tactics manuals warned that the antenna elevation display was "a poor guide for actual

FIGURE 4.16. F-4C Radar Set Control Panels. The controls for the radar were all located in the rear cockpit and spread across two panels; the auxiliary radar set control panel was usually mounted forward of the larger, primary radar set control panel. With all the separate switch actuations and manual tweaking required, Colonel Robin Olds acknowledged that the youngster in the back seat needed a "bit of dexterity" to effectively operate the Phantom's radar. NMUSAF.

antenna positioning" because frequent on-the-line maintenance of the radar usually led to antenna misalignment. The manuals therefore recommended that the GIB always check the elevation alignment of the radar at the start of the sortie so that he could manually correct for any errors when required.[227]

Once set and configured properly, the "gyro out" scan pattern resembled a "fan" approximately eighty-two mils high (4.8 degrees) that swept along the plane of the Phantom's wings 60 degrees to either side of the nose. The radar antenna completed each 120-degree sweep once every second. Since there were no visual indications of the radar settings available to the aircraft commander in the front seat, and since he was likely intently looking outside at the enemy target anyway, the GIB needed to verbally inform the front seater once the radar had been reconfigured in "gyro out" mode. The front seater could then, the manuals explained, "effectively . . . control the radar antenna pattern with the stick. . . . By maneuvering the target into the 'fan' the aircraft commander is providing the pilot [GIB] with a blip to 'lock-on.'" To facilitate a quicker lock-on, the manuals suggested that the aircraft commander verbally update his GIB on the target's "relative position," which would allow the GIB to focus his attention on that sector of his radar scope and better distinguish between the actual target blip and any spurious radar returns.[228]

Once the target blip appeared on the radar scope, the GIB still needed to complete the lock-on. Using the antenna control stick, the GIB maneuvered the radar acquisition symbol over the target blip and commanded automatic tracking by depressing the paddle switch to the second detent. Since there was a high chance the lock might fail to resolve correctly, the GIB then had to verify the radar target parameters. Assessing the target's Vc, as indicated by the gap in the range rate circle, was usually the fastest way to validate the lock. Once the radar lock had been verified, the GIB then had to reach over and switch the radar from "gyro out" back to "gyro nor" to avoid sending corrupt target information to the Sparrow missiles at launch.[229] After all that, the GIB could finally announce over the radio intercom, "Good lock!" Then the aircraft commander could get ready to squeeze the trigger. Trying to complete this laborious process while engaged in an extremely turbulent, close-range dogfight provided an additional level of stress for the aircrew.

The second close-range targeting option available to the Phantom aircrews was called "boresight." A specific mode of the radar, "boresight" was originally designed to provide a limited "auto-acquisition" capability for the aircraft commander in the front seat of the Phantom. If the GIB selected "BST" (boresight) on the radar set control panel in the rear cockpit, the radar antenna would immediately snap to the aircraft boreline (or at least that approximate position due to the potential misalignment of the antenna) and begin staring straight ahead. If the radar was also set to a ten-mile scope range, the aircraft commander could then depress the auto-acquisition switch on his left throttle to command the radar to start scanning for targets dead ahead out to twenty thousand feet and back again. The entire range scan took two seconds. If the radar detected a target, it was supposed to automatically lock on, but the aircrews discovered that the auto-acquisition system was at best unreliable. The boresight mode could also be used to launch a Sparrow missile without a radar lock, but the front seater would then need to keep the enemy target directly in front of the Phantom and within the narrow, 4.8-degree spotlight view of the fixed radar antenna for the missile's entire time of flight. This too was judged to be unreliable.[230]

A better method of using the boresight mode was developed by the Navy and outlined in the May 1965 PACAF tactics bulletin and subsequent 8th TFW tactics manuals. Olds and Clifton had used this modified boresight procedure to obtain their radar lock on their first MiG target during Operation Bolo, and the aircrew in Mink 4 on July 10, 1965, had also tried to use the boresight mode before they realized that their aircraft's radar had failed.[231] Usually, the aircraft commander would tell his GIB that he was going to maneuver for a boresight lock by announcing "Go Boresight!" over the intercom. At that point, the GIB needed to confirm that the radar was configured to a ten-mile scope range, with "wide" gate and clutter "override" selected, and then rotate the radar mode knob on the radar set control panel from "RDR" to "BST." As the aircraft commander aggressively maneuvered the Phantom to position the enemy target dead-ahead and within the small spotlight view of the radar, the GIB stared at the radar scope between his knees waiting for the target blip to appear. As with the "gyro out" mode, once the GIB saw the target blip, he needed to use the acquisition symbol to command automatic tracking. After checking the target's Vc to verify the lock, the GIB then had to reach over and return the radar to its normal "RDR" operating mode. This step unlocked the radar antenna and allowed it to track the target over the full gimbal range of the antenna, freeing the front seater from having to keep the aircraft aimed directly at the target. The tactics manuals warned GIBs to be alert because it was common for the radar to break lock during this transition from boresight to normal radar mode. Once again, the GIB would announce the completion of the process by exclaiming "Good lock!" over the Phantom intercom. If time remained, the GIB might also call out the distance to the target so that the front seater would know if he was within range to fire a Sparrow missile.[232]

In both the "gyro out" and "boresight" short-range scenarios, the key to successfully getting a radar lock and employing the Phantom's primary Sparrow armament was, according to Olds, "a bit of dexterity on the part of the youngster in the back seat."[233] Olds forgot to mention the importance of a good communication channel that allowed the front seater and his GIB to coordinate their actions. While the switchology was cumbersome and inconvenient, the GIB could rehearse the movements and develop the necessary habitual response patterns, in essence training himself to become an automaton ready to flip switches and rotate dials upon his front seater's command. Unfortunately, a reliable communications channel to transmit the front seater's command was not always available. The 8th TFW's tactics manual noted that "communications between the aircraft commander and the pilot are difficult when engaged in combat due to the cluttered strike [radio] frequency." The manual consequently warned the Phantom aircrews to be prepared for potential confusion and miscommunication whenever attempting a short-range lock.[234]

Had the Phantom crews routinely flown together, it is possible that many of the communication difficulties experienced in the air, as well as the general tension within the cockpit, might have been mitigated. Senior officers like Olds usually flew with only a handful of GIBs, which allowed the two officers to develop a rapport and better anticipate each other's actions. However, for most flyers, the pairs of aircraft commanders and GIBs were cobbled together each day as required to fill the flying schedule.[235] Already immersed in a potentially tense cockpit environment, and with each aircraft commander possessing his own preference for "gyro out" or "boresight" mode, the GIB was often placed in the unenviable situation of never quite knowing what to expect and when. One GIB lamented that "you get into a situation where it's react right now or get killed and you don't know exactly what to do, what this guy wants from you in the front seat."[236] Another GIB complained after an April 1966 mission that he "was not contributing anything to the engagement" because he "did not know [the] next action planned by [the] front-seat pilot."[237]

Despite these challenges, the GIBs' procedures during a dogfight scenario were largely mechanical and could be rehearsed. Conversely, operating the radar in the long-range environment required more artistry. Even though the crews knew that most engagements would terminate in close-range battles due to the stringent VID requirement, they continued to use their radars to scan the airspace at long range, hoping to detect any enemy aircraft and gain valuable situational awareness for the ensuing intercept and dogfight.[238] Within this realm, most understood that "finding another aircraft with the F-4 radar was not a simple task."[239]

There were two critical radar adjustments that the GIB needed to refine constantly if the Phantom's radar was going to detect an enemy target at range. The first was the antenna's elevation angle. The 8th TFW's tactics manuals provided guidance for each GIB in a four-ship flight on how best to configure his aircraft's narrow, pencil-beam radar in elevation to maximize the entire formation's chances of detecting a target at any likely altitude. However, the guidance remained somewhat nebulous due to the lack of reliable antenna elevation cues in the Phantom cockpit. For example, the GIB in the flight leader's aircraft was instructed to set his antenna elevation using a one-bar scan on a fifty-mile scope and then adjust his antenna elevation so that "approximately ½-inch of ground return appears at the top of the scope."[240]

Even with a properly set antenna elevation angle, the radar would still miss a target if the GIB failed to attend to the second critical element of long-range radar operations. The Phantom incorporated automatic gain control into its radar circuitry for tracking operations, but in search, properly adjusting the radar gain was the responsibility of the GIB. He had two controls, "coarse" and "fine," both located on the radar set control panel (figure 4.16). Sokol noted that there weren't any suggested settings for the gain, like "eight, or ten, or six"; it was "just a knob." Consequently, the GIB simply "turned it until you saw something you liked."[241] Since the radar return on the scope was a function of varying altitude, terrain, and atmospheric conditions, the GIB had to tweak the gain throughout the flight.[242] Complicating matters, the scope was never as clear as the flight manuals depicted (figure 4.8). "You always had noise on the scope," Sokol recalled. "If you want to see a little airplane, then you had to turn the gain out. [But] turn it up too much, [and] then it kind of starts blossoming with all kinds of [stuff] on there. And there's nothing that filters out the ground returns."[243] Welch described the GIB's job as staring at a screen full of "white dots" trying to "mentally integrate how all those white dots are behaving, because the one that's behaving differently is not the Earth. The one that's behaving differently is something moving over the Earth." The GIBs' ability to do so successfully earned the future Air Force chief of staff's praise: "It was an art to picking an airplane out of all that noise."[244]

All told, to be effective in the Phantom's back seat, the GIB needed to be able to read both his front seater's mind (for close-range engagements) and the figurative mind of the aircraft's radar (in the long-range environment).[245] Seeing an enemy aircraft appear off the wingline, the GIB might conclude his front seater would soon be asking for a "gyro out" lock; if the GIB couldn't see a nearby enemy aircraft, perhaps because it was under the Phantom's nose, then he might anticipate that his front seater was maneuvering for a boresight lock and could mentally prepare for the ensuing "Go Boresight!" command. Similarly, in the long-range environment, the absence of enough "fuzzy blips" on the GIB's scope might suggest that it was time to readjust the radar's gain, whereas too many "fuzzy blips" might indicate that the antenna elevation angle had drifted too low toward the ground. While sometimes described as merely the front seaters' "hands and fingers back there," ready to throw switches and turn dials on command, the young pilot-GIBs certainly had the potential to play a greater role.[246] Whether the GIBs achieved that potential was a function of both the human interpersonal dynamics within the aircraft and their proficiency mastering the set of human-and-machine dynamics localized in the rear cockpit.

Improvising in the Front Seat

The pilots in the Phantom's front seat had their own set of unique concerns. Some, like Bakke, considered the F-4 a "stable and honest aircraft."[247] John Burns, a former P-47 Thunderbolt pilot in World War II and eventual two-star general in charge of the Air Force's Test and Evaluation Center, labeled the F-4 "the most wonderful airplane I had flown."[248] Others routinely boasted that the Phantom was "an extremely easy airplane to fly."[249] Sokol, for one, noted that the Phantom was so smooth that it could lull its pilot to sleep, a fact that was revealed to him one day when his front seater dozed off during a cross-country flight over the United States.[250] There were others like Anderegg, however, who remembered that the F-4 could be a handful. It had its nasty "adverse yaw demon" that would suddenly appear at high AOA with the slightest misapplication of the flight controls. During one particular "eye-watering" out-of-control situation, Anderegg was thrown so hard into the canopy rail that it snapped one of his pocketed pencils in half.[251] The Phantom also exhibited a pronounced "Mach tuck" that occurred whenever the aircraft slowed from supersonic to subsonic speeds and its stabilators suddenly dug in, lurching the aircraft's nose up and potentially overstressing the airframe.[252]

The apparent inconsistency in the pilots' assessments of their aircraft reveals the subtle shift in evolving conceptions of what *flying* a fighter entailed. "It's easy to fly," one pilot noted of the F-4, "it's a little more difficult to fly well."[253] As they would all discover, it was even more difficult to fly well in combat. A seminal shift in fighter aviation was transpiring. In the past, *flying* a fighter was generally synonymous with *employing* a fighter; do the first well and the other naturally followed. Now, the two concepts were beginning to diverge. Welch reflected on the distinction. When he flew his F-84, "the things that allowed you to succeed in the cockpit were . . . instinctive. . . . It was all about flying the airplane." In contrast, when he began flying the F-4, Welch discovered that he had to concentrate on developing his "ability to operate the systems in the airplane. . . . You began to move from [flying the airplane] to being a systems operator."[254]

The dizzying array of switches in the Phantom's front cockpit was one manifestation of the shift. The McDonnell engineers might have placed most of the radar controls in the back seat of the Phantom, but the front cockpit still retained its fair share of cumbersome switches and dials. Echoing others, Burns complained of "too many different switches" arranged in a "terrible" layout. Moreover, Burns noted that once engaged in combat, "there was so little time to get all [the necessary switchology] done."[255] Hence, like the GIB seated in the back, the pilot in the front had to spend a fair amount of time getting acquainted with and rehearsing the intricate switchology.

Perhaps the most notorious switch in the Phantom's cockpit was installed on the missile control panel, located just above the front seater's left knee (bottom left of figure 4.17).[256] The three-position RADAR–HEAT–HEAT REJECT switch was not functionally complex. The top RADAR position selected a radar-guided Sparrow missile for launch; the middle HEAT position selected an IR-guided Sidewinder for launch; and the bottom spring-loaded HEAT REJECT position cycled through the on-board Sidewinder missiles to select the missile that would be launched next.[257] In combat, though, operating the missile select switch proved problematic. During one engagement on April 23, 1966, a Phantom front seater successfully maneuvered behind a MiG-17, but when he tried to lean forward to switch from RADAR to HEAT so he could shoot a Sidewinder, he couldn't. His shoulder harness had locked. "I had difficulty releasing the inertial lock so I could reach the panel and change the switch," he ex-

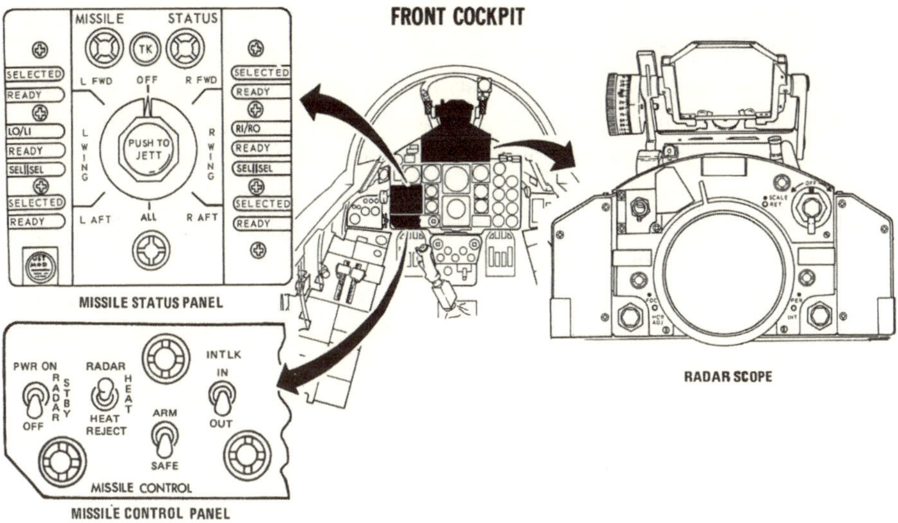

FIGURE 4.17. F-4C Air-to-Air Missile Controls. The missile control panel located in the front cockpit just above the front seater's left knee housed one of the Phantom's most despised controls, the three-position RADAR–HEAT–HEAT REJECT switch, which was used to select the type of missile to be fired. Often, the front seaters had a hard time locating the nondescript switch without glancing inside the cockpit to look at it, something they could not do during a fierce, close-range battle, so many improvised by attaching a piece of plastic tubing to the troublesome switch. NMUSAF.

plained after the mission. "Since the MiG was starting to evade, I elected to remain in the radar position and fire another AIM-7D Sparrow." Fortunately for him, the radar-guided missile worked, and the enemy MiG was last seen in a "near vertical dive" trailing "thick grey and white smoke." It was the Air Force's first victory with the Sparrow.[258]

Being able to reach the switch was a critical design detail. So too was being able to find it in the cockpit. When engaged in a tight dogfight, a front seater could not divert his visual attention from the thrashing bandit, even for a second, to glance at the switch inside the cockpit. Instead, he had to fumble around feeling for the nondescript missile select switch. Many complained about the arrangement. After some time, the front seaters began fitting a self-procured, two-inch-long piece of plastic tubing around the switch. The cheap, expedient modification allowed them to blindly swat at the switch—up or down—to change between missile types. The front seaters also began placing their worst Sidewinder, the one with the weakest tone, at the front of the firing sequence. That way, the front seaters didn't have to worry about stopping in the switch's middle position; they could simply slap the missile select switch all the way down to the bottom HEAT REJECT position, which would shuffle the bad missile to the back of the firing sequence before the switch sprung back to HEAT position. Regrettably, the difficulties the front seaters encountered with the missile select switch were not unique. Anderegg noted that "several other cockpit switches were just as hard to manage."[259]

On top of managing the multitude of misplaced and poorly designed controls, the front-seat pilots also had to become adept at properly estimating their different missiles' launch parameters. In the case of a Sparrow, which first required the GIB to get a radar lock, there were several cues displayed on the radar scope intended to help the crew. These included the Ra and Rmin range strobes and the aim dot. However, none of the cues were particularly effective in the skies over North Vietnam because the weapons display calculations were based on a nonmaneuvering, 1-G target, something the wily MiGs were not.[260] Like their predecessors, the Phantom pilots therefore needed to rely on intuition and experience when determining when to fire their weapons. They also had to contend with the same uncertainty and excitement of aerial battle that had afflicted previous generations of fighter pilots. As one senior Air Force officer explained, "The desire to kill a MIG when you are close to him pushes the patience to wait for the ideal condition or to shoot on the chance you will never reach the ideal conditions."[261] Plus, even when a missile was launched under what was assumed to be ideal conditions, a kill was never guaranteed. With Sparrows, the crews were instructed to always fire two to increase the collective probability of a kill. But even that might not help because, as Olds noted, an alert MiG-17 could "literally out turn a Sparrow."[262]

The Sidewinder with its restricted firing envelope was even more difficult to employ properly. First, the front seater had to maneuver his Phantom into the small thirty-degree cone behind the target. The cone, however, was not static. It followed the enemy's three-dimensional maneuvers like a streamer; the Phantom front seater needed to visualize this streamer and then fly to a position on it.[263] Second, the Phantom front seater had to cue his Sidewinder missile to the target and ensure that it was tracking properly. The front seater accomplished this step by pointing his aircraft at the enemy target and, with the missile select switch set to HEAT, listening for an audible "growl" in his headset. Amid the cacophony of inter- and intraflight radio communications and intra-aircraft intercom transmissions, picking out the "growl" could be problematic. Moreover, the "growl" merely indicated that the Sidewinder had found a heat source, not necessarily the desired one. The front seater's inability to distinguish between detected heat sources led to several Sidewinder missiles being inadvertently launched at the sun. The front seater then had to make sure that his Phantom was at less than 2-Gs before he launched the Sidewinder, otherwise the missile's guidance gyros would tumble during launch causing the missile to veer wildly off course. If the front seater released any back-stick pressure to reduce the aircraft G-load before launching the missile, however, the Phantom would usually exit the necessary thirty-degree firing cone behind the turning target. Finally, the crew also needed to estimate the kinematics of the missile, which varied based on the altitude, airspeed, and flight path of both the Phantom and the target. Hence, although the Sidewinder, like the Sparrow, was easy to employ on paper, it took a combination of pilot experience, intuition, and luck to employ it successfully in dynamic air combat.[264]

The tactics manuals offered some general employment guidelines for front seaters struggling with their Sidewinders. For example, the 8th TFW's tactics manual reminded the Phantom crews that the Sidewinder's minimum range would increase or decrease at a rate of three hundred feet for every 0.1 Mach difference in Vc between the target and the Phantom. The manual also noted that the Sidewinder's maximum range would "increase by 2,000 [feet] and the [acceptable] angle-off by 5° for each 10,000 [feet] of altitude."[265] Few actually expected the front seaters to perform these mental estimations and corresponding computations in the air. More often than not, the Phantom did not have a radar lock to provide the necessary target data when the front seater was looking to shoot a Sidewinder. Consequently, front seaters were taught how to estimate their target's range by comparing the size of the MiG's wingspan to the 2-mil pipper displayed in the F-4's optical sight.[266] In other words, the pilots of the Air Force's premier fighter were told to rely on stadiametric ranging, the original "bugaboo" of aerial weapons employment, when using their IR-guided missiles.[267]

Complicating effective Sidewinder employment, the MiG pilots by mid-1966 were routinely exploiting the American missiles' approximately half-mile minimum range. The enemy pilots realized that by staying close to the Phantoms, they not only limited the Americans crews' ability to use their missiles but also maximized their own offensive advantage due to their MiGs' cannon armament.[268] Already dangerously close to being reduced to mere missile managers, the Phantom aircrews were now becoming borderline ineffective ones at that.

In mid-1967, a cobbled-together innovation developed at DaNang AB under the purview of Korean War double-ace Blesse provided both a temporary solution to the Phantom's armament woes and a much-needed psychological boost to its missile manager-pilots. The Phantom aircrews had been using the SUU-16 external 20-mm gun pod—their "pistols"—since the 45th TFS's first deployment to Ubon RTAFB in mid-1965 (figure 4.18).[269] However, the SUU-16 had always been restricted to air-to-ground strafing. Because the F-4C did not possess a lead-computing gunsight like that used decades earlier in the Sabre, the initial testing personnel had concluded that the gun pod on the Phantom offered only a "limited" capability in the air-to-air environment.[270] The high aerodynamic drag associated with the SUU-16, a function of the exposed gun barrels at its front end, and the fact that it occupied the centerline space normally reserved for the Phantom's 600-gallon external fuel tank, which was vital for the long-distance missions into RP VI, further limited its practicality and left most Phantom pilots unenthused about the idea of carrying the bulky external gun into battle against the nimble MiGs.[271]

FIGURE 4.18. SUU-16/A External 20-mm Gun Pod. The SUU-16/A 20-mm gun pod was originally intended to be used by the F-4C in a ground-attack role only. Spurred by Colonel "Boots" Blesse, Phantom crews from the 366th TFW at Da Nang would eventually devise a way to employ the SUU-16 as an air-to-air weapon. The crews admiringly referred to the six-barrel Gatling guns as their "pistols." USAF.

Blesse was undeterred. Having just spent a year studying at the National War College in Washington, DC, followed by an abbreviated F-4 checkout at MacDill, he arrived at DaNang in early 1967 to serve as the 366th TFW's deputy commander for operations. A longtime proponent of the combat value of an air-to-air gun, Blesse was well aware of the Phantom missiles' lackluster performance. With MiG activity on the rise and the 366th's crews assuming a larger share of the resurgent MiG CAP mission, Blesse was determined to find a way to use the SUU-16 in air-to-air combat.[272] One pilot remembered Blesse telling him, "All I want to do is get a gun on there. . . . I don't care if we have to . . . wire a . . . 38-caliber pistol with a string to it, that's what we'll need against those MiGs!"[273]

The Phantom crews in the 366th responded to Blesse's call. They quickly determined that the lack of a lead-computing gunsight in the F-4C, although certainly not ideal, was not the major stumbling block. Fighter pilots in the earlier world wars had used fixed gunsights, and American pilots over Vietnam could do the same, they reasoned, especially since they would no longer be using a pair of .30-caliber machine guns but rather a six-barrel Gatling gun capable of firing nearly six thousand rounds a minute. Despite carrying only about twelve seconds' worth of ammunition, the Phantom front seaters figured that if they could get the MiG to fly through their "very concentrated burst," the "rounds [would] stitch through the wing or cockpit area like a sewing machine." They therefore planned to simply overlead their target, placing the Phantom's fixed sight "about twice as far [forward] as you thought necessary" before slowly walking it back toward the enemy MiG, all the while spraying bullets out in front.[274] The procedure was imprecise and inefficient, but, as Olds noted, "all you need is a few of them to hit and down he goes."[275]

Instead, the 366th crews discovered that the primary technical roadblock to using the SUU-16 in air-to-air combat involved the aircraft's configuration. Following Operation Bolo, the Air Force's Phantoms had begun carrying an ECM pod on all their flights north. Because only the Phantom's outer-wing stations had the necessary wiring to control the jamming pod, the crews had no choice but to continue flying with the ECM pod on an outer-wing pylon in place of a 370-gallon external fuel tank. The Phantom crews had been relying on the extra 230-gallons in their larger 600-gallon centerline fuel tank to compensate for the missing fuel beneath the wing. However, if that centerline station would now be occupied by the SUU-16, then the aircrews had to figure out a way to carry both the ECM pod and external fuel tanks on each wing. Blesse's team at DaNang solved the dilemma by successfully manufacturing a special wiring harness that effectively rewired the Phantom's inboard wing pylon to support the ECM pod.[276]

Encouraged by their early progress with the modifications, Blesse sent a message on May 5, 1967, to his commander at Seventh Air Force, Momyer, in which Blesse reiterated the motivation for the experiment and asked for permission to complete the planned installation. "This Wing has lost minimum seven kills in the past ten days because of a lack of kill capability [against targets] below 2,000 feet altitude and inside 2,500 feet range," Blesse explained. He also called attention to the irony of the situation: "It is interesting to note we are dusting off deflection shooting info published early WWII and Korea for our Mach 2 fighters."[277] Momyer acceded to Blesse's request, and a week later the first gun-equipped Phantoms thundered off from DaNang's runway.[278]

Two days later on May 14, 1967, the crews of Speedo flight successfully downed two MiG-17s with their SUU-16s. The victorious front seaters were ecstatic and extolled the virtues of the gun during their postmission debrief. One of them, James Hargrove Jr., explained how he was able to begin his attack on the MiG from only two thousand feet away, closing in to "cut him in half with the gun. . . . Without the gun . . . I don't see how I possibly could have gotten a MIG."[279] Another front seater agreed, stating that his kill "with the gun . . . could not have been made with a missile."[280] The 366th's official report of the victories noted: "All members of SPEEDO [flight] spoke praise for the SUU-16 gunnery system. We think the results speak for themselves."[281] By the end of the month, Olds's crews in the 8th TFW at Ubon were also carrying an external gun pod on their Phantoms. They claimed their first gun kill on October 24, 1967.[282]

A few months after the 366th's first gun victories, Blesse was summoned to the Pentagon to meet with McNamara. Responding to one of the secretary's questions, Blesse declared the gun "one of our most versatile and effective weapons, air/ground and air/air in spite of the lack of a computing sight capability."[283] That same month, Olds characterized the gun as "an outstanding development. . . . I gnash my teeth in rage to think how much better this wing could have done had we acquired a gun-carrying capability earlier."[284] Other front seaters like Major Bill Kirk, who claimed the 8th TFW's first victory with the gun and earned Olds's praise as "a hell of a good stick," similarly lauded the gun pod as "the finest thing that was ever invented."[285]

Even several years later the gun continued to garner accolades, and not just from the front seaters. One WSO-GIB in an F-4E recalled that after the experience of "trying to fire three AIM-7s, having one of them blow up in the air, having the other one go ballistic and one not even come off the airplane," the nose gun was "something you could depend on."[286]

The praise for the gun, however, overlooked an important fact; while the gun's mechanical performance may have been more consistent than the Phantom's missiles, it wasn't necessarily easier to employ. Major Robert Dilger, one of the initial cadre in the 366th TFW that worked on the gun program, observed that when one of the wing's front seaters who had "never fired a gun before in his life" tried to do so in combat, "his tactics were so gross that you couldn't expect the gun to have done him a good job."[287] Despite his enthusiastic appraisal of the gun in his July 1967 interview, Olds too harbored reservations, stating on another occasion, "I had no intention of giving any of my young pilots the temptation to go charging off to engage MiG-17s with a gun. They would have been eaten alive."[288] Olds further expounded on his concerns in his memoir, "We needed guns, no doubt about it, but we needed pilots trained to use them even more."[289]

In the end, missiles remained the dominant means to score an aerial victory. Over the course of the Vietnam War, the gun on the Phantom (F-4C/D/E) accounted for only 15½ of the 107½ enemy MiGs claimed by the Phantom aircrews. Another five MiGs were downed during "maneuvering tactics," a catch-all category that included those MiGs which had crashed into the ground while trying to defend themselves. The rest were all shot down by missiles.[290] The tallies illustrate the utility of the gun, especially in light of pilots' claims that many of their kills with the gun would have been unattainable with the Phantom's missiles. Still, the pilots' zest for the once-deemed-antiquated gun was wildly out of proportion to its actual combat results.

For the pilots in the front of the Phantom, the significance of the gun extended to the psychological realm. The gun didn't rely on electronic black boxes and mysterious "beeps and squeaks" that the pilots had no control over. Nor did it rely on a GIB; most of the time the back seater couldn't even see the results of the gun attack due to his restricted forward visibility. The beauty of the gun, in the words of Burns, was that the pilots could now simply "stick [it] in the guy's ear. There is no countermeasure for that."[291] In his history of the Phantom, Bugos concluded that the addition of the gun to the F-4E was "a powerful ideological statement that Air Force pilots were less missile system managers than gunfighters, capable of dogfighting and strafing ground units like their predecessors."[292] In fact, the members of the 366th TFW anointed themselves with a nickname based on their prowess with the SUU-16. They weren't missilemen; they were "The Gunfighters."[293]

Actually, they were neither. Like the plastic tubing covering their missile select switch or their strapped-on SUU-16 gun pod, the fighter pilots in the Phantom were an improvisation—a cobbling together of the collaborative, procedures-driven world of complex radars and missiles and the individualistic, seat-of-the-pants flying world central to the myth of the gun-slinging fighter pilot. The reintroduction of the gun and the corresponding embrace of the historic fighter pilot image did not alleviate the demands placed on the fighter pilots to study and learn their intricate systems, but it did reflect the pilots' ingenuity as they recrafted their human-machine relationships to better suit both their technical and psychological needs. Olds took notice of the subtle shift occurring in the front cockpit: "My point to the guys was you never stop learning, you learn, learn, learn, learn, learn all the time . . . then maybe you live."[294]

Tensions between Cockpits

As the Phantom aircrews worked to resolve the interpersonal and human-machine tensions within their cockpits, conflict erupted within the F-4C community over how best to employ a formation of Phantoms in combat. Schooled in tactics by Korean War veterans according to Blesse's "No Guts, No Glory," Phantom aircrews entered Vietnam flying a Korean-era fighter formation optimized for combat between gun-only-armed jets.[295] Serious deficiencies soon emerged. As one Phantom pilot recalled, "We weren't trained in the right tactics or how to employ [the F-4] in an optimum fashion."[296] Nonetheless, throughout the Vietnam War, the Air Force's radar- and missile-armed Phantom fighters continued to fly the same formations perfected for the Sabres decades earlier. The longevity of the particular formation reveals its dual purpose. Formations weren't tied solely to air combat efficacy. They also possessed an important cultural significance, reinforcing the normative social interactions between flight lead and wingman that were embedded within the fighter pilot myth.

When the 45th TFS began flying its first missions over Vietnam in mid-1965, it used the standard four-aircraft "defensive spread" formation known as "fluid four." The spacing between each Phantom was slightly wider than that used by the F-86 Sabres years earlier in Korea (figure 3.19), and it would steadily increase over the course of the war as aircrews sought more room to maneuver their aircraft without risk of immediately colliding with their flight mate, but the geometry among the aircraft never changed appreciably. The two Phantom elements were positioned nearly line-abreast, split by 7,000 to 9,000 feet, as depicted in the March 1967 edition of the 8th TFW's tactics manual (figure 4.19). The wingmen were supposed to fly closer, slightly aft, and slightly below their leaders, between 2,500 and 3,000 feet back.[297]

The arrangement reflected the unchanged flight member responsibilities. With the wingman tucked in close to his leader, the junior pilot was supposedly well positioned to screen the more experienced pilot's vulnerable six o'clock, thereby freeing the leader to concentrate on his attack against the enemy. In fact, if the flight came "under attack" or was "maneuvering to attack," the wingman was supposed to drop even farther aft into a sixty- to ninety-degree cone behind the leader, spaced only 1,000 to 1,500 feet back—a position known as "fighting wing" (figure 4.20).[298] Intended to make

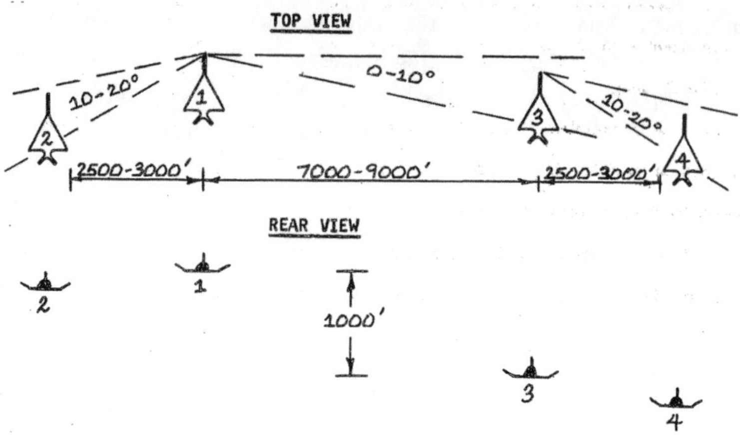

FIGURE 4.19. F-4C Fluid Four Formation. Similar to the formation flown by the F-86 Sabres over Korea decades earlier, the Air Force Phantoms' fluid four formation, as depicted in the March 1967 version of the 8th TFW tactics manual, relied on closely spaced wingman (numbers 2 and 4) and strict flight lead-shooter and wingman-looker responsibilities. The Phantom's greater maneuverability and new guided missile weapons rendered the formation less than optimal in the skies over Vietnam. AFHRA.

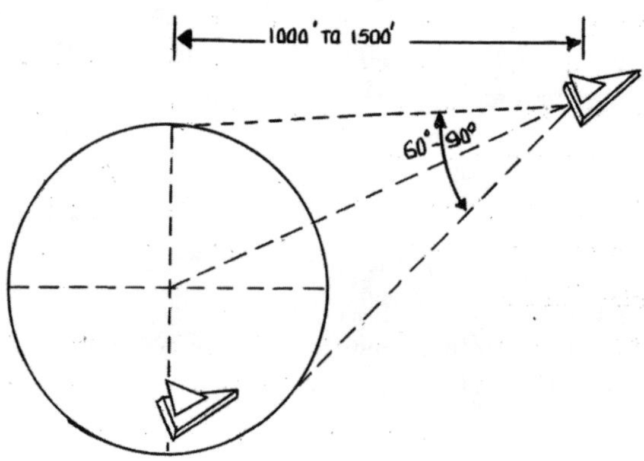

FIGURE 4.20. F-4C Fighting Wing Formation. If the fluid four formation used by the Phantom crews was less than optimal, then the wingman's fighting wing position 1,000 to 1,500 feet behind his leader, which he was to assume whenever his leader engaged a MiG, was totally ineffectual. Optimized for a gun-equipped fighter like the Sabre with excellent cockpit visibility, the fighting wing position did not provide any offensive potential to the missile-armed Phantom wingman; it also rendered him defenseless to enemy attack. AFHRA.

it easier for the wingman to perform his primary duty of protecting his leader, many pilots referred to the "fighting wing" position as " 'welded wing' because the wingman was there to stay, no matter what happened."[299] As before, the flight lead was the designated "shooter"; the element lead, the "secondary shooter." The wingmen were the "lookers."[300]

The "fluid four" and "welded wing" concept may have worked well for the Sabre pilots in Korea when their enemy had to close to within one thousand feet to fire its air-to-air cannons, but it didn't work so well when the Phantom crews squared off against a hard-to-see MiG-21 capable of launching the Soviet's AA-2 Atoll, a Sidewinder-like IR-guided missile, from a mile away. Moreover, those in the Sabre before had enjoyed the benefit of a large bubble canopy that allowed them to see nearly all the way behind them. In contrast, the F-4's faired cockpit design severely restricted the Phantom aircrew's visibility aft of the aircraft's wingline. In this environment, who was going to protect the wingman and clear his six o'clock position? Additionally, with the extreme power of the Phantom and its ability to maneuver rapidly in the vertical, the welded wingman needed to be especially vigilant to maintain formation. If the leader initiated an abrupt turn without warning, the wingman could suddenly find himself out of position or, worse yet, staring at a windscreen full of F-4 and an impending collision.[301] Many wingmen complained that, although they were in fighting wing ostensibly to observe the fight and help their leader, they "never saw anything of the 'fight' because they were just trying to 'hang on.' "[302] The formation might have been "a great plan in Korea or somewhere," Sokol explained, "but [it] was not exactly optimum for the war we were fighting in Vietnam. . . . All you are out there are shock absorbers to suck up missiles."[303]

At least in Korea, the Sabres' close formation allowed the wingman to transition quickly to an offensive position and fire his guns at the enemy when directed. In the missile-only F-4, there was no similar opportunity. Tucked close behind or alongside his flight lead, the Phantom wingman was unable to maneuver his aircraft to obtain a "boresight" or "gyro out" radar lock, and without a radar lock the wingman's four radar-guided Sparrow missiles were useless. The wingman always had his complement of IR-guided Sidewinders that did not require a radar lock, but the Phantoms' formation didn't place the wingman anywhere near the appropriate firing position for that weapon either. Hence, by effectively eliminating the wingmen's opportunity to employ their missiles, the preferred formation tactic cut a four-ship of Phantoms' usable ordnance by half, a troubling statistic in light of the still woefully inadequate performance of the American missiles.[304] One pilot lamented of the wingmen's position in the formation, "You've got a whole damned weapons system back there that is kind of useless as far as employment is concerned."[305]

There were at least two other formation options. One known as "double attack" was developed by Air Force captain Everest Riccioni. A former F-100 pilot, Riccioni had tried for years to free the Air Force fighter pilot establishment of its dogmatic adherence to fighting wing. For Riccioni, the distinguishing feature of his "double attack" system was that there were no preassigned "shooter" or "looker" responsibilities. Flying as a two-ship in a line-abreast formation, split by about one mile, either the leader or the wingman could engage an enemy target; the "shooter" would be whoever was most favorably aligned.[306] Despite his persistent appeals, Riccioni made little headway. Frustrated, in January 1963 while on temporarily assignment to the Air Force Academy teaching astronautics, Riccioni drafted a forty-page treatise extolling the benefits of his "double attack system." He then mailed his report direct to Major General Charles Westover, the vice-commander of TAC. In his cover letter to Westover, recently promoted Major Riccioni made no apologies for his brusque tone, explaining that it was "necessary to shock fighter pilots out of 18 years of lethargic, uncreative thinking."[307] A few months later, Riccioni published his ideas in the *Fighter Weapons Newsletter.*[308] While some aircraft communities within ADC subsequently adopted elements of the "double attack" formation—for instance, the F-106 pilots used a version that they called "six-pac"—Riccioni and his favored formation remained widely shunned within TAC.[309]

The Navy had its own formation, which it termed "loose deuce." Similar to Riccioni's "double attack," "loose deuce" relied on a single, widely spaced element of two aircraft. For its F-4B Phantoms, the Navy recommended an ideal intra-element spacing of "between 6,000–9,000 [feet] and if possible 12,000 [feet]." The Navy also eschewed four-ship formations for its F-4Bs, preferring instead "two 'loose deuce' formations to retain a higher degree of tactical flexibility and for better mutual support between single units." As in Riccioni's "double attack," the two-ship "loose deuce" avoided predesignated "shooter" and "looker" responsibilities in order to take full advantage of each flight member's radar and missile load.[310]

Disenchanted with their service's textbook, fluid four formation-solution, some Air Force Phantom aircrews in Vietnam began experimenting with wider-spaced formations, including the Navy's "loose deuce" concept. One Phantom pilot, Ernest Bedke, remembered a period in mid-1966 when his unit started "spreading our formations and bringing people up to line abreast where we could see them and . . . where we even had a better chance of looking for MIGs coming up at 6 o'clock." However, when a group of stateside Phantom instructor pilots happened to visit Bedke's unit and witnessed the pilots using a non-fighting wing formation, "they were aghast. . . . They threw the damned fighter manuals at us and told us that we had no idea what we were doing."[311]

Even a half-decade later, controversy over the proper positioning and responsibilities of Air Force Phantom crews continued to rage throughout the community. Sokol, by 1971 an F-4 instructor pilot at George AFB, California, remembered one of his fellow Phantom pilots working his Navy connections to score a copy of the sister service's "loose deuce" tactics manual. "We read about it. But we weren't allowed to teach it," Sokol recalled. In fact, "we had to hide that [manual], because we were not allowed to have that book because it was heresy." The pirated manual coincidentally ended up "in the trunk of somebody's car" whenever an inspection team showed up. "It was, by God, fighting wing. . . . You did not deviate from that," Sokol said of the period.[312] Anderegg also remembered the turmoil, explaining that it was "like . . . pulling teeth" getting people to consider anything but fighting wing "because for . . . thirty years, the Air Force [policy] had been [Number] One's the shooter and the wingman is the covering guy."[313] Bedke, who had first encountered the resistance in Vietnam in mid-1966, dryly noted, "It takes a ball bat to the side of the head sometimes to convince [people] that they are antiquated and need to get up to speed and need to start looking ahead."[314]

It wasn't just the old-heads from Korea fighting to retain "fighting wing," though. Richard "Steve" Ritchie, a young captain flying Phantoms in Vietnam in 1972, was a big proponent of keeping his wingman in "tight fighting wing." Ritchie considered the emphasis on "strict, mutual support" an American "advantage over the enemy," even though one of his wingmen had once lost control of his aircraft and was forced to eject while trying to maintain the welded wing position during a violent engagement.[315] Before arriving in Vietnam, Ritchie had served as an instructor at the Air Force's Fighter Weapons School, and his attitude reflected that of his colleagues. In fact, some of the most vocal opposition to anything *but* fighting wing reverberated from the halls of the Weapons School, which was widely recognized as the service's premier source for fighter tactics development and training.[316] In multiple issues of *Fighter Weapons Newsletter* in 1971, the Weapons School instructors railed against the fighting wing alternatives and belittled the pilots who advocated for change.[317] In one piece, the instructors summarily dismissed "loose deuce" as essentially "fluid four" without the wingmen, declaring, "Fortunately that is not our cup of tea as number 2 and number 4 are very valuable to us. . . . We recognize that everyone is not as good as everyone else, and therefore place our people and plan the employment of our tactics accordingly."[318]

The Weapons School instructors' argument smacked of hubris. In affirming their status as the service's premier Phantom pilots and rightful flight lead-shooters, the instructors conveniently ignored the significant increase in offensive potential afforded by the junior wingmen's missiles. The instructors also discounted pilots' experiences since the Korean War and the general acknowledgment that flying as a wingman could be a significantly taxing responsibility, sometimes even more difficult than flying as the flight leader. A far more reasonable concern with the new formation tactics, but one that the instructors seldom invoked, was the assumed already-too-crammed training timelines used to produce F-4 front seaters and the schoolhouses' consequent inability to teach the new wingmen anything other than fighting wing. But, as historian and Air Force fighter pilot Marshall Michel III noted, there was also little incentive to explore potential alternatives: "Since fluid four [and fighting wing] gave the flight leader a great deal of freedom because he knew he was always protected and could ignore his wingmen while pursuing the MiGs, it was not unpopular with the flight leaders," including those teaching at the Weapons School.[319]

The instructors' dogged resistance to alternative formations also ignored some of the significant adjustments that had already taken place during the Vietnam War. One was the Air Force's shift away from strictly coordinated four-ship tactics in favor of more flexible, two-ship element employment. Describing this shift, the 8th TFW's tactics manual explained that, considering the MiGs' patterns of attack, "It will be necessary for the elements to act as independent units until the threat is neutralized and a rejoin effected for mutual support."[320] These new "independent" element tactics allowed one pair of Phantoms to engage the MiGs in a tight battle, thereby keeping the enemy aircraft predictable and anchored in a general location, while the other Phantom element left the battle to gain sufficient separation—two to three miles—so they could turn around, acquire a radar lock, and then fire their missiles. The new tactic required significant pre-mission coordination between the two element leaders to minimize potential confusion during execution. If the two elements mistakenly tried to separate at the same time, or if there was no plan to ensure a clear field of fire for the distant element's missiles, the results could be disastrous. Despite these challenges, the "independent" element tactics became popular among the combat crews in the 8th TFW.[321] The fact that these new tactics were developed and taught on-the-fly in combat also contradicted the Weapons School instructors' argument that two-ship tactics were too complex and too demanding for junior pilots.

The Phantom aircrews' combat experience was also proving that the technical limitations of their hardware, such as the significant variations in aircraft radar performance, could have a pronounced and deleterious effect on their preferred four-ship tactics. Front seaters and GIBs alike began to acknowledge, the former more reluctantly than the latter, that it was unreasonable to assume that two Phantom radars could reliably detect the same enemy aircraft. Because of the known radar display errors, aircrews couldn't even reliably cue another formation member to a detected target. How could the Phantom flight leader fulfill his responsibility to lead the attack if he was unsure of his own ability to detect a specific radar target and, if one was detected, to correlate it with other potential targets detected by his accompanying formation members? To address practical limitations like these, the Phantom aircrews began to develop additional contingency plans. In one scenario, if a wingman obtained a radar lock on, or first sight of, a suspected enemy target but his leader could not, then the wingman could become the de facto leader as the formation closed on the enemy and the "scheduled" flight leader would assume the fighting wing position behind the wingman. If the designated leader was then subsequently able to obtain a radar lock, he could reassume the primary shooter responsibilities and the wingman would have to fall back into the supporting position.[322] The combination of flexible shooter arrangements and independent element tactics contravened the rigid responsibilities outlined in fighting wing. Indeed, several aspects of the new Air Force tactics more closely resembled those from the Navy and their fluid, "loose deuce" formation.

Still, despite these wartime adjustments to the Phantom formations and tactics, fighting wing refused to die. Michel suggested that the Air Force's obstinacy may have stemmed from the fact that "loose deuce" was a "*Navy* formation . . . , and no Air Force officer with an interest in his career would dare suggest the Navy knew better than the Air Force."[323] While that might have been a contributing factor, TAC also proved reluctant to adopt the similar "double attack system" developed by the "curious" but Air Force-blue-wearing Riccioni. Perhaps more significant was the fact that, as Anderegg noted, "having a wingman as a shooter equals partner," and the flight leads weren't looking for "partners" in their wingman.[324] But even this rationale obscures the fact that the wingman and his leader had already entered into a historic partnership to defend against the North Vietnamese SAM threat by using a new, alternative "pod" formation.

Unlike the aircrews' other emerging combat tactics that embraced flexibility, the "pod" formation was extremely rigid. In "pod" each Phantom was to fly precisely 1,500 feet away and 1,500 feet above or below the next fighter. "Modified pod" adjusted the vertical spacing to between 500 and 700 feet to make it easier to engage a MiG if required (figure 4.21). The precise spacing within the formation maximized the total coverage provided by each aircraft's ECM pod, helping to defend the entire formation against North Vietnamese SAMs. If any aircraft got out of position, a gap in the jamming pattern would emerge, and the individual aircraft or his neighbor would become a more inviting target. While the wingman was still responsible for flying off of his flight leader, due to the strict spacing requirements the flight leader also had a responsibility to make sure that his movements were more deliberate and easier for his wingman to follow.[325] "Pod" therefore represented a codependency between leader and wingman that never existed in fighting wing.

Perhaps the transition to "pod" was made easier because the SAMs represented a tangible, technological threat, and the crews' new "pod" formation was similarly designed to enhance a technological ECM-pod counter. The same physicality of these artifacts—the SAM and the ECM pod—were not present in fighter-versus-fighter air combat, and consequently there was no tangible reason to alter the preferred air-to-air tactics. To accept the wingman as functionally equivalent to the flight leader in the air-to-air environment, even just to exploit his offensive capability, was for many "heresy." Acknowledging the necessity of a new formation with new responsibilities would have violated the fighter pilots' assertions that, despite the arrival of radars and missiles, air combat remained essentially unchanged from that conducted in the skies over the Western Front a half century earlier. Popularizing the notion of timeless air combat, the Weapons School instructors invoked the pseudonym "Der Baron" and his oft-cited "anything else is rubbish" mantra to headline their columns in *Fighter Weapons Newsletter* when they lambasted the proponents of "double attack."[326] The

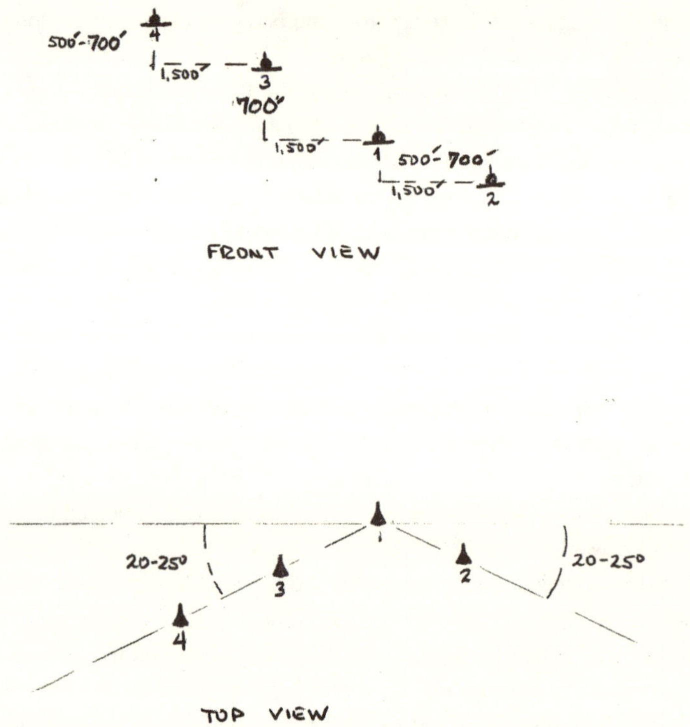

FIGURE 4.21. F-4C Modified Pod Formation. Curiously, while the Phantom aircrews were unwilling to adapt their formation to the new air-to-air environment, they did adjust it to counter the surface-to-air missile threat. The resulting pod and modified pod formations, depicted in the December 1967 version of the 8th TFW tactics manual, were designed to protect an entire formation by optimizing the jamming pattern of each aircraft's individual electronic attack pod. Because survival of all depended on each aircraft maintaining strict spacing, pod formation represented a more equitable partnership between flight lead and wingman that had been historically absent from fighter aviation. AFHRA.

leader did the shooting. The wingman hung on, protected his leader, and watched and learned, proving his worth as a fighter pilot in the process. The fighter pilot myth continued to hold sway over those now flying Mach 2 fighters, and the fighter pilot community writ large was unwilling to forsake its heritage just because today's aircraft now came armed with missiles.

There was one other source of tension between the fighter cockpits that deserves mention—intraflight communications and radio chatter. It was not a new issue. As noted in the previous chapter, the radio was often a source of considerable frustration for the Sabre pilots. With only a few channels available for communicating across scores of airplanes, and with the pilots often unable to restrain themselves from talking during the excitement of battle, the radio over MiG Alley was nearly always saturated. The situation could be maddening for the American pilots, but at least the MiG-15s had to get fairly close to score a kill, which increased the chances of the Sabre pilots detecting them before it was too late. That was not necessarily the case with the Phantom in Vietnam.

Several circumstances conspired to make the F-4's radio a more critical piece of equipment than in wars prior. For one, the MiG-21's longer-range missile capability, coupled with the Phantom's restricted rearward visibility, rendered a timely "Break!" call from a flight mate one of the only ways a crew would know that it was under attack. There was also the requirement to communicate radar target information among the formation members and early warning aircraft like the EC-121. The Phantom aircrews' shift toward element tactics likewise depended on reliable radios. Finally, the radio intercom played a critical role in intracockpit communications between the front seater and his GIB. Although perhaps guilty of hyperbole, Ritchie was not far off when he declared the ultra-high-frequency (UHF) radio "the single, most important piece of equipment over North Vietnam. And then, the next most important thing were pilots and crews that knew how to use it."[327]

Fortunately, the Phantom's radio system was much more capable and versatile than the Sabre's radio. The first F-4Cs used a UHF radio capable of receiving and transmitting on 18 preset or one of 1,750 manually tunable frequencies. The Phantom radio set also incorporated a lower-power auxiliary receiver that could be tuned separately to receive (but not transmit) on one of twenty preset channels, plus an emergency "guard" frequency.[328] Consequently, when you include the radio intercom, Phantom aircrews had to be adept at listening to three different streams of information being piped simultaneously into their ears. Plus, there was also the requirement for the front seater to listen for the all-important and sometimes barely audible Sidewinder missile "growl" during the heat of battle.[329]

The opportunity to select from a wide range of available frequencies was supposed to eliminate too many aircrews crowding on a limited number of radio channels, as was the case in Korea. However, radio saturation continued to be a problem. The issue wasn't the availability of the radio spectrum, but the availability of the information transmitted over that radio spectrum. All the aircrews understandably wanted to be able to listen to the MiG warnings provided by the early warning aircraft. But, according to one EC-121 aircraft commander, "once everyone got on the same frequency the load was just too much. . . . There were delays and sometimes messages [were] not received."[330] The F-4 escort fighters for a time tried to alleviate some of the congestion by assigning one discreet frequency for their communications and another for their accompanying F-105 fighter-bombers; if the two aircraft types needed to communicate with one another, they would simply change their radio channel. There were problems with this arrangement, though. One Phantom front seater explained,

"You hated to [switch frequencies and] miss one fraction of a second of something that may be coming over [the radio]."[331] Once again, the aircrews quickly congregated on a few key frequencies. For the majority of crews, it wasn't uncommon to finish an aerial refueling and push north into Vietnam listening to "fighters who were trying to coordinate their activities within flights and between flights, . . . reconnaissance aircraft up there that were calling off various things, and the air-sea rescue people [who] were operating on the same frequency. . . . In some cases [you] have to have all those people on the same frequency."[332]

As in Korea before, commanders in Vietnam emphasized the importance of radio discipline, keeping "the communications . . . as simple as you possibly can . . . . Keep the noise down; keep the chatter down."[333] However, when the pilots' blood pressures rose during a fierce battle, their voices tended to follow, and any interaircraft "brevity" code or intra-aircraft "descriptive commentary" that was designed to encourage radio discipline was quickly discarded. One Phantom front seater remembered, "Every now and then you'd have some motor mouth . . . and then somebody would soon yell 'shut up.' . . . But that wouldn't always necessarily shut him up."[334] The possibility of having a "motor mouth" in the back seat, garbaging up the intercom and radio, was one of the arguments the front seaters had invoked for keeping a trained pilot in the back seat of the Phantom instead of transitioning to a navigator-GIB.[335]

Even without a "motor mouth" monopolizing the radio time, making sense of the cacophony of radio transmissions could be a challenge for the aircrew, and it added to the complexity of employing the Phantom in combat. Consider, for example, the radio transmissions that a single Phantom—Cadillac 1—had to process during a two-minute span from 13:17:00 to 13:19:00 on July 29, 1972. During the brief 120-second span, there were a total of thirty-three radio transmissions of varying volume, intensity, and duration. They included the intercom communications between the front seater and the GIB as they struggled to locate and target a nearby MiG; their wingman, Cadillac 4, transmitting that he "just took a hit"; another aircraft transmitting on the emergency "guard" frequency as he tried to locate his flight mates; numerous directions from American radar controllers trying to help Cadillac 1 find the elusive MiG target; interflight transmissions from two other nearby flights, Pistol and Trigger, who were also trying to chase down the MiG; and transmissions from another group of F-4s getting ready to drop their bombs.[336] As one front seater noted of the general situation, it could be "very difficult to tell who said what, on what frequency it came in, where to respond to it, or if you should respond to it."[337]

The radios weren't technically challenging, at least not on the order of the Phantom's radar and missile armament, but they were cognitively challenging. Ritchie explained that with "so much information coming in over the radio, the weather, the field, the instruments, the back seater, all this coming in over the headset," he and his fellow Phantom pilots learned "it has almost got to be a computer-type thinking process where you sort it out."[338] Indicative of the importance of these communications, crews started carrying personal, portable cassette recorders into the cockpit with them to capture the radio calls so that, if needed, they could sift through all the information postmission.[339]

The aircrews in the Phantom faced several unique challenges unknown to earlier fighter pilots. As Olds stated when asked to compare flying his Mustang to flying his Phantom, things were "different." While some challenges stemmed from the complexity of the F-4's equipment, not all were technical in nature. Tensions within the cockpit associated with forcing a winged pilot to ride in the back seat, or having a pilot fly in the front seat while listening to someone in the back seat, as well as the pilots' general reluctance to adjust their time-honored and culturally significant formations and tactics are but a few examples of the additional challenges that accompanied the Phantom and its crews into battle. Some were eventually resolved; others never were. But the aircrews still blasted off every day hunting for MiG glory. Still, one important question remained unanswered: with two men aboard the Phantom, who would get the credit for downing a MiG?

Who Gets the Credit?

On October 15, 1972, Captain Gary Rubus was trying desperately to down a MiG-21. Rubus eventually maneuvered his F-4E to within eleven hundred feet of the enemy aircraft and squeezed the Phantom's trigger. His aim was true, and the Phantom's 20-mm rounds began ripping through the center of the MiG's fuselage. Rubus continued to close on the enemy and fired again from eight hundred feet away. Moments later, the MiG's canopy shot off, and the enemy pilot ejected.[340] After the mission, Rubus's navigator-GIB, Captain James Hendrickson, was asked if he should have also received credit for his front seater's gun-kill. Hendrickson replied, "In this particular case, I think it's deserved if you really want to be honest about it."[341] It was a curious appraisal. Hendrickson had no direct involvement with Rubus's gun attack, other than reminding his front seater to "arm up your nose gun" and, when they were about fifteen hundred feet back, shouting over the intercom, "You're in range. Fire." In fact, the three prior Sparrow missile attacks that Hendrickson had direct involvement with had all ended in failure; hence Rubus's decision to forego any additional missile attacks in favor of the more reliable gun.[342]

Without his help, Hendrickson could argue, his aircraft commander would have never found the enemy MiG, a fact that was obscured in Rubus's postmission report that simply stated, "A radar contact was established at a distance of 16 nautical miles, followed by a visual contact shortly thereafter."[343] Hendrickson explained that the MiGs were flying below the clouds and out of sight. It was Hendrickson's skill and dexterity manipulating the radar that had allowed him to detect the MiGs, and then he still needed to coax his front seater down and into position. "They're low. Continue on down," Hendrickson reportedly told Rubus. When the MiGs suddenly started to drift off the radar scope, Hendrickson had to provide further guidance to his front seater, "They're starting to drift to the right, hard right, hard right." Only then did the MiGs finally appear through the clouds, permitting Rubus to take over visually.[344]

Before the Phantom, Air Force fighter pilots had never before had to share their aerial glory with another who had not squeezed a trigger. Indeed, it was the ruthless simplicity and apparent objectivity of the aerial victory credit that made it so appealing and a fixture of the fighter pilot myth. But the Phantom had been engineered specifically to capitalize on the collective contributions of the front seater and the GIB, as well as a host of other automated features, and it remained to be seen if these various contributions to victory could be disaggregated when assigning credit.

Documenting the Phantom's Air-to-Air Performance

The Phantom's recording mechanisms certainly did not facilitate assignment of victory credit to individual components, man or machine. Like the earlier Sabre, the F-4C Phantom included a camera to document the crew's performance. However, whereas the Sabre's camera was mounted in the aircraft's intake to record images of enemy aircraft erupting in flames under a steady stream of well-placed bullets, the Phantom's camera was mounted in the rear cockpit and took pictures of the GIB's radar scope. That was where the action in the Phantom was supposed to take place, and it was an apt reflection of the significance that the engineers attached to the GIB and his role in fulfilling the promise of the Phantom's long-range air combat capabilities. Moreover, the engineers reasoned, a forward-facing gun camera would be of little value trying to record images of missiles impacting an enemy aircraft miles away and not necessarily directly in front of the Phantom.[345]

McDonnell's Direct Radar Scope Camera was touted as a major innovation and a specific "advantage" of the Phantom when it was unveiled. One of the advertisements for the Phantom noted that the "low cost, low volume, reliability, simplicity, inflight reloading, recording capabilities and projector playback characteristics of this device made it ideal for crew training and evaluation."[346] The Navy was so impressed with the 16-mm film system that it asked that the camera be standard equipment on its production models of the F-4B; the Air Force followed suit.[347] But Sokol and Anderegg both remembered that, despite the hype, the system was less than perfect. After the GIB turned the system on, the camera was supposed to snap a picture of the radar scope at the end of each antenna azimuth sweep. Often, however, the door "would come open and it wouldn't run." When it did run, crews still "had to wait four or five days for the developing process. It was just a big pain in the ass." Moreover, Sokol asked rhetorically, "Who wants to watch a film of the [radar]? It was bad enough doing it, let alone watching it."[348]

When the Phantom began participating in close-range air battles over Vietnam, the rear-cockpit radar camera proved wholly inadequate. The radar camera had no ability to record the front seater's employment of the Sidewinder or, after it was added to the Phantom's air-to-air arsenal in May 1967, the external gun pod. Even with a Sparrow, the radar camera could at best record only the disappearance of a "fuzzy blip" following a missile launch—not at all a guarantee of a successful engagement. According to Olds, the lack of a viable camera had cost his Phantoms a potential eighth MiG claim during Operation Bolo.[349] The camera situation was partly corrected in late 1971 when the Air Force began evaluating a new gunsight camera designed to record images through the F-4D/E's optical sight. This new "gunsight camera system," Phantom pilots claimed, would finally "satisfy a long-standing requirement" and allow for proper "film documentation of the techniques employed in both air-to-ground and air-to-air training situations."[350] However, it would take several months before the new cameras began appearing on the Air Force's Phantoms, and they never arrived in sufficient numbers before the end of the war.[351]

Hence, the Phantom would produce no awe-inspiring gun camera imagery to memorialize its role as the Air Force's workhorse fighter of the Vietnam War. Ironically, the iconic air-to-air image of the Vietnam War was instead captured by an F-105 Thunderchief fighter-bomber when it gunned down a MiG-17 at close range in mid-1967 (figure 4.22).[352] More importantly, the Phantom crews lacked the gun camera imagery normally required to validate a pilot's claims of aerial victory. The Phantom crews therefore relied almost exclusively on their flight mates to validate their claims.[353]

The combination of no photographic evidence and all-too-eager wingmen ready to validate a leader's claim might suggest that overclaiming aerial victories in Vietnam would be as rampant as in past wars. However, the issue rarely surfaced. One reason was the greater intelligence gathering capabilities of the Americans, which were often used to corroborate the Phantom pilots' claims. Indeed, it was the ELINT (electronic intelligence) capability that Olds blamed for not allowing his unit to claim an eighth MiG during Operation Bolo.[354] The same national intelligence apparatus also kept close watch over the Phantom aircrews' conduct of the war. Consequently, the Wild West atmosphere that had characterized the battles north of MiG Alley in Korea had by Vietnam largely disappeared.[355]

The intense desire to claim a MiG that had once driven Sabre pilots north of the Yalu River had not dissipated, though. If anything, the relative scarcity of enemy MiGs over Vietnam, and the fits and starts with which they appeared, stoked the Phantom pilots' enthusiasm to be one of the few who scored a victory. Sokol never encountered a MiG during his tours, but to have gotten one, he admitted, would have been like "winning the lottery."[356] Hoyt Vandenberg Jr., son of the first CSAF, noted of his time flying Phantoms in the 366th TFW at DaNang, "Everybody wanted to be a MIG ace. Right? That was the pinnacle in Vietnam, so in order to find MIGs, you had to get up there and mix it up."[357]

FIGURE 4.22. Downing a MiG-17. Although the Phantom was the workhorse of the Vietnam War and earned the majority of air-to-air kills, it never produced the iconic air-to-air film that had thrilled audiences during prior aerial wars. The F-4C had a camera, but reflective of the Phantom's original design, it was installed in the rear cockpit and took pictures of the radar scope. This widely circulated gun camera photograph is from an F-105D fighter-bomber. NMUSAF.

Pushing for the Next Air Force Ace

It wasn't just the Phantom pilots who were enamored with the allure of aerial victory. After heavily marketing the ace figure during the Korean War, the Air Force as an institution eagerly awaited its next round of aces. When the individual aces failed to materialize, the Air Force started applying the "ace" moniker to its squadrons and wings that had accumulated a total of five aerial victories.[358]

Olds came the closest to downing the magic number of five enemy aircraft during Rolling Thunder. After having scored his first MiG with Clifton during Operation Bolo on January 2, 1967, it took Olds another four months before he claimed another MiG on May 4, 1967. Two more MiGs followed less than three weeks later on the twentieth.[359] Olds, however, never got his fifth. Years later, he would claim that he was tipped off that if "I got number five, I would be sent home" for "PR [public relations] purposes." Unwilling to forego his prized Wing commander job, Olds spent his last four months "knowing that if I squeezed my trigger finger, I would lose my job." He ended up letting his wingman "get the MiGs. That was a little hard to take," he admitted.[360] While it's difficult to verify Olds's claim, the fact remains that he completed his tour with only four victories, and no other Air Force pilot had matched it by the end of Rolling Thunder in 1968.

The Air Force might have been able to overlook its lack of an ace except that when air combat resumed in 1972, the Navy produced its own set of aces. On May 10, 1972, Lieutenant Randy "Duke" Cunningham and his RIO, Lieutenant (JG) Bill "Irish" Driscoll, downed three MiGs in a single day. When added to their previous pair of victories, the two crewmembers became the first American aces since Korea. As the Air Force had purportedly hoped to do with Olds, Cunningham and Driscoll were spirited out of combat and back to the States to "transmit first hand and freshly to their colleagues their experiences of aerial combat." Along the way, the two aviators enjoyed "a little exposure and a little reward" courtesy of the Navy PR machine.[361]

The Air Force, if it was to retain its image as the nation's premier air service, desperately needed its own ace. The same day that Cunningham and Driscoll scored their fifth MiG, Ritchie and his WSO, Captain Charles DeBellevue, maneuvered behind a MiG and launched two Sparrows at it. The first missile failed to detonate, but the second missile sent the MiG flaming toward the ground. It was the first victory for both crew members. Over the next four months, Ritchie would use his Sparrow missiles to amass three more kills against North Vietnamese MiG-21s, two with DeBellevue again in his back seat.[362] Ritchie's final and all-important fifth kill came on August 28, 1972, when he and DeBellevue detected a MiG-21 fifteen miles ahead and ten thousand feet above them. Ritchie lit the afterburners and executed a hard, climbing left turn to arrive at the MiG's six o'clock, about four-miles back, just slightly out of the Sparrow's maximum range. During the initial maneuver, Ritchie had fired a pair of Sparrows that he knew were "out of parameters" but that he hoped might "get the MiG to start a turn." The MiG failed to take the bait. After rolling out behind the MiG, the massive engines in the Phantom powered Ritchie closer to the MiG and into a better firing position. Ritchie let loose with his last two Sparrows. "The third missile missed, but the fourth impacted the MiG," which was "seen to explode and start tumbling toward the earth."[363]

Unlike Jabara in Korea, Ritchie embodied the prototypical ace fighter pilot. He was once described by Olds as "brilliant. He is good looking. He is dedicated. He is everything that we wanted."[364] Ritchie was born and raised in Reidsville, North Carolina, home to one of American Tobacco Company's major manufacturing plants, where his father worked as an executive. Selected to attend the Air Force Academy, Ritchie "walked-on" to the varsity football team and played as the team's starting halfback in the 1963 Gator Bowl. He graduated from the academy the same year as Clifton in 1964 and, a year later, finished first in his pilot training class at Laredo AFB, Texas. Deciding midway through pilot training that he "never wanted to fly anything with more than one seat," Ritchie opted for one of the four F-104 Starfighters available to his class. Only later did he realize that the F-104 was a "non-combat assignment," and he spent the next two years at Eglin AFB flying "safety chase on test missions on the range." Looking to transition into TAC and into an F-105, he was instead picked for an F-4D front-seat assignment to DaNang, arriving for his one-year tour in 1968. From there, Ritchie was assigned to Nellis AFB where he completed the Air Force's Weapons School program in the Phantom. He volunteered to stay on and serve as an instructor, becoming one of the vocal opponents to non–fighting wing formations. After his assignment at Nellis, Ritchie opted for a second tour in Vietnam and arrived in January 1972, just a few months before the air war over North Vietnam resumed.[365]

Olds and others would later suggest that Ritchie had been handpicked to become the Air Force's first Vietnam ace, much like Jabara had been in Korea earlier.[366] Ritchie denounced the rumors, explaining that with the unpredictability of the North Vietnamese MiGs, "there's no way they could give you the right missions."[367] But Ritchie did enjoy certain benefits that were unavailable to the earlier aircrews. One, the Combat Tree system installed in the service's new F-4Ds, allowed Phantom aircrews to locate a MiG by interrogating the enemy aircraft's radar transponder.[368] In other areas, Ritchie's intense devotion to his task likely contributed to his success. For example, Ritchie demanded that *his* missiles undergo more frequent diagnostic testing than the regulations specified; his AIM-7E-2 Sparrow missiles achieved an uncharacteristically high success rate of nearly 55 percent as a result. The maintenance troops toiling in the heat and humidity might have balked at Ritchie's request, but the aviator took great care to establish personal relationships with his supporting ground crews. Ritchie also spent time engaging with the radar controllers who he knew might one day vector him toward a MiG and aerial victory. That investment paid off too. On the day that he shot down two MiGs, Ritchie's radar controller recognized the pilot's voice on the radio and, breaking with protocol, reverted to conversational English to alert Ritchie to the nearby MiGs: "Steve, they're two miles north of you."[369] Finally, Ritchie had the benefit of flying with a regular WSO, DeBellevue. As Ritchie explained, he and DeBellevue "had flown [together] enough that Chuck knew exactly the information that I needed. . . . I don't have to say from the front, 'Chuck, is it a good lock?' And he says, 'Yes, it's a good lock.' He knows to tell me at that instant that it's a good lock."[370]

When Ritchie and DeBellevue landed from their eventful ace sortie, their squadron mates from the 555th "Triple Nickel" TFS gathered around the aircraft shouting congratulations. The aircraft's assistant crew chief, Staff Sergeant Frank Falcone, held a stencil of a large star against the aircraft and handed a can of red spray paint to the victorious front seater. As was the practice, Ritchie filled in half of the star and then handed the can over to his GIB to finish it off.[371] Within hours, a message from the secretary of defense, Melvin Laird, arrived for Ritchie: "My warmest congratulations are extended to you as the first US Air Force 'ace' of the Vietnam conflict. You have earned a special salute."[372]

Fittingly, that year Ritchie was awarded the Colonel James Jabara Award for Airmanship, which recognized the Air Force Academy graduate "whose airmanship contributions are of great significance and set them apart from their contemporaries." Ritchie was also a recipient of the National Aeronautic Association's Mackay Trophy, awarded for the "most meritorious flight of the year." There were two other recipients of the Mackay trophy that year—the Air Force's other two aces who followed Ritchie. DeBellevue was one. Captain Jeffery Feinstein was the other. Both were WSOs. Ritchie was the only Air Force ace from the Vietnam War who had ever actually squeezed the trigger.[373]

Lingering Tensions

In spring 1972, before Ritchie downed his first MiG, CSAF General John D. Ryan altered the service's criteria to award an aerial victory credit. Before the change, the Air Force aircrews in Vietnam were held to the same standards and used the same conventions as their Korean and World War II forbears: full victory credit could be awarded only to a single individual; if two individuals contributed to a kill, then credit had to be shared in equal proportion. Unable to disaggregate the Phantom front seater's contribution from that of his GIB, the Air Force originally chose to split victory credit for any downed MiGs between the two crewmembers. Hence, when Olds left combat in October 1967 after having claimed four MiGs, he had amassed only two aerial victory credits. The resumption of air combat in 1972 offered an opportunity to revisit the policy that was stymieing the Air Force's attempts to anoint its next generation of aces. General Ryan seized it, decreeing that front seaters and GIBs would *both* receive a full aerial victory credit for each MiG downed. And he made the policy retroactive to the arrival of the first Air Force Phantoms in SEA, April 1965.[374]

Consequently, there was never any real debate over whether Hendrickson would receive credit for his front seater Rubus's gun kill in October 1972. Nor was there any question when the two GIBs, DeBellevue and Feinstein, achieved ace status. In fact, DeBellevue would score a total of six kills during the war, placing him at the top of the roll of Vietnam aces. But while Ryan's policy obviously appealed to the GIBs, it was generally not well received by those who occupied the Phantom's front seat. Welch recalled "grousing" among the front seaters and sarcastic arguments that if the Air Force was going to extend credit to the GIB, "why don't you give the wingman credit [too]? The wingman is just as responsible for your being in position to kill an enemy airplane as the guy in the back seat. . . . And, by the way, how about the crew chief [that maintains the plane]?"[375] Everest, the commander of the first F-4C training unit at MacDill, also disagreed with the policy but noted that at least the GIB "helped in most cases."[376] Burns was more damning; he declared the policy "not good. . . . I think it confused the hell out of things."[377]

Publicly it didn't matter because few outside the military even took notice. The five Vietnam aces—two Navy and three Air Force—never received the accolades or the headlines that were afforded to their earlier counterparts. Perhaps the public's ambivalence was related to the unpopularity of the tiresome Vietnam War or the fact that the aviator-ace had been supplanted by the next generation of national heroes wearing silver-metallic or white spacesuits. What is clear is that the Vietnam aces were never trotted out in front of a crowd of twenty thousand at Fenway like Jabara. In fact, the *New York Times* made no mention of Cunningham and Driscoll becoming the first American aces since Korea until they arrived in New York two weeks after their historic mission for a swank dinner at the restaurant 21 in Manhattan, "courtesy Grumman"—the manufacturer of the Navy's next fighter, the F-14 Tomcat.[378] Ritchie's accomplishment was recognized in a timelier manner, appearing in print on August 30 two days after his victorious sortie, but it was buried on page sixteen, and it lacked the customary thrilling account of the aerial duel that was a feature of earlier reporting on the Korean War aces.[379] If an individual wasn't already familiar with the cockpit layout of the two-seat Phantom, he or she might miss the qualifying note that accompanied the two-paragraph announcement when DeBellevue achieved ace status as a "backseat weapons systems operator."[380] Feinstein's achievement, which occurred the same day that Rubus and Hendrickson got their gun kill, was buried in a column on the air war in general and made no mention of his position in the cockpit.[381]

One group that did take notice of the anomaly, however, was the American Fighter Aces Association. A private organization established in 1960 to "preserve and promote the distinguished heritage of the American combat fighter ace" and recognize the "special distinction of becoming an ace by destroying five or more hostile aircraft in air to air combat," the Fighter Aces Association maintains a roll of all American fighter pilot aces since World War I. For the Vietnam War, the association recognizes only two aces: Randy "Duke" Cunningham and "Steve" Ritchie.[382] The conspicuous absence of the three ace-GIBs—Driscoll, DeBellevue, and Feinstein—is a reminder of the still-lingering tension over the role of the GIB in fighter combat and the fighter pilots' attempts to preserve their iconic, self-sufficient fighter pilot image. Despite their vital role in victory, the GIBs were still little more than "castrated tomcats" in the eyes of many of the nation's ace aviators.

Conclusion

In March 1915 Roland Garros decided that he preferred the risk of shooting off his own propeller rather than continuing to chauffeur around an observer-gunner who persistently failed to see, much less hit, an enemy aircraft. Almost exactly fifty years later, the observer-gunner was back in fighter combat, except now, instead of looking through oil-smeared goggles and swinging his machine gun around to fire on an approaching enemy, the observer stared at a scope in the back of the Phantom II and let his rocket-powered, radar-guided Sparrow missile do the turning for him. The interpersonal conflict that Garros had conveniently avoided when he abandoned his human baggage at the aerodrome had suddenly resurfaced. In the new Phantom, the lingering interpersonal issues became enmeshed in additional cumbersome human-machine dynamics, all of which would have a significant effect on the combat performance of the Air Force's newest fighter aircraft as it entered combat over Vietnam.

When the first F-4Cs began showing up on TAC flight lines in 1963, the pilots were ecstatic. Although the Phantom may have originally been designed for the Navy, it reflected the Air Force's aspirations for the post-Korean era. The Phantom was the fastest aircraft in the world. It could zoom to near-stratospheric altitudes. And, when combined with its long-range weapons system, it would ensure that no American Air Force fighter pilot would ever have to endure the humiliation of watching enemy MiGs trace lazy contrails in the sky high above him again. While other Air Force aircraft had been pursuing advanced radars, fire control systems, and guided missiles for years, the arrival of the Phantom heralded TAC's first significant foray into the new age of aerial warfare. Hence, the initial cadre of TAC pilots handpicked for the Phantom was understandably excited to take wing in what was widely regarded as the most powerful aircraft in the world.

They soon learned that some aspects of flying the new fighter remained unchanged: pulling back on the stick still made the nose pitch up, and the aircraft still rolled to the left when the pilot pushed the stick in that direction. Other changes that may have initially seemed new simply represented a difference in kind. For example, the pilot now had two throttles instead of one. But there were also more dramatic changes. The large radar scope mounted prominently in front of the pilot was a noticeable deviation from the standard fighter cockpit designs of the past. And then there was the GIB.

In the earlier Sabres, fighter pilots had complained of too much automaticity in the APG-30. With the F-86D's E-4 fire control system, the interceptor pilots enjoyed more control over the radar and its targeting, but "flying instruments" off a television screen mounted between the pilots' knees proved unpopular. The engineers working on the Phantom wanted to provide the desired flexibility in the radar's operation, but their studies showed that a single pilot would not be able to devote the necessary attention to the even more complex radar. Installing a human GIB in the Phantom was the logical, engineering answer. The two crewmembers, it was reasoned, could work together as a team to seamlessly pass critical target information between cockpits using an intercom radio.

The well-engineered design encountered difficulties in practice. Most fighter pilots weren't especially keen about opening their cockpit space to another individual. And, under the Air Force's policy of assigning newly winged pilots to the Phantom's back seat, most GIBs weren't too enthused to *ride in*—not *fly*—the Phantom either. Many of those sitting in either seat in the Phantom would have preferred instead to be flying a single-seat fighter, even if it was an older, less sophisticated, more vulnerable aircraft. The palpable tension within the cockpit threatened to undo the years of intricate planning that had guided the engineers as they divvied up the air combat tasks among the aircraft, the missiles, and the humans. An aircraft commander who told his junior GIB to "shut up" and "go cold mike" violated one of the engineers' basic design assumptions. A GIB who failed to learn the intricate procedures and flexibility of the radar because he was too focused on eventually transitioning to the Phantom's front seat similarly contravened the intended division of tasks. Nor did the reality of air combat over Vietnam conform to the engineering designs. Uneven terrain, VID requirements, and small, nimble MiGs were a few of the challenges that conspired to render the Phantom's radar and missiles ineffective. There were other issues, too. Some, such as a troubling switch, were localized in a specific cockpit. Others spanned the two cockpits, occasionally even reaching across cockpits in the formation. All demanded that the crews develop new innovative tactics and procedures, including, for example, "independent" element tactics or how to use the "gyro out" mode to acquire a radar lock during a dogfight.

In the process, the pilots began to realize that their combat experiences were diverging from that of their predecessors. In the past, Sabre pilots were instructed to be aggressive "tigers" and encouraged to charge headlong into packs of enemy MiGs with guns blazing hoping to score a lucky hit. In Vietnam, commanders like Olds and Blesse still demanded aggressive pilots.[383] But a Phantom pilot wasn't going to hit anything if he didn't remember to listen for the Sidewinder's "growl" before squeezing the trigger, nor would he have a chance with the Sparrow if he couldn't visualize the radar scan pattern in "gyro out" mode and maneuver accordingly so that his GIB could get a radar lock. Indeed, the Phantom crews in Vietnam determined that it was often smarter to send half their four-ship running away from the fight rather than into it. The 8th TFW's March 1967 tactics manual tried to alert the Phantom pilots to the subtle shift. It offered a new definition of *fighter pilot*: "The individual who takes advantage of sound realistic training, who gathers, studies, evaluates, and applies all the information applicable to the mission, who takes advantage of the combat experience within the Wing, who has the forsight [*sic*], who is prepared for all possible contingencies, and who THINKS! These are the people who are effective and confident. These are 'Fighter Pilots!' "[384]

Studying, evaluating, applying information—these terms were anathema to the previous generations of aggressive fighter pilots who boasted of their instinctual "birdman" flying abilities. Despite the new definition, the classic image of the fighter pilot remained. The Phantom, with its powerful radar mounted in its nose and its guided missiles slung beneath its belly and wings, might have been designed to simplify and democratize fighter aviation so that the nation would no longer have to rely solely on the skills of carefully screened and combat-seasoned "birdmen," but it quickly became apparent in the skies over Vietnam that skilled pilots were still needed, and that those necessary skills could often be developed only in combat. Automation procedures and specific "tactics may be learned and recited verbatim," the 8th TFW's tactics manual declared, "but to be of use, this knowledge must be applied in the actual situation so that timing, coordination and teamwork can be acquired, and skills sharpened."[385] Olds similarly observed, "Just because you've got a more modern aircraft, you can't flit about the sky with any degree of impunity. [The enemy's] going to nail you."[386] As they tinkered with the procedural checklists that came with their new equipment and adapted their skills and their tactics to the evolving air combat environment over Vietnam, the Phantom pilots extinguished the notion that they were somehow "flying scientists" or merely black-boxed, button-pushing "missilemen." Despite their missiles and their GIB, they were indeed the proper descendants of the great aviator aces of the past.

The pilots' continued enchantment with the fighter pilot myth, however, had a pernicious side effect. When the tactics or procedures were not derived from new technologies but rather from the historical legacy of fighter aviation, the pilots proved exceedingly stubborn in adjusting them. For example, "fighting wing" had little utility in a battle between *missile*-wielding aircraft; it was designed for gun battles. Yet, because of the cultural significance that the formation position had assumed over the preceding decades, and specifically its role in grooming and "blooding" young fighter pilots, the more experienced pilots steadfastly refused to stray from the time-honored practice throughout much of the war.

The full implications of the system that the engineers and aircrews had collectively wrought became apparent when the fighter pilots started trying to assign credit for aerial victories. In the past, it had been easy to extract the pilot from his aircraft for the purpose of bestowing glory and honor. In the Phantom, however, once the relevant components, including the GIB, had been aggregated in battle, it was nearly impossible to quantify each individual component's relevant contributions to victory. Even in the case of a gun-kill, in which the GIB had almost no role, it was still difficult to draw a distinction between the skill of the individual occupying the front seat and that of his GIB who first detected the enemy from miles away. Nor was it possible to separate the front seater's performance from that of his missiles; many a potential ace could testify to being robbed of rightful glory because of a malfunctioning missile.

General Ryan's 1972 policy of awarding full credit to both Phantom crewmembers offered one means of addressing the controversy, but it did not resolve the existential questions emerging in fighter aviation. In an increasingly complex environment with missiles and radars and GIBs, in which the individual performance of the front-seat pilot could no longer guarantee victory, what constituted heroic action? What feats were worthy of air combat glory, especially when the fighter pilots were spending most of their time not battling MiGs but dropping bombs?

There was another way to address the underlying tension—a return to Garros's original solution. The Air Force would adopt this strategy with its newest, custom-designed fighter, the F-15 Eagle. Designed as a retort to the Phantom experience in Vietnam, the Eagle was a powerful ideological statement for Air Force fighter pilots. Whereas the two-seat, missile-only, and multimission Phantom had compromised the elegant simplicity of the fighter pilot myth, the new Eagle would be a pure fighter, with a single seat and a single mission. It would have a gun, and it would be designed for maximum maneuverability in a close-range dogfight. But then a pair of missile and air combat tests conducted in 1977 intervened, and the fighter pilots once again were forced to confront the question, "What are we?"

CHAPTER 5

Eagles over Nellis

"A Fighter Pilot's Fighter"

What you fly and how you fly it isn't as important as what kind of missile you use and how you use it. . . . Either you fly with long-range kill capability, or you are going to operate at an unacceptable disadvantage.

MAJOR GENERAL FREDERICK C. "BOOTS" BLESSE, USAF (Ret.), 1977

Historically, the skill and cunning of the *men inside* the machines may often have been far more important in determining the outcomes of combat engagements than discrepancies in either the performance or weaponry of their opposing fighters.

MAJOR BARRY D. WATTS, USAF, 1977

On May 3, 1977, a US Air Force McDonnell Douglas F-15 Eagle rocketed down the runway, leaped into the air, and turned north. After a short ten-minute journey, the single pilot took up his assigned position over one corner of "Dogbone Lake," an unmistakable depression in the desert floor below that only rarely accumulated enough water to be worthy of its name. A few minutes later, the pilot detected a small, twin-engine jet fighter maneuvering a few miles away. He slammed his throttles forward to light the afterburners and turned hard toward the center of the lakebed. The Eagle pilot knew that he needed to kill the opposing fighter quickly because it was armed with an advanced heat-seeking missile. Unlike the Sidewinders used by the American pilots earlier in Vietnam, the enemy's new missile was not restricted to being launched from near the target's tail; its infrared-seeker was now sensitive enough to detect other aircraft heat sources that radiated IR-energy out in all directions. It was an *all-aspect* IR-guided missile, and it also had a much greater range than the AIM-9 Sidewinders used by the Americans just a few years earlier.[1]

The Eagle pilot rolled out of his turn. Peering ahead at his foe through the head's up display, the F-15 pilot scanned his aircraft and weapons parameters as the distance rapidly closed. The Eagle was sporting its own advanced missiles for the mission, too. It carried a new version of the radar-guided Sparrow, the AIM-7F, which had traded out the vacuum tubes of the AIM-7E-2 for more reliable solid-state electronics. The updated Sparrow also boasted an effective range almost twice that of the earlier

"Dogfight Sparrow."[2] In addition to its Sparrows, the F-15 was carrying a new, experimental all-aspect IR-guided missile similar to that of the day's adversary. A few moments passed, and then a small "IN RNG" cue flashed up in the HUD, notifying the Eagle pilot that his aircraft's central computer (CC) had determined that he was in a valid missile employment zone relative to his opponent. The Eagle pilot pressed the red "pickle" button located on the top of his control stick, and a second and a half later a deadly missile was sent on its way.

The opposing pilot that day was an expert in Soviet air force tactics, flying an aircraft many considered comparable to the MiG-21 that had proved so troublesome to American aircrews years earlier. He too had detected his adversary over the lakebed and was already aggressively maneuvering for his own missile shot against the F-15. Although his aircraft lacked the Eagle's advertised greater-than-one thrust-to-weight ratio, it could still generate a very tight turn to point its IR-guided missiles at a target. A few seconds later, the enemy aircraft was hit by the Eagle's supersonic missile. But unbeknownst to the F-15 pilot, his missile had arrived one second too late. The opposing pilot had already launched his own heat-seeking missile at the approaching Eagle. The two missiles passed each other in the sky. After a few more seconds, the F-15 pilot received the dreaded radio call: "You're dead."[3]

The engagement was an exercise, one of 463 trials flown by the Air Force's newest F-15 Eagles against a collection of F-5E Aggressor aircraft as part of the joint Air Force and Navy AIMVAL test.[4] No actual missiles had been fired in the desert skies north of Nellis Air Force Base that afternoon, but each pilot had performed every procedure in his cockpit just as if he was carrying real missiles. A revolutionary new training and evaluation tool, air combat maneuvering instrumentation, was collecting all of the data from each aircraft and assessing its simulated missiles in real-time. With ACMI, the combatants didn't scan the sky for smoke and fireballs to alert them of victory (or danger). Instead, they listened for a radio call from test officials on the ground watching the engagement unfold on computer monitors. When the ACMI computers predicted missile success, the observers notified the appropriate aircraft.[5]

The two opposing aircraft for the day's mission each represented a distinct design philosophy. The Northrop F-5E was developed as an extremely agile, single-seat, low-cost fighter airplane. Its designers had shunned the complex electronic black boxes that had been crammed into the earlier Phantom and newer Eagle in favor of simplicity of operation and maintenance. Although the F-5E was equipped with a small radar, it carried no radar-guided missiles as armament, only a cannon and IR-guided missiles.[6] It was also small like a Soviet MiG-21, and its turning performance in a dogfight was considered comparable. Because of its similarities to the predominant threat of the day, the Air Force had recently elected to replace the T-38 Talon trainers in its "Soviet" Aggressor squadrons with the new F-5 fighter (figure 5.1).[7]

The F-15, in contrast, was immense (figure 5.2). A single vertical tail on the F-15 was nearly the size of a MiG-21's wing; the Eagle's wingspan, 65 percent larger; and

FIGURE 5.1. Air Force F-5E Aggressors. The Air Force Aggressor squadrons were established after Vietnam to help aircrews train against more representative threats. Approximately the same size as the Soviet MiG-21 and equally maneuverable, Northrop's F-5E was an ideal platform for the specially trained Air Force pilots to use when mimicking Soviet fighter tactics. The F-5-equipped Aggressors flew as the adversaries during the AIMVAL-ACEVAL tests. NMUSAF.

its fuselage, almost twenty feet longer. At the front of the Eagle, a huge radome shielded the largest air-to-air radar yet developed for a fighter airplane. But, like the F-5, the Eagle had only a single seat in the cockpit, which meant that the F-15 also required an array of electronic gadgets to help the pilot manage the powerful radar and multimissile fire control systems.[8] Engine operation in the F-15 was also complex but largely automated: an air data computer automatically calibrated the position of the hydraulically actuated, variable geometry engine inlets based on air temperature and aircraft speed to ensure proper airflow to the aircraft's pair of Pratt & Whitney F100 engines.[9] From its nose to its tail, the F-15 embodied cutting-edge technology, which was aptly reflected in its nearly $10.5 million unit cost—more than four times that of the F-5E.[10]

The F-15 Eagle was the Air Force's first attempt at building a dedicated air superiority fighter since the F-86 Sabre nearly forty years earlier. For the pilots, their new single-seat "fighter pilot's fighter" promised to restore the sanctity of the fighter pilot myth.[11] Unlike the two-seat Phantom, a "really ugly airplane" in the words of Welch,

FIGURE 5.2. Air Force F-15 Eagle. McDonnell Douglas's F-15 was designed to be the "fighter pilot's fighter." With its single seat, bubble canopy, internal gun, and exceptional maneuverability, the Eagle was optimized for close-range combat. However, it also carried a full complement of guided missiles, four short-range IR-guided Sidewinders mounted under the wings, and four medium-range radar-guided Sparrows slung beneath the engines. Several Air Force officials would question the necessity of the advanced radar and missile weaponry for their newest dogfighter before its 1972 rollout. NMUSAF.

"that didn't do anything really, really well," the sleek F-15 had one purpose, and that didn't include dropping bombs.[12] It would be a dogfighter, capable of not just matching the MiG-21's performance like the F-5E, but of handedly out-turning and outpowering any current or future Soviet fighter. For close-range attacks, the F-15 would sport a new, more-powerful 25-mm air-to-air cannon to pulverize enemy aircraft. For those that tried to flee the gunfight, the Eagle would come equipped with new IR-guided missiles and a powerful radar to guide the latest AIM-7F Sparrows. To aid the Eagle pilot, the aircraft would incorporate separate HUD and HOTAS (hands on throttle and stick) systems, the combination of which would allow the pilot to keep "his eyes on the target and his hands on the stick and throttle, a significant advantage in the dogfight scenario," according to one Air Force press release.[13] Even the Eagle's large bubble canopy, more reminiscent of that on the F-86 Sabre than the faired design on the F-4 Phantom, signaled the F-15's emphasis on classic dogfighting.[14] As Evans had experienced decades earlier, the next generation of Air Force air superiority pilots would be able to look unrestricted over their shoulder, down the Eagle's smooth leading edge, and "imagine all sorts of exciting possibilities."[15]

The results from the Air Intercept Missile Evaluation and Air Combat Evaluation tests, "the biggest fighter-plane shoot-out in peacetime history," upset the Eagle pilots' plans.[16] The F-5E with its new all-aspect IR-guided missile proved exceptionally lethal to the Eagle in the close-range environment. The Eagle scored several victories, too, but not at a rate as lopsided as many had hoped. In the test scenarios, it didn't matter that the F-15 could easily out-power the F-5E, or that the Eagle pilot had a variety of short-range radar auto acquisition modes available to him with the mere twitch of his right thumb. All of the Eagle's technological advantages in the short-range environment could be quickly negated if the opposing pilot simply saw the F-15, turned to point at it, and squeezed off a heat-seeking missile just a second before he was called "dead." In one notorious engagement, four F-15s squared off against four F-5Es, and 112-seconds later everybody was dead, the last two pyrrhic victories coming like those on the third of May when opposing missiles figuratively passed each other in the sky. The infamous four-against-four engagement became known as the "Towering Inferno" mission, a reference to the 1974 Oscar-winning disaster film starring Steve McQueen and Paul Newman, "because," in the words of one analyst, "everybody died."[17] The mission wasn't that abnormal, though. One news report summarized the tests' central finding: "Once a pilot gets into a dogfight, or even comes within range of his adversary's missiles, both pilots are likely to die."[18]

"Boots" Blesse, who retired as a major general two years before AIMVAL-ACEVAL, was invited to observe the trials at Nellis. After his brief visit, Blesse concluded that the new technologies showcased during AIMVAL, including "short-range launch-and-leave missiles, will produce the first real change in aerial tactics since World War I." It wouldn't matter anymore "what you fly and how you fly it," Blesse opined, only whether "you fly with long-range kill capability."[19] It was a curious assessment coming from a combat ace who earlier in his career had fought vigorously against the "missile people" lurking in the Pentagon's halls, and who later led the charge to equip the F-4C with an interim air-to-air gun capability. Perhaps Blesse's opinion was colored by the fact that he was now working as the special assistant to the chairman and president of Grumman Aerospace, the manufacturer of the Navy's F-14 Tomcat, which was also undergoing testing during AIMVAL-ACEVAL. Still, coming from the author of "No Guts, No Glory," Blesse's prediction that "tomorrow's jet ace no longer will have to see an enemy aircraft to destroy it" must have been disheartening for the next generation of Air Force fighter pilots.[20] His assessment might have accorded with the Air Force's and Navy's plans to outfit their forces with high-technology fighter aircraft like the F-15 and F-14, but by assigning primacy to the technology as opposed to the pilot, it belied the central element of the historic fighter pilot myth.

Several challenged Blesse's assertions. Some of the critiques came from industry, especially firms that were trying to market the simple and cheap F-5E as a viable alternative to the more exquisite Eagle and Tomcat. Former Vietnam F-4 pilots also rebuffed Blesse's claims, explaining that they had heard a similar line of reasoning only a decade earlier. Major Barry Watts was one such Phantom aviator. On assignment teaching philosophy at the Air Force Academy, Watts accused Blesse of assuming that "aircraft performance" was and would always be the "determining factor in aerial battle." Invoking Korean War statistics that emphasized the importance of the ace, Watts argued that it wasn't the machinery but "the skill and cunning of the *men inside* the machines" that "historically" determined victory.[21]

The controversy over the F-15, AIMVAL-ACEVAL, and advanced missile weaponry reflects the conflict between the "great machine" and the "great pilot" narratives, as well as the competing imagery of the fighter pilot as "knight" and "scientist."[22] Both aircraft and pilot entered AIMVAL-ACEVAL focused primarily on close-range air combat. They emerged from the tests with fresh requirements for improved technologies, tactics, and procedures that would allow the man and his machine to excel not just at dogfighting but, more importantly, at long-range air combat. It was almost a 180-degree reversal of that experienced by the Phantoms and its aircrews years earlier in Vietnam, and it begged the question of the Eagle pilots, Could their new aircraft still be the "fighter pilot's fighter" if it was no longer going to be primarily a dogfighter?

A *Pure* Air-to-Air Fighter

The genesis of the F-15 Eagle can be traced to a set of nondescript meetings convened in late 1964 at the behest of Major General Arthur Agan. A veteran fighter pilot from World War II, Agan had recently completed an assignment as a SAGE (semi-automatic ground environment) division commander before arriving at the Pentagon that summer to serve as the Air Force's director of plans.[23] He harbored strong opinions regarding the Air Force's contemporary fighter aircraft. His standard was simple: "When you get a fighter with a bomb on it, it's a bomber, not a fighter." Hence, in Agan's estimation, his service's new multirole F-4C was obviously a bomber, and he similarly considered the "big heavy clunker" TFX an attack airplane worthy of an "A-111" designation as opposed to the "F-111" one that it received. Moreover, Agan was furious that the analysts in the Pentagon kept passing him reports and war plans that included the caveat, "Assume air superiority." After his experiences battling the *Luftwaffe* over Europe, Agan considered the caveat tantamount to "assuming you've won the war too." He later remarked that he wanted desperately to "bring somebody's attention to the fact that we [in the Air Force] weren't thinking about beating enemy airplanes in the air with airplanes."[24]

That fall, Agan chartered a small group of recognized air combat experts to help him. Harry Thyng, the combined double ace from World War II and Korea and now a brigadier general, was asked to chair the meetings. Francis Gabreski was invited to participate, as was his fellow World War II triple ace William Dunham. Korean ace Winton Marshall was a member too, along with a few other up-and-coming fighter pilots like John Burns. The aces' subsequent report was rather informal and "generalized," according to Agan, and it languished in the bowels of the Pentagon for months. Civilian budgeteers showed little enthusiasm for its recommendations. Nor did the current chief of staff of the Air Force, the famed bomber general Curtis LeMay. But when General John McConnell replaced LeMay in February 1965, the report resurfaced. Finally, Agan recalled, they were able to get "someone to listen and . . . get a competent effort started toward putting together a requirement for an air fighter."[25]

Like LeMay, McConnell had spent many years as a bomber commander in Strategic Air Command. But McConnell, recalling his time spent flying fighters in the Pacific theater during World War II, also considered himself a fighter pilot.[26] He was aware of the aces' report from when he served as LeMay's vice chief of staff, and he knew about its poor reception within the building. He also had knowledge of another Air Force study completed earlier by Lieutenant Colonel John Bohn Jr. Like Agan's aces, Bohn challenged "the folly of assuming air superiority," especially in light of the new class of Soviet interceptors and fighters then thought to be under development.[27] The sudden loss of a pair of F-105s to North Vietnamese MiG-17s on April 4, 1965, the same day the Phantoms from the 45th Tactical Fighter Squadron departed MacDill AFB for Ubon Royal Thai Air Force Base, offered immediate credence to Agan's and Bohn's concerns. McConnell responded the following month by issuing a "policy statement," likely drafted by Agan, that articulated the "Air Force's requirement 'to win air superiority.'"[28] In a service that hadn't seriously considered achieving air superiority through air battle for several decades, the statement was considered a major coup for Agan and his acolytes. Requirements for a new Air Force fighter known as the F-X were drafted, and in December 1965 a call for proposals and design studies went out to industry.[29]

It would take another three years for the F-X design to stabilize as "a single-pilot, twin-engine turbofan-powered aircraft with a 44,500-lb takeoff gross weight" that, "together with its avionics and weapons," was intended to "be superior in air combat to any present or postulated Soviet fighters both in close-in, visual encounters and in standoff or all-weather encounters."[30] In the interim, the F-X concept experienced considerable turmoil. At one point, plans for the F-X fighter included variable-swept wings, podded-engines, multiple crew members, and a multimission capability. In this configuration, the aircraft was projected to weigh over sixty thousand pounds. Watching the design process unfold, the four-star generals at Tactical Air Command (TAC), Pacific Air Forces (PACAF), and United States Air Forces in Europe (USAFE) grew increasingly distressed. Finally, in February 1966, the three commanders jointly authored a "twelve-star letter" to McConnell expressing their opposition to another multirole fighter like the F-4. The generals got their wish.[31] On December 23, 1969, the Air Force announced that McDonnell Douglas had won the contract for the "F-15 advanced tactical fighter aircraft." The new fighter, the public affairs officials declared, was "designed specifically as an air superiority weapon system."[32]

The first F-15 taxied out from the McDonnell Douglas factory in St. Louis almost three years later on June 26, 1972, to great ceremonial fanfare. With its "engines keening like a banshee" according to one news reporter, Mrs. Iris Rose Hansen, wife of the assistant secretary of the Air Force, christened the Air Force's newest fighter jet "Eagle, and may you reign supreme in your domain." Hansen's husband, Grant, told those gathered that the F-15 would "outclimb, outmaneuver and outaccelerate any fighter threat in existence or on the horizon."[33] Capturing the fighter pilots' excitement for their new fighter, General William Momyer, now serving as the commander of TAC, described "the F-15 as having 'more potential than a pilot can psychologically take.' "[34] The single-seat cockpit was certainly part of the Eagle's appeal to its potential pilots, as was its single, air-superiority mission focus. Yet, even at the rollout, there was still some uncertainty as to how the Eagle would fulfill its single air superiority mission and whether a single pilot, aided by significant automation, could successfully operate the complex aircraft in a hostile environment.

A Single Mission

The refrain "not a pound for air-to-ground" that guided the F-15 design and later thrilled future Eagle pilots allowed the Air Force to develop an uncompromising air-to-air fighter reminiscent of the iconic Sabre, MiG Alley, and ten-to-one kill ratios.[35] The new fighter's singular focus on air superiority was also important for political reasons, for without it and the corresponding specific emphasis on close-combat dogfighting capability, Air Force leaders worried that their new fighter might become indistinguishable from other competing fighter aircraft proposals emerging from within their service and the Navy. Thus, while there were some in the Air Force who were truly captivated by the concept of a fighter optimized for air-to-air combat, notably the three four-star commanders of TAC, PACAF, and USAFE, many others lent their support to the F-X chiefly to avoid the ignominy of having yet another of their service's premiere fighters be sidetracked by outside forces.

Ironically, the Air Force's first step to disarm potential F-X critics and interservice competition required that it adopt another Navy-designed aircraft. In late 1965, the Air Force needed a new close-air-support workhorse. McNamara gave the service a choice: either the new low-cost F-5 or the Navy's A-7. McConnell opted for the A-7 (figure 5.3). An earlier study conducted by Colonel Bruce Hinton, the same officer who had led the F-86s into Kimpo and then scored the Sabre's first MiG kill, had concluded that the A-7 would likely perform the ground-troop support mission better than the F-5. Hinton's assessment happened to match the conclusion reached by the prior Bohn

FIGURE 5.3. Air Force A-7D. Given the choice in the mid-1960s between the new F-5 and the Navy's existing A-7, the Air Force opted for the latter as its next close air support fighter. Some studies suggested that the A-7 would be more effective in the ground-support role, but the service's decision also helped justify the requirement for its favored F-X air-to-air fighter concept that became the F-15. NMUSAF.

study, too. But the A-7 also possessed several other advantages for the Air Force that extended beyond its purported mission efficacy. First, nobody would ever mistake the blunt-nosed A-7 for an air-to-air fighter, whereas they might if the Air Force opted for the F-5. Second, by introducing the A-7 into the Air Force as a dedicated close-air-support platform, the service leaders effectively eliminated the possibility that their new F-X fighter would be saddled with the air-to-ground mission. Finally, because the A-7 was so ill-suited for air-to-air combat, its selection reinforced the service's requirement for a new, dedicated air superiority fighter that could clear enemy aircraft away from the lumbering close-air-support fighter.[36]

Not all bought the single-mission F-X argument, though. Some officials within the Air Force still maintained that the planned air-to-air aircraft was a niche asset, "nice to have . . . but not essential." Others argued that the F-X represented an attempt to assuage the fighter pilots and their "romantic attachment to air-to-air combat, a kind of 'knights of the air' syndrome."[37] The surprise unveiling of the Soviet MiG-25 Foxbat at the Domodedovo Air Show in 1967, which appeared to instantly outclass any American fighter aircraft, tamped down some of the internal criticisms.[38] Still, McConnell felt obliged to inform the Senate Armed Services Committee of the discord during his testimony in May 1968. "We had a very difficult time in satisfying all the people who had to be satisfied as to what the F-X was going to be," McConnell explained. "There were a lot of people in the Air Force who wanted to make the F-X into another F-4 type of aircraft. We finally decided—and I hope there is no one who still disagrees—that this aircraft is going to be an air superiority fighter."[39]

Despite the strong rhetoric and McConnell's Senate testimony, the emerging air-to-air F-15 still retained a "significant" air-to-ground capability. Four months after the CSAF's testimony, an Air Force concept development paper affirmed that the F-X's "potential use in air-to-ground missions, once air superiority is attained, cannot be neglected."[40] Officials wouldn't necessarily neglect it, but they also didn't intend to cultivate it either: any air-to-ground provisions built into the airframe would not be allowed to detract from the aircraft's anticipated air-to-air performance. That was the intent behind the popular "not a pound for air-to-ground" mantra, not that the F-X would possess *no* air-to-ground capability.[41]

In fact, when McDonnell Douglas submitted its proposal for the F-15 contract in mid-1969, an advertised feature of its Model 199 design was that the aircraft's "*air-to-air weapons can be launched with air-to-ground weapons and* [external fuel] *tanks aboard*."[42] The Air Force pilots who put the F-15 through the wringer during testing at Edwards AFB, California, verified the company's claim, and they declared the new air-to-air fighter "an effective air-to-ground weapons platform."[43] However, despite its inherent ground attack capability, the Air Force never could articulate a clear concept of how the F-15 would actually be *used* in an air-to-ground role alongside other Air Force tactical aircraft. In light of the unresolved tactical and doctrinal concerns, the CSAF in October 1975 halted any further air-to-ground tactics development for the Eagle.[44]

Senior Air Force leaders might have finally agreed that the F-X would focus chiefly on aerial combat, but they still had not decided whether the individual victories should be pursued primarily through long-range missile attacks or through short-range air combat. Competition with the Navy would again play an important role shaping the Air Force's decision.[45] Like the Air Force, the Navy was at the time developing a replacement for its fleet of aging F-4 Phantoms. After some fits and starts in its own program, the Navy eventually settled on a VFX design that later evolved into the Grumman F-14 Tomcat (figure 5.4). The Navy's large, twin-engine fighter would, like the Air Force's F-X, focus solely on air-to-air combat. However, the Navy design also emphasized long loiter times and super-long-range missiles to fulfill the requirements of its fleet defense mission. Variable geometry wings and the massive AIM-54 Phoenix missile, along with a second seat for the radar intercept officer who would be needed to operate the complex radar to support the hundred-mile range missiles, consequently became staples of the seventy-thousand-pound VFX aircraft.[46]

When they first embarked on their F-X and VFX programs, the Air Force and Navy both had hoped to avoid the Pentagon's push for commonality that had led to the

FIGURE 5.4. Navy F-14 Tomcat and AIM-54 Phoenix. Designs for the Air Force's Eagle and the Navy's large, twin-tail, twin-engine Tomcat looked suspiciously similar to one another. Lagging the Tomcat's development by more than a year, the Air Force's F-15, although carrying the AIM-7 Sparrow and AIM-9 Sidewinder missiles, was marketed as a close-range dogfighter. In contrast, the Navy's F-14, which added to its available armament the long-range AIM-54 Phoenix—and the back-seat radar intercept officer needed to operate the radar to support it—was optimized for long-range fleet defense. Robert L. Lawson Collection, Naval Aviation Museum.

earlier TFX (F-111) debacle. After arguing their positions for nearly a full year beginning in May 1966, the services finally convinced civilian leadership in the Pentagon that "development of a single, bi-service fighter" was "neither feasible nor desirable."[47] However, the handshake deal did not stop others within Congress or the public from questioning why the Air Force and Navy were both developing separate twin-engine, twin-tail, air superiority aircraft. With the Navy's VFX contract preceding the Air Force's F-15 contract by almost a full year, the Air Force raced to justify its own platform while differentiating it from the Navy's VFX.[48]

A series of internal talking papers produced within Air Force Systems Command in 1969 and 1970 sheds light on the Air Force's strategy. The critical distinction was to be the F-15's superior maneuverability, a product of rigorous aircraft design and calculated trade-offs made according to Colonel John Boyd's energy-maneuverability theory.[49] As one talking paper explained, the Navy's F-14 would enjoy the luxury of flying fleet defense, operating near its carrier in otherwise benign waters. The

Tomcat's mission of lobbing long-range Phoenix missiles at far-off targets didn't require an exceptionally maneuverable aircraft, either. The Air Force's F-15, in contrast, would participate in "offensive missions in an enemy's air space," which would place the F-15 under constant threat of an otherwise undetected attack. "Our fighters," the F-15 proponents therefore argued, "must be able to quickly outmaneuver the enemy to gain the necessary advantage and make the kill. Superior maneuvering capability is the key to success and, therefore, the overriding factor that has been followed in designing the F-15 as a single-purpose, highly specialized aircraft for the fighter versus fighter role."[50]

The Air Force's decision to underscore maneuverability requirements as a means to differentiate its fighter from the Navy's Tomcat influenced several technological choices in the F-15 design. The significance of the F-15's gun was one such manifestation. As noted in chapter 4, the gun exerted a powerful psychological effect over Air Force fighter pilots. One pilot described it as the "fighter pilot's syndrome"—namely, "If I only had a gun, I could shoot down everyone."[51] After having labored to get a gun first *onto* the F-4C/D and then *into* the F-4E, it was probably a foregone conclusion that the service's new F-15 would come equipped with an internal gun. Still, one F-15 briefing delivered to Momyer in 1971 declared explicitly, "The gun is a mandatory item for an air superiority fighter. It is absolutely essential that we have that instantaneous, unjammable, close range capability."[52]

The Air Force didn't want just any gun for its new F-X, it wanted a new and improved air-to-air gun that had a "higher muzzle velocity, flatter bullet trajectory and greater kill capability" than the 20-mm General Electric M-61 Gatling guns currently in use.[53] Its choice, the new GAU-7/A, was designed to use a larger 25-mm caseless round. With a muzzle velocity of four thousand feet per second and an effective range almost 20 percent greater than that of the M-61, the GUA-7/A promised a nearly fourfold increase in lethality over the standard 20-mm round.[54] However, the new gun system encountered significant difficulties during testing. Rather than delay F-15 production, TAC opted to temporarily equip its first Eagles with a slightly improved version of the venerable M-61. The new 25-mm gun never materialized and TAC's temporary solution eventually became permanent.[55] Nonetheless, even when armed with the Vietnam-era M-61, an Eagle pilot was still better off in air combat than his predecessors because the engineers had decided to cant the gun in the F-15 two-degrees up. This simple up-cant essentially provided a head start for a pilot trying to establish the appropriate lead fire against an aircraft target, which, McDonnell Douglas explained, "more than doubles the time-in-firing envelope."[56]

As the Air Force aggressively marketed the superior close-range air combat capability of its F-15, several individuals both inside and outside the Pentagon began to reconsider the utility of the Eagle's powerful yet expensive radar and medium-range Sparrow missile (relative to the Navy's *long*-range Phoenix missile). Early in the design process, F-15 proponents had argued that the radar and Sparrow combination complemented the Eagle's dogfighting characteristics to provide "a carefully balanced spectrum of kill potential."[57] However, even Momyer in late 1969 began to question the wisdom of weighing down what was shaping up to be a remarkable fighter plane with a heavy, complex, and costly fire control system. A year later, the secretary of the Air Force, Robert Seamans Jr., and the new CSAF, General John Ryan, both expressed similar concerns.[58]

Numerous studies were subsequently launched to reexamine the necessity of the F-15's radar and radar-guided missiles. All walked a fine line with their recommendations, urging the service to retain the longer-range weapons but likewise careful to not undo the fastidiously constructed Eagle-as-dogfighter narrative. One, the Ad-Hoc Committee for F-15 Avionics and Armament, justified the Eagle's medium-range armament by couching the value of the Sparrow in its ability to make quick work of any enemy aircraft trying to flee a close-range engagement. Rather than sacrifice the radar and radar-guided missiles for cost's sake, the fighter pilots on the committee instead offered up the F-15's planned Electro-optical Identification and Tracking Set (EOIATS), an updated version of the F-4E Target Identification System, Electro-optical (TISEO).[59] Another study conducted by then-Lieutenant Colonel Welch similarly concluded that the primary value of the F-15's radar was not in long-range Sparrow missile employment but rather in its ability to "provide earlier detection of the opponent and thereby . . . reduce the probability of entering combat with an initial position disadvantage."[60]

A 1971 "spread the word" briefing on the F-15 regurgitated the reports' conclusions. Extolling the Eagle's ability to use its radar and medium-range missiles to dominate the close-range battle, the nearly hour-long sales pitch delivered by a trio of fighter pilots preached that once "the enemy has been found [using the radar], the real name of the game becomes maneuverability and acceleration." Fortunately, they declared, "there is no aircraft in the world that, once in a fight, can disengage from the F-15. There is no maneuver he can perform to get away."[61]

Other Air Force publications reiterated the Eagle's dogfighting credentials. For example, a 1976 F-15 "Fact Sheet" distributed by Air Force Systems Command declared that "agility in the air at medium and low altitudes is a prime requisite to success" in combat. Fortunately, the F-15's "low wing loading and high thrust-to-weight ratio" provided "unprecedented maneuverability," allowing the "F-15 to turn very tightly without losing airspeed." Echoing Welch's argument, the officials in the Office of Information also explained that the F-15 radar's unique ability "to detect, acquire, and track targets at low altitudes and at look-down angles" gave the pilot an important advantage because it would help him "in gaining a favorable position for a close in encounter." Further elaborating on another of the Eagle's distinctive advantages in a dogfight, they advertised: "The F-15 cockpit is designed to help the single crewman achieve air superiority through visual air-to-air combat. A Heads-up-Display (HUD) system provides the pilot with all necessary target and weapons data projected on the cockpit windscreen so that visual acquisition and tracking of the enemy aircraft can be maintained without looking down."[62]

The service's focus on close-range air combat naturally trickled down to the individual F-15 units as the new fighter jets started appearing on the flight line. The first nontest Eagle was delivered to Luke AFB, Arizona, on November 14, 1974, for use as a flight trainer.[63] Sixteen months later in January 1976, the next batch of F-15s officially began arriving at Langley AFB, Virginia, to become part of the 1st Tactical Fighter Wing.[64] After a few months of flying, the Eagle pilots at Langley lodged a complaint with TAC: their regulation-directed "gloss white" helmets were proving to be a definite "handicap in the F-15." Explaining that their shiny helmets made it too easy for their opponents to see which way they were looking, essentially telegraphing their next maneuver to their adversary, the Eagle pilots sought TAC's permission to "camouflage" their helmets gray. The request was approved two weeks later.[65] Although a minor historical footnote, the incident is an apt indicator of the fledgling F-15 community's focus, and it wasn't on the long-range potential of the Eagle's APG-63 radar.

Just as the return of the gun on the F-4 helped stoke the Phantom aircrews' fighter pilot self-image, so too did the singular air-to-air focus of the F-15 for its fighter pilots, especially considering the Air Force's heavy emphasis on dogfighting for its newest fighter. The pilots' "spread the word" briefing made explicit, enthusiastic reference to the fact that they would finally be charged to do what fighter pilots historically had done. Invoking Richthofen, the briefing script proclaimed: "The Red Baron described the air-to-air mission this way—and it is essentially just that—four basic functions: We must go where the enemy is, find him, engage him, and then shoot him down. The F-15 will do all these things better than any fighter in the world."[66]

A Single Seat

Another feature of the Eagle that rekindled the historic glory of fighter aviation was its single seat. According to one Air Force researcher, the single-place cockpit was a "predetermined requirement" of the F-X program that reflected the "general consensus" among Air Force fighter pilots and their preference for individually manned aircraft.[67] The single-seat design would also play an important role as the Air Force sought to distinguish its F-X dogfighter from the Navy's two-seat VFX interceptor. But while a single seat quickly became a staple of the F-X design, it was unclear if a lone pilot would be able to manage all of the tasks demanded of him in modern air combat. The engineers at McDonnell Douglas had planned to incorporate advanced avionics and automation into their design to aid the pilot, but those were also many of the same complex and costly avionics that Momyer and others had begun to question as unnecessary extravagances.

The Air Force's stated preference for a single-seat F-X usually centered on concerns over the additional weight and cost of a two-seat cockpit. For example, an F-X

Development Concept Paper signed in September 1968 warned that the "addition of a second crewman to F-X could result in a 5,000–6,000-lb. takeoff weight increase . . . and could add about $0.5 million to flyaway cost."[68] The estimate drew a wary eye from some on Capitol Hill, and when the singe-seat topic came up during a round of congressional testimony, one staffer challenged the Air Force's proffered rationale: "A man is not that heavy, General." The general testifying retorted, "A man is 200 pounds, but he takes with him all of the oxygen, another ejection seat, more structure, and armor to take care of him."[69] Air Force officials explained that the larger airframe required for a two-seater would also demand more powerful engines and a greater fuel supply if it was to match the performance of a single-seat design.[70]

The Navy was willing to pay this "penalty" with its two-seat VFX, Air Force officials acknowledged, but that was because the Navy's interceptor needed to be able to defend the fleet "in any visibility situation." Doing so required a "second man" who strictly "concentrates on fire control sensor returns, 'looking through' the adverse effects of weather, clutter, and ECM [electronic countermeasures]." In contrast, the Air Force's F-X, focused as it was on dogfighting, would perform "a majority of [its] missions . . . in good visibility." This difference in mission sets, the F-X concept paper asserted, allowed the Air Force to capitalize on the "universally recognized" benefits of "smaller aircraft size, weight and cost for a given level of performance" in a "one-man fighter."[71] Curiously, McDonnell Douglas's winning design proposal nonetheless advertised that the Air Force's single-seat F-15 could easily be converted into a two-seater "with no change required to major aircraft systems . . . and with minimum change to cables, wiring and the environmental control system" (figure 5.5).[72] The two-seat TF-15 was eventually acquired for use during aircrew training, and as McDonnell Douglas predicted, it required only minor modifications. It weighed a mere eight hundred pounds more than its single-seat counterpart.[73]

The F-X's single-seat cockpit not only helped differentiate the Air Force's fighter from the Navy's interceptor while saving some airframe weight, it also circumvented the most loathsome feature of the Phantom—at least according to many of the fighter pilots seated in the front of the F-4. Even after the Air Force started developing professional weapons system operators for its Phantoms in 1969, many front seaters still considered the WSO an impediment to efficient air combat employment. They complained about the Phantom's laborious cockpit crew coordination exercises, especially the requirement to verbalize every mode change during a dynamic air-to-air engagement. The front seaters also continued to criticize the Phantom's restricted rearward visibility, which they typically blamed on the WSO. Olds once explained, "For all-around air superiority, permissive and non-permissive, I prefer a one-man crew. . . . Without a second crewman, [the pilot] should be able to see much more behind him. The major fault of the F-4 is that it's a neckbreaker."[74] Ritchie was even more direct in his reproach. Long after the war, he continued to assert that he lost a MiG victory "on

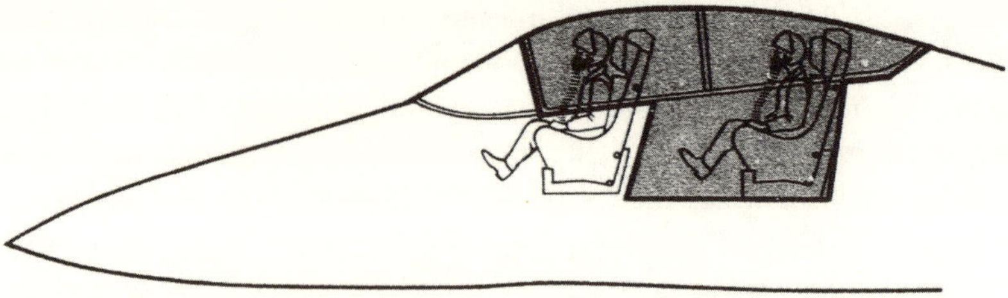

FIGURE 5.5. F-15 Two-Seat Configuration. Preserving the Eagle's single-seat configuration was a priority for the service: it not only stoked the pilots' enthusiasm for their new fighter but also helped distinguish the Air Force's F-15 from the Navy's F-14. Official justification often centered on the additional aircraft structure and corresponding weight required to support a second crew member, but McDonnell Douglas's proposal for the F-15 called attention to the design's ability to add a second seat with only a minimal "area of change" (*shaded area*). Two-seat F-15s would eventually be acquired as TF-15 trainers. Air Force Material Command History Office.

13 June [1972] because I had a WSO in the back seat. . . . If I'd been *by* myself in *that* airplane *that* day, I think I'd have gotten a MIG."[75]

There was also the historic attraction of the single-seat fighter. "Visual counter-air missions have classically employed only one man," the F-X concept development paper explained.[76] Reflecting the pervasiveness of the single-seat bias, 96 percent of fighter pilots declared during an F-X concept survey their preference for a single-seat cockpit in visual air combat. The preference rate dropped only slightly, to 85 percent, if the mission was going to be conducted at night or in poor weather.[77] If the F-15 was supposed to be the proper heir to the Sabre, then it only made sense to the fighter pilots that it too should be a single seater.

The WSOs begged to differ. James Hendrickson, the WSO-GIB who had helped his Phantom front seater Gary Rubus gun down a MiG in October 1972, felt that the second crewmember in the Phantom significantly increased the fighter's capability, especially when "the two people had flown together enough to be able to work together as a crew."[78] Years later, another Phantom WSO, Major William Flanagan, argued that "today's aerial combat task [is] much more complex and demanding," and consequently, "since every fighter pilot cannot be [an ace like] a Hartmann or Bong or Tuck . . . , it makes sense to maximize the efficiency of the average fighter pilot with an optimum crew of two in the cockpit." After all, "the F-15 is not an overgrown P-51 Mustang nor will it be engaging Focke-Wulfe FW-190s," Flanagan asserted.[79] But Hendrickson's, Flanagan's, and the other GIBs' opinions went unheeded. After finally finding a way into a fighter and demonstrating what was achievable with a "professional" GIB, the navigator-WSOs suddenly found themselves excluded from the service's newest, premier fighter. Acknowledging that he was "a little prejudiced," Hendrickson lamented, "I think one of the biggest mistakes we could ever make is build an F-15 with only one seat."[80]

In retrospect, the rationales mustered to justify the Eagle's cockpit configuration and its advanced avionics were curiously circuitous. The F-15 was postured as a dogfighter, which by historic convention meant that it could only have a single seat. But in case the F-15 pilot needed to use a Sparrow missile to chase down a fleeing adversary, the Eagle's proponents argued that it also required a powerful radar, one that in the past required a second crewmember to operate it. Unable (or unwilling) to accommodate the additional crewmember, the Air Force accepted the increased risks associated with developing a highly automated fire control system and a complex and expensive pulse-Doppler (PD) radar to meet the "one-man operability" requirement.[81] But then, even after McDonnell Douglas's winning design demonstrated that a second seat could be accommodated, the Air Force continued down its preferred path.

The new PD radar developed for the Eagle would not rely on a GIB to manually adjust the radar gain or mentally integrate the movements of a multitude of "fuzzy blips." Rather, the PD radar's unique waveforms and automatic electronic processing would perform the gain control and integration automatically, effectively eliminating ground clutter and false returns from the radar scope to provide a true "look down" capability. The F-15 avionics committee boasted that finally fighter pilots would possess a radar "that displays only true targets and not false targets, noise, ground clutter, or large ground point source returns. . . . This clean scope approach" meant that the Eagle pilot theoretically could "scan the radar scope only as frequently as he'd normally scan the rest of his cockpit instruments and still be able to readily recognize a radar target when it appears."[82] Momyer apparently agreed. When given the opportunity to compare the cluttered radar film from an F-4E with the clean images produced by the F-15's radar, Momyer "became convinced that TAC pilots had to have this new pulse-Doppler radar in the F-15 if the objective of one-man operability was to be achieved."[83]

As the prime contractor, McDonnell Douglas was given authority to select the radar for the F-15. Westinghouse, which produced radars for the F-4, and Hughes both submitted bids in response to the Air Force's August 1968 solicitation. The contract called for a radar capable of detecting "small, high-speed objects" from high altitude "down to 'tree-top' level" at long ranges, as well as the capability to automatically acquire targets in the short-range environment. After conducting over one hundred flights with each radar in an airborne test platform, McDonnell Douglas selected Hughes's PD radar in September 1970, one month after the Ad Hoc Committee for F-15 Avionics and Armament released its recommendations. The new Eagle radar became known as the AN/APG-63.[84]

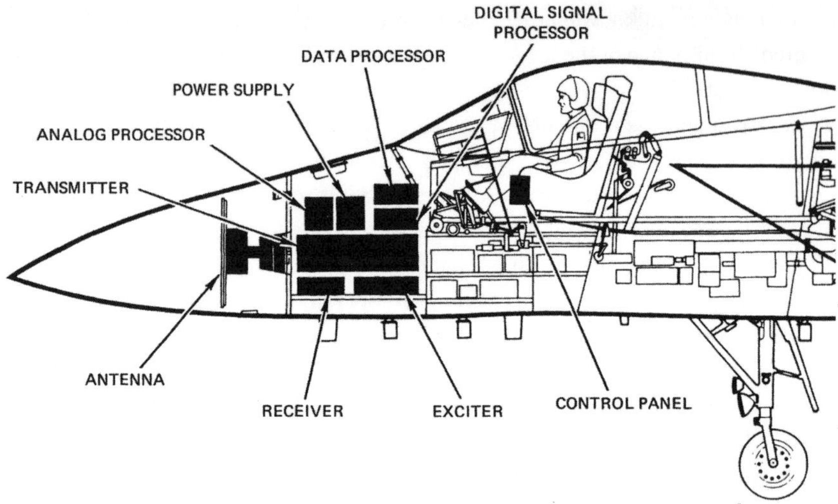

FIGURE 5.6. APG-63 Radar in the F-15. Hughes's APG-63 pulse Doppler radar was advertised as the "Eyes of the Eagle." Unlike the equipment in the Sabre and Phantom, the Eagle's radar components were designed to be easily accessible and independently replaceable. In addition to its massive antenna, the radar's other black-box components dominated the forward section of the fuselage, occupying nearly as much space as the pilot seated in the cockpit. The Eagle's central computer, which linked the pilot to the aircraft's radar and weapons, was located in the compartment beneath the pilot immediately aft of the radar equipment. Reprinted with permission from the Raytheon Company.

Reflecting the Air Force's earlier pronouncements, Hughes advertised its APG-63 as the "Eyes of the Eagle" and "the key to the F-15 one-man operation in all air superiority missions." The APG-63 took full advantage of McDonnell Douglas's original plans for a thirty-six-inch diameter planar radar antenna mounted at the nose of the aircraft, which consequently dictated the massive size of the rest of the airframe. The antenna itself was mounted on a three-axis gimbal to improve its tracking performance during aggressive F-15 maneuvers (figure 5.6). In long-range scenarios, Hughes's radar engineers promised the pilot a "clutter-free radar display" and detection ranges of "more than 100 miles." For close-range combat, the pilot would have the option of using two auto acquisition modes in conjunction with his HUD. Harkening back to the original promises of the A-1C(M), Hughes proudly predicted that the "automatic target acquisition and tracking" of its APG-63 would leave "the pilot free to press the attack."[85]

That promise, however, assumed that the F-15 pilot could successfully navigate the eight separate controls located on the throttles and the additional five buttons located on the control stick that operated the aircraft's radar and weapons, all of which needed to be actuated in separate, precise sequences based on specific engagement parameters.

In the Phantom, Olds had observed that a GIB needed a "bit of dexterity" to properly manipulate the F-4's controls. In the Eagle, the pilots would need to become near-virtuoso "piccolo" players.[86]

And Lots of Automation

"It is true the throttles are designed for busy hands," Captain Robert Hoag acknowledged in his summer 1974 article on the F-15 "Superfighter" in *Fighter Weapons Review*. "However," he continued, "you do not need to be double-jointed or a contortionist to operate any of the controls." Hoag had heard grumbling in the fighter pilot ranks over the multitude of switches and buttons arranged on the Eagle's pair of throttles, and his statement was an attempt to assuage their concerns. "The F-15 cockpit design and pilot procedures are simple and uncomplicated and require no more than average physical dexterity and common sense," he assured.[87] Another initial F-15 pilot also sought to dispel any rumors, explaining that the Eagle's cockpit had been designed "for the man who is going to use it, not to satisfy an engineer sitting behind a desk."[88] Nonetheless, even those like Anderegg who were transitioning from the Phantom to the Eagle quickly discovered that "learning how to use all these switches in concert required much practice." Fortunately for them, flying the new airplane did not.[89]

The engineers at McDonnell Douglas understood that if the pilot was expected to single-handedly *employ* the F-15, he probably would need some help *flying* it. The pilot, they reasoned, had to be able to "confidently maneuver to the limits of his flight envelope" without worrying that a slight misapplication of the flight controls would induce an unrecoverable spin.[90] To provide the necessary stability without degrading the aircraft's maneuverability, the engineers opted to use two separate flight control systems in the Eagle. The first was a classic hydromechanical system that, according to the flight manual, relied on "mechanical linkages and hydraulic actuators to position" the aircraft's standard aileron, stabilator (all-movable horizontal tail), and rudder flight control surfaces.[91]

The F-15's second flight control system was a predecessor to the fly-by-wire systems incorporated in later military and commercial aircraft. The control augmentation system (CAS) used force sensors in the control stick and rudder pedals to generate electrical signals. These signals were then transmitted directly to the individual flight control surface actuators and "superimposed" onto the commands that arrived via the hydromechanical system. By modifying "the control surface deflections," CAS refined the aircraft's response to the pilot's physical movement of the control stick and rudder pedals. Although the flight manual did note that "with the CAS on, the aircraft is fully controllable with the loss of any or all mechanical linkages," the electronic flight control system was not designed for mere redundancy.[92] Rather, it was an essential component of the Eagle's flight control system that worked in combination with the mechanical pushrods to provide "an optimum balance of control sensitivity" at all airspeeds and flight regimes.[93]

However, like Evans's reservations with the hydraulically boosted flight control system in his F-86E decades earlier, several F-15 pilots expressed concerns about relinquishing control to the revolutionary electronic system in their aircraft. "What happens when some berserk electron decides to go on a rampage?," they wondered. Dismissing their fears, Hoag explained that the two-channel, three-axis CAS was "extremely reliable" and incorporated safety features to guard against such random behavior. While acknowledging that the CAS did sound like "black magic" or "something out of *Star Trek*," Hoag considered the smooth flying results indisputable. "CAS-ON is the ultimate in aircraft power steering," he wrote. "Simply put, the nose of the aircraft goes where the pilot puts the stick, with the CAS telling the flight controls how much to deflect and when."[94]

The electronic CAS was only one of many features designed to simplify the task of *flying* the Eagle. The automatic positioning of the engine inlets described earlier was another. Collectively, this automation freed the F-15 pilot to devote more of his attention to *employing* the F-15's fire control system. As noted earlier, considerable emphasis was placed on optimizing the aircraft's radar and weaponry for "one-man operation," and the resulting system was frequently described "as simple as 1-2-3. 1-Look; 2-Select; 3-Fire."[95] While not discounting the remarkable improvements in the APG-63 PD radar and its ability to provide a clutter-free scope, in actuality F-15 long-range radar and weapons employment remained a multistep process essentially unchanged from that of the Phantom. The Eagle pilot still had to set up his radar, see a target appear on the radar scope, maneuver the radar's "acq symbol" over that target, lock it, and then assess his weapons employment options. Although he benefited immensely from more intuitive displays, an improved cockpit layout, and better automation and mechanization in the fire control system, it was still a somewhat cumbersome procedure that demanded significant attention be directed toward the radar scope.

Similar to the Phantom, the F-15 radar in its long-range search mode used a B-scope, which was displayed on the Eagle's four-inch-square vertical situation display (VSD) located in the top left portion of the instrument panel (figure 5.7). However, unlike the Phantom GIB who had to interpret the raw returns on his radar scope, the Eagle pilot instead saw only digitally processed representations of radar targets. There were no "fuzzy blips" on the F-15 VSD, only distinct, synthetically generated target "bricks." Furthermore, only those radar returns that exceeded a variable signal-to-noise ratio controlled by the radar's automatic gain control were displayed. The result was "nothing but moving targets," the Eagle's "spread the word" briefing reported, "no clutter or anything else to confuse the display."[96] Some pilots initially rebuffed the radar's digital processing, complaining that the "highly automated" and "artificially generated" binary targets—"either present or not present" on the display—made it more difficult to "determine signal strength" and limited the pilot's ability to assess the

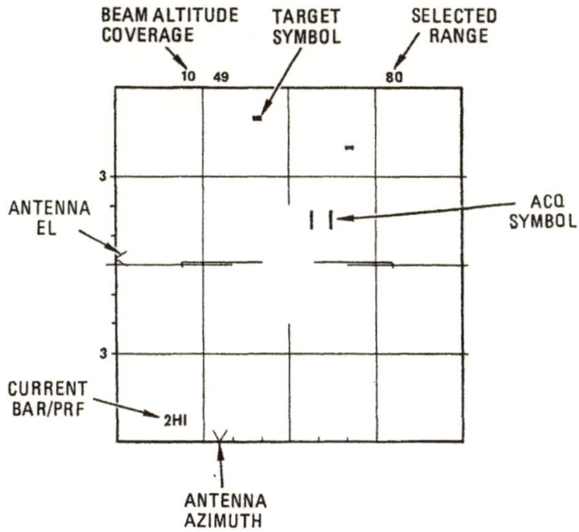

FIGURE 5.7. F-15 Radar VSD in Search. Unlike the raw returns displayed on the Phantom's radar scope, the Eagle's VSD presented digitally processed representations of radar targets. It used a B-scope format as in previous fighter radars, with the maximum scope range (displayed in the upper right corner) selected by the pilot. Additional information, to include the radar's altitude coverage at acq symbol range, was displayed along the VSD borders. In this example, the acq symbol is at fifty miles, and the radar at that range is searching from ten thousand to forty-nine thousand feet in altitude. NMUSAF.

operation and performance of his radar.[97] But Sokol, like many others, appreciated the clarity; comparing the F-15's radar scope to that in the Phantom, he said it was "like the sun just came out."[98]

All told, the F-15's radar set control panel (figure 5.8), which was located on the outboard side of the aircraft throttles, had 35 percent fewer knobs and switches than that used in the Phantom (figure 4.16). Notably, there were no manual gain control knobs for the pilot to tweak. Rather, in marked contrast to the GIB's experience of flipping switches and turning dials on the Phantom's two radar control panels during the heat of battle, the pilot in the Eagle usually set up his single radar control panel on the ground and then rarely touched it again once airborne. If he had set the F-15's radar mode control to its normal AUTO position, the pilot could adjust most radar settings simply by selecting a different weapon using his HOTAS.[99]

Normally during long-range radar operations, the F-15 pilot used a multiple-bar elevation scan that allowed him to search a larger volume of airspace. The radar also automatically interleaved each horizontal sweep of the antenna with alternating high- and medium-PRF (pulse repetition frequency) waveforms to better detect a variety of targets: high PRF was optimized for detecting and tracking targets approaching the

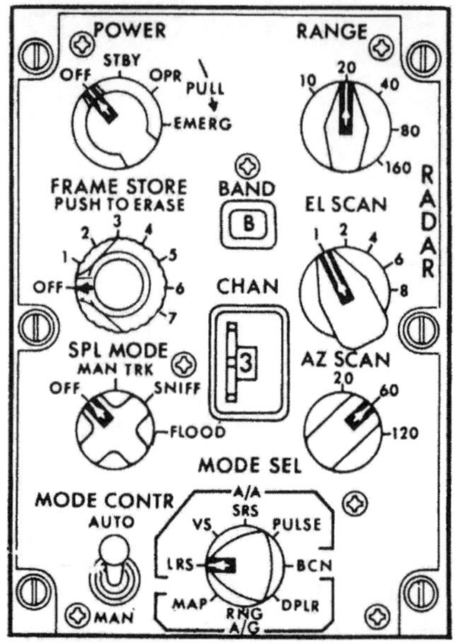

FIGURE 5.8. F-15 Radar Set Control. The Eagle's Radar Set Control panel, located on the left side of the cockpit outboard of the throttles, had fewer knobs and switches than the Phantom's. Critical radar functions could also be adjusted using separate controls mounted on the aircraft throttles and stick. The F-15 pilot usually configured the control panel on the ground and then rarely touched it again once airborne. NMUSAF.

F-15; conversely, medium PRF was better for low-altitude and receding targets. The Eagle's central computer (CC) computed the radar's altitude coverage in search and displayed that information to the pilot, effectively eliminating the mental trigonometric calculation requirements that used to be purview of the GIB. To adjust the radar's altitude coverage, the pilot slewed the antenna up or down using his pinkie finger and the spring-centered elevation control wheel located on the top, far edge of the left throttle (figure 5.9). Unlike the Phantom, the Eagle's antenna elevation carat located on the left side of the VSD was a reliable indicator of the radar antenna position. Additionally, the Eagle's radar display was less cluttered than the Phantom's scope, with no distracting vertical line sweeping back and forth in azimuth, just a small carat moving along the bottom of the VSD.[100]

Once the F-15 pilot noticed that a brick had appeared on the VSD, he needed to move the radar acq symbol over the target to command a lock-on. For that, he used the target designator control (TDC) located on the right throttle. A slightly dimpled disk approximately the size of a penny, the TDC normally rested under the pilot's middle finger. The pilot used the TDC as "an isometric positioning device" to slew the acq symbol across the VSD in azimuth and range. Hoag claimed that the TDC, the "doofer that moves the acquisition symbols around," was so precise that "with a few seconds [*sic*] practice, you can write your name on the scope using this switch."[101] Once the acq symbol was positioned over the target brick, the pilot depressed the TDC to initiate the lock-on process. Notwithstanding Hoag's characterization, pilots quickly learned that it took a fine touch to manipulate the TDC and smoothly command a

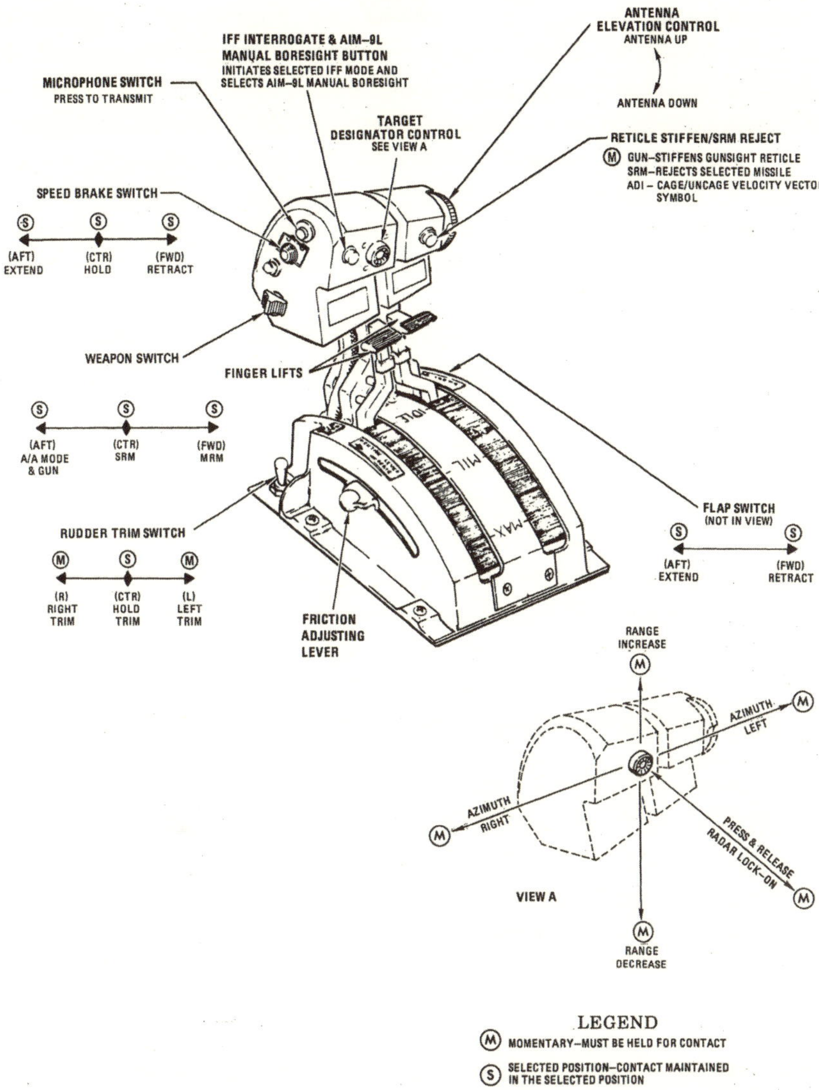

FIGURE 5.9. F-15 Throttle Quadrant Controls. One pilot described the Eagle's controls as requiring only "average physical dexterity and common sense," but others said that operating the maze of control functions was like "playing the piccolo." Just as they had once learned to anticipate their response to a target's maneuvers, successful fighter pilots would eventually learn to anticipate the proper radar and weapons modes, and the necessary switch actuations to activate them. NMUSAF.

lock. Anything other than a straight depression would cause the acq symbols to slide away from the target and disrupt the lock-on process. But, with each target brick synthetically generated, at least the pilot no longer needed to first adjust the radar elevation coverage to achieve maximum target brightness. If a target was on the Eagle's VSD, it could usually be locked, immediately.[102]

Once the radar was in track, the F-15 pilot benefited from a wealth of target information (figure 5.10). Sokol explained that just having the target's altitude readily displayed was a revelation.[103] The target's airspeed, G-loading, and aspect angle (the relative angular position of the F-15 to the target) were all plainly displayed across the top of the VSD. As before, a steering dot and an ASE (allowable steering error) circle, target Vc (closing velocity), and missile range estimations (Rmax1, Rmax2, and Rmin) were also provided to help the pilot navigate to an acceptable weapons launch position. In the past, it would have been the GIB's responsibility to compute, monitor, and communicate these target parameters as the front seater executed the intercept and visually scanned for the target.[104] Now, the single-seat F-15 pilot was responsible for doing everything himself.

Additional automation in the form of a HUD was incorporated into the Eagle's design to aid him. Although the external hardware appeared similar to that used for the optical sights in the F-86 and F-4, the HUD in the F-15 was much more than just a gunsight.[105] In its normal configuration, the HUD displayed the F-15's current airspeed, heading, altitude, and attitude (figure 5.11, left side). It also included a visual indication of the current weapon selected, such as the MRM (medium-range missile) reference circle, and a count of the remaining missiles. Some pilots, including Welch, reported being so overwhelmed by all the information now displayed in front of them that they preferred turning off the HUD's heading, altitude, airspeed, and attitude "scales" to simplify the display (figure 5.11, right side).[106]

Whenever the radar was locked to a target, additional information from the VSD, such as missile employment cues and the target's range and Vc, would be displayed in the HUD.[107] A target designator (TD) box also appeared with a radar lock. One of the pilots' favorite features of their HUD, the TD box visually highlighted the position of a target based on the radar's azimuth and elevation tracking angles. Describing its benefits, Anderegg explained: "To get a tally [visual detection of the other aircraft], all the pilot had to do was point [his aircraft] in the general direction of the target, look through the TD box, and there was the adversary!" Another pilot declared "the TD box the greatest thing since sliced bread."[108] Hoag likened the nondescript square box in the HUD to "a poor man's TISEO"—the optical identification system installed on the F-4E.[109] Still, despite the wealth of data now presented within eyeline of the pilot, not *all* of the target's information could be displayed on the HUD, and the pilot still needed to periodically look inside his cockpit and monitor his VSD.

All of the information presented on the HUD and VSD was computed and processed by the Eagle's central computer, housed underneath the pilot in one of the forward,

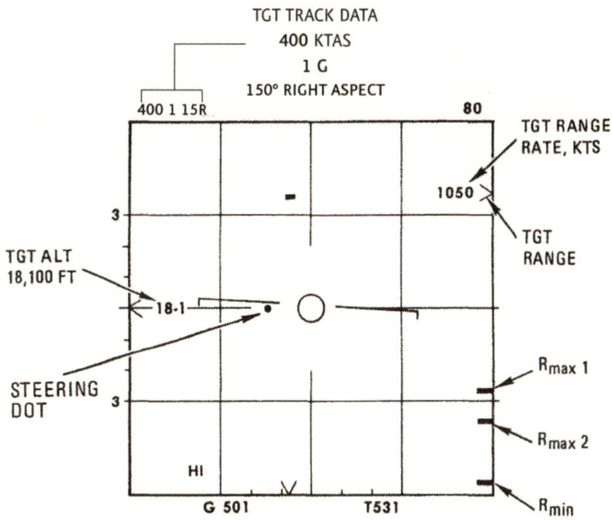

FIGURE 5.10. F-15 Radar VSD in Track. The Eagle's central computer provided a wealth of information about a radar target once the pilot locked it. The target's true airspeed, G-loading, and aspect angle were all displayed along the top left of the VSD; its closure rate on the right side; and its altitude of the left side (in past aircraft, fighter pilots or their GIBs had to mentally compute a target's altitude using trigonometry). The F-15's own ground speed and true airspeed were displayed along the bottom. The target in this example is eight degrees left of the nose, about sixty-three miles away, and nearly level with the F-15 at 18,100 feet. NMUSAF.

left avionics bays. Described as "amazingly sophisticated" and "very intelligent," the 16k, 34-bit, 340,000-instructions-per-second computer also calculated the aircraft's missile launch zones, intercept steering, and gunnery solutions. Equally significant, the CC eliminated the loathsome requirement from the Phantom to verbalize and then fumble with multiple switch actuations to change radar and weapons modes. Instead, the F-15 pilot simply repositioned the weapon select switch at the bottom of his right throttle from MRM to SRM (short-range missile) or GUN. The CC automatically reconfigured the radar for the new mode (when the radar was set to AUTO on the radar control panel) and commanded the aircraft's armament control system (ACS) to prepare the appropriate weapon. For all practical purposes, the human GIB in the Phantom had been replaced by the 48½-pound CC in the Eagle. According to Hoag, "This small, light-weight, high-capacity computer . . . makes the F-15's one-man operation a reality."[110] Just to assuage any pilots' concerns about losing more cockpit authority to yet another electronic black box, Hoag affirmed that "the pilot still owns the *GO—NO-GO* responsibility." The CC could tell the pilot when to shoot—it would display an "in range" cue on the HUD and VSD—but the pilot always had to press the

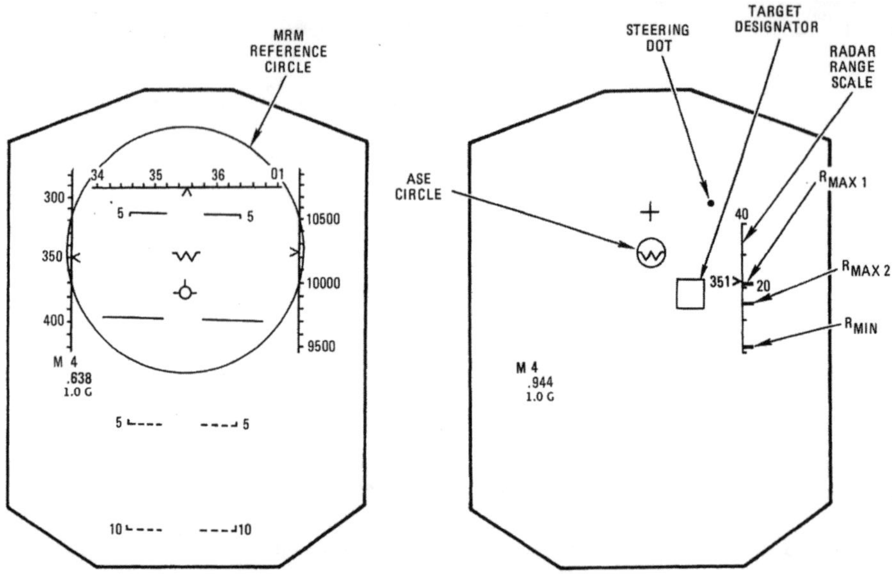

FIGURE 5.11. F-15 HUD in MRM Mode (Radar Search [left] and Track [right]). More than just a gunsight, the Eagle's HUD provided the pilot with a mix of aircraft performance and weapons data in a readily accessible format. One of the pilots' favorite features, the target designator box, indicated where the pilot should look to visually detect a locked radar target. In the HUD image on the left, the aircraft is at 1-G, pointed about two-degrees above the horizon, heading 355°, flying at 350 knots or 0.638 Mach, at an altitude of 10,250 feet. The pilot has MRM selected on the weapons select switch and has four MRMs (AIM-7 Sparrows) aboard. In the HUD image on the right, the pilot has selected symbols reject, which removed the airspeed, heading, altitude, and pitch attitude scales to declutter the display. The pilot has a radar lock on a target located along the azimuth and elevation of the TD box. According to the carat position on the radar range scale, the target is about twenty-miles away, closing at 351 knots. The F-15 is approaching Rmax1 for this target. NMUSAF.

pickle button for a missile to launch.[111] Left unsaid was that, although the pilot retained ultimate control, a failure of the electronic CC would severely cripple the F-15's offensive employment capability.[112]

The automated data processing and improved displays in the Eagle certainly simplified critical elements of the long-range intercept. "Suddenly, your job was to integrate real information," Welch explained, "not to pick a bit of information out of a lot of non-information," as in the Phantom before.[113] However, actual employment of the radar-guided Sparrow remained roughly equivalent to that of the F-4. If the F-15 pilot had obtained a radar lock and had MRM selected, he fired a missile by depressing the weapon release or "pickle" button located on the top of the control stick (figure 5.12). He then needed to maintain the radar lock throughout the Sparrow's entire time of flight, but he now had the benefit of a countdown of the missile's estimated remaining flight time displayed for him in the HUD. If the F-15 pilot ever needed to break lock and return the radar to search, he depressed the auto acquisition switch located on the control stick.[114]

The other functions of the auto acquisition switch, besides releasing the in-flight refueling receptacle, were designed for use in the short-range environment. It was in these close-combat situations that the Eagle's automation and fire control switchology diverged from the Phantom's model to provide true "head out of the cockpit" employment capability.[115] It's also when the F-15 pilot's practice "playing the piccolo" became apparent. The fighter pilot had to learn to anticipate not just the enemy's actions and how he would maneuver to follow them but also which weapon (selected on the throttle) and which auto acquisition radar mode (selected on the control stick) he needed to use to maximize his offensive capability. In short, the intuition that had previously guided a fighter pilot's maneuvers during a dogfight now had to be extended to the switch actuations localized within his cockpit.

For example, if the Eagle pilot visually detected an adversary and did not already have a radar lock, he needed to instantly decide which auto acquisition mode was most appropriate for the situation. He had two options. One was boresight, a mode similar to that originally intended for use in the Phantom, except that it actually worked in the Eagle. The pilot initiated boresight by thumbing forward on the auto acquisition switch. The radar would then begin staring directly ahead of the aircraft, searching in range from five hundred feet out to ten miles. A four-degree reference circle was displayed in the HUD to aid the pilot as he maneuvered to point the aircraft, and thereby the radar, at the target (for reference, the MRM reference circle in the left image of figure 5.11 is twelve degrees). The radar would lock the closest target in range and then automatically transition to full system track, indicated by the sudden appearance of the standard radar track information in the HUD. If there were multiple targets in range and the radar locked the wrong one, the pilot could depress the auto acquisition switch to break the radar lock, adjust his aircraft position slightly, and then thumb forward again to reinitiate boresight. Or, while he still had a lock, he could simply thumb forward on the auto acquisition switch to reject the first radar target and skip to the next closest target in range.[116]

If the pilot did not want to use the boresight radar mode, he could thumb aft on the auto acquisition switch and initiate supersearch. In supersearch, the radar searched an area roughly coincident with the HUD field of view, twenty degrees by twenty degrees, from five hundred feet out to ten miles, the same range as boresight. A twenty-degree reference circle was displayed in the HUD to help cue the pilot. The

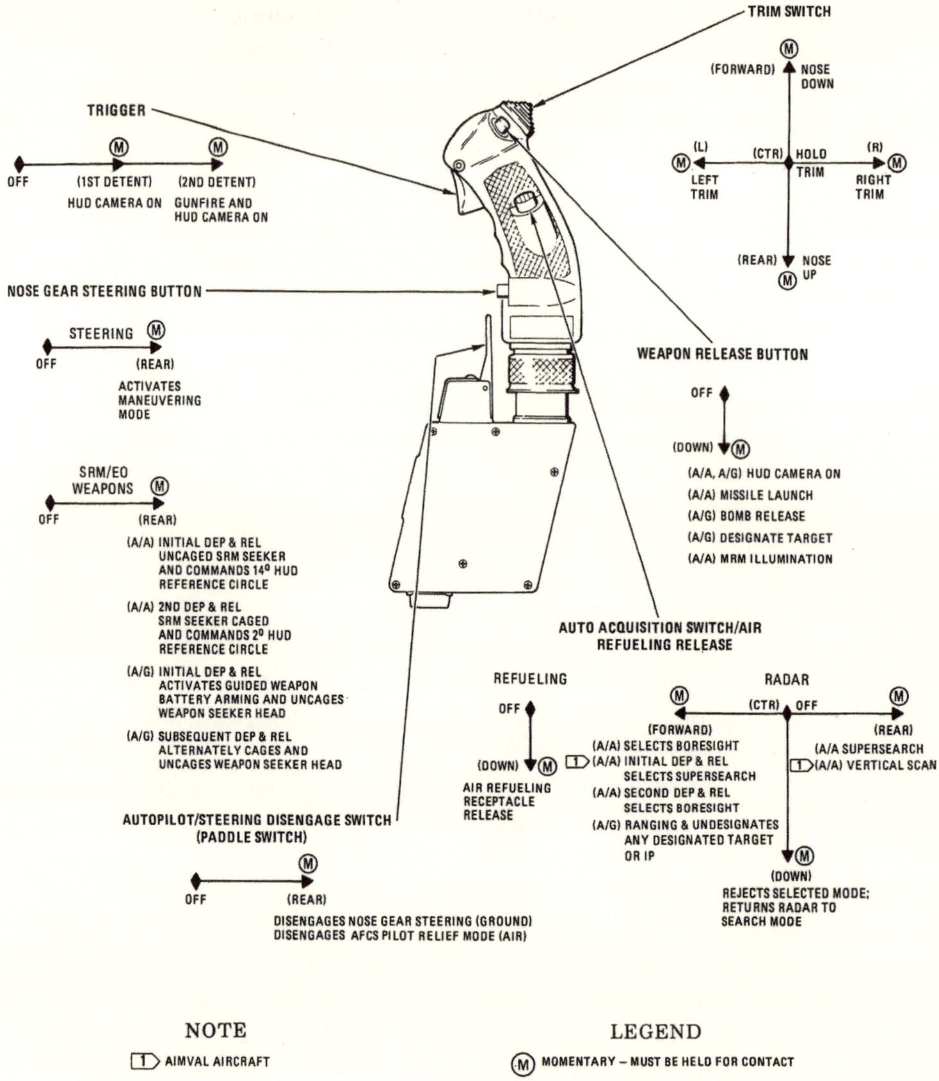

FIGURE 5.12. F-15 Control Stick. The complexity of the controls arranged on the Eagle's throttles (figure 5.9) was matched by those on the control stick. Entries preceded by (A/A) were functions available in air-to-air mode; entries preceded by (A/G) were functions available in air-to-ground mode. The functions identified with an arrowed-1 were added specially for the AIMVAL tests. The large box at the base of the control stick housed the stick force sensor used by the CAS flight control system. NMUSAF.

advantage of supersearch was that the pilot didn't need to point precisely at the target to obtain a radar lock, but it was also therefore less discriminatory and not particularly useful when approaching a swirling dogfight involving multiple aircraft. Additionally, because the supersearch search area was larger than that of boresight, it generally took longer to detect a target and resolve a radar lock. As in boresight, the radar would lock the closest target in range, and thumbing aft on the auto acquisition switch with a lock would cause the radar to break lock and skip to next closest target in range.[117]

There were a few "gotchas," though. If the pilot entered supersearch from normal radar search, depressing the auto acquisition switch to reject the lock or discontinue supersearch would return the radar to normal search. However, if the pilot had thumbed forward to select boresight and then thumbed aft to select supersearch, perhaps because the target aircraft was exiting the four-degree boresight circle, and then he tried to reject the lock by depressing the auto acquisition switch, he would return not to normal search but instead to boresight. Furthermore, although much faster than in the Phantom, the Eagle's radar could still take more than two seconds to resolve a lock and for the CC to update the displays on the HUD and VSD.[118] In that brief interim, a pilot could easily get out-of-sync with his switchology as he raced to get a radar lock in a dynamic and high-stress, short-range combat environment. Purposeful actuations of the various switches and buttons were vital to avoid getting entangled in the maze of HOTAS functionality. Proficiency was critical, too.

As the pilot was working to get a radar lock, he needed to anticipate which weapon would have the highest likelihood of success based on the engagement parameters—Sparrow, Sidewinder, or gun. He selected the weapon type using the three-position MRM-SRM-GUN weapon select switch located on the right throttle. The pilot had to be careful, though, because if he didn't already have a radar lock, adjusting the weapon select switch could also change the radar configuration and alter the VSD display. If he had a lock, then adjusting the weapon select switch would have no effect on radar settings, but it would still alter the symbology in the HUD.[119]

Unlike the radar-guided AIM-7, the AIM-9 could be employed with or without a radar lock on the target. To use the Sidewinder, the F-15 pilot first selected the middle SRM position with the weapon select switch, which caused a two-degree AIM-9 seeker circle to appear in the center of the HUD. The pilot then maneuvered his aircraft to position the seeker circle cue over his intended target. Depressing the nose gear steering button at the bottom of the control stick commanded the AIM-9's seeker to uncage and start tracking the target's heat source. The pilot still needed to listen for the trademark Sidewinder "growl" that indicated that the missile seeker had found and was tracking a target. If he suspected that his Sidewinder missile was malfunctioning, the pilot could switch to a different AIM-9 by depressing the SRM reject button on the left throttle. If the missile seeker failed to track the target, the pilot could recage the missile by reactuating the nose gear steering button and repeating the uncage process on the appropriate heat source. After verifying that he had a good missile "tone" and assessing that he was in range and within proper firing parameters (if he had a radar lock, proper range and firing parameters would be indicated by the CC displaying an "in range" cue in the HUD), the pilot could then launch the missile using the weapons release button on top of the control stick.[120]

Gun functionality was present regardless of the position of the weapon select switch, a reflection of the fighter pilots' traumatic experience in Vietnam and the importance they subsequently ascribed to always having a weapon available for immediate use. If the F-15 was armed, anytime the pilot squeezed the trigger, the 20-mm gun would fire.[121] By selecting GUN with the weapon select switch, the pilot only commanded the CC to display the gunsight in the HUD and begin calculating a gunnery solution (figure 5.13). If the radar was locked to a target, the CC would use the target's range in its computations. If there wasn't a radar lock, the CC would calculate a gun solution using a fixed target range of 2,250 feet; the pilot could adjust that range input to 1,000 feet if he depressed the reticle stiffen button on the left throttle (the same button used to reject a bad Sidewinder when the weapon select switch was in SRM).[122]

In practice, the Eagle's lead-computing gunsight was remarkably similar to the A-1C(M) used decades earlier. The gun reticle moved within the HUD according to the amount of lead required, and the pilot simply needed to maneuver his aircraft to place the target under the 2-mil pipper before he squeezed the trigger. Momentarily tracking the target before firing improved the pilot's chances of scoring a hit. There were some improvements to the gunnery display, however, that provided additional information to the pilot. For example, if a radar lock was available, the range to the target was displayed using a clock position format on the gun reticle (the target in figure 5.13 is at approximately three-thousand-feet range). Later, a small dot was added to the outside of the gun reticle display to indicate the expected maximum effective range of the bullets based on the altitude and airspeed of the engagement. Also, if the CC was in the process of updating the gun solution, a lag line would appear inside the gun reticle, indicating the direction the reticle was moving as it made its corrections.[123]

To complement its internal gun and CC-calculated gunsight, the Eagle also included a gun camera. Mounted inside the cockpit and peering through the HUD, the camera allowed the pilots to capture not only the effects of their gun attacks but also the HUD-displayed weapons parameters from the attack, be it gun or missile (figure 5.14). For this reason, the camera was sometimes alternately referred to as a HUD camera. Anytime the pilot squeezed the trigger or depressed the pickle button, the camera would automatically start filming, and it would continue recording for an adjustable time duration after the trigger or pickle button had been released.[124] The pilots in the 1st TFW at Langley initially had grand plans to use the Eagle's enhanced film documentation to better track their aerial prowess by charting their "weapons firing

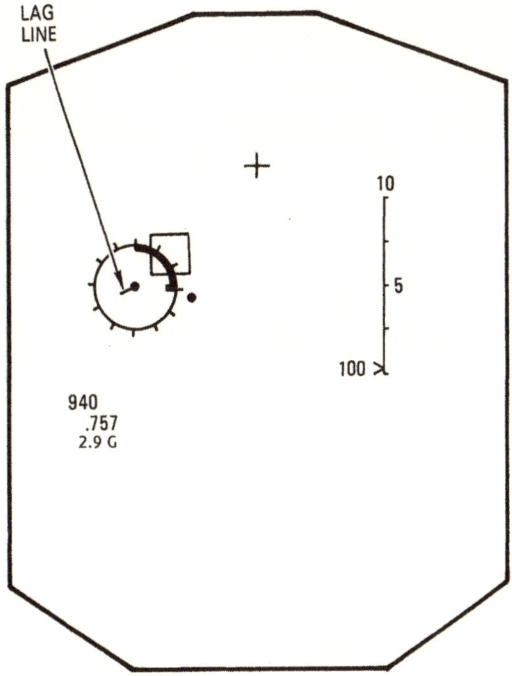

FIGURE 5.13. F-15 HUD in Gun Mode (Radar Track). The Eagle's gunsight display functioned similar to that of the Sabre's A-1C(M); the projected reticle moved in the HUD field of view to command the appropriate lead-fire solution. The pilot maneuvered his aircraft to place the target under the 2-mil pipper at the center of the reticle before he squeezed the trigger. Similar to the HUD image in figure 5.11 (*right*), the pilot has selected symbols reject on the HUD control panel to declutter the display. The "940" in the bottom left corner of the HUD indicates that the pilot has 940 rounds remaining in the gun. NMUSAF.

attempts vs [the probability of] success."[125] But issues with the gun camera remained. First, the aircrews had to get their film developed and processed in a timely manner if it was to be of use. Then, even with the developed film in hand, the images often lacked a critical target parameter needed to judge the weapon's success because not all of the target information from the VSD was displayed on the HUD. Finally, the camera system still had no way to record audio. To help alleviate the last deficiency, F-15 pilots began carrying portable cassette recorders into the cockpit to record the radio transmissions, just as they had done before in the Phantom.[126]

One of the Eagle project officers remarked in mid-1969 that, because "visual combat maneuvering will be with us for a long time," the F-15's designers had aspired to provide a "high degree of automaticity and simplified operation in the cockpit" without excessively "burden[ing] the aircraft with any avionics gimmickry which imposes weight penalties and increases costs."[127] The Eagle's different short-range radar auto

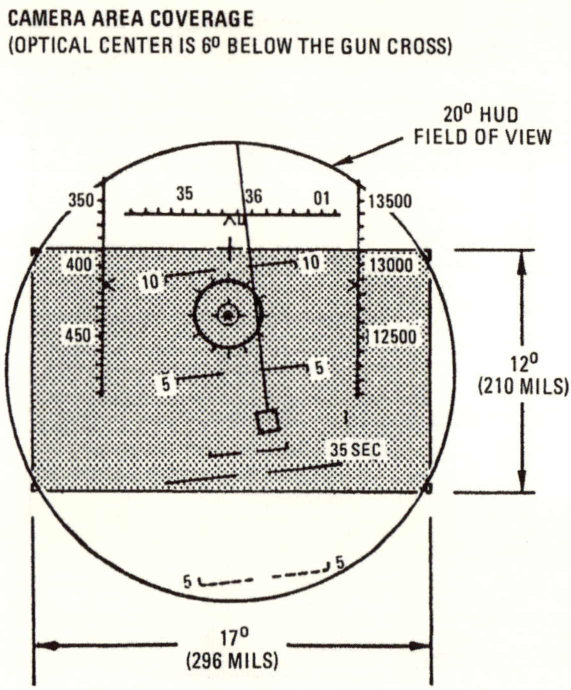

FIGURE 5.14. F-15 HUD Camera Area Coverage. The F-15 included a camera mounted inside the cockpit aft of the HUD. Using the HUD film to assess gun employment postmission was usually straightforward. However, because the HUD did not always display all the relevant target data, assessing missile employment using the HUD film could be more challenging. The recorder also did not have an audio capability, so pilots carried portable tape cassette recorders into the cockpit with them to capture the radio transmissions. This image displays the HUD camera coverage (*shaded area*); the HUD is configured in air-to-ground mode. NMUSAF.

acquisition modes, the HOTAS functionality for selecting them along with the different weapons, the target information that was automatically selected for display in the HUD and VSD, and the camera that recorded the HUD information all reflected the careful balance negotiated by the engineers. The pilots certainly appreciated the attention to detail and the improved capability it brought. Anderegg quoted one Phantom pilot's assessment that only when he sat in an F-15 and "saw what we could do in later years, that's when the travesty of Vietnam came home to me."[128] Anderegg himself remarked that "it probably took me a year of flying the [Eagle] before I stopped saying to myself at least once per mission, 'Well, they got this right.' I mean, all the things that were wrong with the F-4, they were all fixed in the F-15. . . . It was just a joy."[129]

Nevertheless, despite what the glossy brochures had once advertised, few Eagle pilots would ever suggest that employing their new fighter was "as simple as 1-2-3." The F-15 demanded the fighter pilot develop a whole new level of air combat intuition so that he could quickly and properly interact with the aircraft's radar and CC through the various switches and buttons that adorned his control stick and throttles. An Eagle pilot that was out of practice would occasionally find himself accidentally breaking his radar lock while a Sparrow missile was still in flight; or breaking a just-resolving radar lock by impatiently recommanding boresight; or launching a caged Sidewinder missile instead of a Sparrow because he didn't move the weapon select switch from SRM to MRM.[130] To be effective in the air now required considerable, recurring practice processing all the displayed information and manipulating the controls accordingly. Welch suggested that the new skillset reflected the fighter pilot's transition from being a "systems operator" in the Phantom to becoming an "information integrator" in the Eagle.[131] It was likely not the shift the fighter pilots had envisioned when they first heard the claims that the Eagle was going to be *their* fighter.

The F-15 was supposed to be the world's premier dogfighter, but it emerged from the design process a massive aircraft, nearly the size of a tennis court, weighed down by a radar and a suite of avionics so complicated that it took a separate, specialized central computer to help coordinate their activities. Despite the praise that Hoag and others heaped on the F-15, and the justifications mustered to support the Eagle's advanced avionics for its close-combat mission, some fighter pilots wondered how exactly the F-15, other than its single seat, was reminiscent of the small and simple F-86 Sabre of MiG Alley fame. Colonel John Boyd was one of them. Looking at what the F-15 had become, Boyd's biographer explained, "[He] was disgusted. He could tell that his dream for the pure fighter aircraft had vanished."[132] Moreover, a group of defense analysts began to investigate the opportunity costs of building a large, $10 million fighter crammed with expensive and potentially failure-prone equipment. Wouldn't the service and the nation be better off building a smaller, cheaper fighter that had a clearer design connection to history's most successful fighter aircraft? The analysts' question would eventually culminate in the fledgling Eagle's most important trial, not in war over central Europe, but in a pair of tests conducted over the desert north of Las Vegas.

Trial by Test

The joint Air Force–Navy AIMVAL-ACEVAL tests conducted in 1977 constituted the most realistic and comprehensive experimental evaluation of air combat tactics and equipment ever attempted. By the end of the year, the participating pilots had flown approximately sixty-four hundred sorties, and the bill for the pair of tests had eclipsed $150 million, more than triple the original price tag.[133] Promising to "combine real world factors, statistical probabilities, and computer models in [a] single effort to produce a management-oriented evaluation of relative operational utility," Pentagon officials hoped that the two tests would address the emerging "quality" versus "quantity" debate in military aviation.[134] "Quality" advocates had been pushing for extremely capable aircraft like the F-15, knowing that the aircraft's high price tag would preclude procuring significant quantities, but contending that its technological superiority could compensate for the numerical deficiency. In contrast, the "quantity" promoters argued that the Air Force should forsake its unhealthy appetite for technological gadgetry like radar-guided air-to-air missiles and instead steer that money toward acquiring greater quantities of simpler, cheaper aircraft.[135]

Although often unacknowledged today, the AIMVAL-ACEVAL tests shaped the next twenty years of US military investment in tactical aviation.[136] Earlier in the decade, the Air Force had chosen to emphasize close-range air combat when developing and fielding its F-15 Eagle. Following AIMVAL-ACEVAL, the service made an abrupt shift in its procurement strategy and largely abandoned close-range dogfighting weaponry in favor of improved, longer-range radar-guided missiles such as the AMRAAM (advanced medium-range air-to-air missile) and new electronic identification technologies. The lessons of AIMVAL-ACEVAL forced the fighter pilots to adapt their tactics and procedures, too. It was, Blesse observed during his visit to Nellis during AIMVAL, a "strange" set of outcomes from a pair of tests that were originally focused on evaluating "short-range missiles."[137]

Coincidentally, despite the unprecedented level of resources devoted to AIMVAL-ACEVAL, the tests never did resolve the "quality" versus "quantity" conundrum. Indeed, one Senate staffer questioned the tests' utility before they even began, wondering aloud if they would be capable of "really proving the thing they have to prove."[138] His concerns were validated once the evaluations kicked off. There were too many test constraints and the corresponding ambiguous test data could be too easily manipulated to advocate for either the "anti-technology" or "total-technology" positions.[139] Nonetheless, the participants appreciated the tests' ancillary benefits, especially the attention being directed toward developing a better understanding of air combat. Commenting on the significance of the tests, the executive assistant for AIMVAL-ACEVAL, Lieutenant Colonel Robert Fay Jr., told one reporter, "Every perceived principle of air combat I came in here with has changed."[140] Sokol, one of the eight F-15 pilots who flew in the tests, later declared that AIMVAL-ACEVAL accelerated development of air combat tactics in the Air Force by at least five years.[141] Welch, who oversaw a portion of the tests while serving on the TAC staff following his command tour at the 1st TFW, captured both the tests' benefits and issues in his appraisal, explaining that the Air Force and its fighter pilots learned some "pretty darn good lessons out of a . . . very badly flawed, politically motivated, Congressionally directed, horribly expensive test program."[142]

In fact, the early experiences of the Eagle pilots assigned to the 1st TFW at Langley AFB in 1976 and 1977 had already begun to reveal, at least anecdotally, many of the same lessons that were later derived from the data-intensive AIMVAL-ACEVAL program. Foremost was the fact that, after having been promised a modern-day Sabre, the Eagle pilots' best chances at victory and survival were attained not in close-combat but rather when they purposefully avoided the once-iconic dogfight.

Initial Tactics Development

As members of the first F-15 wing not focused exclusively on training new aircrews, the 1st TFW pilots at Langley AFB played an important role developing the Eagle's early air combat tactics.[143] Their sixty-seven-page *Aerial Attack Manual*, released the same time the first few F-15s started arriving at the base in January 1976, set the tone. While the manual's authors acknowledged that the F-15 was "a true fighter pilot's fighter," they also emphasized that "the F-15 is not a magical airplane. It will not automatically maneuver to six o'clock and 'kill.'" Hence, "success in aerial combat" still demanded "preparation" and an aggressive pilot who was "knowledgeable in tactics and procedures." Fortunately for the new pilots, the manual noted that many of the close-range air-to-air maneuvers and the longer-range intercept geometries remained unchanged from those developed earlier for the Sabre and Phantom; "only the techniques to accomplish them are different."[144]

The Langley pilots quickly set about cultivating the new techniques within their ranks. One scenario they used was the "ACE (air combat event) maneuver." Developed to rehearse the "necessary switchology and judgment" required in their new aircraft, the training event had the pilot work the radar and HOTAS to launch a valid (simulated) AIM-7 missile followed by an AIM-9 missile all while simultaneously maneuvering into a position that would permit a stable, "tracking" gun shot against a maneuvering target. The pilot's performance during the exercise was judged on the basis of total elapsed time and assessed weapons validity. The Eagle's new HUD camera helped facilitate the scoring process.[145]

After a few months' training, the 1st TFW pilots had an opportunity to showcase their new aircraft and their fresh skills when the 64th Aggressor Squadron showed up at Langley in April 1976. Welch took special pride in the performance of his wing's pilots and their aircraft during the ensuing training missions. In the past, when the Aggressor squadrons had visited his old F-4 unit, Welch admitted that the specially trained pilots flying their "goddamn little F-5s" simulating Soviet MiG-21s "would wax our ass"—a fighter pilot colloquialism that referenced the opponent's gunsight being all over their aircraft's tail. Moreover, during the mission debriefs, the Aggressor pilots were always "so magnanimous" that it infuriated Welch. During this trip however, facing off against the new F-15s, the Aggressors had their hands full. "We handed them their ass big time," Welch proudly remembered, and "at the out-brief, you cannot believe how magnanimous we were."[146]

Beyond offering a bit of payback for earlier air combat injustices, the training missions against the Aggressors were one of the first opportunities the Langley pilots had to practice employing their Eagles in outnumbered scenarios. The experience yielded an important lesson. Even though the performance of the Eagle far exceeded that of the Aggressors' aircraft, tangling with a skilled pilot in a capable, albeit not equal, fighter aircraft was not always advisable. After the training deployment, the Eagle pilots concluded that if they didn't think they could "kill the bogey" in less than thirty seconds, they were better off exiting the fight to "gain separation, and try[ing] it again."[147] It was a similar conclusion to that reached by the Phantom aircrews years earlier over Vietnam when they began employing their new independent element tactics to compensate for their lack of a viable short-range weapon. The F-15, in contrast, was built for the close-in fight. But the Eagle pilots quickly learned that they could still get shot by an "unseen bogey" if they focused too much attention on gunning down the target in front of them. Rather than engage in a prolonged, swirling dogfight, successful pilots learned to exploit the F-15's superior power and maneuverability to temporarily escape a battle before reentering it from a more advantageous position.[148]

Judiciously deciding when to rely on long-range missiles, when to close for a close-range gun attack, and when to abandon one, combined with the necessary proficiency with the aircraft's radar and weapons systems to seamlessly transition from one tactic to the next, quickly became the hallmark of a good Eagle driver. As one summary report of the training missions against the Aggressors noted, an F-15 pilot needed to be "aggressive, calm, knowledgeable, sneaky, inventive, and patient."[149] While he might have preferred the adrenaline rush of a high-G, sweat-inducing dogfight, ultimately it didn't matter how the pilot killed the bandit, whether by missile or gun—only that he did.

The lesson was reinforced later that year when the Langley F-15s began participating in the large Red Flag exercises being conducted at Nellis AFB. The brainchild of F-4 pilot Major Richard "Moody" Suter, Red Flag was designed to replicate as closely as possible the experience of air combat. Studies had suggested that if a fighter pilot survived his first ten combat missions in Vietnam, he was more likely to survive his entire combat tour. Suter pitched his Red Flag concept as a way to get the young pilots their first "combat" missions in a realistic but controlled training environment. General Robert Dixon, the commander of TAC, agreed. The first Red Flag was held in November 1975, and the postexercise assessments were almost universally positive. TAC began hosting Red Flags several times a year, and they continued to garner rave reviews.[150] After completing a 1976 Red Flag, one F-4 wing commander described it as "the closest thing to actual combat that our aircrews can experience in peacetime."[151]

Langley's Eagles arrived at Nellis in time for Red Flag VI, which began on July 8, 1976.[152] While most fighter units considered themselves lucky if in a year they participated in one or two of the two-week Red Flag exercises, the F-15s from the

1st TFW participated in all four of the remaining Red Flags in 1976 and at least six more the following year. The persistent presence of the Eagles was a benefit to all. The Air Force's non-F-15 units had a chance to witness and appreciate the new fighter's capabilities in simulated combat. For the F-15 pilots, the Red Flag exercises provided a unique opportunity to further develop and refine their tactics. Throughout the exercises, the Langley pilots continued to experiment with different radar settings and weapons employment techniques. They tried flying at various altitudes, stretching from very high to very low, and they also tried out different size formations, ranging from coordinated four-ships down to autonomous singletons.[153]

Whenever the Eagles did fly in formation, the wingmen were not in fighting wing, though. The first pilots who pushed the F-15 through its initial testing had attempted to use the "tried and true" formation, but they quickly concluded that fighting wing was impractical for their new single-seat fighter. For starters, the Eagle wingman couldn't devote sufficient time to operating his radar when he was flying so close to his leader. The problem was exacerbated by the F-15's remarkable maneuverability and its "high turn and acceleration rates. . . . The small time delay due to a wingman's reaction time," the test pilots reported, "often left him well out of the engagement before it began." Plus, when the two aircraft flew close to one another, the wingman's radar transmissions tended to cause "interference" and "induced scope clutter on the leader's radar." Consequently, the early test pilots recommended that future Eagle units adopt a wider "shooter-cover position" or "fluid-two" formation, which was similar to the "loose deuce" concept developed by the Navy for its Phantoms years earlier.[154]

The Eagle's widely spaced formation, which became known simply as "tactical," was soon adopted as the standard within the community (figure 5.15). It could be flown either as a two-ship element or as a four-ship (a pair of two-ship elements). In tactical, the fighters flew nearly line abreast often split by more than a mile, with each pilot responsible for clearing his flight mate's six o'clock position. The wingman was encouraged to take an aggressive vertical "stack" within the formation in order "to increase visual acquisition problems for the enemy." The altitude difference between the fighters also provided options "for immediate split-plane maneuvering" and weapons employment if the Eagles got jumped by an adversary. Forgoing the classic preassigned "shooter" or "looker" responsibilities, both flight leader and wingman were expected to contribute offensively and defensively as the situation presented itself. While there were "inherent maneuvering problems" associated with the miles-wide formation, the pilots developed "specialized, prebriefed procedures" to mitigate the challenges. For example, to turn the formation ninety degrees, the pilot on the inside of the turn was taught to delay his turn until the outside fighter had nearly completed his maneuver; when the turns were properly timed, the two aircraft would roll out on the desired heading in the same line-abreast formation, albeit with the wingman now on the opposite side of his flight leader (figure 5.16).[155]

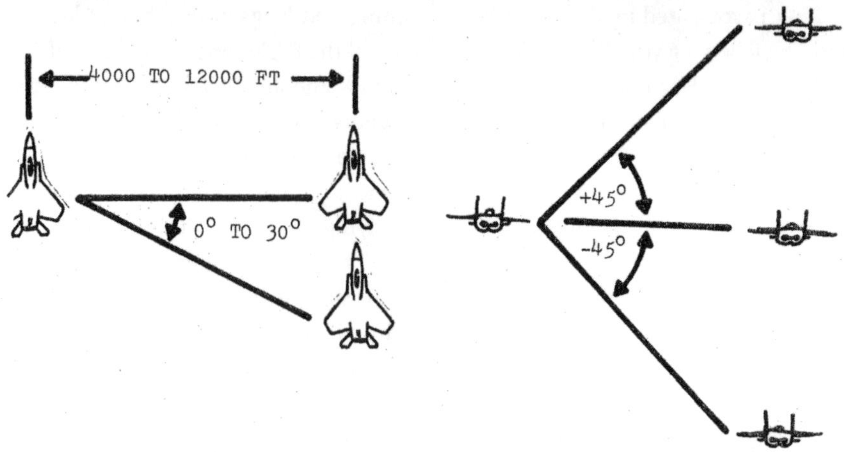

FIGURE 5.15. F-15 Tactical Formation. The first F-15 test pilots tried using the Phantom's fluid four formation, but they quickly determined that it was impractical for their new fighter. Instead they adopted "tactical" as the Eagle's standard formation. The new formation provided additional space and maneuvering flexibility for the wingman to operate his radar. It also eliminated the strict flight lead-shooter and wingman-looker responsibilities; both fighter pilots were now responsible for covering each other's vulnerable six o'clock position while they worked their radars. "Tactical" was functionally similar to the Navy's "loose deuce" introduced years earlier during Vietnam. NMUSAF.

Unlike the rancor that had erupted years earlier when some Phantom pilots expressed interest in adopting a non–fighting wing formation, the transition to tactical in the F-15 was surprisingly affable. Sokol witnessed the adjustment firsthand. Before being assigned to Nellis and then roped into supporting ACEVAL, he was one of the first F-15 instructor pilots at Luke AFB. Sokol remembered fondly, "We had a clean sheet of paper. We didn't have any rules. . . . We didn't have any generals that had been in F-15s and 'knew' how to do it." Consequently, he and the other young instructors, emboldened by the F-15's test reports, decided to throw "that fighting wing crap out with the garbage. . . . The logic was we're spending so much money on all this . . . radar [equipment in the Eagle], . . . so let's let a guy use it."[156] Coincidentally, the F-4 community, finally freed from the exigencies of the Vietnam War, was also undergoing a tactics refresh that eventually led to the demise of fighting wing as their preferred combat formation.[157]

Welch noted that the advantages of the new "tactical" formation were immediately obvious to the pilots flying at Langley. "If we had spread wide enough [apart in formation], there was a good chance [the opponent] would miss one of us," he explained. Plus, it gave both pilots the freedom "to exploit their airplane" and take full

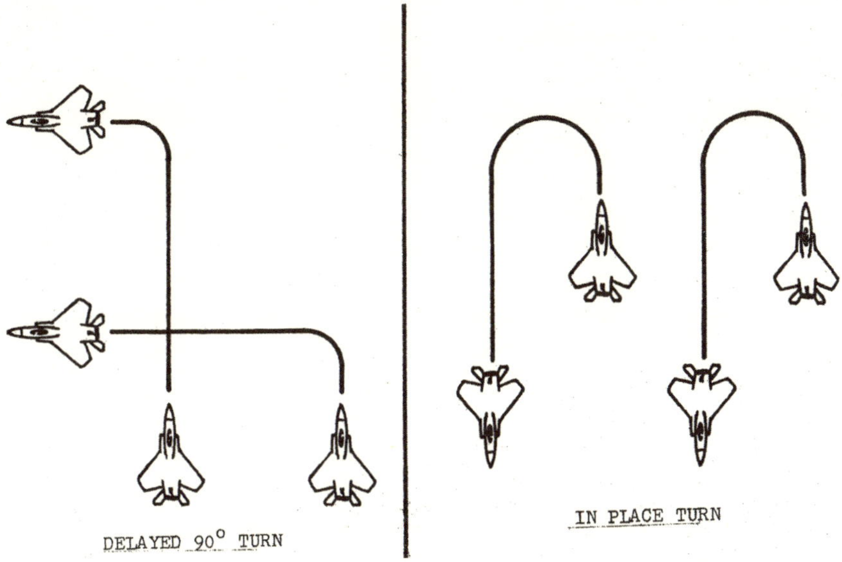

FIGURE 5.16. F-15 Tactical Formation Turns. To overcome the complications associated with maneuvering the wider-spaced formation, the Eagle pilots used "delayed" and "in place" turns. They would later develop techniques to execute the turns "comm out"—without a preemptory radio call. NMUSAF.

advantage of the Eagle's superior maneuverability without having to worry about immediately running into a flight mate.[158] Sokol recognized the implication embedded within the new formation: "Now the task is for *both* guys to use their head *and* their radar."[159]

Whether flying alone or in tactical formation with another Eagle, a consistent lesson learned in nearly all the missions at Red Flag, as well as the training flights off the coast of Langley, was that in a multibogey environment, the F-15 pilot was usually much better off avoiding the turning dogfight for which his aircraft was purpose-built. One post–Red Flag report declared, "Emphasis must be placed on high-speed, minimum-turning [fights]."[160] Another F-15 pilot echoed the conclusion, observing that "high speed, straight dashes, with hard turns (+5.0G) proved to be effective" in the combat-like environments.[161] While the 1st TFW pilots' experiences had not totally invalidated the principle of dogfighting—many an Eagle pilot appreciated the impressive maneuverability of his jet when he suddenly detected an opponent trying to sneak into position behind him—the iconic dogfight was losing its appeal as the preferred method of initiating and succeeding in air combat.[162] It was a remarkable change in expectations, for both the technology and the pilot. A few hundred yards down the flight line from the Red Flag ramp at Nellis, the pilots and analysts working on AIMVAL-ACEVAL were learning a similar lesson.

AIMVAL-ACEVAL

The catalyst for the AIMVAL-ACEVAL tests was the Air Force's and Navy's desire to field an updated IR-guided missile that would outfit their next generation of fighters. By 1975, three separate missile programs had materialized to complement the services' new AIM-7F radar-guided Sparrows. The Air Force was investigating a self-described "low cost, light weight, simple and effective dogfight missile" for its Eagles, appropriately named the CLAW.[163] The Navy, on the other hand, wanted to arm its Tomcats with the technologically exquisite Agile missile, which boasted a wide-angle seeker, thrust-vectoring, and helmet-cued capability.[164] To help "bridge the gap" between their then-current Sidewinders and the futuristic Agile and CLAW, the Navy and Air Force had started to collaborate on a new L-version of the AIM-9. Among other improvements, the AIM-9L incorporated a more sensitive, cryogenically cooled, indium antimonide (InSb) seeker which could detect a target from any angle, thereby eliminating the classic requirement for the fighter pilot to first maneuver to within thirty degrees of his target's tail before employing an IR-guided missile.[165]

However, Congress balked at funding the services' various competing missile programs. It canceled both the Agile and CLAW programs in the 1976 defense budget and instead directed the Air Force and Navy, through the Defense Department's director of defense research and engineering (DDR&E), to explore and then agree upon a common short-range air-to-air missile (SRAAM).[166] DDR&E in turn asked the services to also develop a model of air combat that would help inform the military's planning estimates for future fighter requirements—most agreed that an F-15 or F-14 was more capable than an F-5, but nobody knew how to quantify the advantage (e.g., was one F-15 equal to two F-5s? Three? Four?).[167] The test evaluating the different SRAAM concepts became AIMVAL; the latter force-ratio test, ACEVAL. Both were to be administered and conducted by a joint Air Force and Navy test force operating out of Nellis. They were slated to begin flying in 1976.

The two services recognized immediately that the results from AIMVAL and ACEVAL would define at a minimum the next-decade's worth of investment in air combat capabilities.[168] The services also quickly recognized that while they were ostensibly working together under a congressional and DDR&E mandate to define future air combat requirements, and despite the fact that their Eagles and Tomcats would never be in the air at the same time battling the opposing forces, the coincident matchup of the Air Force's and Navy's aircrews and their latest equipment would inevitably invite public comparisons. More than bragging rights were at stake. Future Air Force and Navy aircraft procurement schedules, as well as potential foreign sales, hung in the balance.[169]

To help balance service interests and quash any accusations of service parochialism, DDR&E split administration of the tests between the two services: a Navy admiral was appointed to serve as the Joint Test Force (JTF) director; the deputy test director was an Air Force general.[170] Nevertheless, concerns about potential military or contractor impropriety were frequently voiced on Capitol Hill. During congressional testimony, JTF officials tried to allay any suspicions by stressing the "complete objectivity" of "the test plan, the test data requirements, and the manner in which the data will be handled and analyzed."[171] The quintessential element underlying the officials' promises of "objectivity" was the ACMI pod that would be mounted on each test aircraft.

Before ACMI, the outcome of a practice air combat engagement was decided according to individual pilot recollections and the occasional review of snippets of gun-camera footage. After the training mission, the pilots from both the "Red" (adversary) and "Blue" (friendly) forces would gather in a conference room and engage in a spirited discussion over who was where when, who shot who, with what type of simulated weapon, and whether that weapon was valid. "We'd have lots of arguments about what happened," Welch remembered of the typical debrief, and they weren't always grounded in reality.[172] According to Jere Wallace, one of the AIMVAL-ACEVAL Eagle pilots, winning the debrief was all about telling a believable story, "If you could lie better than anybody else, they couldn't refute you."[173] In light of Anderegg's assessment that "no one can lie to himself better than a fighter pilot about to pull the trigger," the debriefs could be very contentious, indeed.[174]

Cubic Corporation's ACMI system for the first time offered truth data to replace the pilots' sometimes hazy and often disputed memories. The system served two purposes. First, with ACMI it was possible to determine the relative position and attitude of each aircraft, up to a maximum of eight, in the fast-changing air combat environment: no more arguments about who was where when. When flying over a designated ACMI range, each aircraft was equipped with its own ACMI pod. About the size of a Sidewinder missile, the pod contained an air data sensor, an inertial navigation unit, and a radio transponder. The pods broadcasted their attitude at frequent intervals to a network of ground monitoring stations scattered across the range below. The ground station receivers also measured the range to each pod and triangulated its position. The data were collected, processed, and fed to a central computer, which then plotted the participating aircraft on a "three-dimensional interactive graphic display screen," thereby allowing observers to watch the engagement unfold in real-time (figure 5.17). The data could also be recorded for later playback during the aircrews' debrief.[175]

Each ACMI pod also had the capability to monitor and report its individual aircraft's weapons system settings. Whenever a pod reported a simulated missile launch, the ACMI computers automatically calculated the probability of missile success using the specific engagement parameters, such as range to the target, aspect angle,

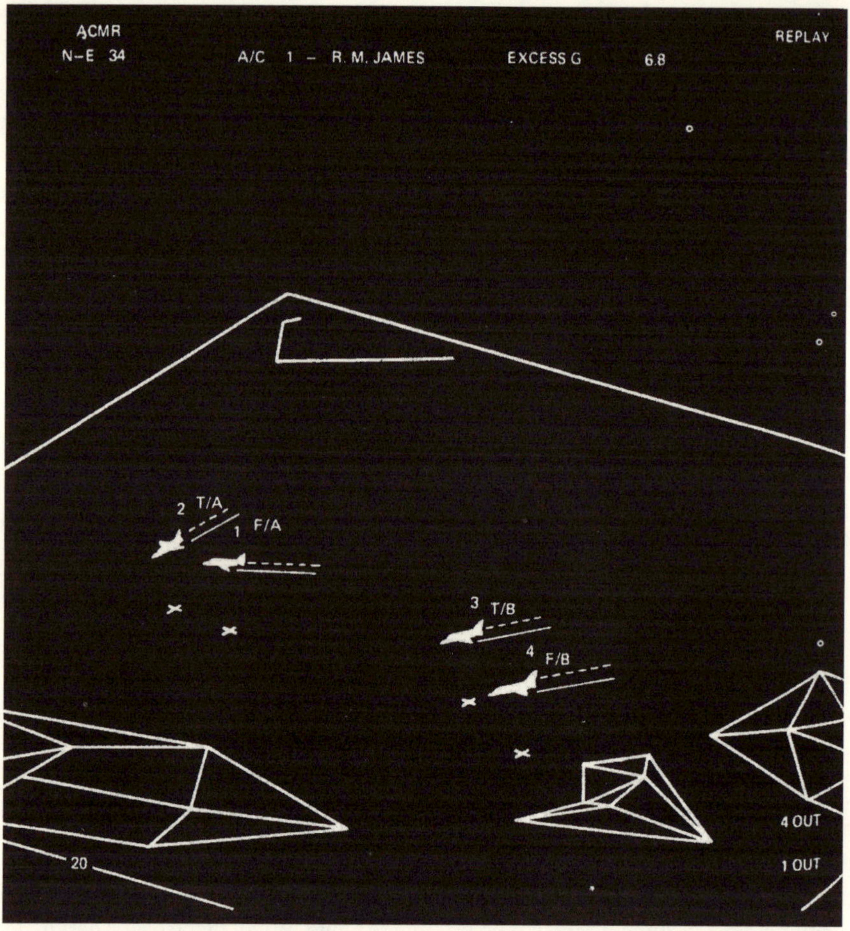

FIGURE 5.17. ACMI Graphic Display. ACMI was a revolutionary training tool. Each participating aircraft (up to a maximum of eight) carried a pod that broadcasted the aircraft's attitude and weapons parameters to a network of ground sites below. The ground sites also triangulated the various aircraft positions. All the data was compiled and processed near-instantaneously, allowing observers in a central control station to watch the fight unfold before them on a computer screen. The data could also be recorded for later playback for the aircrews. This image of the ACMI ground display was included in a Winter 1974 *Fighter Weapons Review* article. MSFRIC.

altitude, and closing velocity. The pods could even measure and report the power level across the seeker in the nose of a Sidewinder—the volume level of its "growl"—to aid assessment. By compiling and merging all the weapons and aircraft position data from the participating aircraft, ACMI could answer the dual questions: Who was shooting at whom and with what type of weapon? If the ACMI system predicted a successful missile intercept, "a schematic coffin appeared over the victim" on the ground display, prompting range controllers to notify the targeted pilot that he had been killed—no more questions about weapons' validity. Even though ACMI necessarily retained a certain degree of artificiality in its weapons simulations, the results it generated were generally accepted by aircrews as an objective, unbiased assessment of the engagement outcome and a significant improvement over the subjective, argumentative approach that had characterized prior air combat debriefs.[176] With ACMI projected to be operational in time for AIMVAL-ACEVAL, test officials proudly declared that the only way to get a more "exact answer" from their tests would be to "fire real missiles."[177]

For all its advantages, ACMI had a significant drawback, though. It only worked over an instrumented range dotted with the special ground receivers that could communicate with the transponders in the pods. At the time of AIMVAL-ACEVAL, there was only one functioning ACMI range, located forty miles north of Nellis, and it covered only a cylinder approximately thirty miles in diameter starting approximately three thousand feet above ground level. Because AIMVAL was designed to evaluate short-range, IR-guided air-to-air missiles, it was determined that the diminutive range space would not significantly bias that test. However, several in the Air Force and Navy worried that the limited range space would significantly hobble the second test, ACEVAL, and its examination of force-ratio effects on air combat. But, after having delayed the two tests for almost a year, there was little stomach in the Pentagon for postponing the tests further to gain a larger ACMI test-space. Congress demanded the test results, and the services needed to deliver them if they hoped to gain the funding that their acquisition plans required.[178]

To ensure that the AIMVAL-ACEVAL engagements would occur within the confines of the ACMI range, JTF officials declared that all test scenarios would include a visual identification requirement.[179] While the experience of Vietnam provided an obvious historical justification, neither the Navy nor the Air Force was too keen to see a VID restriction inserted into the two tests. Many in both services considered VIDs a relic of the strange circumstances of Vietnam and unrealistic for the coming war they expected to wage against the Soviets in the skies over central Europe.[180] Nonetheless, in an effort to quickly bolster its prized fighter's capabilities, the Navy announced its sudden plans to install a new optical television sight system (TVSU) on its F-14s that would extend the Tomcat aircrews' ability to VID a target out to more than five miles. The Air Force, having earlier decided against installing McDonnell Douglas's EOIATS on its Eagles, protested the Navy's unexpected technological advantage, arguing that it would upset the "established balance" between the tests' objectives and constraints. The Navy prevailed, and its Tomcats arrived at Nellis outfitted with the new system.[181] Not to be outdone, the Air Force pilots responded with their own VID-enhancing equipment, installed just in time for ACEVAL. The

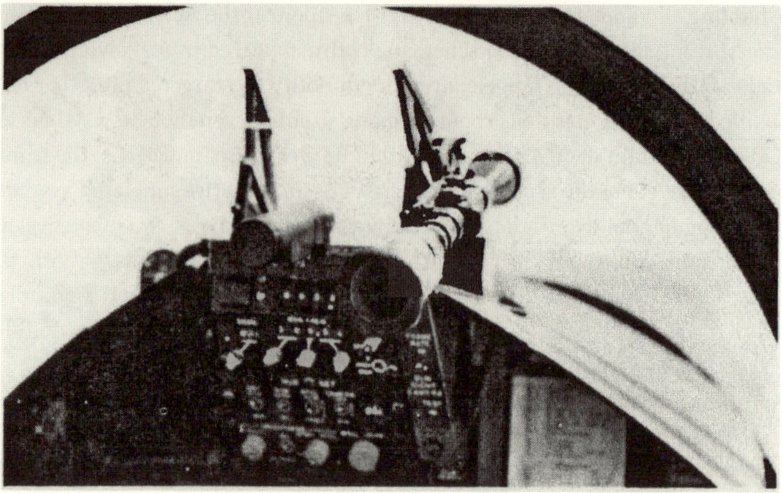

FIGURE 5.18. F-15 Eagle Eye. The Eagle Eye was an off-the-shelf six-power rifle scope bolted next to the F-15 HUD. Hastily procured just prior to ACEVAL, the Eagle Eye theoretically could "more than double unaided visual identification ranges." The innovation was quickly adopted by other F-15 units. (The cylinder mounted just aft of the HUD is the HUD camera.) MSFRIC.

unapologetically simple system known as the "Eagle Eye" consisted of an off-the-shelf six-power rifle scope bolted next to the F-15's HUD (figure 5.18). After the tests, both the TVSU and the Eagle Eye were widely deployed on the Navy's Tomcats and the Air Force's Eagles, respectively.[182]

Interservice squabbling over the tests' restrictions wasn't limited to the VID requirement. Both the Air Force and the Navy repeatedly lodged complaints regarding the tests' orchestration and their statistical design. Some even grumbled that because the tests weren't using real missiles, the aircrews would be at a disadvantage because the pilots wouldn't have the opportunity to see and react against a missile smoke trail heading toward them.[183] One Navy official bemoaned, "The guys who conceived and designed the whole test series clearly didn't know what they were doing."[184]

Unable to satisfactorily compensate for the inherent artificiality of the tests, several officers in both services determined it was "imperative that the impact of the test constraints be highlighted."[185] For example, officials in the Air Force's Studies and Analysis branch issued a cautionary critique: "ACEVAL and AIMVAL are tests involving only close-in, visual air-to-air combat involving flight sizes no greater than four. The test should not be advertised as something it is not."[186] Lieutenant General Alton Slay, the Air Force deputy chief of staff for research and development, shared the sentiment with Congress, reminding the members of the Senate Armed Services Committee in his testimony on March 9, 1977, that, having "theoretically taken a big skyhook and dropped these airplanes into a 30-mile arena," the test was "a canned situation" and should not be taken as an absolute indicator of air combat performance.[187]

Although initially slated to be executed after ACEVAL, AIMVAL was conducted first. It began on January 3, 1977, and took almost five months to complete. The stated objective of AIMVAL was to "compare the operational utility of existing and proposed short-range air-to-air missile concepts [with respect to] sensitivity [and] off-boresight [capability]."[188] During the test, five separate short-range, IR-guided missile designs, ranging from a simple boresight-only missile to an extremely sensitive, high-off-boresight, helmet-cued Agile-like weapon, were evaluated on both Navy F-14 and Air Force F-15 aircraft over a course of 1,033 trials. In addition to the simulated IR-guided missiles, the Blue forces were also armed with simulated AIM-7F Sparrows and a simulated 20-mm gun.[189] To enhance its Eagle pilots' ability to get a radar lock and employ the Sparrow during the anticipated close-range battles, the Air Force added a new auto acquisition mode to the F-15—"vertical scan"—which allowed the radar to automatically search for and lock targets from +5 to +45 degrees in elevation, particularly useful in a tight turning engagement when the F-15 was approaching ninety degrees of bank.[190]

Opposing the F-14s and F-15s were Air Force and Navy Aggressor pilots flying F-5E aircraft armed with a modified, boresight-only version of the AIM-9L Sidewinder, an approximation of a future 1985 Soviet threat.[191] Force ratios during the trials varied from one-versus-one to two-against-four, with the Blue forces never outnumbering the Red forces. The JTF's final report characterized a typical AIMVAL engagement: "Trials commenced at predetermined geographical range start points; both sides were on a fighter sweep mission. The usual engagement pattern was an ingress phase with head-to-head approaches, a large initial exchange of missiles at visual identification, real-time kill removals and re-engagements/disengagements until all aircraft had been killed or the surviving aircrews had exited the range."[192]

Even as AIMVAL was still underway, officials declared the test a "positive influence toward the resolution of common Air Force and Navy needs" for the next SRAAM.[193] Slay later told Congress that rather than simply plugging all the aircraft and missile data into "a computer, kick[ing] that computer and hav[ing] it spit out a roll of tape that tells you what the outcome was," the live-fly test offered an invaluable opportunity to capture the "extremely important . . . human factor" that dramatically influenced the real-world performance of the complex weapons.[194] Of the five IR-guided missile concepts tested, however, none were judged satisfactory; all exhibited difficulty distinguishing between a fighter target and the hot desert background. Additionally, although intended to dictate future *IR-guided* missile requirements, many thought the results from AIMVAL demonstrated a compelling need for a new high-speed, multitargetable *radar-guided* air-to-air missile. Most significantly, AIMVAL vividly illustrated that a relatively simple foe armed with an all-aspect, fire-and-forget missile like the AIM-9L could be lethal to advanced US fighter aircraft.[195]

Although focused on a different question, ACEVAL supported many of AIMVAL's earlier conclusions. Executed between June 2 and November 10, 1977, ACEVAL's stated objective was to "evaluate the factors which affect the outcome of multiple aircraft engaged in close-in air combat between specific aircraft [as a function of] force size/ratio [and] initial conditions."[196] ACEVAL used the same AIMVAL test management structure, much of the same Air Force F-15 and Navy F-14 equipment, and many of the same aircrews. The F-5E-equipped adversary also remained the same as during AIMVAL. Whereas variation in AIMVAL was based on the five missile concepts tested, during ACEVAL all of the F-14s and F-15s used the same simulated armament: AIM-9Ls (AIMVAL Concept C), AIM-7Fs, and 20-mm guns. Instead, variation during the 720 ACEVAL trials was primarily a function of setup parameters: force ratios varied from one-against-one to four-against-four, with Blue forces never outnumbering Red forces; the availability of radar ground controllers varied from both sides having control to only one side (either Red or Blue) having control; and the starting positions of the Red and Blue forces varied among twenty-seven different options within the thirty-mile ACMI range (figure 5.19).[197]

According to Stephen "Shad" Dvorchak, a former F-100 and F-4 pilot who was assigned as a data analyst during the two tests, the ACEVAL findings could be captured in a single sentence: "As the number of fighters in an engagement increases, the exchange ratio trends toward One."[198] In other words, the larger the fight (up to the ACMI limit of eight total aircraft, four-against-four), the more likely everybody died, regardless of their technological sophistication or lack thereof. Moreover, Dvorchak noted that any attempt to view the ACEVAL results strictly in terms of competing technological hardware quickly became "incomprehensible."[199] After spending almost a year trying to quantify air combat, the test administrators could only confidently conclude that they had "produced some data that is worthwhile," mostly the "aircrew-in-the-loop factor."[200]

The public would eventually learn of the uninspiring AIMVAL-ACEVAL results. Following a string of disturbing military failures, including the botched Desert One Iranian hostage rescue mission, an eclectic collection of defense analysts, members of Congress, and even a few within the military coalesced to launch a broad defense reform effort in the early 1980s.[201] One group of "Reformers" located within the Pentagon's Program Analysis and Evaluation shop, successor to McNamara's original "whiz kids" division, suggested that the services' penchant for exquisite, high-priced but low-utility hardware was partly responsible for the military's recent underwhelming performance. The lackluster showing of the Air Force's and Navy's newest, technologically advanced fighters during AIMVAL-ACEVAL naturally attracted the Reform-

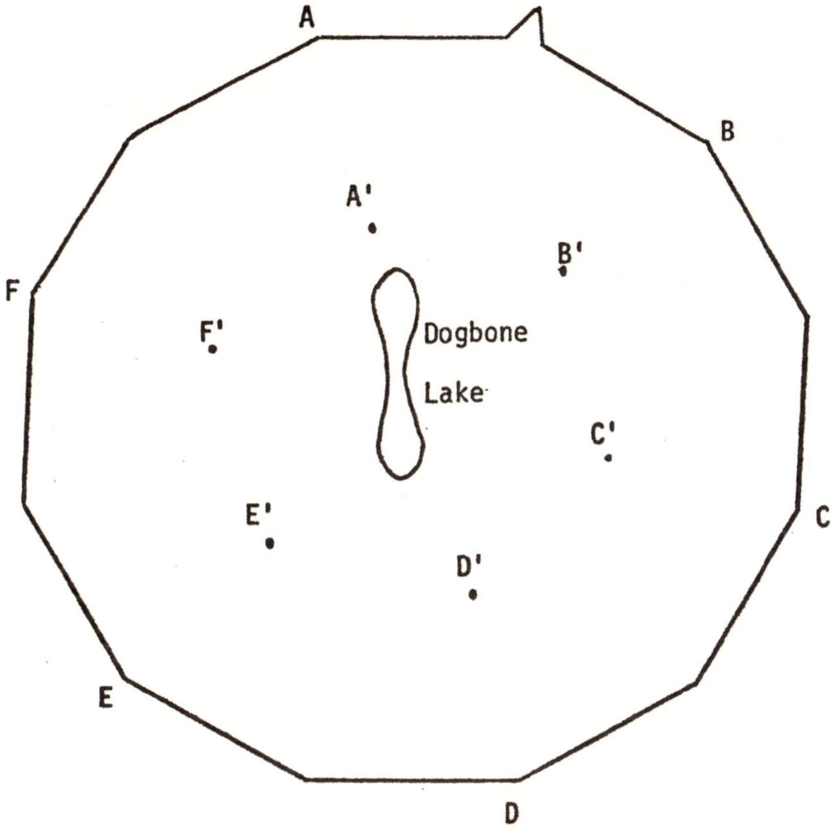

FIGURE 5.19. Nellis ACMI Range Starting Points, ACEVAL. Restricted to a thirty-mile cylinder due to ACMI range limitations, the AIMVAL-ACEVAL test area was centered on Dogbone Lake north of Nellis AFB, Nevada. The small operating space significantly influenced the tests' results. During ACEVAL, aircrews were positioned at the specific reference points on the inner (15-mile, points A′–F′) or outer (30-mile, points A–F) circle according to the test matrix. AFHRA.

ers' attention. Rather than spending tens of millions of dollars for a single Eagle or Tomcat, the Reformers argued that the services should instead be buying smaller, cheaper fighters. After all, "The proud 'air superiority fighters,' F-15s and F-14s," the *Chicago Tribune* declared in a December 1981 summation of the AIMVAL-ACEVAL tests, "had been fought to all but a draw by a comparatively crude $4 million airplane, the F-5."[202]

The Reformers reasoned that F-5-like fighters armed with inexpensive but lethal short-range IR-guided missiles and a powerful gun possessed just enough technology to answer America's tactical fighter requirement, but at a much more affordable price that would finally allow the nation to field a sufficient quantity of aircraft to defeat the Soviet hordes. While the Reformers acknowledged that their preferred aircraft lacked a long-range kill capability, and that the AIMVAL-ACEVAL data excluded longer-range air combat considerations, they countered that Vietnam conclusively demonstrated that the long-range air combat environment was not tactically viable—aircraft inevitably were forced to engage in close-in dogfighting even if they were not designed for it. According to the Reformers, AIMVAL-ACEVAL simply provided more recent, confirmatory evidence of the dominant historical trend.[203]

Others, however, refused to accept that "mutually devastating" mass dogfights should be the modus operandi for American pilots battling for air superiority.[204] For them, the key lesson of AIMVAL-ACEVAL was not that "quantity" suddenly trumped "quality" but rather that a front-aspect, fire-and-forget weapon like the AIM-9L could have a radical, transformative effect on air combat.[205] Welch for one remarked that, after observing AIMVAL-ACEVAL, he suddenly realized that "all" the lessons that he and his fellow Langley pilots had learned during their first years flying the F-15 "went to hell when the [opponent] got the AIM-9L . . . All these tactics, all of our experiences, everything we've taught each other goes out the window."[206] It was a valid assessment, although slightly overstated. As noted earlier, the Eagle pilots at Langley had already begun to acknowledge that their best chances at victory and survival were attained when they avoided a close-range dogfight, and that was against an adversary considerably less capable than the one faced during AIMVAL-ACEVAL.

Following the two tests, both the Air Force and Navy concluded that they had to find a way to keep their fighters out of the increasingly lethal short-range environment. Paradoxically, their longer-range, radar-guided AIM-7 Sparrows were not going to provide the answer. Even when the AIM-7 was fired at its maximum range, the attacking pilot still often found himself less than three miles from his foe when the missile actually impacted the target—a function of the Sparrow's requirement for the attacking aircraft to remain pointed at and illuminating the enemy aircraft with its radar throughout the missile's entire time-of-flight. In the past, it hadn't mattered. Enemy fighters didn't possess an all-aspect-capable IR-guided missile that could detect and track an approaching target; and if the enemy pilot tried to shoot one of his radar-guided missiles (which were shorter range and considered less capable than the Sparrow), as long as the American Sparrow arrived first, the enemy's missile would lose its source of radar guidance and veer off course when the host aircraft was destroyed. Now armed with an all-aspect AIM-9L-type missile, even though the opponent probably would not survive the Sparrow attack, as long as he pressed his pickle button and launched a self-guiding missile before the Sparrow arrived, there was a good chance he would at least even the score.[207]

One AIMVAL-ACEVAL analyst succinctly described the new reality, "If you've got an AIM-7, you're going to die."[208] Because limitations in IR-seekers precluded development of a longer-range, IR-guided missile, the Air Force and the Navy concluded that their aircrews needed a new radar-guided missile, one that didn't rely on the aircraft radar to guide it all the way to the target. The AMRAAM, already under preliminary development, quickly rose to the top of their acquisition lists. To free-up vital funds to develop the new fire-and-soon forget radar-guided missile, the services elected to shelve their plans for an advanced IR-guided missile. Only the AIM-9L Sidewinder remained, which the two services continued to cosponsor.[209]

Even with the additional funding, the AMRAAM would not arrive for at least another decade. In the interim, the Eagle pilots had to figure out a way to transform their aircraft from a close-range dogfighter into a missile-hurling interceptor. And they had to do it while relying on an AIM-7 missile that AIMVAL-ACEVAL had shown was likely to drag them into a close-range environment and get them killed. A new set of radar "sorting" tactics, made possible only with the bounty of information provided by the Eagle's PD radar and CC, and only after the fighter pilots had abandoned the rigid formation assignments associated with fighting wing, was an essential element of that transformation. In the process, the new tactics and the skills required to execute them would profoundly alter the traditional fighter pilot experience.

"Sorting" Things Out

Although the radar set in the Phantom was composed of exquisitely engineered wave guide tubes that transmitted finely tuned electromagnetic energy for periods lasting no more than tens of microseconds at a time, the aircrews' procedures for using their precision instrument weren't very precise. They set their radar's altitude coverage simply by judging the height of the ground return line on their scope. Gain controls were adjusted until they "saw something [they] liked."[210] To determine a target's altitude, they relied on approximate trigonometric relationships that they calculated in their heads. As a result, four Phantoms might be flying together in a formation using their radars to search the same piece of sky, but it was highly unlikely that all four crews would be able to distinguish the same "fuzzy blip" from the array of other "fuzzy blips" on their scope, if the target even appeared at all. Each Phantom crew possessed a slightly different understanding of the airspace in front of it, and nobody knew whose was more accurate.

In the aerial battles over Vietnam, it hadn't usually mattered. Once the Phantoms detected a target with their radar, they were likely going to end up in a close-range battle anyway due to the VID requirement. While they obviously would have preferred to have more information about the number and location of any additional targets in the area, as long as they kept their tails clear, the Phantom crews were not usually at risk of being shot down.[211] They figured they could deal with any MiGs as they saw them, and they were generally correct, *if* they saw them. The same strategy, however, was no longer viable when the adversary possessed an all-aspect, IR-guided missile such as that used by the Aggressors during AIMVAL-ACEVAL. Suddenly, fighter pilots in the Eagle faced a life-or-death requirement: they needed to locate *all* the enemy aircraft with their radars before racing headlong into a close-range dogfight. A failure to account for even just one enemy bandit often resulted in, at best, a tit-for-tat missile exchange and, at worst, several "dead" F-15 pilots.

Fortunately, the F-15 pilots enjoyed a clear advantage over their F-4 brethren. Benefiting from their aircraft's PD radar and CC-controlled displays, all the pilots in an F-15 formation were likely to see the same radar picture during an intercept. The flight leader therefore could now assign specific radar targets to specific pilots in his formation, expecting that his flight mates would both understand the communicated targeting scheme and be able to execute it. This intraflight targeting capability went unexploited before AIMVAL-ACEVAL. After all, the Eagle pilots at that time were focused on short-range air combat and used their radars accordingly—to help secure a positional advantage for an ensuing dogfight, just as they had done in the Phantom before.

Forced to abandon the aerial duel as the favored means of engagement, Eagle pilots had a fresh imperative to develop new coordinated long-range tactics that would capitalize on their common radar picture. With the new tactics came new measures of pilot success. Soon, flight leaders began to consider a wingman with a reputation for always finding his "sort" to be more valuable than one regarded as a good dogfighter. However, the F-15 initially lacked many of the tools needed to assess an individual pilot's precise execution of the radar-intensive tactics. Moreover, the new emphasis on radar interpretation and manipulation was not amenable to the classic trial-and-error training methods historically used by the fighter pilot community. All told, the sudden shift toward long-range sorting tactics following AIMVAL-ACEVAL fundamentally redefined the Eagle pilot's desired skillset, the means of evaluating it, and the techniques for developing it.

New Ways to Use the Radar

The first Eagle pilots considered their new look-down, shoot-down PD radar a godsend. Although they still had to manually adjust their radar antenna's elevation angle and could search only a specific swath of sky at a time, the Eagle pilots could be reasonably sure that if there was an aircraft flying within that volume of airspace, their radar would detect it.[212] The Eagle's radar continued to suffer from a fundamental limitation, though. The synthetically generated target brick that magically appeared on the VSD provided only scant information: range, relative azimuth, and a broad band of possible altitudes. To obtain more precise information, such as the target's aspect angle and its specific altitude, the Eagle pilot had to command a radar lock. A radar lock was also required anytime the pilot wanted to employ a Sparrow missile. If there was only a single target in the airspace, locking it would not be an issue. But if there were potentially multiple targets in the area, then selecting track on a single target meant that the pilot would be unable to search for or monitor any other potential threats. As one AIMVAL-ACEVAL test official explained, going to track was like "look[ing] through a soda straw. . . . You get extremely intelligent about one airplane, and everybody else goes away." The question that therefore confronted the pilots was, succinctly stated, "How the [expletive] can you win a fight looking through a [expletive] soda straw?"[213]

Imagine a single F-15 facing two Aggressors flying in a line-abreast formation. The Eagle pilot detects the two aircraft with his radar and commands track on one of them, say the target on the right. He then launches a Sparrow missile at that target. A few seconds later, the F-15 pilot observes the target aircraft turn away forty degrees. The target's slight maneuver might be enough to rob the Sparrow missile of its ability to complete the intercept. Then again, the missile might still have enough "smash" to hit the target and destroy it. There was no way for the pilot to tell. More importantly though, what about the other target that was on the left? Did it also turn? If it didn't, then it would close the distance to the F-15 faster than the target currently being tracked, and the F-15 pilot would come under attack sooner than expected. But if the F-15 pilot elected to break his radar lock and look for the target that was originally on the left, he would certainly trash his Sparrow missile currently in-flight against the target on the right.[214] One solution to the F-15 pilot's quandary would be to give him a semiautonomous missile so that he could break his radar lock without effectively discarding the missile in flight and then use his radar to search for the other target. Another possibility would be to provide the pilot with a radar mode that could simultaneously search for *and* track targets. Both innovations would eventually emerge—the new missile being the AMRAAM and the new radar mode being TWS (track-while-scan)—but not for several years.[215]

A third possibility that did not rely on a new missile or a new radar mode would be to trust another F-15 flying in formation to target the other enemy aircraft. The process of assigning different radar targets to different flight members became known as "sorting." The precision in the Eagle's radar and the quality of its displays meant that for the first time there was reasonable assurance that a sorting plan could be successfully executed. Both F-15 pilots were likely to see the same targets on their VSDs, and after locking their assigned target, the pilots could use the information that was readily displayed as "positive feedback" to determine if they were properly sorted. For example, if one Eagle pilot said that his target was a "16-Left" and the other pilot said that his VSD indicated a target with an aspect angle of "15-Right," then the two pilots would know that they had in fact locked different targets. Alternatively, the pilots could use the target's digital altitude readout on their VSD as a discriminator. While the sorting tactics had obvious shortfalls if there were more enemy targets than F-15 radars, in those situations there was an even stronger imperative to sort correctly to ensure that the F-15s were not inadvertently "double targeting, or double shooting, a single target."[216] As Wallace explained, with properly executed sorting, "you were making your firepower much more effective."[217]

In addition to a reliable radar and clear displays that provided useful, digital information about the target, as well as a widely spaced formation that allowed the wingman to devote sufficient attention to monitoring the VSD inside his cockpit, sorting was predicated on a high level of mutual trust between flight leader and wingman. Both flight members had to be confident that their fellow pilot was capable of working his radar during a stressful and confusing intercept. Both also had to trust that the other would resist the powerful urge to lock the first target that appeared on his VSD—an "I've got mine, you go find something else" attitude. That led to double targeting and "dead" F-15s, Anderegg explained.[218] Finally, even if the F-15 pilots adhered to the targeting plan and successfully sorted to their assigned target, they still had to trust that their flight mate would complete his task and actually kill the target. A flight leader could die just as easily from his wingman's target as his own; the reciprocal was also true.

During ACEVAL, the Aggressor pilots developed carefully scripted maneuvers purposefully intended to confuse the Blue pilots' targeting schemes. One early Aggressor tactic known as "swap and turkey" or "rope a dope" proved particularly effective. Beginning with two widely spaced formations, the Aggressors would send one formation charging at the Blue fighters. After closing to a certain distance, where it was assumed that the opposing fighters had all gone into radar track and launched their volley of Sparrow missiles, the charging Aggressors would then abruptly turn around and lead the chasing Eagles or Tomcats into a position where the other Red formation could pounce. When it was first executed, the tactic claimed three of four attacking Tomcats. The Eagles experienced similar losses.[219]

Some complained that the Aggressors' tactics were not representative of the expected Soviet threat, that the American "Red" pilots were "too good" and their tactics too imaginative.[220] The test's directors noted, however, that the "ingenuity" was not limited to the Red forces; the F-14 and F-15 aircrews were also constantly developing more flexible tactics of their own to blunt the Aggressors' scripted plans. The officials ultimately concluded that such "tactics change for change's sake was a sound tactical principle," and that the intense competition it fostered helped contribute to the "realism" of the ACEVAL test environment.[221] This battle of wits and the rapid-fire development of air combat tactics and countertactics that it stimulated would become the most valuable legacy of ACEVAL.

In particular, the Aggressors' success at spoiling the Eagles' sorting plans exposed a critical weakness of the latter's new tactic. As one F-15 pilot explained, PD radars made only "one airplane smarter"; somebody still needed to coordinate the multiple radars in the fighter formation, and that was typically accomplished verbally over the radio.[222] Early during the intercept, each pilot would be assigned a different sector of the sky to scan. If a pilot detected a target, he would call the contact out over the UHF radio. Then as the Eagle formation closed on the target(s), the flight leader would communicate the common, assembled radar picture and direct the flight members' targeting. Each flight member would subsequently lock his assigned target, but the formation members then needed to verify their sorts by communicating appropriately distinguishable radar information about each target. Unfortunately, the pilots' radio transmissions throughout the process were usually wordy and imprecise.[223] Dvorchak recalled that the targeting and sorting process for a four-ship of F-15s during an ACEVAL mission often required upward of one hundred separate intraflight radio transmissions. Confusion reigned, and out of the roughly 140 four-against-four trials flown during ACEVAL, Dvorchak couldn't recall a single instance where all the flight members sorted perfectly.[224]

Part of the problem was that for decades, fighter pilots were taught to shun radio communications. As the previous chapters illustrated, there had never been a shortage of excess chatter since radio's first use in combat during World War II. But the fighter pilots had often preached that the solution to the incessant babble was for everyone but the flight leader to simply shut up. Even by the time of the F-15, it was still a common maxim that the only time a wingman should key the microphone was to call: "Two"; "Mayday"; "Bingo" (running low on fuel); or, "Lead, you're on fire."[225] But now with sorting tactics, everybody had to talk on the radio, and nearly all at the same critical time during an intercept.

Despite the difficulties, the ACEVAL pilots declared that the "name of the game is sorting. If you sort, you win."[226] The test results agreed. On the occasions when the Eagles failed to sort and instead all locked on to a single target, "the *best* they ever did," the analysts remembered, "was lose three [F-15s]. . . . When they locked on [to] three out of four [Aggressors], they kicked ass."[227] To facilitate proper sorting, a common fighter pilot vocabulary and a corresponding format and cadence were needed to maximize the flow of information across the formation. The Phantom pilots had already laid the foundation with their intracockpit "descriptive commentary" and intra- and interflight "brevity comm," and there had been continuing development of the radio procedures since. The 1st TFW's *Aerial Attack Manual* included nearly 50 percent more "brevity comm" terms in its glossary as compared to the instruction manuals used by the Phantom aircrews a decade earlier.[228] But the cumbersome and prolific radio exchanges from ACEVAL showed that there was still much work to be done. There was simply too little time during a fast-moving intercept for each pilot in a four-ship to fumble with their words, monopolizing precious radio time with imprecise blathering such as, "I got the one to the left of the blip, which is right over the top of whatever it is."[229] The concept of radio discipline had to change, from simply not saying anything to saying everything simply and precisely.

In addition to refining their communications practices, the Eagle pilots also recognized the need for a simple, universal targeting plan that could be discussed during a premission brief and then executed in the air with minimal radio communications. The pilots had worked on developing such a plan in the buildup to ACEVAL, but it was usually quickly dismantled by the Aggressors' unconventional tactics. "The formations we saw," explained Jeff Cliver, one of the Eagle pilots during the test, "were so diverse and exotic. . . . We had a general sort plan, if we saw a four-ship. But we hardly ever saw a good ol' four-ship; somebody's always 'rope-a-doping' or dragging [turning around] or doing something."[230] Nevertheless, against a more conventional opponent, the ACEVAL pilots believed that a standardized targeting plan would be useful.

Serving as the chief instructor for F-15 intercept procedures at the Weapons School following ACEVAL, Sokol was responsible for codifying the standardized targeting plan and disseminating it to the growing Eagle community. To illustrate the new manual, Sokol spent hours drawing "little squares" that represented the target bricks on an Eagle's VSD, and then he annotated which flight member was responsible for which radar target for a variety of intercept situations. After clearing the new text with the officials at TAC, Sokol distributed the manual to the Eagle units at Langley and other bases.[231] "Lead [takes] left, low, lead" quickly became the prevailing targeting mnemonic in the Eagle community, indicating that the flight leader was responsible for the target on the left if there were two side-by-side, the low target if they were flying at different altitudes, or the lead target if they were flying in trail. The wingman was responsible for the opposite target in each scenario.[232]

Despite the new emphasis on communications and standardized targeting plans, there was still ample opportunity for miscommunication and confusion. For example, what if the two radar targets were not strictly side by side? The flight lead might interpret the enemy presentation as two targets in range and lock the lead target on the right side, but if the wingman thought the enemy presentation still looked like a side-by-side formation, then he too would lock the target on the right, and the two Eagles would find themselves double-targeted. Additionally, while much better than the earlier radar in the Phantom, the Eagle's PD radar still occasionally generated false targets that appeared real, even in radar track. Cliver recalled, "The loneliest feeling in the world is . . . staring through the [TD] box at nothing."[233] And then there was always the urge that every pilot experienced during a confusing intercept to just lock something and start shooting.[234]

Although they generally enhanced the pilots' chances at success and survival, the new coordinated long-range radar tactics represented a substantially more complex employment scheme. Miscommunication of the desired targeting plan, hiccups in individual aircraft radar performance, pilot errors selecting and prosecuting the attack against his assigned target, and persistent missile failures all could render the new tactics ineffective. The additional complexity, however, also introduced flexibility and resiliency into the Eagles' employment since it usually presented opportunities for the flight members to recognize and potentially recover from a rapidly deteriorating situation. The kill (or simulated kill) still remained the ultimate measure of mission success in the air-to-air environment. But the pilots also soon realized that simply tallying the total number of kills after a sortie wasn't helping their younger pilots learn how to recognize and then capitalize on the fleeting opportunities that arose during an engagement.

New Ways to Measure Performance

The myth of aerial victory achieved through individual flying prowess had always been based on half-truths. When the Sabres were battling MiGs near the Yalu, it was the wingman who played the critical but often unacknowledged role, freeing the flight leader to focus his attention on aligning his aircraft behind a target and squeezing the trigger. A decade later in the Phantom, it was the addition of the GIB and the new guided missiles that, along with the wingmen, were typically overshadowed by the glory of the individual pilot. Now in the Eagle, the wingman had gained new prominence, and the GIB had been eliminated, but there were still several steps that needed to be successfully accomplished if the pilots were to achieve victory using their newfound radar-sorting tactics. First, the flight leader needed to develop a targeting plan and communicate it to the flight members, who in turn all had to understand it. Second, each pilot then needed to manipulate his individual radar to accomplish the assigned targeting. Third, missiles had to be launched by the pilots at the appropriate times and under the appropriate conditions. Fourth, the missiles then had to work, still never a guaranteed prospect.

The F-15's HUD camera, although considered a remarkable improvement over previous gun cameras because it also captured the weapons symbology displayed in the HUD, could document only the third step of the process, launching missiles in parameters, and on rare occasions parts of the fourth, if the missiles actually worked and if the pilot was within range to film the ensuing fireball in front of him. But the short segments of HUD film showing weapons parameters at launch did not necessarily verify that the pilot was shooting at his assigned radar target, and as ACEVAL had demonstrated, simply lobbing missiles against any enemy target, even if they resulted in a fireball, was not always effective.

Two other recording systems helped supply critical mission data. The first was the portable cassette recorders that the Eagle pilots, like their F-4 predecessors, carried with them into their aircraft. The audiotape of the radio communications was particularly useful whenever there was confusion regarding the flight leader's targeting direction, as well as when the flight members were clamoring to verify their different radar sorts. The second recording system was ACMI. Like in AIMVAL-ACEVAL, ACMI provided a windfall of information that aided the postmission debrief. By revealing the positions of all the adversary aircraft in relation to the Eagles, the appropriateness of the flight leader's initial targeting direction could be assessed. Additionally, watching the ACMI playback of the mission allowed the pilots to understand the Red forces' maneuvers that were otherwise obscured once the Eagle pilots took their final locks and launched their Sparrows against their individual targets. Finally, ACMI helped the pilots judge the effectiveness of their missile shots. But two significant limitations remained: only a small fraction of F-15 training engagements were flown over an ACMI range; and, even when using ACMI, there was no means to assess the critical second step in the targeting process—the individual pilot's manipulation of his radar according to the information displayed on his VSD.[235]

If the Eagle's radar always worked perfectly and automatically, then there would have been little reason to focus attention on this remaining intermediate step. The simple proof of a successful shot against an assigned target would have been sufficient testimony to a pilot's skill, and, conversely, absence of a valid missile shot against the target demonstrative of his lack thereof. But every pilot knew that the F-15 radar was never perfect and that launching a missile against the correct target entailed more than simply pushing a button. Moreover, AIMVAL-ACEVAL revealed that different techniques for using the radar yielded substantially different outcomes, even when locking the correct sort.

For example, Cliver earned a reputation during ACEVAL for being the first pilot to get a radar contact but the last pilot to ever get a radar lock. He also on average had the shortest time from when he took his lock to when he fired his first missile. It proved to be a recipe for success; Cliver was one of the most lethal Eagle pilots in the test. In contrast, the worst performing Eagle pilot was on average the last to get a radar contact, was first to go into radar track, and spent the longest time in track before firing his missile. The correlation, apparent only after the ACEVAL data were analyzed, was later associated with the rapid decline in a pilot's situational awareness once he entered radar track on a single target.[236] This realization led the pilots to develop additional tactical standards that dictated when they should lock a target and when they should shoot their missiles.

During AIMVAL-ACEVAL, the specially instrumented test F-15s were equipped with systems that could monitor all the pilots' radar actuations, plus two separate videocassette recorders that continuously captured the information displayed on the HUD *and* VSD.[237] Access to these extra data streams facilitated the more nuanced understanding of radar employment and single target track; without them, Cliver's success during the test might have remained a mystery. Regrettably, Eagle pilots in the operational units did not have access to the same additional radar data sources and instead continued to rely strictly on HUD film and hazy memories of what they might have seen or didn't see on their VSDs. Sokol recalled the frequent frustrations he encountered in debriefs trying to reconstruct an engagement without the necessary data:

Flight Leader: "We had two guys on this guy, nobody on that guy. How'd that work? How'd that happen?"

Number 2: "You know, I was looking into the sun. Couldn't see my leader."
Number 4: "I locked on to somebody."
Number 3: "I was focused on something else other than this, and just didn't have time to sit there with my head down in the scope."[238]

Struggling to institute the new long-range sorting tactics across the Eagle community without access to critical radar data postmission, pilots begged for an additional video recorder for their VSDs. Earlier, such a recorder was judged an unnecessary extravagance for a fighter designed for close-range air combat, but TAC finally acceded to the request. It would take several years before all the Eagle units received their additional recorders, but once installed, they became an essential tool for documenting and assessing pilot performance.[239] Wallace explained that after his squadron began flying with the new VSD recorders, his flight leaders could finally "really tell what the guy's doing . . . on [the] radar. If he's spinning the el-wheel from one extreme to another and not settling in and not taking time to search his area of responsibility, you can hammer [him] pretty good."[240]

Wallace's recollection disclosed another significant shift in the expectations of an F-15 pilot. In the Phantom, pilots commended their GIBs' individual artistry that was critical to long-range radar operations. The Eagle's tactics, in contrast, allowed for none. Pilots now emphasized the *mechanics* of working the F-15 radar and the importance of being almost "robotic" when executing the assigned search and targeting responsibilities. Specific radar procedures were detailed during each premission brief: set the altitude coverage to this; search for this many radar sweeps; then reset the altitude coverage to this; search there; lock only when the targets are inside of this range; heed the targeting plan.[241] Pilots who failed to adhere to the contract were, as Wallace said, scolded appropriately during the mission debrief.

At first glance, the strict radar procedures and the emphasis on recording and reviewing them for unauthorized deviations seemed to signal further deterioration of the once-prized individuality of the fighter pilot and a tenuous shift toward finally becoming the push-button missileman that the fighter pilots had long feared. But on closer inspection, the new automaticity in the Eagle that led to "robotic" radar mechanics and strict targeting contracts reflected the familiar pattern observed in the previous two chapters. The new radar equipment had not obviated the operator and his skills; it had redefined them. Whereas in the Phantom it was an avowed art to get the radar to detect and display a target on the scope, in the Eagle the art now resided in knowing what to do with the target once it appeared on the VSD.[242] Not surprisingly, this new skillset was not amenable to the old methods of documenting pilot performance, and the importance of the new VSD recorder, cockpit cassette recorder, and ACMI all reflected the shift.

Nor was the new skillset compatible with the traditional methods of educating a fighter pilot. In the past, an F-86 pilot like Evans was simply thrown into the cockpit during "Clobber College" and told to follow his leader; he'd pick up the rest as he steadily accumulated more hours in the jet. A young pilot in the Eagle, however, wasn't intuitively going to learn how to work the radar or get his sort just by spending more time in the air.

A New Fighter Pilot Experience?

"When I was flying an F-84, . . . I'd fly thirty hours a month, and maybe five to six hours of that was really demanding," Welch recalled. "Fast forward to the F-15; I'd fly twenty-five hours a month, and twenty-one hours of that was demanding." It wasn't the new jet itself. Welch, like countless other Eagle pilots, considered the F-15 an easy aircraft to fly.[243] Granted, employing the F-15 well was another matter, but that didn't mean that employing an F-84 in the mid-1950s was any easier for that pilot, just different. Rather, Welch's observation alluded to the dramatic shift in the quality of training and instruction that coincided with the arrival of the F-15. In Welch's F-84, as in the F-86 and even the early F-4C, fighter pilots had primarily developed their skills through repetition and self-evaluation. Moreover, simply accumulating flight hours often helped cultivate proficiency, since the difference between *flying* a fighter and *employing* it had not yet fully emerged. The same wasn't true in the F-15. An Eagle pilot was not going to develop his radar employment skills by simply flying a few extra, random sorties a month. And, despite now being college-educated, he wasn't likely going to be able to dissect a confusing radar intercept and identify his particular failures—human or machine—without specific, personalized, and reasoned instruction from a more experienced pilot. The symbiotic confluence of these three phenomena—more realistic training, college-educated pilots, and the recognition that self-education and simple repetition were no longer viable techniques to develop young fighter pilots—threatened to upend the traditional fighter pilot experience.

Having witnessed it firsthand as Air Force fighter pilots, both Anderegg and Michel wrote about the metamorphosis in fighter pilot training that transpired after Vietnam.[244] Before the war, the service's leaders had tended to prioritize the preservation of their increasingly expensive aircraft over aircrew training. Blesse once complained that during the decade between Korea and Vietnam, "Safety became more important than the tactics, more important than gunnery, more important than anything. Safety was king. Safety took over." Pilots were routinely prohibited from flying against different types of aircraft, and when they flew against their own type, they were often limited to specific, conservative flight regimes and maneuvers.[245] On other sorties, pilots were reduced to simply flying "around the flagpole" of their bases, successfully logging flight hours but practicing little of consequence.[246] After taking command of TAC in 1973, Dixon acknowledged that the service's focus had been misguided. In his words, the Air Force had sent its fighter pilots to combat having practiced only "what amounted to calisthenics—[doing] the same thing every day in a very unreal atmosphere." New "realistic training" opportunities were sorely needed.[247]

The fighter pilots were not oblivious to the deficiencies in their training, but most did not appreciate how gross the failings were until after they started flying missions over Vietnam. Emerging from the war, many felt betrayed by the service's leaders who had sent them and their squadron mates into combat unprepared, and often against borderline-worthless targets to boot.[248] In Michel's estimation, the pilots' frustration crystallized as a new "combat culture," which "emphasized flexibility, individual responsibility, little top-down control, and a 'get the job done' attitude that was entirely different from the strict supervision and emphasis on flying safety in the stateside Air Force."[249] Combat-seasoned "iron major" fighter pilots like "Moody" Suter embodied this postwar fighter pilot ethos, and when they linked up with Dixon and his new emphasis on realistic training, the Red Flag and Aggressor programs sprouted and then flourished to the benefit of all Air Force pilots.

Another change in the pilot force contributed to the success of "realistic training." For the first time in the history of the Air Force, most of its pilots held a college degree. With that degree, Anderegg noted, the pilots brought an "academic approach to problem solving" that historically had not been valued in fighter aviation. In the past, "when a guy had a problem doing something in an airplane," Anderegg explained, "he just did it over and over and over and over and over and over until he got it right."[250] It was the training method that Evans had encountered in "Clobber College." There were, after all, only a limited number of possibilities, and the outcomes of each aerial trial, whether following a flight lead through a series of maneuvers or shooting machine guns at a towed target, were usually readily observable by the pilot. Moreover, with fighter pilots still regarded as the elite of naturally gifted aviators, it was understood that a pilot with the requisite skills would eventually stumble upon the solution; those that didn't were obviously not in the proper career field and were best left behind.

The shift to college-educated Air Force pilots had begun in the mid-1950s when, following the Korean War, the service began curtailing its Aviation Cadet and Officer Candidate School programs. Both commissioning programs had historically provided a means for those without a college degree to earn their lieutenant's bars and a chance at a set of pilot's wings. While it would make sense that the growing sophistication of the service's aircraft, missiles, and similar technological equipment in the 1950s and '60s had contributed to the increased demand for college-educated officers, in actuality the personnel shift was usually discussed "in the context of officership" and the service's desire for broadly educated leaders conversant in topics like internal relations and business administration. In fact, throughout the transition many maintained that "a college education had little effect on one's ability as a pilot" because *that* career "was based on gross motor skill development and hand-eye coordination."[251] Several former aces agreed with the assessment and openly questioned the Air Force's new college-graduate requirement of its pilots.[252]

There was, however, significant value in the pilots' college education, regardless of whether it was in chemistry, electrical engineering, or history. The "academic approach" that Anderegg referenced guided the fighter pilots' thinking as they evaluated their Vietnam experiences and then rationalized their need for more realistic training programs. When the pilots were then thrust into the melee of the Red Flag exercises that they had constructed, their college-cultivated critical thinking and deductive reasoning skills proved invaluable, especially when they were trying to dissect a complicated multi-aircraft engagement during a debrief. Technical systems knowledge gained through extensive flight manual study—an activity itself once widely shunned by the fighter pilot community—continued to be an important contributor to mission success, but now pilots increasingly had to augment their book studies with detailed mission analysis and synthesis. In the process, the fighter pilots quickly realized that their massive simulated wars were not amenable to the simple trial-and-error and brute force methods of learning commonly used in the past.[253] A new imperative suddenly emerged, Welch remembered, to ensure that "every sortie" flown was a "productive" one.[254]

The quest to maximize productivity extended to the mission debrief. In the past, there was typically no formal postmission analysis other than what was required to validate a flight leader's kill. The lack of a detailed debrief was partly mitigated by the fact that a Sabre or Phantom flight leader could usually assess his wingman's performance merely by looking over his shoulder: if the wingman was still in the welded wing position, all was well; if he wasn't, then the flight leader could quickly adjust his own tactics while providing instantaneous verbal feedback to the wayward young pilot. However, in the context of the new combat environment, characterized by more demanding scenarios, technologically sophisticated aircraft and weaponry, and complex mutually dependent tactics, flight leaders quickly realized that they couldn't assess their wingman's performance from afar and they couldn't simply wait for the youngsters to figure it out on their own. As Sokol explained, only half-joking, "You got Lieutenant [So-and-so] out there in Number 4, he may have forgotten to turn on the radar."[255] Recognizing that a young wingman who didn't know how to use his radar properly or who couldn't find his assigned sort was a liability to everyone in the formation, the experienced flight leaders had a strong, self-preservation motive to get their young wingmen up to speed as fast as possible.

Consequently, the wingman position in the F-15 community evolved from the traditional fighter pilot rite-of-passage to an apprenticeship program conducted under the watchful tutelage of an experienced flight leader. Gone were the days of simply showing up at the duty desk, filling out a mission card, and walking out to the jets on a personal quest to prove who was best.[256] Instead, detailed premission briefs were given, during which radar responsibilities were outlined and targeting plans and radio communications procedures reviewed. Even more significant was the emphasis on instruction and critical feedback after the pilots had landed.[257] "Suddenly, the debriefings got very, very brutal," Welch recalled of his F-15 time. "We went from, you're really kind of careful [criticizing] your flight; to no, we're going to talk about what you did wrong; to no, it's going to be the whole truth. And it's going to be brutal. And you're going to do better on your next mission or the next time you're going to hear about why in the hell didn't you fix that."[258] Simply converting JP-4 fuel to jet noise was no longer acceptable. Flight leaders and wingmen had to train together and learn from one another on every mission.

Anderegg noted that as the flight instruction and debriefs became more intense, pilots began directing additional "discipline, . . . motivation, . . . [and] intellectual curiosity" toward their ground studies.[259] Granted, a pilot didn't need to know how to build the Eagle's radar and the missiles it carried, or the scientific theory and mathematical calculations that made them work, but he did need to become an expert in their operation and employment across a variety of dynamic scenarios. Pilots no longer had time to think through the complex "decision matrix" that confronted them during an air-to-air intercept, and they couldn't always wait for the CC-computed displays to appear in their HUD or VSD. They needed to learn to anticipate the eventual computer output just as they had learned to anticipate which radar mode they needed to select with their HOTAS during a dogfight. "You have to have already studied that," Anderegg explained. "You have to already know the intricate working details and the interrelationships of all those systems before you ever put the engine in afterburner and raise the landing gear. . . . The days of throwing your scarf around your neck, . . . and pulling your goggles over your eyes, and pointing over the side and saying, 'Let's go after them, boys!', those days are gone."[260]

The effects of the new training scenarios, analytic thinking, detailed instruction, and additional book study were profound. Since the earliest days of military aviation, fighter pilots thought they could identify who among them was the best and most likely to become a future ace. It showed every time they flew, and there was often a wide gulf that separated the top pilots from the rest of the pack. Welch noted that, in his two years as the Wing commander at Langley flying the first operational Eagles, that gap had effectively disappeared. While acknowledging that there were probably some "air-to-air guys who were held up as better than others," Welch felt that pilot performance in the F-15 was remarkably consistent. The F-15, in his words, had become "a great equalizer."[261] Wallace agreed with the assessment, admitting his own difficulty when he tried "to identify who was good and bad in a tactical aspect. . . . There wasn't enough definitive that I saw [where] you could really say, 'OK, this guy's a top gun.' Everybody was a top gun."[262]

The post-AIMVAL-ACEVAL experiences of the Eagle pilots had the potential to shatter the myth of the fighter pilot. As Blesse had predicted after his brief observation of AIMVAL, with almost every pilot now regarded as a "top gun," the technology of radars and longer-range missiles appeared to have finally supplanted the pilot's traditional stick-and-rudder skills. The top fighter pilots were no longer restricted to those who had been blessed at birth with "good hands"—good fighter pilots could now be made, not just found. Some elements of the still-popular myth persisted, though. The air-to-air kill retained its allure as an intense, individual affair; it just so happened that that lone pilot could now be either a flight leader or a wingman. The gun too continued to provide a tangible connection to the classic fighter pilot mystique. Even after the shift in emphasis from close-range dogfights to longer-range intercepts, the presence of the Eagle's gun was repeatedly invoked to affirm the vital role of the individual human pilot seated in the cockpit. Retired brigadier general Robinson Risner, a decorated Sabre ace from Korea who went on to fly F-105s in Vietnam before closing out his career supervising preparations for AIMVAL-ACEVAL, explained, "It is still the human being that makes the airplane perform. . . . If it is an air-to-air fight, you fire your missiles; you try to kill him off at long-range, and when some get through, you bring him under fire with your Gatling gun. That requires pilot ability. There is no technological thing that you can use to get behind that airplane. It is pure intuition and training."[263]

But there were also plenty of other situations when the pilots' distinctly human capabilities, his "intuition and training," proved essential to effective employment of the F-15. It didn't matter whether it was firing the gun during a dogfight or working the radar and shooting missiles, pilot aggressiveness and tenacity, when properly coupled to the capabilities of the aircraft, remained vital to mission success. Even Risner, although continuing to underscore the close-range dogfight as the ultimate test of a pilot's skill, acknowledged the necessary symbiosis between human and machine that extended across all combat regimes, "You still have the requirement of pilot capability, just pure old pilot capability, hooked to the technological capabilities of the aircraft."[264] Welch offered a similar appraisal. For him, "the real distinction" in the Eagle "had to do with . . . whether or not you seemed to have the leadership and the savvy to lead a flight to a successful mission."[265] The new technologies and more focused instruction might have helped normalize pilots' employment of their weapons, but they had not obviated the role of heroic pilot in the air.

Anderegg offered a glimpse into what those heroics would entail in the years after AIMVAL-ACEVAL: "When you're leading a flight of four F-15s at 35,000 feet, supersonic, closing on ten adversaries, . . . [and] you're managing [your] radar, [and] you're managing the radar of your three other fight mates, [and] you're managing your radar warning receiver, [and the] warning indications from your other three flight mates, and you're able to employ just on your airplane two or three very different kinds of missile systems, you need to have your shit together before the fight starts."[266] But then again, every successful pilot, whether he was flying a Sabre, a Phantom, or an Eagle, needed to have his "shit together" too; it just amounted to different "shit."

Conclusion

Declared by some to be the "world's most successful jet fighter," the twin-engine, twin-tail F-15 would accumulate an unmatched 104–0 combat record over the next four decades.[267] From the outset, the Eagle was intended to be a classic dogfighter and proper heir to the storied F-86 of MiG Alley fame. It would be highly maneuverable, focus solely on air-to-air combat, be armed with a gun, and have only a single seat. It represented not only everything that the F-86 was but also everything that the F-4 Phantom was not. Except, of course, for one crucial element. Like the Phantom, the Eagle would carry guided air-to-air missiles and use a powerful radar to support them. Neither the Phantom's radar nor its missiles had earned a stellar reputation in the early stages of the Vietnam War, but the need for the advanced armament represented one of the key lessons the Sabre pilots had learned as they watched MiG-15s trace lazy contrails just above them. The Eagle's engineers might have planned for their aircraft to have unparalleled maneuverability, but they still wanted it to have a level of redundancy and design flexibility unachievable in a machine-gun-only-armed fighter. Hence, rather than renouncing the Phantom-model of air combat to embrace that of the revered Sabre, the Eagle was in reality a compromise between the two.

It took time for many to come to this realization. Several, including the TAC commander Momyer, questioned why the Air Force's newest dogfighter, supposedly patterned after the Sabre, needed such complex and costly avionics. The criticisms sent those working on the F-15 program scurrying to justify the requirement for the Eagle's radar. Their supporting arguments generally refrained from emphasizing the potential value of guided missiles in combat; instead they outlined the advantages the radar would provide to a pilot as he maneuvered for a close-range dogfight. Once the need for the radar was acknowledged, then it was obvious, they claimed, that the radar had to be a new, advanced PD-type, because only a radar that could effectively eliminate ground clutter was capable of being operated by a single pilot. Why only a single pilot? Cost, aircraft performance, the interpersonal conflict experienced in the two-seat Phantom, and the simple fact that the air superiority mission historically had been accomplished by single-seat fighters were all proffered as answers.

When the Eagle finally rolled out of McDonnell Douglas's assembly hangar in 1972, it was nearly double the size of the Sabre and "a good deal larger than the F-4."[268] It had a massive thirty-six-inch PD radar antenna buried in its nose, and pilots began to hear rumors that they would need to be "double-jointed" to operate its controls. Hoag told his fellow Air Force fighter pilots not to worry, "The entire air-to-air function is a simple, automatic process which frees the pilot to fly."[269] It was a pronouncement that the World War II and Korean fighter pilot veterans had heard many times before. And, like most past utterances, it proved to be false. But the statement's speciousness was not tied wholly to lingering engineering oversights in the weapons system's automation; it was misleading because it implied that the most demanding and important task a fighter pilot accomplished was the flying itself.

When the myth of the fighter pilot first emerged, stick-and-rudder flying had in fact consumed the majority of the pilot's attention as he struggled to keep his aircraft aloft. However, the demands on the pilot's flying skills steadily diminished over the ensuing decades as the act of flying became merely a means to the eventual ends of aerial victory. By the time of the Sabre, flying skills remained important, but piloting the aircraft was no longer in and of itself especially challenging. The divergence widened with the Phantom and then eventually reached chasmlike proportions in the Eagle. Flying the F-15 was simple and brought exhilaration, but it did not usually bring glory; that would be reserved for those who had mastered the Eagle's radar and weapons, and the maze of HOTAS functionality that controlled them. However, this steady transformation of the fighter pilot's skills went unacknowledged as the public and pilots alike continued to associate the terms *flying* and *pilot* with their historical connotations. The label "fighter pilot's fighter" that was attached to the Eagle relied on these dated definitions, steeped as they were in the persistent myth of the fighter pilot.

The increasing divergence between *flying* a fighter and *employing* a fighter was revealed once the Eagles started populating Air Force flight lines. No longer content to merely accumulate flight hours, the pilots began arranging complex, realistic training exercises that would allow them to practice their demanding long-range tactics. Then, to maximize their and their flight mates' learning during the challenging sorties, the pilots began to rely on a broader array of data sources to help assess their performance, including HUD film, ACMI, and eventually VSD recorders.

AIMVAL-ACEVAL exacerbated the flying–employing dichotomy. By most accounts, the yearlong set of tests failed to accomplish its stated purpose but proved invaluable in the process. As Dvorchak explained, every "good thing about ACEVAL and AIMVAL, either one of them, was by mistake. We learned a hell of a lot, because we didn't know the questions to ask when we went in."[270] Some of the lessons were focused on technology. The value of a new, longer-range, semiautonomous and multitargetable radar missile like the AMRAAM, as well as a new track-while-scan radar mode, became clear in the skies north of Nellis AFB. More important than updated hardware capabilities, though, was the pilots' realization that against an anticipated Soviet threat armed with an all-aspect, fire-and-forget, IR-guided missile, an F-15 pilot couldn't simply rely on his flying ability to ensure aerial success anymore. If he did, he was likely going to get himself killed pretty quickly. It wasn't the "best hands," Anderegg declared, but the "best head" that now defined a good fighter pilot in the Eagle.[271]

Despite the shift to long-range radar tactics that emerged post–AIMVAL-ACEVAL, the F-15 pilots were not reduced to the pushbutton missilemen their forbears had once feared. The democratization of fighter aviation, a goal since a group of young Korean aces had argued on behalf of radar-ranged gunsights in 1952, had taken a significant step closer to realization with the Eagle and long-range air combat. While there would always be a place for natural flying talent in fighter aviation, the gates were now open to those individuals not specially endowed with the attributes that had historically predetermined one's worthiness to become a fighter pilot. Still, no one would suggest that a successful Eagle pilot lacked fighter pilot skills. They were just now different fighter pilot skills.

"Abby" Sveden's air combat engagement during Desert Storm that was described in the introductory chapter profoundly illustrates the metamorphosis in fighter aviation that accompanied the post–AIMVAL-ACEVAL Eagle. With a radar and an AIM-7 missile, Sveden with only a few hundred hours' experience in the F-15 was rendered as lethal as his Weapons School graduate flight lead, "Gunga" Dingee. Neither Dingee nor Sveden's element leader, Prather, had voiced a concern about the young wingman flying out in tactical formation, locking his assigned sort, or launching his missiles from miles away. He had done exactly as they had trained him, and he earned sole credit for his kill because of it. Granted, Sveden was flying a newer C-model of the Eagle that had been slightly modified to incorporate the technological lessons of AIMVAL-ACEVAL, but most of the systems in the jet, including the HUD and HOTAS, remained the same as when the jet was first optimized for employment in "close-in, highly maneuverable combat."

In addition to the flexibility of the airframe and the technologies therein, the details of Sveden's victory reveal the flexibility of the human operators as they adapted with their technological hardware. Without the new formations, new training methods, and new coordinated radar tactics that were developed following AIMVAL-ACEVAL, Sveden likely would not have earned his victory that January afternoon. The new tactics represented the pilots' best attempts to use their new machinery in an optimum fashion. They were distinctly social phenomena that, although related to the arrival of the F-15, had not been predetermined when the engineers first drew up the Eagle's design. Cliver, specifically referencing the AIMVAL-ACEVAL experience but also alluding to fighter aviation more generally, explained, "That's what's beautiful about having people in airplanes. Guess what? You're going to figure out how best to use what you are provided."[272]

Cliver's comment indicates an appreciation for the human-and-machine perspective, but it also conflates the operator's physical presence with the coevolutionary process. The Eagle flight leader's use of coordinated radar tactics to extend his control across multiple radars in his formation was one indication that physical presence in a particular cockpit was not in and of itself vital. That seemingly minor detail would have profound implications over the coming decades when new technological developments started pushing pilots out of the cockpit and into remote ground control stations.

CHAPTER 6

Conclusion

> War shall yet be, and to the end;
> But war-paint shows the streaks of weather;
> War shall yet be, but warriors
> Are now but operatives; War's made
> Less grand than Peace,
> And a singe runs through lace and feather.
>
> HERMAN MELVILLE, "A Utilitarian View of the *Monitor's* Fight,"
> *Battle-Pieces and Aspects of the War*, 1866

The Union navy's USS *Monitor* of the American Civil War was a peculiar vessel—part iron raft, part submarine. Only the ship's conspicuous revolving turret, which shielded two eleven-inch Dahlgren guns behind nearly a foot of armor plating, extended much higher than the waterline. Not quite sure what to make of the curiosity, several American literary authors of the period, to include Herman Melville, selected the *Monitor* as their subject. Many of them chose to cast the iconic ironclad vessel as a harbinger of the age of industrial warfare, when mathematical calculations and "engineering skill" would replace the "passions" of the fighting men. According to historian David Mindell, Melville's outlook was particularly unique, though, and his "Utilitarian View" the "richest" because he was able to capture the irony that others often overlooked.[1] That irony did not escape the ship's acting assistant paymaster, William Keeler. Destined to serve in its dank but otherwise well-appointed and well-shielded officers' cabin, Keeler wrote his wife upon seeing his new vessel, "There is not danger enough to give us glory."[2]

Glory would come to the crew of the *Monitor*, but not according to the classic "war-paint" and "lace and feather" tradition that had marked previous naval heroics. After the *Monitor*'s thunderous four-hour fight-to-a-draw battle against the Confederate ironclad *Virginia* in the waters of Hampton Roads on March 9, 1862, President Abraham Lincoln and the Union navy were reluctant to risk their prized ship in another battle. Thereafter relegated to riverine and blockade duty, Keeler complained that his ship was placed in a "glass case," its crew imprisoned "in the bowels of our iron monster."[3] "The only fear I have," Keeler lamented, "is of getting eaten through by rust."[4] Despite their inability to showcase their valor on the open sea, the men aboard the *Monitor* nevertheless "experienced the full flush of military glory." It was, in Mindell's words, the "paradox of the war machine: political imperatives of developing, selling, and deploying a complex technology gave new life to the heroic aspect of war by promoting the *Monitor* and its crew, even as the machine's operations reduced their status from warriors to operatives."[5]

Mindell notes, however, that this transformation did not eliminate the opportunity for the *Monitor*'s "operatives" to demonstrate their heroism. The leaky, smoky ship nearly suffocated its crew during its first voyage out of New York harbor, and the ship's "basic lack of seaworthiness," which culminated in her loss in a gale on New Year's Eve, 1862, became well known within the navy. As Keeler acknowledged in a later letter, "anyone could fight behind an impenetrable armor," but volunteering "to undertake the trip and go into the fight in an untried experiment and in our unprepared condition" was itself worthy of "the credit, if any is due."[6] Thus, while the members of the *Monitor* crew initially may have worried that their new mechanical curiosity would rob them of their role in warfare, Mindell argues that "a place for human skill, and hence for heroism," remained. It was a "*contingent* heroism," though, inextricably linked to the performance of their war machine and, therefore, "always in question and always in need of maintenance."[7]

In concluding his study of the *Monitor* and its crew, Mindell hypothesizes that the fighter pilots of the First World War might have had much in common with the "iron coffin" sailors of the American Civil War. For the new aviator aces, as for the sailors a half century earlier, "heroism increasingly meant mastering the machine as much as destroying the enemy." Indeed, a pilot like Richthofen or Rickenbacker could not accomplish the latter without first developing the skills demanded by the former. And therein, Mindell suggests, lay some of the irony of the new flying combatant; even the acknowledged masters could never escape the "inherent danger of the machine." The uncomfortable reality was regularly reinforced whenever a pilot saw his friend's wings suddenly crumple or his engine erupt in flames. The aces' relationship with their aircraft cut both ways: it could be "simultaneously protector and weapon, potential savior and potential tomb."[8]

The Irony of the Fighter Pilot

The irony did not suddenly disappear when the first military pilots returned home from the Western Front. Nor was it solely restricted to the fighter pilot's prospect of sudden fame or gruesome death based on his machine's ability to withstand the rigors of air combat. Rather, an equally ironic and potentially more informative element centers on how the collective image of the fighter pilot and aerial heroism remained relatively static in the ensuing decades, despite the dramatic transformation in the skills that the profession demanded. Granted, from its outset the myth of the fighter pilot—an individual who marshaled an aggressive and tenacious spirit and superior stick-and-rudder flying skills to conquer opponents in valorous aerial battle—had always sacrificed historical "richness" and "contingency" in favor of its simpler, more symbolic representation. But the discrepancy between the myth and the actual fighter pilot experience was exacerbated following the introduction of advanced fire control systems and their associated missile weapons after 1950. These new technologies threatened to replace the knightly, "warrior" image of the fighter pilot with that of a scientist or, worse yet, a push-button missileman "operative." The editors at *Life* had surmised as much in early 1952, and they publicly wondered what was to become of the fighter pilot?[9]

Addressing this question requires us to move beyond the often captivating but one-dimensional "great machine" or "great pilot" narratives that dominate the historiography and instead reassess the history of fighter aviation from a human-and-machine perspective. By constructing a cognitive history of the fighter pilot, we can begin to appreciate the complex and ever-evolving interactions that linked the pilot with his machine. This approach also reveals how changes within an individual cockpit profoundly altered the pilots' social conventions and relationships outside the cockpit.[10] In the end, understanding how the fighter pilots of the past responded to their new fire control technologies can inform today's discussions about increasing automation in our workspaces.

The different formations that the fighter pilots flew in combat were one manifestation of their shifting interpersonal relationships. During the First World War, pilots realized that coordinated, multi-aircraft action maximized their chances of collective survival and success, and they consequently began arranging themselves into mutually supportive formations whenever they took to the air. The soon-agreed-upon convention placed the more senior pilots out in front where they could concentrate on offensive action; the junior pilots in contrast focused on defensive lookout. World War II reinforced this division of tasks. By the advent of jet combat in MiG Alley, the distinct leader and wingman responsibilities were firmly ingrained within the American fighter pilot experience. Corresponding social conventions had also emerged, which at times demanded the wingman's near-total subjugation to his flight leader. Even the wingman's contributions to his leader's victories were often purposefully obscured. While there were still obvious tactical reasons for designating a "shooter" lead and "looker" wingman when battling MiGs over Korea—not the least of which was the speed at which a speck on the canopy could transform into a fireball-spewing MiG—the wingman experience had by then assumed an ancillary purpose, serving as an institutionalized rite of passage for a young, budding fighter pilot in the US Air Force.

By the time of the F-4, the experience of being "blooded" as a wingman flying in "fighting wing" had supplanted the classic formation's tactical utility. Snuggled up close to his leader's aircraft, the Phantom wingman was unable to employ any of his missiles if his flight leader needed support. Additionally, because the F-4 was much more powerful and maneuverable than previous fighters, the fighting wing position also regularly compromised the wingman's ability to perform his defensive, lookout task, since he had to constantly focus his attention on following his leader's abrupt maneuvers. Some fighter pilots advocated for a change, but they encountered stiff resistance. They quickly learned that adjusting the tactical formation to take better advantage of the F-4's missiles required not just a geometric shift in the wingman's position but also a psychological one in the minds of the veteran fighter pilots. Moving the wingman to a more line-abreast position represented a more equitable partnership between the two aircraft. Senior flight leaders would no longer be able to devote most of their attention to offensive employment; they would now be as responsible for their wingmen's survival as their wingmen had historically been for theirs. Moreover, the newly proposed formations threatened the flight leader's historic prerogative of seizing the glory of aerial victory for himself. Consequently, despite potentially viable alternatives, the welded wing formation and the social hierarchy it represented remained largely unchanged throughout the Vietnam War.

The Phantom's second seat generated another significant interpersonal challenge for the fighter pilots. The introduction of a GIB dedicated to operating the Phantom's complex radar shattered the idyllic image of the fighter pilot as self-sufficient, lone eagle. The impact could be seen in the young, newly winged pilots' scorn for the Phantom's second seat. Despite the F-4's record-setting speed, few wanted to spend their first years as an Air Force pilot riding in the back of a fighter staring at a radar screen. What they should have been doing, they thought, was what they had been trained to do: fly in the *only* seat of a proper fighter like the F-100 or F-105. Moreover, the disdain in the cockpit was often mutual. The front seaters, already reluctant to acknowledge their evolving partnership with somebody flying alongside them, were even more recalcitrant in acknowledging their dependence on somebody who had invaded the sanctity of their aircraft. The fact that the offending individual wore the same set of Air Force wings on his flight suit only exacerbated the tension.

The F-15 Eagle, advertised as the "fighter pilot's fighter" and heir to the fabled Sabre, promised to revive the image of the single-seat fighter pilot and the cherished autonomy it represented. Because the Eagle wasn't expected to perform most of its combat at long range or in adverse weather, there was no need, or any desire, for another GIB. However, once the Air Intercept Missile Evaluation and Air Combat Evaluation demonstrated the lethality of the short-range environment, the Eagle pilots were forced to adjust their tactics and, consequently, their expectations of their fellow aviators. The pilots already had discarded the fighting wing position of Korea and

Vietnam in favor of a wider-spaced "tactical" formation, implicitly accepting the more equitable partnership between leader and wingman that it represented. But, to maximize the effectiveness of their pulse-Doppler radars and their guided missile armament, the Eagle pilots also had to develop new communication strategies that would allow them to seamlessly share information across the formation. These new long-range sorting tactics, which made little distinction between leader and wingman, signaled the end of the strict "lead-shooter, wingman-looker" responsibilities that had been so central to the historic fighter pilot experience. Accompanying the new tactics was an unprecedented emphasis on providing detailed instruction to all flight members in order to maximize their learning during the increasingly complex training scenarios. With that, the wingman's traditional experience of proving his worth by figuring things out on his own was largely abandoned, and his position transformed into an apprenticeship pursued under the watchful eye of a more experienced and more interested master.

In addition to the shifting interpersonal relationships, there were several intracockpit changes that altered the fighter pilots' historic experience of air combat over the ensuing three decades. When the first A-1C(M)s began arriving in Korea, the pilots were told that the new gunsights would automatically solve the challenging mathematical calculations that defined a successful gun attack; the pilot would simply need to maneuver to place a mechanically aimed pipper over his radar-ranged target before squeezing the trigger. The combat debut of the new gunsight, however, was a flop, a function of poor design, unreliable electronics, and faulty assumptions of air combat invoked when developing the system. As the pilots were apt to point out, the final aiming and squeezing of the trigger accounted for only 10 percent of the total air combat problem. The new A-1C(M) not only failed to help with the other 90 percent, it actually made its accomplishment all the more difficult as the Sabre pilots struggled to maneuver their heavier aircraft into position behind their target and within range of the same .50-caliber bullets that many of them had used years earlier in their P-51s and P-38s.

While the initial deployment of the A-1C(M) was a disappointment and did not appreciably alter the tasks of the fighter pilot, it signaled many of the changes that would accompany later fighter armament and fire control systems. Most significantly, once the Sabres' radars and gunsights had been restored to factory condition through Project Jaybird, the pilots realized that, to use their radar-gunsight system effectively in air combat, they had to actually study and become knowledgeable in its operation. Today, the requirement seems commonsensical, but two preconceptions conspired against the pilots in Korea. First, for many pilots it was their initial exposure to a fire control system advertised as being completely automatic. The engineers at MIT had in fact earlier discouraged the pilots from developing their own intuitive understanding of the A-1C(M) for fear that they would deviate from the gunsight's direction. Second, fighter aviation was still seen as the province of naturally gifted aviators. While the pilots' innate flying instincts might have gotten them airborne in a Sabre with only a bare minimum of training, the pilots had to come to the realization that those same natural abilities did not extend automatically to their new fire control systems. At first disconcerting, the differentiation of skills was eventually embraced. Indeed, when controversy erupted over the future of radar-ranged gunsights in fighter aircraft, a key argument in favor of the complex fire control systems was that they didn't rely on the same set of "birdman" skills, honed by years of air combat experience, that had characterized successful fighter pilots in the past. The new systems would, in effect, help democratize fighter aviation, opening it to individuals who possessed a broader range of intrinsic flying acumen.

The F-4 Phantom with its all-missile ordnance embodied this democratization principle. The Navy-designed aircraft therefore constituted a significant threat to the classic tasks of the fighter pilot. In the past, Sabre pilots employing machine guns needed to deftly manipulate their aircraft's stick and rudder to maneuver into a position close to their adversary and near his six o'clock before opening fire—the closer and nearer, the better. Flying intuition would help the Sabre pilot get there, and then luck (or maybe an operable A-1C[M]) would help him bag his prey. In the Phantom, flying to that same position behind the adversary would eliminate *any* chance of aerial victory; the Phantom's Sparrow and Sidewinder missiles simply could not be employed from that position. Rather, the F-4 pilot was supposed to leave the aggressive maneuvering to his missiles. The new arrangement, it was believed, would render fighter aviation simpler, at least from the perspective of the pilot.

The circumstances of the Vietnam War and the pilots' inability to employ their missiles from longer ranges upset the plan. The F-4 pilots eventually realized that they needed to continue cultivating the flying skills that would allow them to dogfight a smaller, nimbler MiG. Simultaneously, they had to develop new intuition that would guide them as they tried to employ their missiles from a mystifying offensive position that constantly morphed and shifted based on the changing geometry between attacker and target. For those flying Phantoms in Vietnam, additional requirements were thrust upon them due to the panoply of air-to-ground missions that they regularly executed. The net result, in the words of Welch, was that the F-4 pilot was transformed into a "systems operator," responsible for managing not only his aircraft but also his semiautonomous missiles, various air-to-ground weapons, and (through his GIB intermediary) his radar.[11]

The single-seat, single-mission focus of the F-15 was designed to alleviate some of the systems-level responsibilities that had been levied on the Phantom pilots. Simply flying the Eagle was made easier with the introduction of a dual, hydromechanical and electronic flight control system. There was neither a GIB to manage nor any air-to-ground munitions to worry about. And, while the aircraft was armed with Sparrow missiles and equipped with a powerful radar to support them, the F-15's designers had exercised great care to optimize the fire control system for the Eagle's planned close-range battles. Fighter pilots, the Eagle's proponents advertised, would finally be able to exercise their flying intuition without distraction.

Their promise, however, misrepresented the tasks of modern fighter aviation. It suggested that *flying* a fighter and *employing* one were still synonymous. They weren't. The importance historically ascribed to manipulating the aircraft's stick and rudders had been supplanted by the fighter pilot's ability to effectively use his advanced fire control system. AIMVAL-ACEVAL and the subsequent emphasis on long-range radar tactics illuminated the widening chasm. While some of the Eagle pilot's tasks resembled those in the Phantom, others were vastly different. The fighter pilot in the Eagle no longer had to hunt for a particular "fuzzy blip" among a sea of "fuzzy blips." Instead, he worked with discreet target bricks that appeared on a synthetically generated, digital representation of a radar scope. Additionally, the individual intuition that had previously guided the Phantom pilot's employment of his missiles had now been programmed into the Eagle's high-speed, automatic central computer, and the resulting solutions prominently displayed for the pilot in the head's up display. These and other advancements addressed some of the employment difficulties that had plagued the Phantom, and they in turn freed the Eagle pilot to devote significant attention to processing all the information now available on the HUD and vertical situation display. The ability to integrate this abstract information, coordinate action across the formation, and then execute the proper tactic became the hallmark of a skilled F-15 pilot.

Despite the remarkable thirty-year transformation within their cockpits and across their formations, the pilots still maintained their powerful, romantic connection to their individual aircraft. The changing technologies had not fundamentally altered that oneness which was central to the fighter pilot experience since the Great War. It was present even in the Phantom. Reminiscent of Salter's description of his Sabre, Olds reflected:

> Man merges with machine; he doesn't simply use it. You don't climb into an aircraft and sit down. You strap the machine to your butt, become one with it. Hydraulic fluid is your blood; titanium, steel, and aluminum, your bones; electrical currents, your nerves; the instruments, an extension of your senses; fuel, the food; engine, the power; the control surfaces, the muscle. You are the heart, yours is the will, yours the reasoning power. You are something more than earthbound man. You are augmented and expanded by the miracle of the machine. You are tied to it physically and you are part of it emotionally.[12]

In the Eagle, pilots now also had a powerful PD radar to serve as their eyes and a myriad of HOTAS (hands on throttle and stick) switches and buttons—new neural connections—to link their brain with the CC brain of their aircraft. The increased automation introduced with each successive generation of fighter aircraft had not, in the fighter pilots' opinion, subverted their autonomy, because in the end, the pilots' intuition and skill were never subsumed by their new equipment. A pilot was always needed to make the system work, to provide the "heart, . . . the will, . . . the reasoning power."

Some concluded that the pilot's persistent and essential role managing the new fire control systems demonstrated the inherent limitations of automation within fighter aviation. Air combat, they suggested, was too dynamic and the decision matrices too complex to be programmed into mechanical or electronic computers. Each attempt to automate a fighter pilot's tasks—be it stadiametrically ranging a target, maneuvering into position to employ weapons, or manipulating and interpreting reflected radar energy—had never matched initial expectations. In the end, the once-touted technologies had only saddled the pilot with additional tasks as he compensated for the automated systems' failings. Indeed, a common refrain of the "Reformers" during the 1980s was that engineers and service leaders needed to stop burdening fighter pilots with these costly, complex, and failure-prone devices; the pilots should be left alone to do what they did best—fly.

The ironic thing is that while flying continued to distinguish the pilot as "something more than earthbound man," it was never as all-encompassing a task as the myth of the fighter pilot suggested. The first military pilots and the engineers that responded to their requests were able to look past the tales the publicists spun and recognize that the act of flying—manipulating the aircraft's stick and rudder to stay aloft—was but a means to an eventual end. Consequently, there were from the earliest days attempts to transfer some of the pilot's flying responsibilities to his machine. Fighter pilots longed for an aircraft that possessed just enough stability so that they could shift their attention to scanning the sky or clearing a jammed machine gun without worrying their fighter would suddenly go out of control. But the pilots also wanted to retain the option to force their aircraft to the edge of its flying envelope, flinging it about the sky as needed to either prosecute an attack or thwart one. Henceforth, a fighter pilot's skill would increasingly be defined by his ability to rapidly reallocate his attention, deciding when, where, and how to intervene in order to exercise more direct command over the machinery that surrounded him.

The history of fighter aviation is therefore also a history of the struggle to understand automation in its broader sense. Only decades later would Thomas Sheridan and others help illuminate the various gradations within the term *automation* itself. Previously, designers, engineers, and operators all struggled to articulate how their various fire control systems fit within the automation rubric. The initial design of the A-1C(M) represented automation applied in the strict sense. It granted the pilots little agency, effectively reducing them to the role of mere "operative," and it stoked their fears of being further deskilled by future automated systems. But within months of the new gunsight's introduction in combat, pilots and engineers realized that this strict separation of tasks between pilot and machine was inefficient. A more balanced and flexible approach was needed. In the case of the A-1C(M), the pilot was subsequently encouraged to understand and optimize the performance of his fire control system for particular situations, which he exercised on the basis of new procedural and technical knowledge and novel control mechanisms such as Jenkins's range limiter. The shift in expectations was akin to recognizing the difference between "full automatic control" and "human supervisory control," a distinction later popularized by Sheridan.[13]

Lacking the specific *supervisory control* terminology, however, did not hinder development of later fire control systems that embraced the concept. While many continued to advertise the benefits of automation in their advanced fire control systems, notably their systems' purported ability to free the pilot to perform other essential flying tasks, there was a growing appreciation among engineers and pilots alike that the new automated systems also came with their own set of critical tasks, and that these tasks could often be performed only by the human pilot seated in the cockpit. One of the great ironies within fighter aviation during the period 1950 to 1980 is that the systems the pilots once feared would push them *out* of the cockpit in fact became increasingly reliant on skilled pilots remaining *in* the cockpit. And, as long as they remained in the cockpit of fighter aircraft, the operators would continue to be known as fighter pilots. The airmen realized this, and the anxiety they revealed in the pages of *Fighter Weapons Newsletter* in the 1950s (e.g., figure 1.3) eventually subsided.

But, while the new fire control systems validated the pilot's necessary presence in the cockpit and therefore his place in the fighter pilot myth, the new equipment also dramatically transformed the skills required of him. By the time of the F-15, the fighter pilot's flying intuition had been largely supplanted by radar intuition, and his skill manipulating the control stick overcome by his skill operating the HOTAS. Moreover, these new fighter pilot skills were not readily transferable between aircraft. Whereas in the past a pilot could transition from one aircraft to another with relative ease and often without formal training, it became increasingly difficult to do so in modern fighters: an F-15 pilot was as uncomfortable initially flying an F-4 as an F-4 pilot was trying to figure out an F-15's HOTAS.[14]

This unique pairing of specific pilot skill with particular aircraft was one manifestation of the evolving human-*and*-machine *system* within fighter aviation. The introduction of highly complex, technologically advanced, and procedurally driven systems had not eliminated the requirement for hands-on expertise and intuition. Supervising systems and integrating information may not have conformed to the fighter pilot myth as the means by which heroic pilots instilled their "heart" and "will" into their aircraft, but it nonetheless more accurately reflected their role in the modern cockpits. Importantly, the pilots eventually realized that this shift did not lessen or cheapen their fighter pilot experience. It happened to be the same lesson learned by the sailors aboard the *Monitor* a century earlier.

Regrettably, the historical "richness" and "contingency" within fighter aviation is often obscured by the still-popular appeal of the fighter pilot myth. "Great machine" and "great pilot" narratives, and the ahistorical interpretations they perpetuate, continue to flourish. Even today, many fighter pilots neither understand nor appreciate the evolving nature of their tasks, their changing relationships with their technologies, and the implications therein. The effects can be pernicious.

A Lesson for Future Automation

Although this study terminated with the F-15A, the push toward greater automation in fighter aviation did not abruptly end when the C-model version of the McDonnell Douglas Eagle started rolling off the production line in St. Louis in 1979. Semiautonomous, fire-and-soon-forget air-to-air missiles like the AMRAAM, bombs steered by satellite signals, and more complex aircraft and fire control systems that facilitated the weapons' employment continued to propagate throughout the Air Force. For example, the Air Force's newest fighter, the F-35 Lightning II, requires some 9.5 million lines of computer code for flight operations, three times the amount needed in the service's next-newest fighter, the F-22 Raptor.[15] Commenting on the remarkable pace of development, Welch noted that "today's fighter pilot is orders of magnitude more effective than I ever was. . . . We give them the opportunity to do with one airplane what used to take a couple of squadrons a week to do." But, Welch also admitted, "I would never have been interested in becoming today's fighter pilot," and the former chief of staff of the Air Force suspected that today's fighter pilots, often seemingly relegated to "simply transporting ordnance," are "probably not terribly happy about it" either.[16]

Welch's assessment is not atypical. It reflects the common tendency for fighter pilots to emphasize their own particular set of skills, optimized for their particular type of aircraft, which they feel most closely connects them to the classic fighter pilot myth. This same mindset also often causes fighter pilots to discount the skills and any potential personal satisfaction that might come with mastering the newer equipment outfitting the latest aircraft. Confronted with ever-more fighter aircraft automation, the veteran pilots' connection to their cherished myth would be further strained in the 1990s and 2000s.[17]

The Lockheed Martin F-22 Raptor, one of the Air Force's newer fighter aircraft that Welch referenced with his comment, provides an example of the dominant trend.[18] In the Eagle, pilots like Welch prided themselves on being information integrators, analyzing data from a variety of federated sources, including the four-inch square, monochrome VSD radar display, the HUD, the RWR (radar warning receiver) display (located on the opposite side of the instrument panel from the VSD), and extensive radio communications. In the Raptor, the pilot's information integration task was largely automated through a process known as "sensor fusion." Describing the process, one industry news report explained that Raptor pilots have "all the available data on the airspace fused and displayed on a single, easy-to-read screen."[19] If the Raptor's

FIGURE 6.1. F-22 Raptor Cockpit. The Raptor's center eight-inch-square and three surrounding six-inch-square color displays dominate the cockpit. All are pilot-customizable, but the community standard is to configure the center screen for tactical situational awareness, the left screen for defensive information, the right screen for attack information, and the bottom screen for weapons status and aircraft performance information. The F-22 ejection seat has been removed in this picture. Lockheed Martin photo by John Rossino. Reproduced by permission.

computers determined a target to be hostile, the aircraft symbol on the F-22's center, eight-inch-square color display appeared in red; friendlies appeared in green; unknowns in yellow (figure 6.1). When the F-22 was first fielded, the pilot had very little insight into which sensor detected the target and caused it to appear on the display; his or her only concern was that it had appeared and its current color.[20] Several industry publications attributed the "F-22's unprecedented combat effectiveness" not to its radar-stealth properties but to the aircraft's "ability to perform fusion of multisensor information," which, echoing the earlier promises of automation in fighter aviation, "significantly reduc[ed] the pilot workload during battle conditions."[21] Even voice communications had been significantly curtailed in the Raptor through an advanced fighter-to-fighter datalink.[22] The Eagle pilots, however, saw the F-22's advanced automation and sensor fusion as an ill-advised attempt to steal more control away from the pilot. When a twelve-ship deployment of F-22s encountered a nearly catastrophic software error while attempting to fly across the International Date Line in 2007, the Eagle pilots howled with laughter and "I told you so's."[23]

Yet even with the high degree of automation built into the Raptor's cockpit, its pilots were never reduced to ordnance transporters and simple push-button missilemen and missilewomen. Freed from the information integration tasks of the Eagle, the F-22 pilots could now shift their attention to planning in real time their air and ground attacks into heavily defended airspace, regions that aircraft like the F-15 or F-16 typically avoided due to the challenging threats. The tasks required to operate in these areas were distinct from those demanded of pilots in earlier fighters. One Lockheed executive noted the shift, explaining that junior pilots plucked to fly the Raptor fresh from pilot training were performing exceptionally well, largely because "the new pilots have far fewer habits to unlearn" than the older, more experienced pilots coming from other airframes like the F-15.[24] Moreover, contradicting Welch's appraisal and even that of some newer Eagle pilots, few pilots in the Raptor expressed misgivings that their aircraft's automation was robbing them of their fighter experience. They did, however, occasionally acknowledge that their advantage at long-range air combat limited their chances to engage in "fun" dogfight maneuvers.[25]

Far more unnerving to fighter pilots than the automated F-22 experience was the Air Force's increasing inventory of RPAs (remotely piloted aircraft). The emergence of novel control technologies and high-bandwidth, low-latency satellite data transmissions meant that, in the case of RPAs, pilots no longer needed to be seated in the same cockpit as their equipment; they could now perform their supervisory tasks from the other side of the globe (figure 6.2). With their human flesh no longer physically present in the sky much less over the battlefield, and their cockpits more resembling a poorly designed computer workstation, significant controversy erupted over the professional identity and proper recognition due the growing ranks of RPA operators.

Two incidents stand out. In June 2012, Air Force major Dave Blair published a short essay in the Air Force's professional periodical, *Air & Space Power Journal*. In it, Blair articulated an updated definition of *combat risk* and argued that, despite their physical separation from the battlespace, RPA operators should be entitled to receive combat medals for the "direct actions" they take that "affect the outcome [of battle]."[26] The response to the article on the journal's website was both swift and, to a large extent, severe. Several current service members and retired veterans were appalled. "You really can't be serious, can you?" questioned one respondent.[27] Reflecting a common theme in the responses, a retired lieutenant colonel explained, "One must experience [combat] firsthand," which, according to another Desert Storm veteran, entailed the "imminent

FIGURE 6.2. MQ-1/9 RPA "Cockpit." The cockpits for the Air Force's Predator and Reaper RPAs have undergone significant modification over the past two decades. One indication is the array of new computer monitors, supplying additional data and communication streams as well as helpful computer-based tools, that have sprouted from the main console. USAF.

threat of losing life or limb . . . while under direct fire from SAMs and AAA fire with nothing but thin sheet metal and a seat cushion in between."[28]

A few months later in February 2013, outgoing secretary of defense Leon Panetta announced the creation of the Defense Department's Distinguished Warfare Medal (DWM). Acknowledging that "technological advancements have, in some cases, dramatically changed how we conduct and support combat and other military operations," Panetta directed that the new medal would recognize "extraordinary achievement, not involving acts of valor, directly impacting combat operations . . . regardless of . . . the member's physical location." The DWM was to be ranked hierarchically just above the Bronze Star.[29] Panetta's announcement triggered a backlash similar to that which Blair encountered after publishing his essay. The Military Order of the Purple Heart declared in a press release that the new medal was "degrading and insulting to every American Combat Soldier, Airman, Sailor or Marine who risks his or her life and endures the daily rigors of combat in a hostile environment."[30] The Veterans of Foreign Wars likewise rebuked the Pentagon leadership, declaring that "medals that can only be earned in direct combat [like the Bronze Star and Purple Heart] must mean more than medals awarded in the rear."[31] One British commentator questioned if the new medal would "feature an armchair and a Coke machine or two crossed burgers," and many in the United States wondered the same.[32] The new secretary of defense, Chuck Hagel, abruptly cancelled the medal.[33]

The uproar over the DWM and Blair's essay revealed the stigma that many in both the public and the military attached to RPA operators who remained isolated from the physical dangers of combat. Even before the events in 2012 and 2013, it was common for RPA operators to be told in conversation with other military pilots that "flying your UAV [unmanned aerial vehicle] is like a big video game." Others declared the RPA job ideally suited to the incoming "X-Box generation" of young officers.[34] On one occasion, Blair encountered a "particularly vociferous (and inebriated) F-22 pilot" who explained that the problem with RPAs, and by extension RPA operators, was that "fighting a war via video teleconference isn't very honorable." It was a remarkably peculiar assessment coming from a pilot who flew cocooned in a nearly invisible stealth fighter and performed cockpit tasks that bore little resemblance to pilots' tasks from even a few years before.[35] But the nameless Raptor pilot's attitude wasn't unique. The director of the American Fighter Aces Association noted that many of his organization's members also regarded RPA operators as "less than true pilots."[36] A spokesperson for a Vietnam veterans' organization of fighter pilots explained, "When you tell a guy who flew an F-105 over Vietnam that this guy [in an RPA] is stressed out, he doesn't see it."[37]

But stressed out they were, albeit in different ways. When the Air Force began fielding RPAs in the 1990s, the aircraft's technologies and the corresponding tactics that governed their use were designed to minimize human involvement. The RPA pilot, seated in an isolated ground station, was basically tasked with "babysitting the autopilot" during the extended intelligence-gathering missions. Over the ensuing two decades, however, new technologies had emerged, the aircraft had assumed additional mission sets, and the RPA crews had moved beyond the mere "automaton" role they were once assigned. Tim Cullen, a former Air Force fighter pilot who studied the RPA transformation, observed that, once freed from the physical constraints of an aircraft cockpit, the RPA operators were uniquely empowered to "manipulate, coordinate, and integrate information from a constellation of people and automated tools . . . to create virtual worlds, and operators used the virtual worlds to identify patterns and to discover discontinuities they could correlate with friendly and enemy behavior."[38]

These were complex tasks that no machine algorithm could accomplish. These were also tasks that were never assigned to the human operators on the basis of "a master plan or conscious reflection about the role of humans and RPA in war." Rather, they emerged through an iterative process of human-machine collaboration, negotiation, and renegotiation. Although the RPA pilots admitted that their new information-integration tasks were "not as fun as flying fighters, bombers, or transport aircraft," the missions were "personally rewarding," Cullen observed. "As a mission commander of a RPA, they saw more battles, saved more lives, and killed more insurgents and terrorists than their airborne peers, and they cherished the relationships they developed with supported ground forces."[39]

The RPA crews' unique skills and the unprecedented level of coordinated engagement they facilitated over the battlefield, albeit virtual, came at a cost, though. An insatiable demand for RPAs soon emerged within the military, and the consequent "unrelenting pace of operations" stretched the service's deployed-in-garrison RPA crew force to the breaking point. The fact that the crews used ground stations that were just down the road from their homes to fly combat missions half-a-world away produced its own set of unanticipated complications, too.[40] Lacking any detailed understanding of what its RPA operators actually did when they walked into their ground control stations, the Air Force's attitude toward its RPA aircraft and the people flying them vacillated depending on the audience. At times, the service emphasized the RPA's ability to achieve incredibly unique effects in the battlespace—a "great machine" narrative. Other times, the service argued that its RPA pilots were no different from the pilots of its other manned platforms, similarly skilled and subject to similar personal and professional stressors—a version of the "great pilot" narrative. In reality, the machine and its pilots could not be disaggregated; the Air Force's RPAs achieved unique effects because the crews that flew them were unique. But lacking an appropriate understanding and coherent narrative to explain why that was the case, the Air Force never could develop a comprehensive strategy to guide its burgeoning RPA force, human or machine.[41]

In late summer 2015, the Air Force finally launched a Culture and Process Improvement Program (CPIP) to investigate some of the systemic problems that had plagued the RPA enterprise for years. At the outset of their study, CPIP officials acknowledged that the service's RPA operators did not enjoy "the same level of holistic quality of life and professional development" as other Air Force aircrews, a remarkable admission that itself was a long time coming. "A lot of assumptions were made over the years," one CPIP program official explained, "and people don't realize how stressful and overworked the [RPA] field is."[42] The pace was unrelenting indeed. General Hawk Carlisle, the commander of Air Combat Command, told Congress in early 2016 that RPA aircrews had been operating at a "surge" level for several years, working seven ten-hour days, six spent flying and one spent attending to nonflying duties, before earning two days off.[43] At least the grueling work also often went unappreciated. Senior leaders might have occasionally praised the contributions of their RPAs over the battlefield, but recurrent perceived bureaucratic slights and overt derision of the RPA force, like that encountered in the dust-up over the DWM, decimated morale in the RPA squadrons.[44] CPIP officials reportedly heard "over and over again" during their interviews that the RPA operators felt like "second-class citizens" in Air Force aviation.[45]

Regrettably, CPIP's proffered solutions may offer no more than symptomatic treatment. For example, to ease the demands on the RPA force, the Air Force has significantly ramped up RPA pilot production to more than 150 pilots a year. Carlisle expressed his hope that as the new scores of RPA pilots graduate from training, the proportion of traditional aviators temporarily assigned to RPA units will decline from its present third to only 10 percent. The elimination of the involuntary RPA assignments for traditional aircrews would serve two purposes: it would help the service address an emerging but critical shortfall in Air Force manned aviation; and it would also eliminate, it is thought, a key source of poor morale within the RPA units.[46] However, bolstering the RPA flight training program alone may further institutionalize the "second-class citizen" stigma of RPA pilots because many traditional aviators consider it to be an abbreviated and simplified version of the intense pilot training program that they once endured.[47] Additionally, while rarely thrilled about the assignment, the traditional "manned" aviators who temporarily fly or have flown RPAs are perhaps the best equipped to appreciate and then articulate what the RPA crews do, both in their cockpits and in the airspace over the battlefield. Curtailing this already small population of pilots with experience in both platforms, absent any larger organizational change, will likely impede the service's efforts to address the cultural divide that exists between the two communities.

But there is also no guarantee that the gulf can even be bridged successfully. The myth of the fighter pilot and the myths of autonomy both conspire against the RPA pilots.[48] So too does the Air Force's inability to articulate what its RPA operators actually do, settling instead for one-dimensional "great machine" or "great pilot" narratives. In this environment, the professional isolation of the RPA community will likely persist.[49] It is unfortunate, because while there are several ways to distinguish a traditional pilot from one that flies an RPA—the unmistakable one being that the former establishes a *physical* presence above the ground, whereas the latter establishes only a *virtual* presence there—arguing like many do on the basis of skill or lack thereof is misguided. Mindell reminds us that the critical issues when discussing autonomy center not on dichotomous "manned versus unmanned" distinctions but rather on the answers to four questions: "*Where are the people? Which people are they? What are they doing? Where are they doing it?*"[50]

Indeed, a critical lesson of the history of fighter aviation is that, despite the myth of the fighter pilot and corresponding notions of his timeless skills, different fighter aircraft have historically demanded different things of the pilot. Just as the F-86 pilots' skills were not wholly transportable to the F-4, nor the F-4 pilots' to the F-15, so too would an F-22 pilot today find his or her skills inappropriate if he or she was suddenly called upon to operate an RPA. In each historical case, certain pilot skills that were once considered challenging and central to the fighter pilot experience were adjusted and sometimes even rendered wholly irrelevant depending on the aircraft and the automated fire control technologies within it. New tasks steadily emerged in their place, and often they were as challenging as the now outmoded skills, if not more so. Hence, despite the myth of the timeless fighter pilot, the individual and social experience of fighter aviation has been constantly evolving. RPAs are but the latest iteration.[51]

This pattern of human-and-machine coevolution set within powerful narratives that tend to privilege either the operator or the machine is not unique to military aviation. Commercial aviation, robotic-assisted surgery, and even education have all seen the introduction of new technologies that threaten the historic identity of the user-practitioner. For example, some claim that using an autopilot to land a commercial jet isn't flying;[52] others argue that surgeons need to get their hands bloody to earn their title;[53] still others, that a teacher needs to be standing in front of students in a lecture hall.[54] The history of Air Force fighter aviation, automated fire control systems and the myth of the fighter pilot may therefore offer lessons for other sociotechnical systems as we continue to strive for the optimum blend of human-machine collaboration in our future enterprises.[55]

Knights or Scientists?

The editors at *Life* in February 1952 wondered, When confronted with new, cosmic fire control technologies and their associated weaponry, would fighter pilots be able to preserve their flying knight image, or would they be transformed into flying "scientists" preoccupied with the "electronic brains" in their new equipment? Moving past the myths, the history of fighter aviation from 1950 through 1980 illustrates a remarkable shift in the practical knowledge and skills required of a fighter pilot. Pilots did necessarily become conversant in scientific topics like gigahertz radar frequencies, pulse repetition frequencies, and Doppler shifts. But the pilots also never surrendered the mystique of the traditional goggle-clad, silk-scarf-wearing fighter ace. And they were able to maintain their deep, personal connections with their aircraft, even as those connections were increasingly mediated by the new fire control technologies. Thus, it was probably never an either-or proposition. Fighter pilots have always occupied the gray area between heroic, passionate knight and detached, critical scientist. Even the great aces of World War I fame were analytically minded as they struggled to learn how best to use their new machinery. Similarly, the Eagle pilots in AIMVAL-ACEVAL still possessed an intensely competitive spirit reminiscent of Richthofen.

In some respects, the "knight" versus "scientist" distinction seems only to propagate the "great pilot" and "great machine" dichotomy, with knights associated with the former and scientists the latter. Unfortunately, these one-dimensional narratives often obscure important context and confound our ability to use the lessons from aviation history to inform current discussions about automation and autonomy. The most interesting and useful revelations tend to emerge when we study the humans and their automated machinery working not in isolation or in competition but rather in partnership.

Perhaps a better label for the successful Air Force fighter pilot, instead of *knight* or *scientist*, is *tiger*. Unlike the fighter pilot's skills, which had to evolve with the changing technologies, the value of an aggressive and tenacious spirit, once expressed in Korea as "Every man a tiger," is technology-agnostic.[56] The intense desire of fighter pilots to use their equipment as best they could is a common theme that runs through the history of Air Force fighter aviation. Granted, this tiger-like attitude is not unique to fighter pilots. "We have all heard the old story, 'It takes a special breed of cat to be a fighter pilot,'" one fighter ace explained during an interview. "It takes a special breed of cat to be almost anything—an interviewer, a writer, a personnel guy, a fighter pilot. . . . There's nothing magic about being a fighter pilot." But, to be a *successful* fighter pilot, the aggressive and tenacious attitude was not optional, and the ace later clarified his remarks. "Call it dedication," he said.[57] Olds echoed the assessment, "You have to want it. You have to study. You have to think."[58]

What the pilots wanted rarely changed over time, but what the pilots studied and how they thought about it did. In the F-86, it was about learning where to find the enemy and then how to outmaneuver him. In the F-4, it was about understanding the Phantom's different radar modes, the limitations of its missiles, and developing a partnership, albeit reluctantly, with the GIB. In the F-15, it meant practicing the HOTAS switchology, radar manipulation and interpretation, and interflight communications. While the skills of an F-86, F-4, and F-15 pilot were not readily interchangeable, the pilots' tiger-like attitude, when present, provided sufficient drive and determination for the pilot to navigate the transition successfully.

Faced with ever-increasing automation in their cockpits, fighter pilots repeatedly demonstrated that they could be both *knights* and *scientists*. The successful ones, though, were always *tigers*. They will be in the future, too.

NOTES

Chapter 1. Introduction

1. Kilduff, *Richthofen*, 44–50; Wohl, *Passion for Wings*, 220–25; Fritzsche, *Nation of Fliers*, 80–82.

2. *Dicta Boelcke* was a set of seven precepts for air combat. They "became the standard reference for all other Jasta pilots, and their basic principles applied to a new generation of fighter pilots during World War Two, for whom they appeared in printed booklet form." Franks, introduction, 4–5.

3. Wohl, *Passion for Wings*, 225–27; Fritzsche, *Nation of Fliers*, 80–82.

4. Quoted in Kilduff, *Richthofen*, 193.

5. Quoted in ibid., 194.

6. Driggs, "Aces among Aces," 574; Wohl, *Passion for Wings*, 250. Driggs quoted C. G. Grey.

7. Kilduff, *Richthofen*, 199–205; P. Hart, *Aces Falling*, 166–77; Persico, *Eleventh Month, Eleventh Day, Eleventh Hour*, 218; Driggs, "Aces among Aces," 574. The tribute was inscribed on a wreath placed on Richthofen's casket. Driggs's London correspondent also reported, "Only a few days ago one of the best of our airmen expressed the hope that he and von Richthofen might survive the war, so that they might compare notes." On Richthofen's death, Hart admitted that he "always sentimentally preferred to believe that it was indeed [Captain Roy] Brown" in a Sopwith camel that downed the famous German ace, "but no-one will ever know and one cannot be dogmatic at this distance in time" (174).

8. Singer, *Wired for War*, 363; Wohl, *Passion for Wings*, 250.

9. On the mechanistic horrors of the Great War and the war's challenge to the notion that a nation's technological superiority was the "surest proof of their civilization's superiority," see Adas, *Machines as the Measure of Men*, chap. 6. On the flying ace as a restorative tonic for the Western psyche, see Wohl, *Passion for Wings*, chap. 7; Fritzsche, *Nation of Fliers*, chap. 2.

10. Fritzsche, *Nation of Fliers*, 88. The discrepancy between the public image of gallantry and fairness and the sordid reality of sneaking up on lumbering reconnaissance planes, as Richthofen had done on his seventy-fifth victory, is a topic for chapter 2. See also P. Hart, *Aces Falling*.

11. Richthofen's "duty of the fighter pilot" is an oft-cited aviation expression. As with most popular historic quotes, several versions exist, and no authoritative source can be found. The version cited here is from D. English, *Slipping the Surly Bonds*, 89. For a sampling of other versions of the quote, see Boyne and Handleman, *Brassey's Air Combat Reader*, 6; Middleton, *Air War—Vietnam*, 210; Hettena, *Feasting on the Spoils*, 26; Wilcox, *Scream of Eagles*, 185.

12. Pisano et al., *Legend, Memory, and the Great War*, 29. Pisano et al. cited a wartime speech by David Lloyd George celebrating the aviators: "The heavens are their battlefield; they are the Cavalry of the clouds. High above the squalor and the mud, so high in the

firmament that they are not visible from the earth, they fight out the eternal issues of right and wrong. . . . Every flight is a romance; every report is an epic. They are the knighthood of the war, without fear and without reproach. They recall the old legends of chivalry, not merely the daring of their exploits, but by the nobility of their spirit, and amongst the multitudes of heroes, let us think of the chivalry of the air."

13. P. Hart, *Aces Falling*, 45. See also Kilduff, *Richthofen*, 238.

14. *4th FIG Tactical Doctrine*, July 22, 1951, ii. The organizational hierarchy within the Air Force typically adheres to the following pattern (from top to bottom): HQ Air Force, Major Command (either functional or geographic), Numbered Air Force, Wing, Group, Squadron.

15. *Resume of Our Tactics*, 16–17. "Every man's a Tiger!" or some variation thereof, was a common refrain within the F-86 units. Lieutenant General William E. Brown Jr., a 125-mission F-86 pilot in Korea, remembered it as "Every man a tiger" in Hallion, "Air Dominance," 43.

16. D. Evans, *Sabre Jets over Korea*, 217.

17. "Here's a Fighting Man"; D. Evans, "I Fight the Red Jets"; D. Evans, *Sabre Jets over Korea*.

18. D. Evans, *Sabre Jets over Korea*, 130–31, 234; D. Evans, "I Fight the Red Jets."

19. D. Evans, *Sabre Jets over Korea*, 131; D. Evans, "I Fight the Red Jets," 90.

20. *Resume of Our Tactics*, 12.

21. D. Evans, *Sabre Jets over Korea*, 105. Just maintaining formation could be challenging; see Blesse, "No Guts, No Glory," 10–12.

22. D. Evans, *Sabre Jets over Korea*, 131–32.

23. Ibid., 133.

24. Ibid., 163.

25. S. Davies, *F-15C Eagle Units in Combat*, 57–59; Deur, "Wall of Eagles," iv, 19.

26. Deur, "Wall of Eagles," 19; S. Davies, *F-15C Eagle Units in Combat*, 57. The "Rambo" call sign was pirated from Sylvester Stallone's 1982 movie of the same name, in which the title character, John Rambo, singlehandedly lays waste to an entire small town. Another popular Eagle call sign had similar cinematic roots; "Conan" refers to Arnold Schwarzenegger's 1982 movie *Conan the Barbarian*. The two call signs were (and still are) routinely used by F-15 flights at the USAF's Weapons School at Nellis AFB, Nevada. For the January 19, 1991, mission, the Eagle flight lead, a recent USAF Weapons School graduate, found the higher-headquarters-assigned call sign for his formation—JJ—lacking, so he renamed his and the accompanying formation Rambo and Conan.

27. Deur, "Wall of Eagles," 19–20; S. Davies, *F-15C Eagle Units in Combat*, 58.

28. S. Davies, *F-15C Eagle Units in Combat*, 58–59; Deur, "Wall of Eagles," 19–20. Sveden (Rambo 4) recalled the radio exchange between Rambo 1 and 3 as they took their initial locks at long range: "In my view, No. 3 then demonstrated the best comm and execution of any of the F-15 kills—as cool as you could possibly be, he called, '"Rambo 3" has a side-side breakout on the nose for 20.' No. 1 came back and said, '"Rambo 1," same.' Then No. 3 responded, '"Rambo 3," sorted side at 7000,' followed by '"Rambo 1," same, 9000.'" Prather (Rambo 3) emphasized that the radar procedures were something "we do every day in training." Prather also explained the concept of *sorting*, "Wingmen are supposed to look for other contacts, and then their goal in life is to lock someone different than their flight lead, so 'Abby' [Rambo 4] locked a different guy than I had."

29. The turn to the west toward the Iraqi fighters had also placed Rambo flight looking into the setting sun, a clear violation of *Dicta Boelcke*'s second rule: "Try to place yourself between the sun and the enemy. This puts the glare of the sun in the enemy's eyes and makes it difficult to see you and impossible for him to shoot with any accuracy." Boelcke's *Dicta* is included in Franks, introduction, 5.

30. Quoted in Deur, "Wall of Eagles," 22.

31. Toward the end of the war, Eagle pilots started racking up victories against Iraqi jets that were fleeing to Iran. Most of these kills were achieved at close range from a stern position. Of all the Eagle engagements during Desert Storm, only two resembled the close-range, swirling dogfights of prior aerial battles. During one such battle on January 19, 1991, Captain "Cherry" Pitts and Captain "Kluso" Tollini tangled with a pair of MiG-25 Foxbats. At one point during the confusing battle, Tollini was chasing an aircraft that had its afterburners lit. Just to make sure he was behind an enemy aircraft and not a friendly Eagle, Tollini queried his flight to see if anybody was using afterburner. That same day, Captain "Mole" Underhill and Captain "Rico" Rodriguez encountered two Iraqi MiG-29s. Underhill destroyed the first MiG at range, but following some precautionary defensive maneuvers, the pair of Eagles found themselves visually engaged with the second MiG. The Eagles started a climbing turn to engage the lone MiG. The MiG started a descending turn to engage the Eagles. A few moments later, the MiG crashed into the desert floor below. Ibid., 16–19, 40–41; Laslie, *Air Force Way of War*, 138–40.

32. Quoted in Deur, "Wall of Eagles," 15.

33. "From Struts to Jets," 62–63.

34. Representative of this perspective is the argument that, following the dismal performance of the Air Force's favored aircraft during the Vietnam War, the service rededicated itself to getting "back on the right track to field superior technology and a world-beating force," which naturally led to the brilliant success of Desert Storm. Werrell, *Chasing the Silver Bullet*, 7. See also most of the popular aviation texts that focus on specific aircraft, for example, W. Thompson and Dildy, *F-86 Sabre vs MiG-15*; Werrell, *Sabres over MiG Alley*; Thornborough and Davies, *Phantom Story*; Thornborough, *USAF Phantoms*; P. Davies, *USAF McDonnell Douglas F-4 Phantom II*; P. Davies, *USAF F-4 Phantom II MiG Killers, 1965–68*; S. Davies, *F-15C Eagle Units in Combat*. The military's alphanumeric aircraft designation format likely contributes to this deterministic perspective, as more sophisticated aircraft that possess higher numbers (e.g., the F-15) replace inferior aircraft with lesser numbers (e.g., the F-4). The deterministic undertones and the sheer volume of aviation "great machine" narratives have contributed to attitudes, expressed in Builder, *Icarus Syndrome*, chap. 14, that, more so than any other American military service, the Air Force "has long worshipped at the altar of technology." On the competing perspectives of technological determinism and socially constructed technologies, see M. Smith and Marx, *Does Technology Drive History?*; Bijker, Hughes, and Pinch, *Social Construction of Technological Systems*. On the historiography of technology, see Kranzberg, "The Newest History: Science and Technology"; Usselman, "From Sputnik to SCOT: The Historiography of American Technology." On the oft-overlooked social and cultural elements of aviation history, and the corresponding preponderance of "picture books and hagiographic accounts of particular pioneers" that dominate the historiography, see Hansen, "Aviation History."

35. Westrum, *Sidewinder*; Bugos, *Engineering the F-4*.

36. Women pilots were barred from flying USAF fighter aircraft until the early 1990s; hence, all of the references to the fighter pilots within this 1950–80 period are exclusively masculine.

37. See, for example, Yeager, *Yeager*; Mahurin, *Honest John*; Trest, *Once a Fighter Pilot*; Blesse, *Check Six*; Handley, *Nickel on the Grass*; McCarthy, *Phantom Reflections*; Olds, *Fighter Pilot*. For additional information on fighter pilots' motivations, characteristics, and their seemingly uncanny ability to overcome technological and bureaucratic impediments, see Wolfe, *Right Stuff*; Andrews, "To Fly and Fight"; Michel, "Revolt of the Majors." Laslie's recent *Air Force Way of War* offers another example of the "great pilot" narrative. In it, Laslie rebuffs the "great machine" narratives that emerged post–Desert Storm and contends instead that the pilots' revolution in training produced victory: "Superior technology provided only an opportunity. That opportunity was successfully exploited because many, if not most, of the hundreds of American and coalition aircrews had attended Red Flag or similar exercises" (135).

38. The Air Force and DARPA (Defense Advanced Research Projects Agency) launched a set of comprehensive studies in the mid-1970s to define the "right stuff" that made successful fighter pilots. However, both studies tended to overlook the role of new technologies and their effect on evolving fighter pilot traits and skillsets. The DARPA study's final report made little distinction between a successful fighter pilot from World War II and one from Vietnam, thereby obscuring any potential effects of changing technology on the fighter pilot's necessary skillset. Youngling et al., *Feasibility Study to Predict Combat Effectiveness*, sect. 1, p. 1. The same was true for the Air Force's Corona Ace Oral History program. This oversight is not confined to those within the defense establishment. Sherwood attributed the demise of the Korean fighter pilots' raucous "flight suit attitude" not to any corresponding changes in technology but to the fact that today "there are plenty of other pilots who want to fly F-15s and are willing to behave themselves for the privilege." J. Sherwood, *Officers in Flight Suits*, 164.

39. The F-86A/E and F-4C/D/E all saw extensive use in combat. The F-15A, however, did not; its slightly modified successor, the F-15C, served as the predominant air-to-air fighter during Desert Storm. Unfortunately, because the F-15C is still actively used by the Air Force, many of its systems remain classified, as do the specific F-15 combat reports from Desert Storm. I chose to terminate my study with the F-15A to avoid these classification issues. To compensate for the lack of combat data, I rely on a pair of tests conducted in 1977 that were lauded as being the most realistic air combat tests ever devised.

There were obviously other technologically advanced aircraft fielded by the Air Force during this thirty-year period; for example, the Air Force's century series of fighters, which included the F-100, F-101, F-102, F-104, F-105, and F-106. However, none of these aircraft experienced *air-to-air* combat (or simulated combat) like the F-86, F-4, or F-15. Additionally, interceptor aircraft like the F-102 and F-106 were assigned to Air Defense Command (ADC) and were flown by "interceptor pilots," not "fighter pilots." The fact that the former wore orange flight suits, as opposed to the latter's olive-drab flight suits, was a visible distinction separating the two. See the discussion in chapter 4.

Finally, I focus on the air-to-air aspect of fighter combat, as opposed to the bomb-dropping air-to-ground component, because it remains directly linked to the celebrated experiences of the early great aces like Boelcke, Richthofen, and Rickenbacker. See the discussion in chapter 2.

40. With regard to the term *weapons systems*, I use it to refer generically to the fire control systems—the combination of aiming and targeting systems and the projectiles they used, be they bullets or guided missiles—that were installed in the fighter aircraft.

41. *A-1 Series (Gun-Bomb-Rocket) Sight*, 2; "New Radar Sight Guides Jets' Guns." The *New York Times* proclaimed that the "secret radar sight . . . does everything but fly the plane."

42. Bugos, *Engineering the F-4*, 27–28. Although technically the Phantom II—the original FH-1 Phantom first flew in January 1945 (see Bugos, 11–12)—most references to the latest incarnation describe it simply as the Phantom. Later it became popular to vilify the engineers and acquisition desk jockeys for their faith in the unproven missile armament, but at the time many American pilots, still haunted by memories of MiGs soaring high above them near the Yalu and out of reach of their .50-caliber machine guns, supported the shift from anachronistic guns to modern missiles. Michel, *Clashes*, 13.

43. Anderegg, *Sierra Hotel*, 153–54; Werrell, *Chasing the Silver Bullet*, 70; Anderegg, interview, April 9, 2013; Wallace, interview.

44. *Eyes of the Eagle*, 2.

45. The image of modernity has always been a critical component of the mystique of aviation. See, for example, Schatzberg's concept of an "ideology of progress" guiding aviation's transition from wooden aircraft to metal aircraft. Schatzberg, *Wings of Wood*.

46. D. Evans, *Sabre Jets over Korea*, 234.

47. McCarthy, *Phantom Reflections*, 16; Handley, *Nickel on The Grass*, 66n29.

48. Trotti, *Phantom over Vietnam*, 28. Trotti was a Marine Corps Phantom pilot.

49. Quoted in Anderegg, *Sierra Hotel*, 150.

50. Hoag, "Superfighter," 23.

51. Yeager, *Yeager*, 41–43, 53–54, 58–59.

52. L. Welch, interview, October 1, 2013.

53. *USAF F-15 Fighter Summary*, 7; E. Lewis, *History of the F-15 Eagle*, 48, 85.

54. Olds, Corona Ace Oral History Interview, 65–66.

55. Hutchins observed that "to study activity as an integrated system, we need tools that allow us to look at multiple sources of data about the activity in a synchronized and integrated way." Hutchins et al., "Integrative Approach to Understanding Flight Crew Activity," 358. The fighter pilots came to this realization as more and more disparate data streams were generated by their varied equipment, all of which could contain essential bits of information at any given time.

56. See, for example, the gunnery footage in the following World War II newsreels: *Gen. MacArthur Leads Attack on Admiralty Islands*; *Allied Drive On in Italy*.

57. See discussions on towed target shooting in "A-1 Sight and Air to Air Gunnery"; "Gun Camera"; "How to Set Up a Film Assessing Room"; "Lead Assessing"; Giordano, "Dart Target," November 1956; Giordano, "Dart Target," March 1957; R. Bond, "Elusive Dart."

58. Harmer and Anderegg, *Shootdown of Trigger 4*, 5; Anderegg, interview, April 9, 2013; Sokol, interview. For an example of typical radio chatter, see Futrell et al., *Aces and Aerial Victories*, 20–22.

59. Sokol, interview; Wallace, interview.

60. Taylor, "Air Combat Maneuvering Instrumentation"; Turner, "ACMI Update"; Waddell, "ACMI Improves Pilot Combat Training."

61. Bugos, *Engineering the F-4*, 26.

62. Wohl, *Passion for Wings*, 207–8, 214; Creech, Oral History Interview, 54; Wallace, interview.

63. Milling, *Air Tactics*, sect. 1, p. 9.

64. Doctrine has been described as the "glue" that holds a human-machine system together, mediating between the two components to prescribe how a specific task is to be accomplished. O'Mara, "Socio-Technical Construction of Precision Bombing," 345. For a different definition of *doctrine*, see Posen, *Sources of Military Doctrine*. Within the military, there are multiple layers of doctrine. At the highest level, *strategic* doctrine outlines how the military plans to use its forces to win a war. *Tactical* doctrine operates at a lower-level, outlining, for example, how a handful of fighter pilots should work together as a formation team to accomplish a mission.

65. Michel, *Clashes*, 169–72; Anderegg, *Sierra Hotel*, 59–60; Anderegg, interview, April 9, 2013; Anderegg, interview, October 2, 2013; Sokol, interview; Wallace, interview.

66. Of note, O'Mara explored the impact of changing aircraft technologies on the tactical formations flown by B-17s in World War II. O'Mara, "Socio-Technical Construction of Precision Bombing."

67. D. Evans, *Sabre Jets over Korea*, 170.

68. Quoted in Wildenberg, "A-1C(M) Gunsight," 35; Werrell, *Sabres over MiG Alley*, 30.

69. *T.O. 1F-86E-1*. The F-86E armament equipment information is located on pages 4-22 through 4-34. *T.O. 1F-4C-1*; *T.O. 1F-4C-34-1-1*. The -34 series manuals did not appear until 1964, a reflection of the growing technical knowledge requirements associated with the more complex aircraft and the wider variety of armament they carried. Winkler, "Look at the -34 Aircrew Non-nuclear Weapons Delivery Manual." On the F-86 pilots' reluctance to use their checklists, see Werrell, *Sabres over MiG Alley*, 58.

Chapter 2. The Myth of the Fighter Pilot

1. Blesse, *Check Six*, 89–92. Blesse's commanding officer, Colonel Bruce Hinton, the commander of the first Sabre unit deployed to the Korean War and a jet ace himself, described Blesse as "an alert, creative mind . . . who did almost everything well" (92).

2. Ibid., 91; "Major General Frederick C. Blesse."

3. Blesse, "No Guts, No Glory"; Blesse, *Check Six*, 57.

4. Blesse, *Check Six*, 3–5, 11, 56.

5. Olds, *Fighter Pilot*, 5–6. Olds's father, Robert, was a pioneer in early American military aviation; he served as aide to General William "Billy" Mitchell and later became director of the Air Corps Tactical School. Hence, according to Robin, "the pulp-fiction heroes of *G-8 and His Battle Aces* were also the real men that moved through my daily life."

6. Mahurin, *Honest John*, 18.

7. McCarthy, *Phantom Reflections*, 7.

8. *Pursuit Text*, 17; Blesse, *Check Six*, 56.

9. Barthes, *Mythologies*, 11, 109, 117–18, 129, 151. See also Campbell, *Hero with a Thousand Faces*; Campbell and Moyers, *Power of Myth*.

10. While others, notably Robert Wohl and Peter Fritzsche, have already investigated the public and private purposes of the fighter pilot myth as it emerged in World War I (albeit without using the *myth* label), I am specifically interested in the inconsistencies that arose between the myth and the pilots' actual experience of air combat. See Wohl, *Passion for Wings*,

chap. 7; Fritzsche, *Nation of Fliers*, chap. 2; P. Hart, *Falling Aces*; Morrow, *Great War in the Air*, xiv–xv.

11. On the similarities between humans' relationships with things and with people, see Sennett, *Craftsman*.

12. Pisano et al., *Legend, Memory, and the Great War*, 35; Wohl, *Passion for Wings*, 203, 304n2; Crouch, *Wings*, 121, 157.

13. Pisano et al., *Legend, Memory, and the Great War*, 35–36; Wohl, *Passion for Wings*, 304n2; Fritzsche, *Nation of Fliers*, 74; Crouch, *Wings*, 157; Gurney, *Five Down and Glory*, 15. There was no logical precedent for the initial five-victory requirement. Pisano et al. note that Pégoud was bestowed with the title of *l'as* after his fourth victory; they likewise claim that Immelmann had only four victories when he became "the first German pilot to be designated with this honor." There is some discrepancy, however; see, for example, Crouch and Fritzsche. Furthermore, the five-victory requirement was not static during the war. Wohl noted that as the number of aircraft in the skies over the Western Front multiplied, the French Air Service upped the requirement in 1917 to ten victories. There were plans to increase the requirement once more in 1918 to twenty victories, but war's end intervened.

14. Wohl, *Passion for Wings*, 243. See also Wohl's earlier chapters on the European enthusiasm for aviation prior to the Great War.

15. Fritzsche, *Nation of Fliers*, chap. 2; P. Hart, *Aces Falling*. Fritzsche primarily focused on the German experience; Hart, on the English.

16. Adas, *Machines as the Measure of Men*, 368–69; Fritzsche, *Nation of Fliers*, 3, 82; Pisano et al., *Legend, Memory, and the Great War*, 29. Fritzsche observed, "If machines were the measure of men in the modern era, as Michael Adas argues, airplanes and airships were the measure of nations at the beginning of the twentieth century" (3). To extend the argument, the pilots who flew the airplanes and airships were in turn seen as measures of the virtues of the citizenry.

17. Wohl, *Passion for Wings*, 203–10; Morrow, *Great War in the Air*, 91–92; Crouch, *Wings*, 157. Garros's invention quickly proliferated throughout the service. Crouch cited Pégoud's description of shooting a machine gun through the specially equipped propeller's arc: "About every sixth shot hits the propeller and bounces back at me. Often I hear it whistle past my ear. Sometimes I fear I am more apt to shoot myself down than I am to shoot down my intended victim."

18. Quoted in Wohl, *Passion for Wings*, 210.

19. During the war, the Americans used the term *pursuit* for their aircraft designed and equipped "primarily to destroy hostile aircraft in flight." The French referred to theirs as *avions de chasse*. Hence, World War I fighter pilots were sometimes referred to as pursuit pilots or *chasseurs*. *Pursuit Text*, 3. For consistency, I simply refer to these individuals as fighter pilots and their aircraft as fighters.

20. Quoted in Werner, *Knight of Germany*, 114–15, 126–27, 135; Wohl, *Passion for Wings*, 214. The Germans discovered Garros's innovation when he was forced down behind enemy lines in April 1915. Within months, the Dutch designer Anthony Fokker had developed a mechanical interrupter gear that synchronized the machine gun's firing with the spinning propeller. When paired with a small but fast 80-hp monoplane, the Fokker E, the new armament system helped inaugurate a nearly six-month period of unmatched German air superiority, bemoaned behind Allied lines as the "Fokker Scourge."

21. Quoted in Wohl, *Passion for Wings*, 226, 239, 244–45. Wohl also cited Richthofen's memoirs in which he likened the thrill of air combat to that of hunting big game.

22. Quoted in Fritzsche, *Nation of Fliers*, 88.

23. Quoted in ibid., 87–89; Driggs, "Aces among Aces," 574; Wohl, *Passion for Wings*, 250. Fritzsche noted: "Böhme looked forward to 'fencing' with a 'fair Englishman'" (87). It is ironic that within two weeks of penning his letters, Böhme would inadvertently bring down Boelcke in a disastrous midair collision—apparently it wasn't always "equal weapons and equal chances."

24. Pisano et al., *Legend, Memory, and the Great War*, 31. Pisano et al. were quick to point out, however, that the "ace's combat role was not as vital as that of the 'mighty Generals behind the lines.'"

25. Quoted in Wohl, *Passion for Wings*, 229–33.

26. Quoted in ibid., 233–36 (emphasis in original). Guynemer's name was subsequently inscribed in the pantheon as a "symbol of the aspirations and enthusiasms of the Nation."

27. Ibid., 215–16, 221–22. Wohl, citing one of Boelcke's eulogies, commented that the ace flyer's meteoric success and tragic fall likewise became a powerful testament to Germany's indomitable spirit: "Up to his death he was unconquered. Now he is not only someone who was great, now he will remain great forever" (222).

28. Hart, *Aces Falling*, 48; Kilduff, *Richthofen*, 107, 198; Wohl, *Passion for Wings*, 225–28. Wohl noted that Richthofen's wartime memoir, *Der rote Kampfflieger*, was carefully crafted to achieve maximum morale-boosting effect. Hart cited one of Richthofen's passages: "I should consider myself a despicable creature if, now that I am loaded with fame and decorations, I should consent to live on as the pensioner of my own dignity and to preserve my valuable life for the nation, while every poor fellow in the trenches, who is doing his duty just as much as I am, has to stick it out." The Red Baron's humility is also made apparent in another passage quoted by Kilduff: "I am no record keeper, [and] in general, records are far from our thoughts in the flying service. One only fulfills one's duty."

29. P. Hart, *Aces Falling*, 2.

30. "Enemy Aircraft Destroyed by US Army Air Service." The circular continued: "The work of observation and bombardment is considered equally as hazardous as that of Pursuit, but due to the fact that the Observation and Bombardment pilots are not called upon merely to destroy enemy aircraft, it should not be allowed to aid in establishing a popular comparison of results merely by relative victories."

31. Briscoe and Stannard, *Captain Ball*, 220–21, 291; P. Hart, *Aces Falling*, 2, 48.

32. W. D. Lewis, *Eddie Rickenbacker*, 1, 13–20, 56–58, 80–86; Glines, "Charmed Life of Captain Eddie Rickenbacker"; Mass, "Eddie Rickenbacker"; Driggs, foreword, v. Driggs claimed that "Rickenbacker was the idol of the automobile racing world at the moment when America entered the war."

33. W. D. Lewis, *Eddie Rickenbacker*, 95–98, 102–4, 115–20; Driggs, foreword; Rickenbacker, *Fighting the Flying Circus*, 1–2.

34. Rickenbacker, *Fighting the Flying Circus*, 338–39. Rickenbacker claimed, "I do not recall a single violation of this principle by any American aviator that I should care to call my friend."

35. Driggs, foreword, vi–vii.

36. W. D. Lewis, *Eddie Rickenbacker*, 213–14, 228–29, 234–35. Rickenbacker would continue to exercise careful control over the tales of his aerial battles. See, for example, the

account of one of his battles in Gurney, *Five Down and Glory*, 53–55. For descriptions of the same battle in the other sources, see Rickenbacker, *Fighting the Flying Circus*, 113–16; W. D. Lewis, *Eddie Rickenbacker*, 154, 163–64.

37. Driggs, foreword, vii. Driggs makes explicit mention of Rickenbacker's good sense to avoid getting killed during the war, "saving himself for the continued service of his country." On Rickenbacker's later life, which included stints as the owner of the Indianapolis 500, general manager of Eastern Airlines, and nearly three weeks spent bobbing in a raft in the middle of shark-infested South Pacific waters during World War II, see W. D. Lewis, *Eddie Rickenbacker.*

38. Wohl, *Passion for Wings*, 239, 245, 249; "Gallant Feats in the Blue," 58.

39. Quoted in Wohl, *Passion for Wings*, 244.

40. Fritzsche, *Nation of Fliers*, 73. Morrow suggested that the flyers' control over their aircraft offered a new symbol of "technocratic heroism." Morrow, *Great War in the Air*, 26.

41. Quoted in Wohl, *Passion for Wings*, 232.

42. P. Hart, *Aces Falling*, 21, 159, 222–23; Crouch, *Wings*, 166. Crouch reported that British aviator Cecil Lewis had a more pessimistic estimation of his fellow flyers' life span fighting over France in summer 1916—only three weeks.

43. P. Hart, *Aces Falling*, 4. For Hart, the pilots' despondence can be seen in their photographs: "An aura of doomed youth still clings strongly to the images that have survived of many of the aces. Looking at their photographs the natural effects of overwhelming tiredness and stress are transmuted in our imagination into the 'thousand-yard stare' of legend. . . . Even the greatest of aces was as grist to the all-consuming mill . . . that was the Western Front in 1918."

44. Ibid., 197–99. Burning to death was a prime concern of the early pilots. See discussions in Rickenbacker, *Fighting the Flying Circus*, 95–97.

45. P. Hart, *Aces Falling*, 200–204. A few weeks before his death, Mannock wrote a friend, "Things are getting a bit intense just lately and I don't know how long my nerves will last out. I'm rather old now, as airmen go, for air fighting. Still, one hopes for the best" (200).

46. Kilduff, *Richthofen*, 174; Wohl, *Passion for Wings*, 228.

47. Quoted in Wohl, *Passion for Wings*, 240.

48. Fritzsche, *Nation of Fliers*, 90.

49. Quoted in W. D. Lewis, *Eddie Rickenbacker*, 142; Fritzsche, *Nation of Fliers*, 93.

50. *Pursuit Text*, 24–25. The 1929 text noted that, in the Great War, "it was extremely rare that the great aces met each other in combat. The novices in fighting airplanes frequently fell easy victims to the greater experience of the famous fighting pilots. It was the world old game of survival of the fittest." Of Richthofen's eighty confirmed victories, less than half were scored against similarly equipped single-seat fighters.

51. Quoted in Fritzsche, *Nation of Fliers*, 90–91; P. Hart, *Aces Falling*, 45; Kilduff, *Richthofen*, 238. Maneuvering to achieve surprise and "attack[ing] when the enemy least expects it or when he is pre-occupied" was one of the tenets of Boelcke's *Dicta*, included in Franks, introduction, 5.

52. Deullin, "Pursuit Work in a Single Seater," 2–3; Deullin, "Chasseur Patrols," 1. Translated copies of Deullin's reports made their way to the Americans, although the French ace's name in the copies was misspelled as "Albert B. Deuillin S.73 [Spad *Escadrille*.73]."

53. Rickenbacker, *Fighting the Flying Circus*, 262–63.

54. Deullin, "Pursuit Work in a Single Seater," 1; *Air Service Pursuit and Combat Manual*, 2. Richthofen likewise warned his pilots, "The greatest danger for a single-seater is the surprise attack from behind." Quoted in Kilduff, *Richthofen*, 237.

55. Wohl, *Passion for Wings*, 240; Fritzsche, *Nation of Fliers*, 87–89.

56. Quoted in P. Hart, *Aces Falling*, 159. McCudden recalled being surprised "at times when I have passed quite close to a Hun machine and have had a good look at the occupant. . . . 'By Jove! there is a man in it.'" Quoted in Wohl, *Passion for Wings*, 306n106.

57. Quoted in Werner, *Knight of Germany*, 240–41.

58. Quoted in P. Hart, *Aces Falling*, 202.

59. Rickenbacker, *Fighting the Flying Circus*, 122–26. Rickenbacker's approval of Campbell's tactics, and indeed his satisfaction with his own surprise attacks, is particularly intriguing in light of his assertion (cited earlier) that he would never attack a disabled Hun and that he knew of no friend that had ever tarnished the principle of fair play.

60. Once the enemy aviators were disarmed of their machine, they could be afforded the customary courtesies granted military officers. For example, Boelcke during an engagement on January 7, 1916, successfully downed a British two-seater. The German pilot landed nearby and ran to the "enemy machine. . . . I went straight up to the Englishmen, shook hands with them and told them I was delighted to have brought them down alive. I had a long talk with the pilot, who spoke German well. When he heard my name, he said with a grin: 'We know all about you!'" Quoted in Werner, *Knight of Germany*, 161–62.

61. Deullin, "Pursuit Work in a Single Seater," 1. Deullin for one declared that a single-seater pilot "should be before everything else skillful in maneuvering. He can never practice too much aerial acrobatics."

62. Quoted in Kilduff, *Richthofen*, 236–39. A few paragraphs later in a section on "Individual Combat," Richthofen stated, "One does not need to be an aerobatic artist or a trick shooter; rather, to have the courage to fly right up to the opponent."

63. Quoted in P. Hart, *Aces Falling*, 53, 91. Hart cited another British pilot who was "quite unnerved" following a large engagement on March 18, 1918, and landed muttering to himself, "Frightful affair, frightful affair."

64. Wentworth, "Lectures by Pursuit Pilots," 2.

65. Ibid., 4; P. Hart, *Aces Falling*, 59.

66. P. Hart, *Aces Falling*, 42, 51, 54–55, 59. Of course, roaming with others could also lead to a midair collision, as Boelcke and Böhme learned.

67. *Pursuit Text*, 27½.

68. Werner, *Knight of Germany*, 198–99. Before Boelcke's Jagdstaffeln, either pilots flew as isolated combatants or they were told to accompany and protect nonfighter aircraft such as reconnaissance or bomber aircraft. On the significance of the Jagdstaffeln, see Fritzsche, *Nation of Fliers*, 70–71; Franks, introduction, 3–4. On the "triangular" formations, see *Pursuit Text*, 41–43; Deullin, "Pursuit Work in a Single Seater," 7; Deullin, "Chasseur Patrols," 1. Deullin alluded to the initial transition in his "Pursuit Work" text: "After some fruitless and even painful attempts, our hunters were forced to admit that the time of the solitary single seater has passed and that we had to look for something else." In "Chasseur Patrols," he noted that pilots followed their leader about "500 yards behind and 100 yards above."

69. This description by Captain Leonard Rochford, RNAS, is cited by P. Hart, *Aces Falling*, 103.

70. Wentworth, "Lectures by Pursuit Pilots," 2.

71. Ibid., 1. This "minimize[d] in so far as possible the undesirable necessity of [the new pilot's] solving situations as seem best to him at the moment. They are based on the fact that similar situations have been encountered before, the possibilities, analyzed and the method of dealing with them incorporated in the essentials."

72. Boelcke's *Dicta* is included in Franks, introduction, 5.

73. Quoted in P. Hart, *Aces Falling*, 53–54.

74. Wentworth, "Lectures by Pursuit Pilots," 4. Wentworth warned the Americans that "a pilot breaking from his formation to give combat . . . jeopardizes the safety and efficiency of all" (1).

75. Ibid.

76. Deullin, "Pursuit Work in a Single Seater," 7; Wentworth, "Lectures by Pursuit Pilots," 3.

77. Quoted in P. Hart, *Aces Falling*, 53–54. Douglas's description of battle was not published until 1963, but Deullin's 1917 report provides historical support for Douglas's stated rationale: "The guid [*sic*] gives the signal for the attack [and] his comrade follows him, keeping his distance and protects his rear." Deullin, "Pursuit Work in a Single Seater," 7.

78. Deullin, "Pursuit Work in a Single Seater," 7.

79. Wentworth, "Lectures by Pursuit Pilots," 1–2. Richthofen also granted significant freedom to his accompanying pilots once the battle had begun, telling them, "From the moment the commander dives on an enemy squadron, each pilot should be intent on being the first to engage the enemy." Quoted in P. Hart, *Aces Falling*, 46. In the event the formation came under sudden, surprise attack, the initial defensive maneuvers were, like those for the attack, also largely predetermined; see Wentworth, "Lectures by Pursuit Pilots," 2; *Air Service Pursuit and Combat Manual*, 6.

80. Milling, *Air Tactics*, sect. 6, p. 2. See also *Pursuit Text*, 87.

81. Morrow, *Great War in the Air*, xv–xvi, 60, 300–302, 315–16. On the superiority of the Fokker D.VII, Morrow noted that it possessed "an apparent ability to make a good pilot out of mediocre material" (300). See also Crouch, *Wings*, 168.

82. Quoted in P. Hart, *Aces Falling*, 38. Hart teased that the Camel's unpredictability could be of benefit in combat: "If the pilot barely knew what was happening how could his opponent?"

83. Morrow, *Great War in the Air*, 217–18; Wohl, *Passion for Wings*, 228; Holley, *Ideas and Weapons*, chap. 7; "Report of the Chief of Staff," 137. In a letter penned to a friend in mid-1917, cited by both Morrow and Wohl, Richthofen bemoaned of his old Fokker D.V: "I can assure you that it is no longer any fun being leader of a fighter unit. . . . Our airplanes are inferior to the English in a downright ridiculous manner. . . . Our fighter pilots, though quite good, are consequently lost! The D5 is so antiquated and laughably inferior that we can do nothing with it. . . . You would not believe how low morale is. . . . No one wants to be a fighter pilot any more." Senior leaders recognized the importance of technological superiority (or at least equivalency) in the air. The American army chief of staff in his 1917 report to Congress complained of the difficulties his service experienced getting suitably modern aircraft into the war, lamenting "a machine to-day is obsolete to-morrow."

84. *Pursuit Text*, 107.

85. Quoted in Kilduff, *Richthofen*, 237. See also P. Hart, *Aces Falling*, 47.

86. Deullin, "Pursuit Work in a Single Seater," 8. See also P. Hart, *Aces Falling*, 46–47; *Pursuit Text*, 156–57.

87. P. Hart, *Aces Falling*, 38–39.

88. *Pursuit Text*, 108, 152–58. Following the war, the Air Corps' 1929 *Pursuit Text* lauded the "considerable advantage" of the newly emerging .50-caliber guns (153).

89. P. Hart, *Aces Falling*, 61–62.

90. Deullin, "Pursuit Work in a Single Seater," 7–8. Furthermore, Deullin acknowledged that "it is utterly impossible in practice to do more than approximately estimate speeds and distances." On the advantages and disadvantages of using tracer ammunition, see *Pursuit Text*, 179–80.

91. *Pursuit Text*, 164–65.

92. *New Developments*, 29–30.

93. *Pursuit Text*, 163. The text noted, "At such [close] ranges elaborate sighting devices are unnecessary since the direction of flight and the trajectory of the bullets are almost coincident."

94. Ibid., 164.

95. Deullin, "Pursuit Work in a Single Seater," 7. Deullin, however, noted that one could "do too much sprinkling, and in one's anxiety and haste, to use the machine gun as a sort of hose with which to spray the sky."

96. Quoted in P. Hart, *Aces Falling*, 62, 196. Of course, the pilot could also always resort to ramming the enemy plane, as American Wilbur White elected to do on October 10, 1918. Hart cited Rickenbacker's recollection: "White rammed the Fokker head on while the two machines were approaching each other at the rate of 250 miles per hour! It was a horrible yet thrilling sight. The two machines actually telescoped each other, so violent was the impact. Wings went through wings and a first glance both the Fokker and Spad seemed to disintegrate" (335–36).

97. Ibid., 61. Contributing to the problem, Hart noted that "quarrels over claims were avoided and many squadrons had a cheerful 'You scratch my back, I'll scratch yours' attitude to the formal confirmation of kills." See also Hart's description (59–61) of a hypothetical mission and the confusion over aerial victories that logically resulted when pilots followed their leader during an attack.

98. Ibid., 204.

99. On the larger contributions of aviation within the Great War, see Morrow, *Great War in the Air.*

100. Quoted in P. Hart, *Aces Falling*, 138.

101. Quoted in ibid., 119. Flying over the maze of trenches, pilots had a difficult time distinguishing enemy from friendly troops; flying low enough to do so rendered the pilot "more vulnerable."

102. Quoted in ibid., 121–22.

103. *Pursuit Text*, 25; Crouch, *Wings*, 164–65. According to *Pursuit Text*, the sixty-five leading aces from the various belligerent nations accounted for 2,069 destroyed aircraft, with the remaining 609 aces of the war downing an additional 7,279 aircraft. Hart noted that with each nation possessing its own constellation of star-aces, and with their individual capabilities often mitigated by periods of waxing and waning technological superiority, the aces' collective effect over the battlefield was usually "redundant" and offsetting. P. Hart, *Aces Falling*, 204.

104. Fritzsche, *Nation of Fliers*, 85. Fritzsche cited a German cartoon that "aptly represented fliers as 'the upper ten thousand,' a class as remote and exempt in war as the *haute bourgeoisie* had been in peace."

105. Burrows, *Richthofen*, 141; *Pursuit Text*, 25; Wohl, *Passion for Wings*, 225.

106. Notable films included Howard Hughes's "multi-million dollar air spectacle" *Hell's Angels*, which premiered on May 27, 1930. *Dawn Patrol*, released a few months later, likewise incorporated spectacular aerial footage. Neither, though, matched the excitement and spectacle that was *Wings*, released three-years earlier in August 1927. *New York Times* film critic Mordaunt Hall wrote of *Wings*: "This feature gives one an unforgettable idea of the existence of these daring fighters—how they were called upon at all hours of the day and night to soar into the skies and give battle to enemy planes; their light-hearted eagerness to enter the fray and also their reckless conduct once they set foot on earth for a time." Hall, "Screen: The Flying Fighters." *Wings* went on to win the first-ever Academy Award for Outstanding Picture. "Hell's Angels"; "Dawn Patrol"; "Wings"; Hall, "Screen: Sky Battles." See also Wohl, *Spectacle of Flight*, 114–32, 139.

107. Corn, *Winged Gospel*, 20–26; Wohl, *Spectacle of Flight*, 29–37.

108. Wohl, *Spectacle of Flight*, chap. 4. The works of Aéropostale pilot and budding French literary genius Antoine de Saint-Exupéry also appeared during this period.

109. Corn, *Winged Gospel*, x, 12–13, 16–17, 30, 136–38.

110. Quoted in ibid., 75. Corn suggested that it was "the female pilot [who] became the antidote for the intrepid birdman stereotype." See also pp. 74–88. More recent scholarship suggests that the "antidote" was not as effective as Corn first alleged; see A. Meyer, *Weekend Pilots*, 3, 9.

111. The December 1940 *Saturday Evening Post* article is cited by A. Meyer, *Weekend Pilots*, 9. For more on the Ercoupe, see pp. 90–97.

112. Ibid., 3–4, 30–33.

113. *Pursuit Text*, 23, 27.

114. Ibid., 24, 46.

115. On the postwar issues of the American military aviation establishment, especially during the air mail crisis in early 1934, see Nalty, *Winged Shield, Winged Sword*, 122–27. See also D. Johnson, *Fast Tanks and Heavy Bombers*.

116. Quoted in Jeon, "Technologies of the Operator," 148. Jeon cited a meeting on the "Selection and Training of Aircraft Pilots," May 25–26, 1941, noting that "the military did not look 'merely for the men to fly but for men who *know how to take it* after they know how to fly.' . . . A pilot was someone 'who can stand the gaff' " (emphasis in original).

117. Quoted in Cameron, *Training to Fly*, 395. Although citing a Navy pilot, Cameron noted that the Army and Navy flight training regimens were comparable in this respect.

118. Rippon and Manuel, "Report on the Essential Characteristics," 411–12. The authors explained their rationale for the pilot "knowing little or nothing" of his aircraft: "The less the fighting scout pilot knows about his machine from a mechanical point of view the better. From the very nature of his work he must be prepared to throw the machine about, and at times subject it to such strains that did he realise how near he was to the breaking-point, his nerve could go very quickly." The medical profile also noted that the pilot "has little or no imagination, no sense of responsibility, keen sense of humour, able to think and act quickly." See also Morrow, *Great War in the Air*, 315; Cameron, *Training to Fly*, 131; W. Mitchell, *Winged Defense*, 164.

119. J. Flanagan, "Selection and Classification Program for Aviation Cadets," 231 (emphasis in original). See also Jeon, "Technologies of the Operator," 104, 110.

120. Cameron, *Training to Fly*, 539. Cameron noted the paradox: "The air arm . . . glorified the individual, specifically the warrior-pilot who flew against extremely dangerous odds, in training as well as in combat. The flight training program formulated rules and regulations in part to defy air force culture, in that training procedures attempted to curb the eccentric, the dangerously individualistic, the tendency for airmen to rely only on themselves and each other" (2).

121. Quoted in ibid., 403–4.

122. Olds, *Fighter Pilot*, 15–16.

123. Quoted in Cameron, *Training to Fly*, 272. Yeager rationalized the "gruesome weeding-out process. . . . Those who were killed in Nevada [during training] were likely to have been the first killed in combat." Quite simply, some people were made to be fighter pilots, and the others who weren't and who died trying, according to Yeager, rightfully earned disrepute "for destroying expensive government property as stupidly as if they had driven a Cadillac off a bridge." Yeager, *Yeager*, 15.

124. O'Mara, "Socio-Technical Construction of Precision Bombing," 52–54; Biddle, *Rhetoric and Reality*, chap. 3.

125. Quoted in Cameron, *Training to Fly*, 400.

126. Mahurin, *Honest John*, 109, 115–16.

127. Ibid., 116, 118–19, 132. Before achieving ace status, Mahurin distinguished himself in a more infamous manner. Still new to the theater and looking for a thrill, Mahurin closed to within a few feet of a lone B-24 bomber and exchanged waves with the crew. Planning to descend below the large bomber and then accelerate away from it, Mahurin's tail was instead clipped by the bomber's propellers. Mahurin's plane was destroyed (he jumped out of it using his parachute). He was ordered to pay a one-hundred-dollar fine for "use of governmental equipment prejudicial to the best interest of the United States" (123–24).

128. Mahurin, "Major Walker M. Mahurin," 23–25. Reprinted in Mahurin, *Honest John*, 130–32. Mahurin's advice, along with that of the other leading aces, would not be distributed until May 1944. It was accompanied by a note from the Eighth Fighter commander that the "outline of the technical phases of long range fighter escort" did not diminish the pilots' "determined, courageous, and skillful battle." Kepner, "Major General W.E. Kepner," 3. Olds likewise recalled the influence of the media on his image of a fighter pilot. After downing a FW-190, Olds "buzzed" and then did "two victory rolls" over the German pilot; "then I felt like an ass doing such a silly, damned-fool, kid thing like that. Obviously I'd read too much of Hogan's *G-8 and His Battle Aces* and watched too much of *Wings* and *The Dawn Patrol*." Olds, *Fighter Pilot*, 83.

129. Mahurin, *Honest John*, 118, 124–25, 131; Mahurin, "Major Walker M. Mahurin," 25.

130. Mahurin, *Honest John*, 125–27. Olds similarly remembered of his combat experience, "The airplane was a target, not a man. It was a surprise when a body would suddenly come flying out of the stricken bird, arms and legs flailing, chute blossoming." Olds, *Fighter Pilot*, 92.

131. Quoted in Shaw, *Fighter Combat*, 17.

132. Tilley, *Fighter Tactics*, 3.

133. Mahurin, "Major Walker M. Mahurin," 23; Mahurin, *Honest John*, 130. American World War II pilots referred to the surprise attacks as "bounces." See references in *Long Reach*.

134. On one occasion, Mahurin's flight got separated from the larger group formation, and he shot down two aircraft. Still, his commander, Colonel Zemke, was livid and threatened to court-martial the young pilot. Fortunately for Mahurin, General Kepner, the commander of Eighth Fighter Command, elected to intervene on his behalf. Mahurin, *Honest John*, 134–35.

135. McCollom to Commanding Officer, 66th Fighter Wing, "Narrative Report on Bomber Escort by P-47"; Mason to Commanding Officer, 66th Fighter Wing, "352nd Fighter Group Combat Formations"; "Tactics and Techniques of Long-Range Fighter Escort." On the importance of staying close to one another, the Eighth Fighter Command warned that "excessive spreading out of the formation makes it difficult for the leader to maintain control of his squadrons, makes friendly aircraft appear as unidentified aircraft, and subjects individual planes to attack. In addition, excessive power must be used by pilots in order to stay in formation when a turn is made." "Tactics and Techniques of Long-Range Fighter Escort," 18.

136. Mahurin, "Major Walker M. Mahurin," 24–25. Mahurin offered his assessment of flying in formation: "It is up to the wing man to cover his element leader no matter what. Sure, I know its [*sic*] tough to sit back and tell a guy that he is clear behind so that he can shoot down a Jerry. But," Mahurin lectured, "sooner or later" the wingman would have a chance to become a leader himself, then he would "get the chance to shoot."

137. "Tactics and Techniques of Long-Range Fighter Escort."

138. McFarland and Newton, *To Command the Sky*, 5; Futrell, *Ideas, Concepts, Doctrine, 1907–1960*, 153–54. According to MacFarland and Newton, it was this "symbiotic" combination of bombers delivering strategic attacks and fighters battling for air superiority that allowed the Allies to seize the initiative in the skies over Europe and to ensure the success of the Normandy landings.

139. Mahurin, *Honest John*, 132, 136.

140. *Pursuit Text*, 166–69. The text forewarned of a critical problem pilots would face when using their radios in combat: "Stringent regulations within each flight would be necessary to prevent jamming." See also predictions in Wentworth, "Lectures by Pursuit Pilots," 4–5.

141. "We Must Seek Out the Hun," 6; Tilley, *Fighter Tactics*, 3–4.

142. For examples of the gun camera footage used in newsreels, see *Gen. MacArthur Leads Attack on Admiralty Islands*; *Allied Drive On in Italy*.

143. Mahurin, *Honest John*, 128–29.

144. Olds, *Fighter Pilot*, 1.

145. Mahurin, *Honest John*, 128–29.

146. "Jet Plane Explosion Kills Major Bong."

147. Mindell, *War, Technology, and Experience*; Mindell, *Iron Coffin*.

148. "It's the Size of the Fight in the Man!," 77 (ellipses and emphasis in original).

Chapter 3. Sabres over Korea: "Every Man a Tiger"

1. In aviation parlance, a G refers to one unit of earth's gravity; hence, a 2G turn means that the pilot and his aircraft are subjected to an acceleration equivalent to two times earth's gravity.

2. D. Evans, *Sabre Jets over Korea*, 34, 126, 206, 225. Pilots frequently complained about the bulk and weight of their flight gear, but they were also reluctant to forsake any of it lest they got shot down over the mountains of North Korea. See, for example, Mahurin, *Honest*

John, 34; Hinton and Wingo, "Analysis of Operations at Kimpo," 3; Hise, "F-86E versus MiG-15 in Korea," 21; Baker, "Report on F-86 Operations," 9.

3. *T.O. 1F-86E-1*, sect. 1, pp. 27–29. To improve transonic handling, North American Aviation engineers working on the F-86E elected to discard the F-86A's classic, hydraulically boosted elevator. In its place, they installed a solid, controllable horizontal stabilizer powered by a redesigned, irreversible hydraulic system. The new hydraulic system also powered the ailerons. While the new design allowed the pilot to easily move the flight controls at all airspeeds, it also limited his ability to feel (through the control stick) the air pressure against the fight control surfaces. To compensate for the loss of the "conventional stick feel," engineers added an "artificial feel system" using "spring bungees." See also Wagner, *North American Sabre*, 54.

4. D. Evans, *Sabre Jets over Korea*, 78, 95–96, 198. For other assessments of the E-model's flight control arrangement, see Jones to Commanding Officer, 4th Fighter Interceptor Wing, "F-86E Combat Evaluation," 2; Mahurin, *Honest John*, 22; Thyng, "Operation of the 4th Fighter Wing in Korea," 3; *Operational Suitability Test of the F-86E*.

5. D. Evans, *Sabre Jets over Korea*, 122. Evans's account contains an inconsistency: in the new A-1C(M) gunsight, the "diamonds" were replaced with a solid, illuminated circle. The diamond-marked reticle from the earlier Mark 18 would eventually be reintroduced in the A-4, which was fielded beginning in late 1952 as a replacement for the A-1C(M), but the timing of Evans's account coincides with the arrival of the A-1C(M), not the A-4. Despite the discrepancy, Evans's account is useful because it is reflective of the level of instruction, or lack thereof, that pilots typically received regarding their new gunsights. On the A-4, see "A-4 Sight"; or, as a component of the MA-3, Tucker, "Fire Control."

6. D. Evans, *Sabre Jets over Korea*, 122–23.

7. Salter, *Hunters*, 62–63; Salter, *Burning the Days*, 153, 163, 196. Horowitz elected to write under the name Salter, a name "as distant as possible from my own," because he considered it "essential not to be identified and jeopardize a career [in the Air Force]—I had heard the sarcastic references to 'God Is My Copilot' Scott."

8. Blesse, *Check Six*, 58, 60; Mahurin, *Honest John*, 31; Hallion, "Air Dominance," 43. The "every man" quote is from Lieutenant General William E. Brown Jr., a 125-mission F-86 pilot in Korea, cited in Hallion. See also references in *Resume of Our Tactics*, 16–17.

9. "Harmonization and Firing Techniques," 1. There was a pressing need for the improved capability, the manual also noted, because of the "higher speed, greater armor, and increased retaliatory firepower" of enemy aircraft.

10. *Accelerated Comparison Test*, 1–2.

11. "Radar Gunsight," 66.

12. Baker, "Report on F-86 Operations," 11. See also a similar assessment in Thyng, "Operation of the 4th Fighter Wing in Korea," 7. Blesse likewise noted that "air to air gunnery is only about 10 or 15 percent of the overall problem of destroying an enemy aircraft." Blesse, "Air Tactics Problem," 15.

13. Stuart, "Salute to Combat Leaders," 24.

14. Mahurin, "Major Walker M. Mahurin," 25; reprinted in Mahurin, *Honest John*, 131–32.

15. L. Davis, Oral History Interview, 1–14; Wildenberg, "A-1C(M) Gunsight," 30–32. Davis was supposed to begin flying A-36s, a ground-attack version of the famed P-51 Mustang (although in his 1973 interview, Davis incorrectly referred to the planes as A-31s). Rationaliz-

ing the outdated equipment, one Army representative at a conference on fighter fire control in August 1941 explained that "from 1918 until about two years ago the Army had no funds available for activity in this field." *Control of Plane-to-Plane Fire*, 2. In fact, the Army had made some progress during the interwar years with its N-series reflector gunsights. Nevertheless, the ring-and-bead sight remained a fixture in fighter cockpits because it was durable and inexpensive, important attributes for a cash-strapped air service. See Whistler, "Brief History of USAF Reflector Gunsights."

16. L. Davis, Oral History Interview, 15–16. Davis complained that the .50-caliber folks "wanted to get that 1,500 rounds per stoppage up to 1,800 rounds per stoppage," a negligible improvement in Davis's opinion.

17. "Lieutenant General Leighton I. Davis"; Wildenberg, "A-1C(M) Gunsight," 33.

18. On the aerial gunnery problem, see Ehrenfried, *General Operating Principles of the A-1*; *A-1 Sight for the Control of Gunfire*; "Harmonization and Firing Techniques," 2–7; *Type A-1C(M) Gyro Computing Sight*, 23–25; "Velocity Jump"; "Fire Control Problem," 21; Shaw, *Fighter Combat*, 7–10.

19. Ehrenfried, *General Operating Principles of the A-1*; *A-1 Sight for the Control of Gunfire*; "Harmonization and Firing Techniques," 128–36; *Type A-1C(M) Gyro Computing Sight*.

20. "Harmonization and Firing Techniques," 129–32.

21. Mahurin, "Major Walker M. Mahurin," 25; Mahurin, *Honest John*, 131.

22. Shaw, *Fighter Combat*, 9. Shaw, a veteran Navy pilot, summed up the aerial gunnery problem: "Given all the foregoing factors that come into play, it's amazing that an air-to-air gun kill is ever recorded. . . . Little wonder the most effective technique is often to 'fill the windscreen with target and let 'er rip.'"

23. L. Davis, Oral History Interview, 10–12.

24. Mindell, *Between Human and Machine*, 220–23; *Gun Sight: Mark 14*.

25. As will be explained in more detail later, dampers attached to each gyroscope limited the effect of these abrupt movements, smoothing the commanded lead angle over time.

26. *Gun Sight: Mark 14*, 21 (emphasis in original).

27. Ibid., 11. Inputting the range into the Mark 14 was only one of the range setter's responsibilities. According to the manual: "The Range Setter sets the range on the Sight, watches for targets, helps the gunner locate the targets, and watches so the other half of the team doesn't fire at friendly ships and planes."

28. Ibid., 30–39; Mindell, *Between Human and Machine*, 221. In the instance when the target was attacking the ship, the range would be collapsing so fast that the operator's manual recommended setting the range knob at eight hundred or twelve hundred yards and leaving it alone; errors in the calculated lead angle due to the erroneous range input would naturally decrease as the target got closer.

29. Mindell, *Between Human and Machine*, 221.

30. Quoted in ibid., 220–21. The Navy initially purchased twenty-five Mark 14 sights in mid-1941; by 1945, the service had more than thirteen thousand. *Antiaircraft Action Summary*, 12.

31. "Navy Identifies Battleship 'X.'" At the time, the *South Dakota* was referred to in the press as Battleship X. After the war, the Navy figured that "guns controlled by [the Mark 14] shot down more aircraft than any other." *Antiaircraft Action Summary*, 12.

32. L. Davis, Oral History Interview, 12–13. Davis's frustrations with the service persisted; see L. Davis to Chief, Armament Laboratory, "Aircraft Sights."

33. Bush to Spaatz, March 11, 1942.

34. Werrell, *Sabres over MiG Alley*, 27; *Control of Plane-to-Plane Fire*; *Conference on Plane-to-Plane Fire Control*. Britain had begun working on LCGs for aircraft use in 1939, part of its frantic attempts to bolster its fighter defenses in the buildup to World War II. Writing after the war of the Office of Scientific Research and Development's efforts, Boyce noted that "under the stress of war the versatility of the fighter plane had flourished, but instrumental aids to the pilot were virtually nonexistent. The problem of rendering help in this situation was especially difficult because of the stringent requirement that the sight had to *help* the pilot, and not give him so much of a task of sight manipulation that he couldn't fly his plane" (emphasis in original). Boyce, *New Weapons for Air Warfare*, 58–59.

35. Werrell, *Sabres over MiG Alley*, 10, 14, 27. Werrell noted, "There is no explanation why [the USAAF] picked the navy gunsight over its AAF sibling, the K-14," for the new F-86. The first flight of the F-86 (then designated the XP-86) occurred on October 1, 1947. Rickenbacker's old squadron, the 94th Hat-in-the-Ring, was the first squadron to receive the new fighter in February 1949.

36. "Harmonization and Firing Techniques," 31–37.

37. *Pursuit Text*, 164. More formally, stadiametric ranging is based on the trigonometric principle of similar triangles whereby one milliradian (mil) approximately subtends a length of one foot at a distance of one thousand feet.

38. *T.O. 1F-86A-1*, sect. 4, pp. 18–19; "Harmonization and Firing Techniques," 31–37; "Gun Sight: Mark 18"; Wagner, *North American Sabre*, 25; Werrell, *Sabres over MiG Alley*, 26.

39. *T.O. 1F-86A-1*, sect. 4, p. 19. The three-step procedure was accompanied by five additional "operational notes," the first three of which concerned the use of the sight in combat: "1. Keep sight operating whenever its use is probable. 2. When maneuvering into position for attack, keep sight set at shortest range (large-diameter gyro reticle image circle) and decrease diameter to correspond to size of target. 3. Always track target before firing. By operating ranging control, continually frame the target while tracking for a minimum period of one second; then fire. Only after such tracking will the lead be correctly compensated."

40. The requirement to know "the size and angle off of the target" was identified as "an obvious disadvantage" of stadiametric ranging. A. C. G. Mitchell, "Airborne Fire Control," App 205. One group of Air Force officers suggested that the foreshortening issue could be mitigated if the gunsight and reticle were recalibrated to span the target fuselage instead of the wingspan. Their recommendation was never adopted. "Bailey Pipper."

41. Recognizing the gunsight's limitations, designers allowed the pilot to toggle between the lead-computed reticle and a fixed aiming cross. The pilot could even select both to be displayed at the same time, providing a clear visual indication of the prediction angle computed by the gunsight. "Gun Sight: Mark 18." See also Jenkins, "Some Notes."

42. D. Evans, *Sabre Jets over Korea*, 85. Describing one engagement in which he "made a head-on" pass against a MiG, Evans remarked, "What a closure rate! Just too fast for a calm range estimation peering through a gun sight" (182).

43. *Accelerated Comparison Test*, Inclosure 2:3; "Harmonization and Firing Techniques," 31, 129.

44. Werrell, *Sabres over MiG Alley*, 27.

45. Greene to Commanding General, Continental Air Command, January 31, 1950, 2.

46. Gurelac, *Radar in World War II*, 253–65, 271–73; Buderi, *Invention That Changed the World*, 27–28, 33, 37, 43, 46, 51, 70–71; Bowen, "Tizard Mission." According to Buderi, Bowen

later described his secret parcel as "a gift from the gods" for the Americans. On Britain's earlier attempts at developing AI radars for its night fighter aircraft, see Gurelac, *Radar in World War II*, 145–52, 159–64. On the magnetron in airborne radar, see ibid., 185–87, 224–31; Stimson, *Introduction to Airborne Radar*, 18–19.

47. McFarland, *Conquering the Night*, 7–12; Skolnik, "Fifty Years of Radar," 184; "Harmonization and Firing Techniques," 40–41. For more information on airborne radar principles, see Stimson, *Introduction to Airborne Radar*; Gurelac, *Radar in World War II*, chap. 2.

48. *Test of A-13B Ball Turret*, 3; Bloss to Commanding Generals, "Status of Development," 1; Gurelac, *Radar in World War II*, 329–32.

49. DuBridge et al., *Radar and Communications*, 11:22.

50. *A-1 Series (Gun-Bomb-Rocket) Sight*, 2.

51. "Fixed Price Contract." The contract was worth $268,500. Even by October 1944, Davis was still encountering resistance to developing more advanced LCGs for Army aircraft. In one memorandum, Davis questioned his superiors on why the service was continuing to buy "outmoded [gunsight] systems for applications where dynamic performance is the criteria [*sic*]? One thing we *are* sure of is that the airplane is not going to stop and let us compute the problem on a static or even on a 'steady state' basis." L. Davis to Chief, Armament Laboratory, "Aircraft Sights," 4 (emphasis in original).

52. *A-1 Sight for the Control of Gunfire*, iii–vi.

53. Wildenberg, "A-1C(M) Gunsight," 33–34. The A-1C(M) was first mass-installed in the F-84E. In addition, the last twenty-four F-86As that rolled off the production line were also equipped with the A-1C(M). Earlier produced F-86As were eventually retrofitted with the A-1C(M) and AN/APG-30. See also Bloss to Commanding Generals, "Status of Development," 4; "Standard Aircraft Characteristics: F-86A," December 8, 1950, 3; "Standard Aircraft Characteristics: F-86A," September 21, 1951, 3; "Standard Aircraft Characteristics: F-86E," 3; *T.O. 1F-86A-1*, sect. 4, p. 20; Werrell, *Sabres over MiG Alley*, 23–24, 27–28.

54. On the multi-weapon capabilities of the A-1 sight, see *A-1 Series (Gun-Bomb-Rocket) Sight*, 2–3; Ehrenfried, *General Operating Principles of the A-1*, 1; *A-1 Sight for the Control of Gunfire*, 1, 5–6.

55. "Harmonization and Firing Techniques," 12, 31.

56. *A-1 Series (Gun-Bomb-Rocket) Sight*, 25; Ehrenfried, *General Operating Principles of the A-1*, 3; "Harmonization and Firing Techniques," 18, 24. The eight components of the A-1C(M) were the amplifier, range servo, rocket setting unit, sight head, dimmer control, manual range control, bomb target unit, and computer. The decision not to include the radar in the A-1 system reflected the Air Force's desire that the A-1 be universal, capable of working with any radar. However, this arrangement also meant that no one entity coordinated development of the total fire control system, which directly contributed to significant problems with the system's reliability and maintainability. Years later, the Air Force would finally link the two components together under a new J-series Fire Control System designation: J-1 to designate the A-1C(M) and APG-30 combination, J-2 to designate the new A-4 sight and APG-30 radar. See *Accelerated Comparison Test*, Inclosure 1:1; Long, *USAF Fighter Aircraft Fire Control Systems*.

57. Wildenberg, "A-1C(M) Gunsight," 34; *Automatic Range-Only Radar*. The AN/APG-30 was designed specifically for fighter aircraft and the A-1 gunsight. Before APG-30 production, some early A-1B sights were paired with the AN/APG-5 radar which was originally designed for use directing B-17 and B-24 gun turrets.

58. Installing a radar in the nose of a high-speed fighter proved challenging in its own right. The radar needed to be covered by a "nonmetallic material" that would "not interfere" with its operation. Initially, engineers at North American tried a "lacquered glass covering," but they soon discovered that when an F-86 flying at 650 miles per hour encountered rain, the covering "wore away." A "makeshift covering" was developed using "keg-liner, the material used to line the inside of beer cans," before a neoprene radome was introduced as a "more permanent solution." Campion, "But Why Are We Losing," 68.

59. R. Wells, "Tips and Info for AN/APG-30 Radar"; "Harmonization and Firing Techniques," 42–44; Blake to McCormick, "Request for Communication and Electronic Equipment Data," Inclosure 4; *T.O. 1F-86A-1*, sect. 4, p. 20.

60. *T.O. 1F-86A-1*, sect. 4, p. 25; R. Wells, "Tips and Info for AN/APG-30 Radar"; "Harmonization and Firing Techniques," 42–44. Another T.O.-specified technique to verify the radar lock was to watch the variably sized reticle (discussed later): "As range decreases, the reticle should grow larger to span target continuously."

61. Ehrenfried, *General Operating Principles of the A-1*, 20–21; "Harmonization and Firing Techniques," 23–24, 26–27; *Type A-1C(M) Gyro Computing Sight*, 30–32, 34–38; *A-1 Series (Gun-Bomb-Rocket) Sight*, 23–26. Adjusting for air density was critical to refining the gunsight's time of flight prediction for the bullet. A unique, gunsight-specific, factory-set calibration was also applied to the sensitivity current. See *Operational Suitability Test of the A-1CM*, 8, 10, 12.

62. Ehrenfried, *General Operating Principles of the A-1*, 18; Jenkins, "Some Notes"; "Harmonization and Firing Techniques," 19–20, 25; *Type A-1C(M) Gyro Computing Sight*, 4–5, 26–30, 33, 44–45; *A-1 Series (Gun-Bomb-Rocket) Sight*, 17–25; *T.O. 1F-86A-1*, sect. 4, p. 22.

63. Ehrenfried, *General Operating Principles of the A-1*; *Type A-1C(M) Gyro Computing Sight*, 4–5, 26–34, 44; "Harmonization and Firing Techniques," 18–20, 26–27; *A-1 Series (Gun-Bomb-Rocket) Sight*, 17–25. Because the dampers directly affected the prediction angle computations, they needed to be precisely controlled, including their operating temperature (specified as between 160° and 180°F). The air temperature in the prediction computer needed to be even more precisely regulated to maintain 125°F ± 5°F.

64. *Type A-1C(M) Gyro Computing Sight*, 5–6, 38–39; "Harmonization and Firing Techniques," 21–23; *A-1 Series (Gun-Bomb-Rocket) Sight*, 8, 26. One light bulb provided light for the variable diameter reticle, its size automatically controlled by an adjustable range cone in the sight unit. The other bulb provided the light source for the pipper and, when activated, the fixed diameter reticle.

65. *A-1 Series (Gun-Bomb-Rocket) Sight*, 22, 26; *Type A-1C(M) Gyro Computing Sight*, 5–7, 44–45; "Harmonization and Firing Techniques," 21–23.

66. *A-1 Sight for the Control of Gunfire*, 2; "Fixed Price Contract," Exhibit A; *A-1 Series (Gun-Bomb-Rocket) Sight*, 3–4. When the aircraft was flown at altitudes below about five thousand feet, the radar became unreliable due to ground clutter. The radar likewise was rendered unusable when flying in clouds because the radar signal was reflected by the moisture in the air. *Accelerated Comparison Test*, Inclosure 1:3; Wildenberg, "A-1C(M) Gunsight," 33–34. For more information on ground clutter, see Stimson, *Introduction to Airborne Radar*, chap. 22.

67. *T.O. 1F-86A-1*, sect. 4, pp. 22, 25; *Type A-1C(M) Gyro Computing Sight*, 38, 45; "Harmonization and Firing Techniques," 23, 27. The pilot was also reminded to use the electrical caging

switch as required during the manual ranging process. The wingspan selector wheel was not used in the normal radar-ranging mode.

68. *T.O. 1F-86A-1*, sect. 4, p. 22; *Type A-1C(M) Gyro Computing Sight*, 33, 39, 44; "Harmonization and Firing Techniques," 22–23, 25–27.

69. *T.O. 1F-86A-1*, sect. 4, pp. 24–25; *T.O. 1F-86E-1*, sect. 4, pp. 28–29. There were some minor differences in the order of steps between the F-86A and the F-86E flight manuals. Other A-1 manuals stressed the importance of the initial tracking period and ensuring "that the reticle be held firmly on the target with a minimum of movement." Ehrenfried, *General Operating Principles of the A-1*, 20–21; *Operational Suitability Test of the A-1CM*, 10, 12.

70. Ehrenfried, *General Operating Principles of the A-1*, 22–23; *Operational Suitability Test of the A-1CM*, 8; "Harmonization and Firing Techniques," 15, 128–32. For example, if the fighter was flying an *uncoordinated* maneuver (such as a slip or skid) in violation of the third assumption, then the bullet's velocity-jump due to centrifugal force would not be in the same plane as the elevation gyro and could not be measured by the attached unbalanced mass, rendering the A-1's computed prediction angle incorrect.

71. *Operational Suitability Test of the A-1CM*, 2.

72. Ehrenfried, *General Operating Principles of the A-1*, 3; *A-1 Series (Gun-Bomb-Rocket) Sight*, 2; *Type A-1C(M) Gyro Computing Sight*, 3; Mindell, *Between Human and Machine*, 221.

73. *Operational Suitability Test of the A-1CM*, 10.

74. Getting, "Conference on Fighter Fire Control," I–5; Olsen to Holland, Schwarze, and Rhame, "Evaluation of Results." After observing pilots use the A-1 gunsight during the 1950 Air Force Gunnery Meet, MIT engineers expressed their hope that the pilots had realized that, "as with any precision instrument used in flight, snap decision corrections or judging will not improve results obtainable with the A-1C sight over those possible if correctly employed."

75. "New Radar Sight Guides Jets' Guns."

76. Hone, "Korea," 459; Y'Blood, *MiG Alley*, 1; Futrell, *United States Air Force in Korea*, 7; Stratemeyer, *Three Wars of Lt. Gen. George E. Stratemeyer*, 37. At the start of hostilities, the NKAF possessed sixty-two Il-10 ground attack aircraft, seventy Yak-3 and Yak-7B fighters, and twenty-two Yak-16 transports. In contrast, the entire South Korean Air Force consisted of sixteen aircraft. The NKAF attacks against the airfields at Kimpo and Seoul damaged a control tower, destroyed a disabled American C-54, and crippled seven of ten South Korean aircraft parked on the ground.

77. Futrell, *United States Air Force in Korea*, 9–13.

78. Ibid., 67–70, 98–103. FEAF, initially led by General Stratemeyer, represented the USAF component of Far East Command, which was led by General Douglas MacArthur. Three numbered air forces were assigned to FEAF: Fifth Air Force, headquartered at Itazuke AB, Japan; Twentieth Air Force, based out of Okinawa, Japan; and Thirteenth Air Force, operating from Clark AB, Philippines. Because of its proximity to the action, Fifth Air Force would provide the preponderance of tactical aircraft for use in the Korean War. Rather than being assigned to a numbered air force, the B-29s on-loan from the Strategic Air Command remained under direct control of FEAF. See pp. 2–4.

79. Hone, "Korea," 462; Y'Blood, *MiG Alley*, 6; Futrell, *United States Air Force in Korea*, 102–3. Futrell quoted Stratemeyer's assessment, "As it happened, the air battle was short and sweet. Air Supremacy over Korea was quickly established." But in his diary entry dated

October 14, Stratemeyer cautioned himself, "The principal lesson learned [is] not to draw wrong conclusions from Korean War because of absence of aggressive hostile air." Stratemeyer, *Three Wars of Lt. Gen. George E. Stratemeyer*, 232. Still, the battle had not been bloodless. For example, on July 12, 1950, the NKAF downed an American B-26, B-29, and L-4 (a military-version of the Piper Cub). Warnock, *USAF in Korea*, 7; Werrell, *Sabres over MiG Alley*, 71.

80. Futrell, *United States Air Force in Korea*, 244; Crane, *American Airpower Strategy in Korea*, 24–26; Stratemeyer, *Three Wars of Lt. Gen. George E. Stratemeyer*, 101.

81. Appleman, *South to the Naktong*.

82. Stratemeyer, *Three Wars of Lt. Gen. George E. Stratemeyer*, 252; Futrell, *United States Air Force in Korea*, 219, 244–45; Y'Blood, *MiG Alley*, 10; Thompson and Nalty, *Within Limits*, 21–22. The MiG-15 was known to American intelligence officers since its unveiling at the July 1948 Moscow airshow. For more information on the MiG-15, see Werrell, *Sabres over MiG Alley*, 61–65. There may have been early signals indicating that China had taken possession of Stalin's vaunted fighter, but they went largely unnoticed. On the Soviet-Chinese-Korean relationship during the Korean War and its influence over the Communist air strategy, see Zhang, *Red Wings over the Yalu*.

83. Bruning, *Crimson Sky*, 161; Futrell, *United States Air Force in Korea*, 220–30; Futrell, *USAF Operations in the Korean Conflict, 1 Nov 1950–30 Jun 1952*, 7–9.

84. Futrell, *United States Air Force in Korea*, 223, 244; Boatner, "Letter Report on Relative Aerial Combat Capability." Despite its disadvantages, the F-80C did enjoy occasional success against the MiG-15. The November 8, 1950, engagement between a MiG-15 and an F-80 piloted by Lieutenant Russell J. Brown has historically been celebrated by the Air Force as the "first jet-to-jet aerial victory in history." Warnock, *USAF in Korea*, 23; Stratemeyer, *Three Wars of Lt. Gen. George E. Stratemeyer*, 269. However, recent scholarship suggests that there was no enemy downing that day and the honor should instead be awarded to US Navy Aviator William Amen for destroying a MiG on November 9; see Zhang, *Red Wings over the Yalu*, 88.

85. Stratemeyer, *Three Wars of Lt. Gen. George E. Stratemeyer*, 175, 267–68, 272, 277; Futrell, *United States Air Force in Korea*, 246–48; T. Sherwood, *4th TFW Chronology*, 29–30; *History of the 4th Fighter-Interceptor Group, Oct–Dec 1950*, 5; Werrell, *Sabres over MiG Alley*, 76–77; Futrell, *USAF Operations in the Korean Conflict, 1 Nov 1950–30 Jun 1952*, 17, 122. Senior Air Force officials in Washington had repeatedly refused FEAF's and Stratemeyer's prior requests for new aircrews and new aircraft, citing limited inventories of F-86s and pressing "air defense" requirements in the United States. The Fourth had been flying the F-86A Sabre since the end of March 1949.

86. Hinton and Wingo, "Analysis of Operations at Kimpo," 1, 8; *4th FIG Tactical Doctrine*, July 22, 1951, 1. The December 15 mission "was uneventful, serving only to familiarize the pilots with routes to be traversed on missions of future dates." The following day, bad weather grounded the Sabres. The F-86 unit at Kimpo reached full strength on December 24. In the *4th FIG Tactical Doctrine*, July 22, 1951, the group's primary mission "while on temporary duty with the Far Eastern Air Force" was described as "the destruction of enemy Air Forces, protection of United Nations aircraft engaged in operations against the enemy, and to establish and maintain air superiority over a given area at a given time" (1).

87. Hinton and Wingo, "Analysis of Operations at Kimpo," 8–10. See also "Enemy Air Activity, 17 Dec 1950"; Limpantsis, "FEAF Intel Form 1, 17 Dec 1950."

88. *FEAF Report on the Korean War*, 2:2–3; Futrell, *USAF Operations in the Korean Conflict, 1 Nov 1950–30 Jun 1952*, 18–19, 42–43; Stratemeyer, *Three Wars of Lt. Gen. George E. Stratemeyer*, 262–67. The Joint Chiefs of Staff doggedly refused to grant MacArthur approval to bomb Chinese targets north of the Yalu. While many aviators complained of the Communists' aerial sanctuary in Manchuria, UN forces also enjoyed a de facto sanctuary out over the water; UN pilots frequently observed their MiG foes abruptly disengage when the fight drifted past the Korean coastline. The Americans reasoned that neither the Chinese nor the Soviets wanted the true identity of the MiG pilots revealed, and their suspicions seemed to be confirmed when they observed a flight of MiGs strafe one of their own pilots who had gotten shot down and was floating in the Yellow Sea awaiting rescue by an American helicopter. See also Mahurin, *Honest John*, 92–93.

89. Quoted in Futrell, *Ideas, Concepts, Doctrine, 1907–1960*, 171–72. For a similar postwar assessment on the necessity of "air superiority," see Tedder, *Air Power in War*; Momyer, *Airpower in Three Wars*, 125–29.

90. Futrell, *United States Air Force in Korea*, 279, 293–96. Suwon was recaptured at the end of January 1951, but it remained "a waterlogged, bomb-pitted, concrete runway in the middle of a sea of mud" through early March. Unable to support full operations, it was used as an intermediate staging base by the Sabres beginning on March 6, 1951. Anticipating the return of the Sabres, FEAF sent its B-29 bombers back north five days *prior*, but the results were catastrophic—ten Superfortresses damaged, three severely.

91. Preston to Commanding Officer, 4th Fighter Interceptor Wing, "Analysis of F-86 Fighter Encounter," 5; Hinton, "MiG-15 versus F-86A," 18–22; "Special Subject: Two Years of MiG Activity," 11; *FEAF Report on the Korean War*, 2:8; Crane, *American Airpower Strategy in Korea*, 83–89; Futrell, *USAF Operations in the Korean Conflict, 1 Nov 1950–30 Jun 1952*, 114–16. Sabre pilots repeatedly expressed concern about their inability to protect the lumbering B-29s as they attacked targets near the Yalu during daylight. For example, one September 1951 memorandum by Preston emphasized "that it would be virtually impossible to completely stop a determined attack against a formation of B-29s by aircraft as fast as the MiG-15."

92. Peterson, "F-86A Operation as a Fighter-Interceptor," 2; *FEAF Report on the Korean War*, 2:2; Mahurin, *Honest John*, 37; Gabreski, "(End of Tour) Report," 4; McMillin et al., "FEAF Intel Form 1, 18 Jun 1951." FEAF officials complained, "The entire air war in Korea was characterized and influenced by several conditions which rendered this conflict unique and unprecedented in the annals of military history." Mahurin described the frustration as being "exactly as though one professional boxer were [*sic*] asked to fight another with one hand behind his back." The exasperation is palpable in the postmission summary from June 18, 1951.

93. Futrell, *United States Air Force in Korea*, 411–12, 416–18; Crane, *American Airpower Strategy in Korea*, 90–92.

94. *4th FIG Tactical Doctrine*, December 28, 1952, 23.

95. Peterson, "F-86A Operation as a Fighter-Interceptor," 2. See also Hinton, "MiG-15 versus F-86A," 4–6. In its final report on the Korean War, FEAF identified "five distinct phases" of air combat between the MiG-15 and the F-86, "each apparently resulting from a change in enemy tactical concept." *FEAF Report on the Korean War*, 2:2–4. Cf. "Special Subject: Two Years of MiG Activity"; "Growth of MiG Alley;" Thyng, "Operation of the 4th Fighter Wing in Korea," 8–9. A "cyclic rise and fall in air activity" and pilot proficiency was also observed within each phase, eventually leading many intelligence analysts to conclude,

correctly, that whole Chinese and/or Russian units were being rotated in and out of Korean combat. See Hise, "Markings on MiG-15 Type Aircraft"; Hise, "F-86E versus MiG-15 in Korea," 16; Hise, "Analysis of the Tactics of Aerial Warfare," 2; Zhang, *Red Wings over the Yalu*; Futrell, *United States Air Force in Korea*, 418–19.

96. Hinton, "MiG-15 versus F-86A," 6–14; Futrell, *USAF Operations in the Korean Conflict, 1 Nov 1950–30 Jun 1952*, 108–9.

97. D. Evans, *Sabre Jets over Korea*, 89, 119; Hise, "Analysis of the Tactics of Aerial Warfare," 1, 13–15; "Growth of MiG Alley," 53; Beckwith to Deputy for Intelligence, 5th Air Force, "Re-evaluation of MiG-15 Tactics"; *FEAF Report on the Korean War*, 2:4; Futrell, *United States Air Force in Korea*, 402–4; Futrell, *USAF Operations in the Korean Conflict, 1 Nov 1950–30 Jun 1952*, 109–10; Futrell, *USAF Operations in the Korean Conflict, 1 Jul 1952–27 Jul 1953*, 53; Y'Blood, *MiG Alley*, 26; Thyng, "Operation of the 4th Fighter Wing in Korea," 8–9; *Resume of Our Tactics*, ii. Evans expressed his displeasure with the situation: "I get mad as hell when I think of the odds we have to contend with, plus the superior climb, altitude, and acceleration ability of the MiGs." In December 1951 American pilots reported seeing eight times as many MiGs as they had six months earlier, including 366 spotted on a single day. "Special Subject: Two Years of MiG Activity," 10–11.

98. *FEAF Report on the Korean War*, 2:4; "Special Subject: Two Years of MiG Activity," 11–12.

99. Howard and Hise, "FEAF Intel Form 1, 18 Nov 1951"; Hinton, "MiG-15 versus F-86A," 5–6.

100. Quoted in Hallion, "Air Dominance," 40. Harbison continued, "When they decided to engage, they came down to attack us in numbers. Once they did that, we could mix it with them, but, if they wanted to, they could stay out of trouble."

101. Futrell, *United States Air Force in Korea*, 391, 402–4, 413–15, 421; Futrell, *USAF Operations in the Korean Conflict, 1 Nov 1950–30 Jun 1952*, 82, 88–89; *United States Air Force Statistical Digest: Fiscal Year* 1953, pt. I; "Special Subject: Two Years of MiG Activity," 12; Werrell, *Sabres over MiG Alley*, 84–86, 125–26; Y'Blood, *MiG Alley*, 24, 28–29; Crane, *American Airpower Strategy in Korea*, 83–89; Mahurin, *Honest John*, 38. Werrell noted that Vandenberg's decision represented a sudden reversal from an earlier memo dated September 20, 1951, announcing that no more F-86s would be sent to Korea.

102. Futrell, *United States Air Force in Korea*, 250, 296, 419; *FEAF Report on the Korean War*, 2:15; Thyng, "Operation of the 4th Fighter Wing in Korea," 4; D. Evans, *Sabre Jets over Korea*, 91, 98, 209; J. Sherwood, *Officers in Flight Suits*, 83; Blesse, *Check Six*, 65–66, 72–76. Often, a pilot would suddenly find himself engaged with MiGs while dangerously low on fuel, as Evans did on October 5 when he landed with less than twenty gallons in his tank. Evans later joked that because of his fellow pilots' tendency to return home low on fuel (or out of fuel), they should rename "the outfit to the 4th Glider Group." After the war, FEAF reported that "jettisonable fuel tanks represented the largest single operating cost of an F-86 fighter interceptor wing."

103. Futrell, *United States Air Force in Korea*, 403, 412, 421; Futrell, *USAF Operations in the Korean Conflict, 1 Jul 1952–27 Jul 1953*, 50–51; Schinz, "Brief Resume of the Experiences," 4; "Enemy Activity in Korea: Air to Air," November 15, 1952, 9. On the Soviets' decision to send their premier jet fighter to combat, see Zhang, *Red Wings over the Yalu*, 91.

104. Hone, "Korea," 468–70; Werrell, *Sabres over MiG Alley*, 19–23; Y'Blood, *MiG Alley*, 33; Futrell, *United States Air Force in Korea*, 253, 651, 696–99; *FEAF Report on the Korean War*,

2:173–74. To address the armament deficiency, FEAF conducted Project Gun Val, discussed later in this chapter.

105. Mahurin, *Honest John*, 33–34, 37; D. Evans, *Sabre Jets over Korea*, 109, 196–98; Werrell, *Sabres over MiG Alley*, 88; Futrell, *USAF Operations in the Korean Conflict, 1 Jul 1952–27 Jul 1953*, 50–51. Evans recorded in his diary entry dated October 30, 1951: "Again I say what an airplane that MiG is." Cf. G. Smith to Commanding General, Far East Air Force, "Combat Evaluation of the MiG-15"; Hinton and Wingo, "Analysis of Operations at Kimpo," 29–31; Hinton, "MiG-15 versus F-86A," 3–4; Hise, "F-86E versus MiG-15 in Korea," 1–5. The American press also published varying assessments of the F-86 versus the MiG-15; see, for example, "MIG's Top Sabres"; "Jet 'Superior to' MIG-15"; "Ramey Holds F-86 Better than MIG"; "Gabreski Tells Truman MIG-15 Is Good as F-86."

106. Jones to Commanding Officer, 4th Fighter Interceptor Wing, "F-86E Combat Evaluation"; Futrell, *United States Air Force in Korea*, 402, 509, 512, 651; Futrell, *USAF Operations in the Korean Conflict, 1 Nov 1950–30 Jun 1952*, 122; Futrell, *USAF Operations in the Korean Conflict, 1 Jul 1952–27 Jul 1953*, 66–69; Baker, "Report on F-86 Operations," 5; Werrell, *Sabres over MiG Alley*, 24–26, 87–88. Jones reported: "The inability of the F-86A and F-86E to climb, zoom, out-turn, and accelerate with the MIG-15 at any altitude, particularly above 25,000 feet, makes the need for a greater power plant obvious. It is believed that the full combat potential of the F-86E cannot be realized until an engine of substantially greater thrust is installed."

107. *FEAF Report on the Korean War*, 2:5; Y'Blood, *MiG Alley*, 44. The Korean War Sabre-MiG kill ratio continues to be a source of controversy, and it has been revised downward on several occasions. Cf. Futrell, *United States Air Force in Korea*, 695–96; Werrell, *Sabres over MiG Alley*, 141–43; Zhang, *Red Wings over the Yalu*, 201–4; Sayers, "Red Baron Reports."

108. D. Evans, *Sabre Jets over Korea*, 103; Peters, "Air Force, Korea, and Kosovo," 19; W. Thompson and Nalty, *Within Limits*, 28; Hone, "Korea," 468, 475, 488; Werrell, *Sabres over MiG Alley*, 51, 79, 221; Futrell, *United States Air Force in Korea*, 246, 698–99; Crane, *American Airpower Strategy in Korea*, 89; J. Sherwood, *Officers in Flight Suits*, 77. The obvious corollary to the favored rationale was the ineptitude displayed by the majority of the MiG pilots.

109. Werrell, *Sabres over MiG Alley*, 26, 32–33; *FEAF Report on the Korean War*, 2:16. Werrell also suggested that "the poor marksmanship of the MiG pilots could be attributed to . . . an outdated gunsight." See similar assessments of the gunsight in J. Sherwood, *Officers in Flight Suits*, 75; W. Thompson and Nalty, *Within Limits*, 28; Y'Blood, *MiG Alley*, 15; Wildenberg, "A-1C(M) Gunsight," 34.

110. D. Evans, *Sabre Jets over Korea*, 11. Buck Rogers was a space-age, science-fiction hero popular in the 1930s and '50s.

111. Ibid., 181. Hone, "Korea," 470–71. Hone's assessment echoes that of a 1952 report from the Fourth FIG, which stated: "Basic air to air fighter tactics used in Korea are fundamentally the same as they always were, except for certain modifications imposed due to high rate of fuel consumption, loss of visual acuity at high altitudes, and the much larger air space required for maneuver at high speeds." Hise, "F-86E versus MiG-15 in Korea," 5.

112. J. Sherwood, *Officers in Flight Suits*, 94; *FEAF Report on the Korean War*, 2:16.

113. D. Evans, *Sabre Jets over Korea*, 138–40; Salter, *Burning the Days*, 88. On the hierarchy among different aircraft types and its role in Air Force pilot "elitism," see J. Sherwood, *Officers in Flight Suits*, 67.

114. Maurer, *USAF Credits: Korean War*; Werrell, *Sabres over MiG Alley*, 144; "Air Force Aces," 128; Haulman, "Aerial Victory Credits"; "American Fighter Aces." The aces in Korea accounted for 310½ of the Air Force's 896½ aerial victory credits. Werrell concluded that approximately "four percent of the pilots that flew Sabres in the Korean War accounted for forty percent of the victories." However, discrepancies over the number of aces and total victory counts exist among the cited sources, some due to Cleveland's late addition to the ace rolls. While the specific totals and quantities are subject to debate, the significant and disproportionate contribution of the aces to the victory totals is not. On Cleveland, see Seamus O'Connor, "Board Recognizes Korean War Pilot"; Bates, "Retired General Becomes Air Force's Newest Fighter Ace"; Trest, *Once a Fighter Pilot*, 88–90, 234–42; Cleveland, Oral History Interview, 36–41.

115. "These Seventeen Americans," 35. *Life* included a special feature on the first jet ace, Major James Jabara, and the F-86 Sabre in earlier editions: Dille, "Jets' First Ace"; "US Sabres Splash Red Jets."

116. Sherwood, *Officers in Flight Suits*, 8, 73. On the other Air Force missions and their contribution to the air war, see: Y'Blood, *MiG Alley*, 44; Hallion, "Korea: The Forgotten (Air) War," 2–3.

117. D. Evans, *Sabre Jets over Korea*, 157, 177; Mahurin, *Honest John*, 42, 44, 87; Blesse, *Check Six*, 65; J. Sherwood, *Officers in Flight Suits*, 79.

118. Blesse, *Check Six*, 65, 69–70.

119. Quoted in Trest, *Once a Fighter Pilot*, 88–96. Although missing the all-important fifth MiG kill, Cleveland had nevertheless racked up an impressive tally: four destroyed MiGs, four damaged MiGs, and two probable victories. Blesse referred to the young lieutenant as "'The Ivory Ace,' 99 44/100ths percent pure," mimicking the popular soap advertisement. Blesse, *Check Six*, 69.

120. Salter, *Burning the Days*, 153.

121. Mahurin, *Honest John*, 30. Blesse expressed a similar opinion, observing after one eventful flight, "Relying on other people in this business can kill you." Blesse, *Check Six*, 47.

122. D. Evans, *Sabre Jets over Korea*, 13.

123. Salter, *Hunters*, 193. Salter's description was only the latest (and perhaps most eloquent) attempt at describing the feeling of flying a single-seat fighter. For example, a 1922 text on air combat offered the following: "The whole psychological medium differs. The pilot, even in formation, is an entity, controlling his machine by voluntary effort. True, he can see his comrades to right and left and observe the turn of the fight. But there is no elbow touch, no complete submergence of his own individuality in that of the mass." Milling, *Air Tactics*, sect. 6, p. 4. See also D. Evans, *Sabre Jets over Korea*, 195; J. Sherwood, *Officers in Flight Suits*, 71.

124. Gabreski, "(End of Tour) Report," 4. Schinz begged that the retrained pilots not be sent to Korea until they had completed a yearlong tour stateside. Schinz, "Brief Resume of the Experiences," 5.

125. Hise, "F-86E versus MiG-15 in Korea," 22. The report noted, "We feel that it is almost an impossible task to take a pilot who flew transport or heavy bombers in WW II, who got out of the service and grew old mentally and fat physically for several years, recall that pilot, and expect him to fly a quick course in F-80s and become a replacement pilot for F-86 flying. In most cases it just simply won't work."

126. D. Evans, *Sabre Jets over Korea*, 235 (emphasis in original).

127. Ibid., 13–28. References to the World War II incarnation of "Clobber College" can be found in Olds, *Fighter Pilot*, 124.

128. Brown's comments are quoted in Hallion, "Air Dominance," 31. Olds recalled that flying formation in the Sabre at forty-four thousand feet required "total attention and a delicate hand on both stick and throttles [*sic*]." Olds, *Fighter Pilot*, 202.

129. Cleveland, Oral History Interview, 25–27. Cleveland described the gunnery training: "The pilots in the fighters that were attacking [the banner] would come up 3,000 or 4,000 feet above it and off to one side of it and about even with it; turn into it, fly with it, and try to get the rag in the gunsight and shoot." On banner gunnery training, see "A-1 Sight and Air to Air Gunnery"; "Radar Ranging"; "Square Corner."

130. Harbison recalled his experience in Clobber College, "I must say that I thought I was experienced on the F-86—and indeed I was—but [the combat experienced instructors] really flew the airplane to its limits, exploring corners of the envelope that I hadn't seen." Quoted in Hallion, "Air Dominance," 40. See descriptions of typical dogfighting training in Blesse, *Check Six*, 52–53; Blesse, Corona Ace Oral History Interview, 61–63.

131. Later, Sabre units began experimenting with "camera gunnery," which combined gunnery training with dogfight training: the attacker didn't fire any bullets but instead recorded the engagement with his gun camera. Unfortunately, because the gun camera did not record the gunsight information, there was no way to validate whether the engagement would have been successful. See J. Meyer to Commanding General, Air Defense Command, "F-86 Activities Report," 3; J. Meyer to Commanding General, Air Defense Force, "F-86 Activities Report (Special)," 4–5; Schmid, "Operations of the 4th Fighter-Interceptor Wing," 1; Hesse, "Flying: Continuation Flying Training"; *History of the 334th Fighter Interceptor Squadron, Apr 1952*, 3.

132. Tactics, according to one pilot, involved discussions of "where the MiGs were coming across the border, what altitude they were flying at, . . . what was the most effective and quickest way to get on their tail." Risner, Oral History Interview, 26–27.

133. Obviously, the attitude was not shared by all. For example, Hagerstrom sought to prepare himself for Korean air combat by "taking courses over and over again on the A-4 gun sight, and reading all the intelligence reports he could get his hands on." J. Sherwood, *Officers in Flight Suits*, 70.

134. Werrell, *Sabres over MiG Alley*, 58. Sometimes the exigencies of combat necessitated a brusque aircraft introduction. DeArmond recalled how he arrived in Korea having never flown the F-86; "in fact, I had never seen one. Our captain stood on the wing of a Sabre, told me how to start it, and off I went. We got three or four rides, and I went into combat. . . . The first mission in combat, I was so 'clanked up' and so tight I couldn't remember where the gun switch was when we crossed the bomb line [into enemy territory]." Quoted in Hallion, "Air Dominance," 34. Recounting his tendency to jump in unfamiliar aircraft and take to the skies, Olds explained, "We didn't care much about the niceties of emergency procedures in those immediate postwar days, nor did we pay much attention to a lot of bothersome flight regulations either." Olds, *Fighter Pilot*, 96–99, 156–58. Another pilot of the period recalled that "it took several days for my flight commander to locate" an aircraft flight manual for him to "study prior to having to check out in the airplane." Umstead, Oral History Interview, 10–11.

135. D. Evans, *Sabre Jets over Korea*, 170 (emphasis in original). See also pp. 160, 228. Recall that the air services and medical professionals had earlier encouraged this technical ambivalence; see Rippon and Manuel, "Report on the Essential Characteristics," 411–12.

136. Werrell, *Sabres over MiG Alley*, 58; *Summary of F-86E Aircraft Accidents*, 2–3. Air Force safety officials ultimately concluded that from January 1951 through August 1954, "pilot error" was the "most frequent primary cause factor in F-86E accidents," and "in the majority of cases, improper operating procedure was the first step in the sequence of events leading to the accident."

137. "How Old Can You Jet?" The surgeon general of the air force had recently issued a report concluding that "the combat utilization of pilots for jet type aircraft ceases at age 30, and for conventional [prop] type fighter aircraft at age 32." The experience of World War II seemed to confirm the assessment, with men between the ages of twenty-one and twenty-six reportedly making the best fighter pilots; see the assessment in "Tactics: V & XIII Fighter Commands," 5.

138. J. Meyer, Oral History Interview, 1–3; Werrell, *Sabres over MiG Alley*, 152–55; Dille, "Jets' First Ace," 135.

139. Mahurin, *Honest John*, 49, 52. Mahurin's anguish was as much for his wingman as his own reputation. "I was crushed because I hated to establish myself as a guy who had no concern for others in his flight." Blesse took pride in the fact that he had only one wingman "who didn't come back in his own aircraft." Blesse, *Check Six*, 72.

140. Quoted in Werrell, *Sabres over MiG Alley*, 155.

141. Ibid., 153–54; Dille, "Jets' First Ace," 139. Dille's account of Jabara's engagement is interesting in its description of the lost wingman, posing it as a testament to Jabara's fighting spirit rather than a breakdown in the established formation team: "Jabara climbed through this melee, back up to 20,000 feet, where he spotted six more MIGs. But Jabara's wing man was not with him. Kemp had tangled with some MIGs on the way up. Few pilots would consider tackling a flight of MIGs without a wing man to keep an eye out while he is shooting. But Jabara bore in on one of his newfound MIGs anyway."

142. "World's First Jet Ace Downs 2 MIG's within 10 Minutes"; "Two New 'Kills' Give Yank Title of First Jet Ace"; "Kansan Becomes History's First Jet Fighter Ace"; "Kansan Downs 2 MIG Planes." The editors at *Life* devoted six pages to Jabara in its first issue after his achievement: Dille, "Jets' First Ace." As expected, *Air Force* also devoted several pages to Jabara in its June edition, including a self-authored piece: Jabara, "We Fly MiG Alley." Curiously, all the newspapers claimed that Jabara was an ace in World War II, yet the Air Force record credits him with only 1½ kills, a discrepancy possibly stemming from the temporary policy during World War II of awarding credit for enemy aircraft destroyed on the ground; see Haulman, "Aerial Victory Credits." Furthermore, with new access to Russian archival materials, Jabara's claims and his status as the first jet ace have come under increased scrutiny; see Werrell, *Sabres over MiG Alley*, 154. The day after his ace victory, Jabara was spirited out of combat and into the hands of the Air Force's public relations officials. "Jet Ace Out of Combat"; "Fenway Crowd Sees 1st Jet Ace Awarded DFC"; "Jabara, Jet Ace, Set for Mid-East Good Will Tour"; "Ace to Tour Jet Bases"; "First Jet Pilot Ace Back from Europe, Middle East." Jabara would eventually return to combat, more "the modest herotype," according to Meyer, and finish the Korean War as the second leading American ace with fifteen kills. J. Meyer, Oral History Interview, 1; "First Korean Jet Ace Eager to Fight Again"; "Jabara to Fight Again."

143. "13 Red Jets Shot Down in 2 Battles; Ace Gets 4."

144. Parrott, "Sabres Bag 6 MIG's." Davis's MiG tally ended at fourteen after he was shot down, which also earned a front-page story: "Top U.S. Jet Ace Lost in Korean Air War."

145. "Fernandez Bags 13th MIG, Becoming the Top Jet Ace"; "First Triple Jet Ace Helps Bag 9 MIGs."

146. "First Triple Jet Ace Helps Bag 9 MIGs."

147. "'Old Pros' of MiG Alley."

148. Futrell, *United States Air Force in Korea*, 655; "Air Force Aces," 128–29.

149. "Air Force Aces," 122–28. But indicative of what was to be finally learned in Korea, almost 50 percent of the *top* World War II aces scoring fifteen or more kills were field-grade officers.

150. Futrell, *United States Air Force in Korea*, 698.

151. *FEAF Report on the Korean War*, 2:7. Another report noted, "Pilots who fought in every theatre in WW II generally agree that the combination of high speed and inability to see far enough at altitude create a fighting situation that requires more judgment and skill than any other air-to-air fighter flying they have experienced." Hise, "F-86E versus MiG-15 in Korea," 17.

152. Quoted in Hallion, "Air Dominance," 41. Discussions of pilots' age and experience can also be found in "First Korean Jet Ace Eager to Fight Again"; Gabreski, "Our Top Ace Talks Shop," 41; Jabara, "We Fly MiG Alley," 65; Dille, "Jets' First Ace," 135; Thyng, "Operation of the 4th Fighter Wing in Korea," 2.

153. Quoted in Hallion, "Air Dominance," 44. Brown observed, "One ace is worth a whole lot of journeyman fighter pilots. . . . They have a great sense of situational awareness." Spick identified *situational awareness*—"the ability of the pilot to keep track of events and foresee occurrences in the fast-moving, dynamic scenario of air warfare"—as the "mysterious ingredient" in successful fighter pilots. Spick, *Ace Factor*, vi. Situational awareness has also "entered the mainstream of human factors" research; see, for example, Wickens, "Situation Awareness." Watts cited statistics that "a mere 4.8% of the 800 Sabre pilots who flew at least twenty-five counterair missions are credited with no less than 38.2% of the kills." B. Watts, "Changing World of Air Combat."

154. Werrell, *Sabres over MiG Alley*, 137. Werrell figured that pilots flying in the lead position "registered 82 percent of the total MiG claims, a percentage that varied little over the course of the war." Additionally, "the men who finished as aces scored most of their kills as leads, 93 percent compared with 75 percent of those pilots who scored but did not tally five kills." The relationship between flight position and probability of claiming a kill is also discussed in Youngling et al., *Feasibility Study to Predict Combat Effectiveness*, sect. 1, pp. 8–10, sect. 3, pp. 3–5, 12–14.

155. *4th FIG Tactical Doctrine*, December 28, 1952, 7. See also: Futrell, *USAF Operations in the Korean Conflict, 1 Nov 1950–30 Jun 1952*, 111–12; Mahurin, *Honest John*, 45; Thyng, "Operation of the 4th Fighter Wing in Korea," 3, 7, 10; Hinton, "MiG-15 versus F-86A," 19–21; Hise, "F-86E versus MiG-15 in Korea," 19; *FEAF Report on the Korean War*, 2:7–8, 11. Complicating matters, the Sabre pilots frequently had issues communicating with the B-29s bombers; see Gray to Commanding General, 5th Air Force, "Transmittal of Pilot's Statements"; Beckwith to Commanding Officer, 4th Fighter Interceptor Wing, "Analysis of Aerial Engagements, 23 Oct 1951."

156. *Resume of Our Tactics*, 4.

157. Ibid., 5–8. See also *4th FIG Tactical Doctrine*, December 28, 1952, 7–8, 22–23; *FEAF Report on the Korean War*, 2:8–9; Futrell, *USAF Operations in the Korean Conflict, 1 Nov 1950–30 Jun 1952*, 112.

158. *4th FIG Tactical Doctrine*, December 28, 1952, 16–17.

159. *Resume of Our Tactics*, 12–13; *4th FIG Tactical Doctrine*, July 22, 1951, 9; *4th FIG Tactical Doctrine*, December 28, 1952, 16–17; *4th FIG Tactical Doctrine*, June 17, 1953, 15–19; Hise, "F-86E versus MiG-15 in Korea," 11; Hinton and Wingo, "Analysis of Operations at Kimpo," 20. By 1953, the 4th FIW allowed its element leader to fly "approximately 400 yard[s] out, slightly back and stacked either slightly below or slightly above the lead element, depending largely upon the position of the sun."

160. Stocker, "Wing Man Has to Have Eyes," 32; *Resume of Our Tactics*, 12–13; *4th FIG Tactical Doctrine*, July 22, 1951, 9; *4th FIG Tactical Doctrine*, December 28, 1952, 16–17; *4th FIG Tactical Doctrine*, June 17, 1953, 15–17; Hise, "F-86E versus MiG-15 in Korea," 11; Hinton and Wingo, "Analysis of Operations at Kimpo," 20; Mahurin, *Honest John*, 49; Salter, *Burning the Days*, 144.

161. *Resume of Our Tactics*, 12–13; *4th FIG Tactical Doctrine*, July 22, 1951, 9–10; *4th FIG Tactical Doctrine*, December 28, 1952, 16–17; Blesse, *Check Six*, 59; Mahurin, *Honest John*, 50. Complaints about wayward wingmen are a staple of Korean pilots' memoirs; see, for example, Blesse, *Check Six*, 62; D. Evans, *Sabre Jets over Korea*, 186; J. Sherwood, *Officers in Flight Suits*, 87–88; Salter, *Hunters*, chap. 19. There were three reasons for a wingman to fire on a MiG: to save his flight lead; if his flight lead ran out of ammunition and directed him to; or if he was conveniently aligned and could do so while still protecting his flight lead; see descriptions of each in "FEAF Intel Form 1, 11 Jul 1951"; "Enemy Activity in Korea: Air to Air," October 25, 1952, 4; Howard et al., "FEAF Intel Form 1, 16 Oct 1951 (0401)." For a brief period, the Sabre pilots experimented with six-ship flights, but they soon abandoned the formation as too unwieldy; see Futrell, *USAF Operations in the Korean Conflict, 1 Nov 1950–30 Jun 1952*, 112; Hise, "F-86E versus MiG-15 in Korea," 11; *4th FIG Tactical Doctrine*, July 22, 1951, 18; Hinton, "MiG-15 Versus F-86A," 16.

162. Stocker, "Wing Man Has to Have Eyes," 35.

163. *4th FIG Tactical Doctrine*, December 28, 1952, 17.

164. Stocker, "Wing Man Has to Have Eyes," 35.

165. *4th FIG Tactical Doctrine*, July 22, 1951, i–ii; Mahurin, *Honest John*, 49. After having progressed to element lead, Evans had no desire to fly as a wingman again. But on one mission, his squadron was short of aircraft and in the hasty reorganization of the flight, he "ended up on wing again—*#*&%*!!" D. Evans, *Sabre Jets over Korea*, 92. See also Giraudo, Oral History Interview, 85–87.

166. Hallion, "Air Dominance," 30.

167. Blesse, *Check Six*, 60–61; J. Sherwood, *Officers in Flight Suits*, 124–25, 139. Cleveland promptly schooled the squadron pilots in ping-pong, all except for Blesse.

168. J. Sherwood, *Officers in Flight Suits*, 125–29. Alcohol is a central component in most Korean fighter pilot memoirs. According to one fighter pilot cited by Sherwood, "Korea was the easiest place in the world to become an alcoholic." Many Sabre pilots flew drunk or still suffering the aftereffects of the previous night's boozing—the effects of alcohol on aerospace physiology, or even driving a car, had not yet been thoroughly studied. For example, Mike DeArmond, later a Brigadier General, was shot down when his flight lead, who had

won the prior night's martini drinking contest, failed to turn tight enough after his formation came under attack. Hallion, "Air Dominance," 34. See also D. Evans, *Sabre Jets over Korea*, 105.

169. Commenting on the young lieutenants, Harbison noted, "very soon they caught the fever. The competition was fierce." Quoted in Hallion, "Air Dominance," 41. For other opinions of the young pilots, see Gabreski, "(End of Tour) Report," 4; Schinz, "Brief Resume of the Experiences," 2; Thyng, "Operation of the 4th Fighter Wing in Korea," 2. Most agreed with Thyng's assessment: "These youngsters . . . are good, they are able, . . . they are eager, courageous, and the United States Air Force has no worries about this young generation."

170. Gabreski, "(End of Tour) Report," 4.

171. Salter, *Hunters*, chap. 19.

172. Low and Blesse are quoted in Werrell, *Sabres over MiG Alley*, 180–81. See also "Pilot Is 17th U.S. Jet Ace in Only 6 Weeks in Korea"; Dorr, Lake, and Thompson, *Korean War Aces*, 61; Gabreski, "Our Top Ace Talks Shop," 37.

173. *Resume of Our Tactics*, 6–7.

174. Hallion, "Air Dominance," 47. Beginning in late 1952, the Americans were able to use a radar and tactical air-direction center on tiny Cho-Do Island, some 140 miles northwest of Seoul in the Yellow Sea, to help alert them to any nearby MiGs. The Cho-Do station, using the call sign "Dentist Charlie," could also provide Sabre pilots with the same detailed, GCI (ground controlled intercept) instructions that the MiG pilots had been enjoying for months. Before the emergence of Dentist Charlie, F-86 pilots received only occasional alerts of airborne MiGs by radar operators stationed at Kimpo, call sign "Dentist," but the ground-based radar coverage and reliability from that location was marginal.

Beginning in May, five months before the UN radar installation at Cho-Do was fully operational, American intelligence officers and ground controllers began eavesdropping on the MiGs' inflight communications, which allowed them to alert the Sabre pilots to the MiGs' intentions. The Americans were able to glean significant tactical warning from the radio transmissions because that same month, the Chinese began providing detailed GCI instructions to some of its MiGs. The Americans' "electronics assistance" mimicked the "Y-service" that was used heavily by the Allies in the later stages of World War II. Futrell, *United States Air Force in Korea*, 423, 514; Mahurin, *Honest John*, 58–67. On Y-service in World War II, see "Tactics and Techniques of Long-Range Fighter Escort," 17.

175. *FEAF Report on the Korean War*, 2:7. In his description of World War II combat, Beeson offered a superb account of the importance and mechanics of visual detection. Beeson, "Capt. D. W. Beeson," 71. FEAF echoed the earlier assessments, noting that in Korea, "To see the enemy first is a great advantage and to develop this ability to its fullest requires hard work and constant practice. This factor has become increasingly critical as the speed and altitude of aerial warfare have increased. Many potentially good fighter pilots have been unsuccessful simply because they cannot learn to search the sky effectively."

176. *Resume of Our Tactics*, 14. On the medical efforts to study the effects of high-altitude flight on the Sabre pilots' visual acuity, see references in *History of the 51st Fighter Interceptor Wing, Jul–Dec 1952*, 19.

177. D. Evans, *Sabre Jets over Korea*, 94. See also pp. 41, 128, 202. Evans commented, "Planes sometimes seem to vanish or dissolve in the blinding blue depths of very high altitude."

178. Hallion, "Air Dominance," 37; Thyng, "Operation of the 4th Fighter Wing in Korea," 6; Baker, "Report on F-86 Operations," 10; J. Sherwood, *Officers in Flight Suits*, 84; Blesse, "No Guts, No Glory," 73–74; Salter, *Burning the Days*, 151.

179. J. Sherwood, *Officers in Flight Suits*, 84; Werrell, *Sabres over MiG Alley*, 185. Hagerstrom apparently bragged "that he could look at a 500-watt light bulb with these glasses and read the writing on the bulb." When the optometrist warned Hagerstrom that wearing the special glasses might permanently damage his eyesight, the fighter pilot reportedly replied, "I don't give a shit."

180. Hallion, "Air Dominance," 30. Brown remembered, "Some of our leaders were old guys in their thirties, and I was a young twenty-four-year-old. So the difference in eyesight was significant." One exception was Baker, whom Risner described as having "eyes like an eagle"—Baker became one of the top Korean aces. Risner, Oral History Interview, 12.

181. Mahurin, *Honest John*, 82–83. The 4th's tactics manual offered another example of a correct call, "Awning Red Leader, Awning White Two here, bandit's six o'clock high closing fast." *4th FIG Tactical Doctrine*, July 22, 1951, 26.

182. Raebel, "Communications: Frequency Assignments." Two radio channels were used by the control tower, two were used to help vector the Sabres to landing, and one was used for emergencies. The radios couldn't be readily retuned on the ground either; maintenance crews had to requisition new crystals before a different frequency could be used. *History of the 51st Fighter Interceptor Wing, Apr 1952*, chap. 2, "Communications." Salter, with his elegant prose, described his experience with the radios: "The planes sit out in the weather and dampness affects them, the radios become unreliable. *Break!* they are crying in a fight and you hear nothing. The silence is uncanny. *Break right!* they are shouting, *break right!* For some reason you look back and there, behind you, is an intake the size of a locomotive." Salter, *Burning the Days*, 158–59 (emphasis in original). A "break" call indicated that an *immediate* defensive maneuver was required for survival.

183. D. Evans, *Sabre Jets over Korea*, 73; *FEAF Report on the Korean War*, 2:10. On fighter sweep missions, three or four dozen F-86s would be launched on a single patrol, all listening and transmitting on a single radio channel. The chatter between the two pilots in *The Fighting 51st* was probably typical for its extraneous banter—"Uh, Rog, I'm goin' in now." "OK, you're clear. Go ahead." "Rog, I'm gonna fire now." "Uh, you're still clear" "OK, he's down. You got him too."—but atypical in that there was no one else trying to use the radio at the time. *Fighting 51st*.

184. *Resume of Our Tactics*, 23. See also *4th FIG Tactical Doctrine*, July 22, 1951, 26; *4th FIG Tactical Doctrine*, December 28, 1952, 20.

185. D. Evans, *Sabre Jets over Korea*, 42, 96, 184–86. Resembling Evans's experience, on a November 30, 1951, mission, Able 3 and 4 became separated. Able 4 thought he spotted his leader and tried calling him on the radio, asking Able 3 to "wag his wings" to confirm it was him. However, Able 4's "radio plug had become disconnected, and when he established radio contact, he heard Able 3 calling for help." Able 3 had been attacking an enemy formation, thinking his wingman was with him "until he leveled and observed great balls of fire above his canopy." Howard et al., "FEAF Intel Form 1, 30 Nov 1951."

186. Roberts, Corona Ace Oral History Interview, 27; Hone, "Korea," 492; Futrell, *USAF Operations in the Korean Conflict, 1 Nov 1950–30 Jun 1952*, 113; D. Evans, *Sabre Jets over Korea*, 71–72. Futrell cited a statistic that "often less than 3 seconds elapsed between first sighting through engagement to the break-off."

187. Salter, *Burning the Days*, 156.

188. Blesse, *Check Six*, 71.

189. Quoted in Trest, *Once a Fighter Pilot*, 91.

190. D. Evans, *Sabre Jets over Korea*, 125; Futrell, *USAF Operations in the Korean Conflict, 1 Nov 1950–30 Jun 1952*, 117; *FEAF Report on the Korean War*, 2:7–8; *Resume of Our Tactics*, 15; *4th FIG Tactical Doctrine*, December 28, 1952, 10; Hise, "F-86E versus MiG-15 in Korea," 17. Mahurin identified the "three advantages" regularly enjoyed by the "Communists": "superior speed, superior altitude and advantageous position." Mahurin, *Honest John*, 37.

191. D. Evans, *Sabre Jets over Korea*, 89, 107; Hinton, "MiG-15 versus F-86A," 12; *4th FIG Tactical Doctrine*, December 28, 1952, 10; McMillin, "FEAF Intel Form 1, 17 Jun 1951"; Kenyon and Beckwith, "FEAF Intel Form 1, 9 Sep 1951."

192. Futrell, *USAF Operations in the Korean Conflict, 1 Nov 1950–30 Jun 1952*, 113; Hinton, "MiG-15 versus F-86A," 10, 17–18; D. Evans, *Sabre Jets over Korea*, 116–17.

193. Quoted in Hallion, "Air Dominance," 38. Risner told of trying to chase down a MiG when "an F-86 went right across my head. [The other pilot] said, 'I have this MiG.' . . . So I said, 'You get out from behind that MiG, or I'll shoot you down.'" Neither caught the MiG. Risner, Oral History Interview, 22.

194. Mahurin, *Honest John*, 73–74; Werrell, *Sabres over MiG Alley*, 121–24. Only at close range did the MiG's trademark high T-tail and midmounted wing become discernable. Most famously, Jabara accidently shot down his wingman, Richard Frailey, in the commotion of one battle. See also Hallion, "Air Dominance," 53–54. The Sabres' "jet stream" tactic exacerbated the identification problem because, unlike mass formations that all arrived together, new F-86s were constantly flowing into the area. *4th FIG Tactical Doctrine*, December 28, 1952, 8.

195. *Resume of Our Tactics*, 15–16; Blesse, "No Guts, No Glory," 72–73.

196. On one mission, Baker Lead closed to within "50 feet" of his MiG target firing "long bursts," but "no strikes" were observed. Howard et al., "FEAF Intel Form 1, 16 Oct 1951 (0401)."

197. Salter, *Burning the Days*, 146.

198. Futrell, *USAF Operations in the Korean Conflict, 1 Jul 1952–27 Jul 1953*, 46, 63. A MiG pilot defector explained that "Red pilots knew that to break would expose their cockpits to [the Sabres' gun]fire, whereas, by absorbing the [gun]fire in the engine and armor plate behind them, the MIG pilots found that they had an excellent chance to recover, or at least safely to bail out after the F-86's had broken off."

199. *FEAF Report on the Korean War*, 2:6. See other descriptions in Futrell, *USAF Operations in the Korean Conflict, 1 Nov 1950–30 Jun 1952*, 118. For example, another "experienced Sabre pilot fired twelve hundred rounds at a MIG-15 at estimated ranges as low as 150 to 200 feet, and, although the MIG smoked, it still was able to cross the border under control."

200. Blesse, *Check Six*, 65–67; Dille, "Jets' First Ace," 136, 139; Werrell, *Sabres over MiG Alley*, 153–54. Later, Blesse would tell pilots without equivocation, "If you have a 'hung' external tank, leave the combat area." Blesse, "No Guts, No Glory," 55.

201. D. Evans, *Sabre Jets over Korea*, 98; Mahurin, *Honest John*, 54–56.

202. Blesse, *Check Six*, 74–75.

203. D. Evans, *Sabre Jets over Korea*, 210; Werrell, *Sabres over MiG Alley*, 156–57; Bruning, *Crimson Sky*, 181–88; Blesse, *Check Six*, 55–56. Werrell noted that some pilots thought Davis had "become more brazen, more aggressive, and more willing to take risks in Korea than he was during World War II." Bruning cited one pilot's assessment that Davis had grown "quite

contemptuous of the MiG pilots. . . . [He] considered himself immortal." The pilots had a name for the "affliction"—"*MiG Madness.*"

204. Quoted in Trest, *Once a Fighter Pilot*, 71–72. See also Cleveland, Oral History Interview, 40–42.

205. Mahurin, *Honest John*, 68–69.

206. Ibid., 69–71, 78, 80, 83–85. Mahurin offered another rationale for their flights north: "A nation should not handicap its fighting men so that their chances of survival in battle against the enemy are reduced substantially." Mahurin admittedly was in "an intense competition with Gabby" Gabreski, his former commander in the 51st, remarking, "I would have given my eye teeth to shoot down all the Migs in China—especially if it contributed to bringing the war to a close. At the same time I would almost have given my right arm to beat Gabby in every way possible." There was one restriction that most pilots adhered to, though: they would not strafe any MiGs on the ground—that would be too obvious a violation of Manchurian sanctuary (and too difficult to cover up). Instead the Sabres orbited over the Communist airfields waiting to see dust clouds kicking up behind the MiGs as they started down the runway. Then the Sabres would swoop down and try to catch the MiGs just as they left the ground. See also Blesse, *Check Six*, 67–69; Cleveland, Oral History Interview, 41.

207. Mahurin, *Honest John*, 70, 84; Werrell, *Sabres over MiG Alley*, 132. Mahurin's statements cast doubt on the veracity of intelligence summaries like that published after a mission on October 25, 1952, which read: "The leader of the first Sabrejet element closed to 2,000 feet on his target, scored hits on the left wing, and the Red aircraft immediately decelerated. Two more bursts resulted in many more hits on the MiG, and the Red pilot bailed out 12 miles southeast of Sinuiju. *The pilotless plane continued on a west heading, finally crashing on the Manchurian side of the Yalu River, approximately two miles north of the Antung Airfield*" (emphasis added). "Enemy Activity in Korea: Air to Air," October 25, 1952, 3–4. The gig was finally up in late January 1953. When questioned about his sortie by Colonel David Jones, Captain Dolph Overton elected to tell the truth; the next day the Air Force's newest ace was grounded, stripped of his medals, and ordered home. He didn't receive credit for his five victories for almost a year. Most border violations ceased after Overton was sent home. However, some like ace Lonnie Moore kept flying north under the impression that the Fifth Air Force commander "says screw the Yalu." In June 1953 Moore was caught flying in Manchuria and was also promptly shipped home. See Werrell, *Sabres over MiG Alley*, 133–36.

208. Maurer, *USAF Credits: Korean War*; Haulman, "Introduction: Korean War Aerial Victory Credits"; Futrell, *USAF Operations in the Korean Conflict, 1 Nov 1950–30 Jun 1952*, 126; Werrell, *Sabres over MiG Alley*, 137. The half-credit policy mimicked that of World War II. In contrast, during World War I, pilots and observers could both be awarded a full credit for a single downed enemy aircraft. For a brief period, Fifth Air Force also promised to award credit for an "aircraft [that] was observed to explode or burn while *on the ground*," but that criterion was quickly rescinded (emphasis added).

209. See concerns expressed in Mahurin, *Honest John*, 23; D. Evans, *Sabre Jets over Korea*, 141.

210. Occasionally a pilot in a news story would credit his wingman. See, for example, "Jet Flier Shy of 'Ace' by 1 in 5 Fights." Cleveland also acknowledged the contributions of the ground personnel in Trest, *Once a Fighter Pilot*, 83. The importance of the ground crew personnel to the mission was also emphasized in *Fighting 51st*.

211. Werrell, *Sabres over MiG Alley*, 178.

212. "Standard Aircraft Characteristics: F-86E," 3; "Harmonization and Firing Techniques," 82–84; Strauss, *Analysis of Korean Air-to-Air Combat Film*, 4; Hollstein to Banfill, "Comments of Study," 2; "You've Got to Prove It!"; "Enemy Activity in Korea: Air to Air," November 15, 1952, 8; Alkire, "Gun Camera Brief," 4–7. Evaluating the gun camera installation on the F-80 and F-84E, Hollstein noted, "It is obvious that the role of combat recording photography has been neglected since the termination of hostilities in World War II." Still, one *FEAF Intelligence Roundup* report declared in 1953 that "the most convincing and fool proof means of confirming a kill is by the gun camera carried in each Sabrejet." "Gun-Camera Records Kills." Even by 1955, the *Fighter Weapons Newsletter* was still complaining about the N6 gun camera, calling it "a bundle of trouble." "GI Gun Cameras," 62. When it worked, the gun camera footage could also be used as a source of intelligence on the MiG; for example, combat film analysis seemed to indicate that the MiG carried fuel in its wings, contradicting earlier assessments that all its fuel was carried solely in the fuselage. See Beckwith to Commanding General, 5th Air Force, "Wing Fuel-Load in MiG-15."

213. Quoted in Hallion, "Air Dominance," 36. Hinton reported similar occurrences: "In other cases, pilot reports and film verification showed a large number of strikes on the MIG-15, with the latter continuing his flight with little or no reduction in speed. As a result, it was always necessary to follow the MIG to the ground, often at speeds in excess of the normal Mach, to insure [*sic*] his destruction." Hinton and Wingo, "Analysis of Operations at Kimpo," 30.

214. Porter to Director of Reconnaissance, "Report of Staff Visit," 4; James to Director of Reconnaissance, "Report of Staff Visit"; *History of the 4th Air Base Group, May 1952*, 6; *History of the 4th Air Base Group, Jun 1952*, 6; *History of the 51st Fighter Interceptor Wing, Mar 1952*, 13; Hise, "F-86E versus MiG-15 in Korea," 20–21. There were additional challenges associated with developing the film in an austere combat environment. Initially, the only film development site was located 250 miles away from Kimpo. When the bases finally built their own on-site film facilities, they frequently encountered problems obtaining reliable electricity and sufficient quantities of clean water to process the combat film. See J. Meyer to Commanding General, Air Defense Command, "F-86 Activities Report, 2 January 1951 thru 13 January 1951," 1; J. Meyer to Commanding General, Air Defense Force, "F-86 Activities Report," 3; *History of the 4th Fighter-Interceptor Wing, Jul–Dec 1952*, 163–67.

215. Werrell, *Sabres over MiG Alley*, 277n74; *History of the 334th Fighter Interceptor Squadron, Oct 1951*, 4; D. Evans, *Sabre Jets over Korea*, 113, 134. Evans was ecstatic when his gunnery film "came out good for a change." Naturally, only the best gun camera footage found its way into *Fighting 51st*.

216. Mahurin, *Honest John*, 71–72, 83–84. Mahurin recalled his pilots "coming back from missions with some of the damndest pictures. Migs with wheels, flaps and dive breaks extended in preparation for landings; Migs just breaking ground after take-off; Migs upside down, sideways, in formation, out of formation and so-on—with most of the pictures taken just around the enemy-held air bases." All "that film had to be screened."

217. Trest, *Once a Fighter Pilot*, 81. Cleveland ended up claiming a probable on his July 11, 1952, mission. Werrell described another typical encounter: DeArmond remembered his flight lead, the Wing commander Thyng, asking him before they stepped into the debrief where he had seen Thyng's MiG go down. Knowing that they had been deep inside Manchuria before

engaging the MiG, DeArmond responded, "somewhere around the mouth of the Yalu," to which Thyng replied, "Son you have a bright future in the Air Force." Werrell, *Sabres over MiG Alley*, 132.

218. J. Sherwood, *Officers in Flight Suits*, 88. Sherwood described one engagement that illustrated the importance of having a wingman to verify a claim: under attack, an American flight lead somehow managed to whip his jet around and shoot down the attacking MiG moments before ejecting from his own damaged Sabre, but because his wingman had panicked and already left the fight, the MiG claim was denied. Years later, still frustrated by the experience, Turner remarked: "As far as the Air Force is concerned I never got a MiG, but as far as I'm concerned, I got three." Hagerstrom demonstrated that there was another way to verify a MiG kill without a wingman or gun camera film when he returned with a piece of a MiG aircraft stuck in his Sabre.

219. D. Evans, *Sabre Jets over Korea*, 210; Bruning, *Crimson Sky*, 187–88.

220. Mahurin, *Honest John*, 87–88. The dispute was obviously not featured in the ensuing news report: "New England Flier Becomes 16th Jet Ace."

221. Werrell, *Sabres over MiG Alley*, 155. Giordano noted that in the ensuing discussion, Jabara "being the Major and Executive Officer and me being a punk second lieutenant, I lost the argument." Jabara was more modest and complementary toward his wingman in his *Air Force* article: "[Being a wingman] is a tough assignment and as far as I'm concerned half my victories should go to good wingmen. . . . To me he's worth his weight in .50 caliber ammo." Jabara, "We Fly MiG Alley," 27.

222. Mahurin, *Honest John*, 84.

223. Salter, *Burning the Days*, 141, 153. Salter noted, "It was victory we longed for and imagined. You could not steal or be given it. No man on earth was rich enough to buy it and it was worth nothing. In the end it was worth nothing at all."

224. Mahurin, *Honest John*, 84. Cleveland similarly shrugged off any concerns, blandly observing, "Every kill counts." Quoted in Trest, *Once a Fighter Pilot*, 72.

225. Mahurin, *Honest John*, 30–31.

226. Hagerstom is quoted in J. Sherwood, *Officers in Flight Suits*, 87.

227. Mahurin, *Honest John*, 86–87, 125; J. Sherwood, *Officers in Flight Suits*, 90.

228. Mahurin, *Honest John*, 50–52. After the war, Mahurin learned that his wingman survived the ejection, emerging from "prison camp at the end of the war, alive and well."

229. DeArmond is quoted in Hallion, "Air Dominance," 34. See also Werrell, *Sabres over MiG Alley*, 150; Mahurin, *Honest John*, 48, 127.

230. Quoted in Werrell, *Sabres over MiG Alley*, 170; Hallion, "Air Dominance," 37.

231. Blesse, *Check Six*, 65.

232. *4th FIG Tactical Doctrine*, July 22, 1951, ii. "Many times the tide of an air battle has been turned by apparently suicide attacks on superior numbers of enemy aircraft and injecting enough confusion and panic into the situation to cause the larger force to take the defensive and attempt to make its 'operations elsewhere.' This often results in major damage to the larger force inflicted by the numerically inferior but *aggressively superior* attacking force" (emphasis in original).

233. Mahurin, *Honest John*, 84.

234. *Accelerated Comparison Test*, 1, Inclosure 1:1–6; Werrell, *Sabres over MiG Alley*, 29–32; *FEAF Report on the Korean War*, 2:11–12; Futrell, *USAF Operations in the Korean*

Conflict, 1 Nov 1950–30 Jun 1952, 119–20; Wildenberg, "A-1C(M) Gunsight," 34–36. Curiously, the Air Proving Ground's report identified September 4, 1952, as the date of the CSAF's tasking, but FEAF's final *Report on the Korean War*, Werrell, and Wildenberg all fixed the date as September 8.

235. Wildenberg, "A-1C(M) Gunsight," 35–36; *A-1 Series (Gun-Bomb-Rocket) Sight*, 2.

236. Wildenberg, "A-1C(M) Gunsight," 36; Dorr, Lake, and Thompson, *Korean War Aces*, 61; "Pilot Is 17th U.S. Jet Ace in Only 6 Weeks in Korea."

237. Quoted in Hallion, "Air Dominance," 51.

238. *Accelerated Comparison Test*, Inclosure 1:6.

239. Blesse, *Check Six*, 64–65, 72. Blesse was able to clear his windscreen by accelerating above 350 knots.

240. Hinton and Wingo, "Analysis of Operations at Kimpo," 31; Broad, "FEAF Intel Form 1, 21 Dec 1950"; Peterson, "F-86A Operation as a Fighter-Interceptor," 2; "Enemy Air Activity, 30 Dec 1950"; J. Meyer to Commanding General, Air Defense Force, "F-86 Activities Report (Special)," 7–8. On airspeed's effect on aircraft gun boresighting, see "Harmonization and Firing Techniques," chap. 4.

241. Blesse, *Check Six*, 57, 65. See also Evans's accounts of his first engagements in D. Evans, *Sabre Jets over Korea*, 84–85.

242. *Accelerated Comparison Test*, Inclosure 2:4; Thorne, "Major James H. Thorne," 21.

243. *Accelerated Comparison Test*, 1.

244. Ibid., Inclosure 2:4; Mahurin, "Major Walker M. Mahurin," 23.

245. *Resume of Our Tactics*, 16. On tracer ammunition, see also: Futrell, *USAF Operations in the Korean Conflict, 1 Nov 1950–30 Jun 1952*, 118–19; Thyng, "Operation of the 4th Fighter Wing in Korea," 7; Blesse, "No Guts, No Glory," 71–72; Commanding General, 5th Air Force to Commanding General, Far East Air Forces, "Evaluation of the A1CM Gun Sight," 3.

246. Commanding General, 5th Air Force to Commanding General, Far East Air Forces, "Evaluation of the A1CM Gun Sight."

247. Ibid.; Jenkins, "Some Notes," 51.

248. Jenkins, "Some Notes," 51; *FEAF Report on the Korean War*, 2:11; *History of the 4th Fighter-Interceptor Wing, Jul–Dec 1952*, 70; Futrell, *USAF Operations in the Korean Conflict, 1 Nov 1950–30 Jun 1952*, 119–20; Wildenberg, "A-1C(M) Gunsight," 34–35; *History of the 6400th Air Depot Wing, Feb–Jun 1952*, I:230–31.

249. G. Smith to Commanding General, Far East Air Force, "Combat Evaluation of the MiG-15," 2. Smith was unsure if the results were due to the MiGs' "low volatile fuel," "high performance," or "ability to absorb gun fire."

250. Hinton and Wingo, "Analysis of Operations at Kimpo," 30; Futrell, *USAF Operations in the Korean Conflict, 1 Nov 1950–30 Jun 1952*, 118.

251. Nuckols, *SNL A-57 Gun Machine Cal .50 M3*; J. Meyer to Commanding Officer, 4th Fighter Interceptor Wing, "F-86-A5 Armament"; Eagleston to Commanding General, 4th Fighter Interceptor Wing, "Recommendation concerning Armament Equipment," 2. See also Alper and Caywood, *Weapons for Fighter-to-Fighter Engagement*, 16.

252. Hise, "F-86E versus MiG-15 in Korea," 19–20 (emphasis in original); Preston to Commanding Officer, 4th Fighter Interceptor Wing, "Analysis of F-86 Fighter Encounter," 6; Gabreski, "(End of Tour) Report," 7; Thyng, "Operation of the 4th Fighter Wing in Korea," 7; Baker, "Report on F-86 Operations," 10; Schinz, "Brief Resume of the Experiences," 5.

253. Werrell, *Sabres over MiG Alley*, 19–23; Futrell, *United States Air Force in Korea*, 651; Futrell, *USAF Operations in the Korean Conflict, 1 Jul 1952–27 Jul 1953*, 63–64; *FEAF Report on the Korean War*, 2:173–74; *History of the Fifth Air Force, Jul–Dec 1952*, 1:203–8; *Final Report on Combat Suitability Test of F-86F-2 Aircraft with T-160 Guns*; CG FEAF to CG AF FIVE, "OP-RQMT GUN-VAL"; CG FEAF to CG AF FIVE, "OP-RQMT T-160 INSTL"; Baker, "Report on F-86 Operations," 10; Schinz, "Brief Resume of the Experiences," 5. FEAF cautioned in its *Report on the Korean War* that the "higher percentage of enemy aircraft destroyed or damaged" during Gun Val "probably was due in part to the skill and experience of the test pilots, and to the favorable combat positions flown on the test missions." During the test, the 20-mm guns had only 115 rounds of ammunition each, whereas the Sabre's standard ammunition load for the M3 was 300 rounds each. Schinz also noted that the introduction of the T-160 would demand increased "fighter gunner proficiency," placing a heavy burden on the stateside F-86 training program. For footage of the Gun Val test, see the silent film *Project Gun-Val.*

254. Mahurin, *Honest John*, 41.

255. D. Evans, *Sabre Jets over Korea*, 73, 133, 169, 190.

256. McClure to Commanding General, Continental Air Command, "Unit Operations and Analysis Report on A-1-C-M," 3. The unit flew F-84E Thunderjets.

257. Greene to Commanding General, Continental Air Command, January 31, 1950; Olsen to Holland, Schwarze, and Rhame, "Evaluation of Results." Despite the issues in the January 1950 test, the MIT representatives proudly noted, "Experienced pilots state that they have obtained better gunnery results with the A-1CM than with any other equipment they have used."

258. *FEAF Report on the Korean War*, 2:11. See also J. Watts, *Operational Survey of Type A-1CM*, 3–4.

259. J. Meyer to Commanding General, Air Defense Force, "F-86 Activities Report," 2; Futrell, *USAF Operations in the Korean Conflict, 1 Jul 1952–27 Jul 1953*, 64–65; Schinz, "Brief Resume of the Experiences," 2–3; Gabreski, "(End of Tour) Report," 6. The equipment shortages were not confined to A-1 components; however, the A-1 situation was especially egregious. Futrell noted that "spare parts, spare components, and test equipment [for the A-1C] did not reach the Far East until at least six months after the sight was first put into use in mid-1951."

260. Futrell, *USAF Operations in the Korean Conflict, 1 Jul 1952–27 Jul 1953*, 64–65; Wildenberg, "A-1C(M) Gunsight," 34; *Accelerated Comparison Test*, Inclosure 1:2; Werrell, *Sabres over MiG Alley*, 28; Greene to Commanding General, Continental Air Command, January 31, 1950; Hise, "F-86E versus MiG-15 in Korea," 20. Evans described Kimpo's runway as "rough as a cob from all the old bomb and shell crater patching and endless repairs." D. Evans, *Sabre Jets over Korea*, 128.

261. Gabreski, "(End of Tour) Report," 6.

262. "A-4 Sight"; Tucker, "Fire Control"; "Standard Aircraft Characteristics: F-86F," 3; Werrell, *Sabres over MiG Alley*, 29. The A-4 was described in *Fighter Gunnery* as "a dressed up A-1." Regrettably, many of the errors that accompanied the fielding of the A-1C were repeated with the A-4; see McNickle to WCSWF, "Reworking A-4 Sightheads"; *History of the 4th Fighter-Interceptor Wing, Jul–Dec 1952*, 35; *History of the 6400th Air Depot Wing, Jul–Dec 1952*, 1:213–16; Futrell, *USAF Operations in the Korean Conflict, 1 Jul 1952–27 Jul 1953*, 65.

263. Gabreski, "(End of Tour) Report," 6–7.

264. Quoted in Futrell, *USAF Operations in the Korean Conflict, 1 Nov 1950–30 Jun 1952*, 119. On pilots opting to cage their gunsights, see Hallion, "Air Dominance," 36–37.

265. *FEAF Report on the Korean War*, 2:11. See also *History of the 6400th Air Depot Wing, Feb–Jun 1952*, 1:114, 261; Futrell, *USAF Operations in the Korean Conflict, 1 Nov 1950–30 Jun 1952*, 119; Futrell, *USAF Operations in the Korean Conflict, 1 Jul 1952–27 Jul 1953*, 65–66; *History of the 4th Fighter-Interceptor Wing, May 1952*, 51; *History of the 51st Fighter Interceptor Wing, May 1952*, 10; Werrell, *Sabres over MiG Alley*, 28–29; Wildenberg, "A-1C(M) Gunsight," 34. The exact timing of Jaybird is disputed in the cited sources, but all agree that most activity took place between April and June 1952. The F-86 depot unit reported that "605 units were modified and processed" as part of Project Jaybird. Mimicking the necessity and eventual success of Jaybird, Air Materiel Command would initiate Project Sightline in February 1953 to rehabilitate the A-4 sights in the field. See references in *History of the 6400th Air Depot Wing, Jan–Jun 1954*, 1:153.

266. Werrell, *Sabres over MiG Alley*, 29; Schinz, "Brief Resume of the Experiences," 2–3. Occasionally, circumstances beyond human control seemed to conspire against the gunsights: in April 1952 the "Group Radar and IFF Shop [at Kimpo] burned to the ground. As a direct result of the fire, [the maintenance] section was relieved of its one and only 28-volt rectifier." *History of the 334th Fighter Interceptor Squadron, Apr 1952*, 8.

267. *History of the 51st Fighter Interceptor Group, Dec 1951*, 13; Baker, "Report on F-86 Operations," 6; *History of the 4th Fighter-Interceptor Wing, Jun 1952*, 44; *History of the Fifth Air Force, Jul–Dec 1952*, 1:199.

268. *FEAF Report on the Korean War*, 2:11.

269. J. Watts, *Operational Survey of Type A-1CM*, 4–6. Although Watts's report covered F-84E units, the F-86 units faced similar circumstances. Watts estimated that "about 90 percent of replacement pilots arriving in Korea, to date, have had no prior training with the A-1CM gunsight."

270. *4th FIG Tactical Doctrine*, July 22, 1951, i; *Resume of Our Tactics*, 16.

271. D. Evans, *Sabre Jets over Korea*, 122.

272. *History of the 4th Fighter-Interceptor Group, Jul 1952*, Maintenance, Supply, Communications, and Armament; *History of the 4th Fighter-Interceptor Wing, Jul–Dec 1952*, 70–71; *History of the 4th Fighter-Interceptor Wing, Oct 1951*, chap. 6; *History of the 51st Fighter Interceptor Wing, Feb 1952*, chap. 2.

273. Commanding General, 5th Air Force to Commanding General, Far East Air Forces, "Evaluation of the A1CM Gun Sight," 2. See also other favorable pilot reports included in the letter.

274. Baker, "Report on F-86 Operations," 6; "He's a Poor Shot, Says Top Jet Ace." The *New York Times* reporter noted the marked contrast in Baker's appraisal of the system: "Unlike another jet ace who said recently that he had shot down eight planes by using 'Kentucky windage' rather than the gun sight, Colonel Baker classified himself as a rather poor marksman. He said that in making his kills he had to get close to the enemy and use the radar gun sight entirely." See also Thyng, "Operation of the 4th Fighter Wing in Korea," 5.

275. Quoted in Trest, *Once a Fighter Pilot*, 80.

276. *Accelerated Comparison Test*, Inclosure 1:1; Wildenberg, "A-1C(M) Gunsight," 34; Werrell, *Sabres over MiG Alley*, 29.

277. Schinz, "Brief Resume of the Experiences," 3.

278. Gabreski, "(End of Tour) Report," 5–8; Wildenberg, "A-1C(M) Gunsight," 35; Werrell, *Sabres over MiG Alley*, 30. See also Gabreski's comments on the need for "simplicity of design" in Gabreski, "Our Top Ace Talks Shop," 41.

279. General Vandenberg's message is quoted in CG FEAF to CG AF FIVE, "Personal from Smart to Barcus." The problem of complex gadgetry in fighter aircraft was not a new problem. An Army Air Corps training manual from 1929 warned: "Aircraft performance is the essence of all aerial warfare, offensive or defensive. In the air, speed and climb are, in the present state of affairs, infinitely more valuable to pursuit than the number of guns or the weight of projectiles. Only sufficient power to destroy without being destroyed is necessary. When pursuit aircraft are properly designed almost everything is sacrificed to the one essential—performance." *Pursuit Text*, 88. See other voiced concerns about burgeoning equipment requirements for fighter aircraft in Futrell, *USAF Operations in the Korean Conflict, 1 Nov 1950–30 Jun 1952*, 117; Schinz, "Brief Resume of the Experiences," 5.

280. *Accelerated Comparison Test*, 2.

281. Ibid., Inclosure 1:7.

282. Ibid., 5–6. Whisner similarly focused his comments on the poor reliability of the system, conceding that "if the radar was reliable to a percent of 90 to 95, then each of us would say that we would buy it, in spite of the weight, expense, and problems involved." But he also declared his doubts that "we can get adequate reliability into this gunsight system."

283. Ibid., Inclosure 2:2.

284. Ibid., Inclosure 1:5, 8, Inclosure 2:2. Johnston and Green flew in the test but did not sign the report, apparently indicating their dissension.

285. Ibid., Inclosure 1:5–6.

286. Ibid., Inclosure 1:1–3. The report noted: "The only known 'yardstick' for determining the relative accuracies of the various systems are results of firing on banner targets. The shortcomings of this method are well known, the two principle ones being the slow speed of the target and the fact that firing must be accomplished at an angle off in excess of 15 degrees which is not realistic or comparable to fighter vs fighter combat. These limitations must be borne in mind when evaluating the results."

287. Ibid., 1–2; Strauss, *Analysis of Korean Air-to-Air Combat Film*, 7 (emphasis in original). While the University of Chicago study determined that most firing passes were made from outside the stern position and at long range, it also concluded that "the most effective attack, in terms of hits, was from an angle-off of less than 15° and a range of about 600 feet."

288. CG FEAF to CG AF FIVE, "Personal from Smart to Barcus."

289. Hale to Commanding General, Far East Air Forces, "A1CM Gunsight," 2–4; *FEAF Report on the Korean War*, 2:12; Futrell, *USAF Operations in the Korean Conflict, 1 Jul 1952–27 Jul 1953*, 66; *History of the Fifth Air Force, Jul–Dec 1952*, 1:198–99; Werrell, *Sabres over MiG Alley*, 30–32. Hale's report summarized a September 11 meeting of the aces.

290. Long, *USAF Fighter Aircraft Fire Control Systems*; Tucker, "Fire Control."

291. Futrell, *USAF Operations in the Korean Conflict, 1 Jul 1952–27 Jul 1953*, 64.

292. Jenkins, "Some Notes," 51 (emphasis in original).

293. *4th FIG Tactical Doctrine*, July 22, 1951, i.

294. Jabara, "We Fly MiG Alley," 27. Evans likewise proudly stood up one day in front of a group of the Sabre's designers and affirmed, "But isn't the pilot the really decisive factor anyway?" D. Evans, *Sabre Jets over Korea*, 197.

295. *4th FIG Tactical Doctrine*, July 22, 1951, i–iii.

Chapter 4. Phantoms over Vietnam: The "GIB"

1. Clifton, "There I Was . . . Part 1." The Route Pack structure is discussed in more detail later in this chapter. In his recollection of the events, the Wing commander later claimed that Clifton had completed only "eighty-plus" missions. Olds, "How I Got My First MIG," 40.

2. Clifton, "Who Am I"; Clifton, "There I Was . . . 4."

3. Clifton, "There I Was . . . Part 1"; Olds, "How I Got My First MIG," 40; Asay, Covington, and Stone, *History of Operations BOLO*, 6, 10.

4. Olds, "Briefing to the Staff of Project CORONA HARVEST on Operation BOLO," 16–17.

5. The ensuing description of Operation Bolo is based on the following sources: Olds, "Briefing to the Staff of Project CORONA HARVEST on Operation BOLO"; Asay, Covington, and Stone, *History of Operations BOLO*; Olds, "How I Got My First MIG"; Boyne, "MiG Sweep"; Ow, *Mission Bolo*; Olds, *Fighter Pilot*, chap. 17; Attinello, *Air-to-Air Encounters in Southeast Asia: Account of F-4 and F-8 Events Prior to 1 March 1967*, 415–46; Futrell et al., *Aces and Aerial Victories*, 35–42; J. Sherwood, *Fast Movers*, 31–37; Michel, *Clashes*, 73–74.

6. Asay, Covington, and Stone, *History of Operations BOLO*, 39; Olds, "Briefing to the Staff of Project CORONA HARVEST on Operation BOLO," 48.

7. Asay, Covington, and Stone, *History of Operations BOLO*, 41.

8. *T.O. 1F-4C-1*; *F-4 Aircrew Operational Procedures*, 1. Per the F-4C flight manual, the front-seat occupant was the *aircraft commander*; the rear-seat occupant was the *pilot systems operator* or *pilot*. However, Phantom crews often informally referred to the front-seat occupant as the *pilot*, because he flew the plane, and the back seater as the *GIB*. Adding confusion to the terminology, both occupants were, at least until a change in policy in 1969, Air Force–trained pilots—officers who had completed pilot training and who therefore wore pilot wings on their uniform. To avoid confusion and respect the terminology of the time, I generally avoid using the term *aircraft commander* and, when appropriate, instead identify the crew members by their position in the cockpit, that is, front seater or back seater.

9. Futrell et al., *Aces and Aerial Victories*, 43–44; Michel, *Clashes*, 74; Boyne, "MiG Sweep," 51; Attinello, *Air-to-Air Encounters in Southeast Asia: Account of F-4 and F-8 Events Prior to 1 March 1967*, 451–59.

10. Olds, *Fighter Pilot*. Quotes are from the "Advance Praise" section of the dust jacket of *Fighter Pilot*. Sherwood titled his chapter on Olds "Old Lionheart." J. Sherwood, *Fast Movers*, chap. 1.

11. Clifton, "Who Am I"; Clifton, "There I Was . . . 4."

12. Futrell et al., *Aces and Aerial Victories*, 129, 133, 135. Olds and Crocker destroyed two MiGs on May 20, 1967, bringing Olds's final tally for the Vietnam War to four kills.

13. Kirk, Oral History Interview, 24 (emphasis added).

14. Bakke, Oral History Interview, 9 (emphasis added).

15. Olds, *Fighter Pilot*, 303–4. The Air Force's anthology of aerial victories in Southeast Asia tended to refer only to the aircraft commander in its narratives, "for the sake of convenience." Futrell et al., *Aces and Aerial Victories*, 22.

16. Hill, Oral History Interview, 21–22. Anderegg, who also flew as a GIB during Vietnam, was unaware of the denotation associated with the term *gib*, indicating that the correlation, although fitting, may have been unintentional. Anderegg, interview, October 2, 2013.

17. Salter, *Hunters*, 193; D. Evans, *Sabre Jets over Korea*, 127; Creech, Oral History Interview, 54. General Wilbur Creech, a future commander of TAC, recalled fighter pilots teasing

their Phantom brethren, "I've never been against people breathing in my ear, but I didn't want it to come from a back seater on a hot-mike intercom." Implicit in the comment was that it was also a *man* "breathing" in the front seater's ear, an obvious affront to the masculine, fighter pilot ethos.

18. *T.O. 1F-4C-1*. See also references in *F-4 Aircrew Operational Procedures*, 1.

19. "From Struts to Jets," 62–63; "Black Box to Replace Piot."

20. *Accelerated Comparison Test.*

21. *Air-to-Air Guided Missiles*, 1–2. The lecture included a detailed example of the geometric problems associated with the intercept pursuit curves for high-speed aircraft, eventually concluding, "as aircraft speeds rise, it will become more and more difficult for fighters to attack other aircraft." In dismissing the effectiveness of the .50-caliber ordnance, even when aimed by an A-1 gunsight, the lecture assumed an optimistic effective range of "800 yards." It noted, "The way aircraft are being built these days it would be a very lucky round indeed that might destroy another ship." The predictions proved prophetic based on the Korean War experience a few years later.

22. Wildenberg, "Visionary Ahead of His Time," 6; R. McMullen, *History of Air Defense Weapons*, 12; *Air-to-Air Guided Missiles*, 3–4.

23. R. McMullen, *History of Air Defense Weapons*, 44, 47, 88–89; Wildenberg, "Visionary Ahead of His Time," 8.

24. Knaack, *Encyclopedia of US Air Force Aircraft*, 1:69–111.

25. Redmond and Smith, *From Whirlwind to MITRE*, 19–29, 429–33; Westrum, *Sidewinder*, 25–27. On SAGE, see also T. Hughes, *Rescuing Prometheus*, chap. 2; Edwards, *Closed World*, chap. 3.

26. *Pilot's Introduction to the E-4*, 4; Wildenberg, "Visionary Ahead of His Time," 9–11; Long, *USAF Fighter Aircraft Fire Control Systems*.

27. Knaack, *Encyclopedia of US Air Force Aircraft*, 1:69–111; Wildenberg, "Visionary Ahead of His Time," 9–10. One F-86D pilot lamented, "Firing air-to-air rocketry was certainly a lot less exciting than firing air-to-air .50-caliber machineguns." Umstead, Oral History Interview, 23.

28. Wildenberg, "Visionary Ahead of His Time," 9–11; *Pilot's Introduction to the E-4*, 4; Long, *USAF Fighter Aircraft Fire Control Systems*.

29. E. Miller, "Thunderbolt Geometry," 71. See also references in Austin, Oral History Interview, 137.

30. Gabreski, "F-86D Program Deficiencies." Reminiscent of his previous gripes with the A-1C(M), Gabreski complained in a private memo that the Air Force was "so absorbed with the prospects of technical advancement to be achieved that the capability of the Air Force to properly maintain and operate the weapons system was not kept in proper perspective" (1).

31. *Pilot's Introduction to the E-4*, 5.

32. McVeigh and Perry, *Development of the F-102*, 10.

33. *Pilot's Introduction to the E-4*, 5–11; Wildenberg, "Visionary Ahead of His Time," 10. It is interesting to note that the images in the radar operating manuals almost always included only a single, crisp target blip. More likely, the radar scope would have displayed several radar returns, some from valid radar targets, but many others due to clutter. One pilot likened the F-86D to a "flying television station." Austin, Oral History Interview, 137, 140.

34. *Pilot's Introduction to the E-4*, 6–15. See also E. Miller, "Thunderbolt Geometry," 73. During an evaluation of the F-94 and its E-5 fire control system (an offshoot of the F-86D's

E-4), the two-person crews successfully completed only twelve of ninety-four intercepts because they were worried that their fire control system would cause them to collide with the target. *F-94C Squadron Operational Suitability Test*, 194.

35. *Pilot's Introduction to the E-4*, 10.

36. Knowing that the Sabre pilots would never waste their fuel practicing instrument flying when they could be hunting MiGs, the squadrons in Korea were provided with spare T-33 aircraft for the pilots to practice their instrument flying skills. More often than not, the T-33s sat on the ground broke; the pilots didn't seem to mind. See *History of the 4th Fighter-Interceptor Wing, Jul 1951*, 35–36; *History of the 16th Fighter Interceptor Squadron, Mar 1952*, 5; *History of the 16th Fighter Interceptor Squadron, Apr 1952*, 5; Risner, Oral History Interview, 43. On the introduction of instrument flying during the interwar years and its effect on the identity of the pilot, see Jeon, "Technologies of the Operator," chap. 1; Conway, *Blind Landings*; Doolittle and Glines, *I Could Never Be So Lucky*, 120–41.

37. "208 Rockets Fired at Runaway Plane"; Merlin, "Battle of Palmdale"; Westrum, *Sidewinder*, 30. Still in use by 1961, the effectiveness of the unguided rockets remained a crapshoot.

38. A nuclear-tipped version of the Falcon was also developed that had a slightly different missile-body shape than the conventional Falcons. Hughes's involvement in the Falcon can be traced back to the Tiamat and Project MX-570, but specific development of the Falcon stemmed from Project MX-904. For a while, the Falcon was known as XAAM-A-2 (XAAM-A-1 was Ryan's Firebird missile). As the missile entered early testing, it was designated the F-98 to signify its role as an unmanned aerial interceptor. However, that naming practice was discontinued in August 1955, and the Falcon thereafter assumed the standardized GAR (guided air rocket) label. Later in 1963, McNamara implemented a policy that eliminated the GAR designation in favor of AIM (air intercept missile). Sean O'Connor, *Arming America's Interceptors*; Wildenberg, "Visionary Ahead of His Time," 8. On the nuclear version of the Falcon, see R. McMullen, *History of Air Defense Weapons*, 158, 290–96; Schaffel, *Emerging Shield*, 234.

39. Sean O'Connor, *Arming America's Interceptors*; Westrum, *Sidewinder*, 28–29; Luke, *Evaluation Report on GAR-1*, 1–2.

40. "Hughes Falcon."

41. Luke, *Evaluation Report on GAR-1*, 1–2; Sean O'Connor, *Arming America's Interceptors*.

42. Westrum, *Sidewinder*, 28; R. McMullen, *History of Air Defense Weapons*, 157, 277; Knaack, *Encyclopedia of US Air Force Aircraft*, 1:94–96; Luke, *Evaluation Report on GAR-1*, 8–12; Bauman, *Test Report on GAR-1*, 5. Luke's analysis concluded, "The probability of hit for each missile is 0.25 giving a 0.578 probability of hit for a salvo of three missiles." In contrast, Bauman's October 1956 report noted that "out of 48 missiles launched from 1 January 1956 until 1 September 1956, only seven intercepted their target," a 14 percent hit rate.

43. "New Missile Ready"; R. McMullen, *History of Air Defense Weapons*, 278–80, 284; Sean O'Connor, *Arming America's Interceptors*; Luke, *Evaluation Report on GAR-1*, 1–2, 5, 8–12. The follow-on GAR-3 also raised the acceptable aircraft airspeed to launch the missile from Mach 1.3, based on the F-102, to Mach 2.0, based on the Air Force's faster F-106.

44. Luke, *Evaluation Report on GAR-1*, 2, 12; Sean O'Connor, *Arming America's Interceptors*. Luke's report reflected the Air Force's dominant vision of missile-armed interceptors thwarting fleets of invading Soviet bombers. Curiously absent from the assumed advantage that the

infrared missile "will select a target" was the criterion that the missile would actually guide toward the target that it was fired against; apparently any target would do, and there would supposedly be plenty of them in the sky. *Ground clutter* occurs anytime the radar is pointed below the horizon and radar energy reflected off the ground masks the target return; see Stimson, *Introduction to Airborne Radar*, chap. 22.

45. R. McMullen, *History of Air Defense Weapons*, 280; Sean O'Connor, *Arming America's Interceptors*.

46. R. McMullen, *History of Air Defense Weapons*, 283–84; Sean O'Connor, *Arming America's Interceptors*. Carrying the IR-guided missiles internally in the Air Force's high-speed interceptors further complicated development because the missile's seeker had no view of the target until after launch.

47. Long, *USAF Fighter Aircraft Fire Control Systems*; McVeigh and Perry, *Development of the F-102*, 68–71, 74–76; Sean O'Connor, *Arming America's Interceptors*; Wildenberg, "Visionary Ahead of His Time," 10–11. Gabreski, already worried about the "complexity" of the F-86D and its E-4, was dumbfounded that the E-9 planned to use "twice the number of [vacuum] tubes and black boxes as does the E-4." Gabreski, "F-86D Program Deficiencies," 4.

48. Knaack, *Encyclopedia of US Air Force Aircraft*, 1:158–73, 206–21.

49. McVeigh and Perry, *Development of the F-102*, 89, 111; Wildenberg, "Visionary Ahead of His Time," 10–11. The report also noted that the F-102's high-altitude, supersonic dash "would punish the pilot physically."

50. McVeigh and Perry, *Development of the F-102*, 81–83; Sean O'Connor, *Arming America's Interceptors*.

51. Andrews, "To Fly and Fight," 47–48; McCarthy, *Phantom Reflections*, 8.

52. E. Miller, "Thunderbolt Geometry"; Blesse, "No Guts, No Glory."

53. Whitehead, "Supersonic Air-to-Air Gunnery"; Austin, Oral History Interview, 137–38.

54. Kutzscher, "Application of Infrared Radiation." The article made no mention of IR-guided missiles, only that "infrared techniques may be used in search equipment, in detection and tracking systems and also in warning devices" (30).

55. Giordano, "A-4 Sight Check."

56. Tucker, "Fire Control."

57. Boyd, "Fire Control Problems." Boyd's article earned a special notice from the editors: "It is suggested that this article be given more than just a cursory reading." On the shift in the journal's contents, see Giraudo, Oral History Interview, 141–42. See also the pair of mid-1956 articles on computing gunsights and the intricacies of the APG-30 radar: Jenkins, "Some Notes"; R. Wells, "Tips and Info for AN/APG-30 Radar." Of course, not all pilots immediately took to the new technical approach; see, for example, Cook, *Once a Fighter Pilot*, 33–34.

58. Knaack, *Encyclopedia of US Air Force Aircraft*, 1:123. The initial decision to explore use of IR-guided missiles on the F-100 occurred in late 1955. In September 1956 the Sidewinder was selected over the GAR-1B Falcon for use on the F-100D, but the new capability did not appear until after the 184th F-100D rolled off the assembly line.

59. Thor, "GAR-8," 30. The initial article was followed by a more detailed one fifteen months later; Giordano, "F-100D GAR-8 Tactics." One F-100 pilot explained that "firing my first Sidewinder missile . . . was a great experience, but there was really nothing to it once you learned what to do. . . . You just pressed that trigger and away it went and 'blam.'" Knight, Oral History Interview, 119.

60. *F-4C System Package Program*, August 31, 1962, sect. 5, p. 2; *F-4C System Package Program*, September 1963, sect. 12, p. 17; Knaack, *Encyclopedia of US Air Force Aircraft*, 1:112–33, 190–205, 222–35; *History of the Tactical Air Command, Jul–Dec 1961*, 1:17, 202–12; *History of the Tactical Air Command, Jan–Jun 1963*, 1:275–80; Gillespie, "Mission Emphasis," chap. 3. The F-101 Voodoo was likewise based on a 1951 design and never found an enthusiastic following in TAC except as a tactical reconnaissance platform. The F-102 and later F-106 were specially designed as ADC interceptors, not TAC fighters. The Kennedy administration's new emphasis on "flexible response" and "limited," as opposed to "total" or "nuclear," war was also a commonly cited justification for upgrading the quality and quantity of TAC's fighter force.

61. *History of the Tactical Air Command, Jul–Dec 1961*, 1:194–202; *History of the Tactical Air Command, Jan–Jun 1962*, 1:242–44; *F-4C System Package Program*, September 1963, sect. 12, p. 8; Knaack, *Encyclopedia of US Air Force Aircraft*, 1:333; Bugos, *Engineering the F-4*, 105–6, 116–20; John J. Burns, Oral History Interview, June 5–8, 1984, 168–69. Several authors contend that the F-4's ability to be adapted to drop tactical nuclear bombs was a major selling point for TAC officers looking to establish themselves in an Air Force hierarchy dominated by the bomber-generals of SAC. On the "SAC-timization" of TAC, see Creech, Oral History Interview, 41–44; Hildreth, Oral History Interview, 54; L. Welch, Oral History Interview, 71; Crane, *American Airpower Strategy in Korea*, 172; Ziemke, "In the Shadow of the Giant"; Hannah, *Striving for Air Superiority*; Laslie, *Air Force Way of War*. However, the Air Force also had analyzed the Navy F-4H's capability to provide "close support of friendly ground forces" and concluded that the new fighter would increase capability by "more than 100 times per sortie over that of Korea." Gardiner, "Effectiveness of F-4H," 2.

62. Bugos, *Engineering the F-4*, chap. 1; Boyne, *Phantom in Combat*, 10.

63. Bugos, *Engineering the F-4*, 76–79; Westrum, *Sidewinder*, 44. The beam-riding Sparrow had its antenna located in the tail of the missile. When the attacking aircraft locked the target with its radar and fired a Sparrow, the missile maneuvered itself to remain in the radar beam that went from the attacker to the target. Consequently, if the target or the attacker turned while the missile was in flight, the missile needed to adjust its position to remain in the straight-line radar beam between the attacker and the target aircraft, thereby flying an energy-depleting curved flight path to the target rather than a straight-line, cut-off intercept.

64. Bugos, *Engineering the F-4*, 79–80; Westrum, *Sidewinder*, 43–44, 132.

65. Bugos, *Engineering the F-4*, 80–81; Westrum, *Sidewinder*, 44–46, 132.

66. Bugos, *Engineering the F-4*, 27–28, 81–89. Raytheon's Aero 1A Airborne Missile Control System (AMCS) was selected for the Phantom; it included the Westinghouse APQ-72 radar, APA-128 controller, and the AAA-4 auxiliary, nose-mounted IR-tracking system. Initially designed to use a twenty-four-inch radar antenna, the Phantom's dimensions were tweaked to accommodate the larger and more powerful APQ-72 radar with its thirty-two-inch antenna.

67. Ibid., 89–90; Westrum, *Sidewinder*, 129–44, 170–75; Michel, *Clashes*, 13–14. Whereas the other guided missiles procured by the Air Force and Navy were designed by industry on contract, the Sidewinder missile was designed in house at the Naval Weapons Laboratory at China Lake, California, under the supervision of William McLean. The resulting missile performed well, often besting the Sparrow by a factor of two in missile tests, and at only one-half the cost per missile. McLean's simple design would ultimately be stolen by the Soviets, appearing as an AA-2 Atoll; see Westrum, *Sidewinder*, 205–6; Michel, *Clashes*, 41.

68. Bugos, *Engineering the F-4*, 27–28, 82–89; John J. Burns, Oral History Interview, June 5–8, 1984, 170–72. Burns suggested that Navy engineers removed the Phantom's gun to better streamline the aircraft so that it could achieve its extreme airspeeds more efficiently. Reflective of the common attitude within the Pentagon, McNamara reportedly quipped, "In the context of modern air warfare, the idea of a fighter being equipped with a gun is as archaic as warfare with bow and arrow." Quoted in Michel, *Clashes*, 16.

69. Bugos, *Engineering the F-4*, 26–27.

70. Ibid., 26–27, 106–7.

71. Quoted in ibid., 98.

72. Ibid., 105–11.

73. Holsapple to AFSC and AFLC, "F-4H"; *History of the Tactical Air Command, Jul–Dec 1961*, 1:197. The November 7, 1961, memo followed a two-day conference held November 2–3. Of note, McNamara's memo announcing termination of the F-105 and directing the Air Force to acquire the Navy's F-4 was dated February 17, 1962 (included in *F-4C System Package Program*, September 1963, sect. 12, p. 8).

74. Holsapple to AFSC and AFLC, "F-4H"; *F-4C System Package Program*, August 31, 1962; *F-4C System Package Program*, September 1963; Donovan to HQ USAF, "Configuration of the F4H"; HQ TAC to ASD, "Proposed System Package for the F4H"; *History of the Tactical Air Command, Jan–Jun 1962*, 1:247–52; *History of the Tactical Air Command, Jul–Dec 1962*, 1:231–45; Bugos, *Engineering the F-4*, 121–23; John J. Burns, Oral History Interview, June 5–8, 1984, 169–70.

75. Holsapple to AFSC and AFLC, "F-4H."

76. *F-4C System Package Program*, September 1963; *F-4C System Package Program*, August 31, 1962; *T.O. 1F-4C-1*; *F-4 Aircrew Operational Procedures*, 1. The original August 29, 1962, SOR-200 is included in section 12 of the September 1963 revision to the "F-4C System Procurement Package" document. The SOR stated: "The Navy F4H is crewed by a pilot/radar operator combination. The Air Force F-4C will be manned by a crew of two pilots" (sect. 12, p. 20). See also Norman to USAF, "F-110 Radar Pilot Training." Neither ADC's F-89 nor F-94 aircraft included provisions or controls for the back seater to fly the plane.

77. Hill, Oral History Interview, 20–21; John J. Burns, Oral History Interview, June 5–8, 1984, 215–17; Everest, Oral History Interview, 13–14; Cook, *Once a Fighter Pilot*, 90–91. Hill suggested that the Air Force used the F-4 rear seat "as a back door way to get authorization for more pilots." Burns similarly noted that he was told the rationale for using two pilots in the Phantom was that, given the "choice . . . to either turn up the gain on pilot training or navigator training," it made more sense to increase the number of pilots, especially since "we almost always seem to be short of [them] in a crunch period." But Burns also acknowledged that, as the Air Force got into "an era of very expensive airplanes, . . . an argument could be made that by putting the fellow in the back seat of an F-4 . . . he could get his apprentice training." Cook remembered one general telling the new Phantom pilots, "With two big engines and two pilots, there will be no excuses for F-4 accidents." The decision to field pilots in both seats of the F-4C demanded a significant increase in Air Force pilot production. Struggling to meet the new quota, officials in ATC toyed with the idea of shortening the pilot training program by ten weeks. See Briggs to HQ USAF/AFPDC, "Proposed Changes in Pilot Training"; McGehee to ATC (ATXDC), "Flight Indoctrination."

78. Donovan to HQ USAF, "Configuration of the F4H"; *F-4C System Package Program*, August 31, 1962, sect. 6, p. 37; *F-4C System Package Program*, September 1963, sect.6, p. 36; *History of the Tactical Air Command, Jan–Jun 1962*, 1:233. The Navy version of the Phantom included provisions for a removable, stowable flight control stick, but it could not be used during intercept missions. Cook observed of the rear seat equipment, "Pilots don't like flying when they don't have a set of controls." Cook, *Once a Fighter Pilot*, 85.

79. *F-4C System Package Program*, September 1963, sect. 6, pp. 36–37, sect. 12, p. 20 (emphasis added); *F-4C System Package Program*, August 31, 1962, sect. 6, pp. 37–38; Cook, *Once a Fighter Pilot*, 90–91; McCarthy, *Phantom Reflections*, 32–33; Attinello, *Air-to-Air Encounters in Southeast Asia: Account of F-4 and F-8 Events Prior to 1 March 1967*, 197.

80. *F-4C System Package Program*, September 1963, sect. 6, p. 36; *T.O. 1F-4C-1*, Fold Outs 17–19; Anderegg, interview, October 2, 2013; Sokol, interview. Because they were pilots, the GIBs still had to maintain their Air Force pilot currencies, to include passing annual instrument flying checkrides; see additional discussion later in this chapter.

81. Knaack, *Encyclopedia of US Air Force Aircraft*, 1:266.

82. Andrews, "To Fly and Fight," 55–56. For a sampling of the varying accounts, see John J. Burns, Oral History Interview, March 22, 1973, 3, 9, 16–18; Catledge, Oral History Interview, 30–32; Blesse, Corona Ace Oral History Interview, 59–61; Blesse, *Check Six*, 119–20; Cook, *Once a Fighter Pilot*, 208–11. For example, Burns characterized the Pentagon's prevalent attitude at the time as reflecting an opinion that aircraft guns were "anachronisms, throwbacks to earlier, bygone days. . . . That the day of the gun was gone, and that the day of the maneuvering fighter was gone, and that air combat would consist entirely of a radar detection and acquisition and lock-on, followed by a missile exchange"; those, like himself, who argued to the contrary were "voices in the wilderness." See also Bugos's interpretation of the rationale for removing the F-4's gun in Bugos, *Engineering the F-4*, 28.

83. "F-110." On the Phantom's numerous aviation records, see also McDonnell's advertisement, "Feats of a Phantom."

84. Allen, "Meet the F-110A," 20–22. Another test pilot was quoted as saying, "The F-110A is one of the best aircraft I've ever flown," in *History of the Tactical Air Command, Jan–Jun 1962*, 1:245. Cook boasted that he already "knew all the numbers" when he received his assignment to fly the F-4C: "Zero to thirty nine thousand feet in one minute and seventeen seconds. One thousand, six hundred miles an hour." Cook, *Once a Fighter Pilot*, 82.

85. Olds, *Fighter Pilot*, 249.

86. *History of the Tactical Air Command, Jul–Dec 1962*, 1:182–83; *History of the Tactical Air Command, Jan–Jun 1963*, 1:303. The TAC commander wanted to fill his new Phantoms with "only the most highly qualified personnel," and he consequently asked his F-100 wing commanders to send him the names of their top ten fighter pilots for reassignment. Among other requirements, candidate pilots needed to possess "a minimum of 1,500 hours total flying time" and "1,000 hours single engine jet time." Of the names forwarded, fourteen were designated as the initial cadre in the F-110, while another fifteen were sent to bolster the ranks of F-105 pilots. The last five F-4Bs on loan from the Navy were returned on June 3, 1964; *United States Air Force Statistical Digest: Fiscal Year 1964*, xliv.

87. *History of the 15th Tactical Fighter Wing, Jan–Jun 1964*, 1; *History of the 15th Tactical Fighter Wing, Jul–Dec 1964*, 14–15; *History of the 4453d Combat Crew Training Wing, Jul–Dec 1966*, 1:1–2; *History of the 15th Tactical Fighter Wing, Jan–Jun 1966*, 1:8–9; Scholin, "New Phlyers for the

Phantom," 46. At the height of the war, F-4 training was conducted by the 15th Wing at McDill AFB; the 479th at George AFB, California; and the 4531st at Homestead AFB, Florida; only a single stateside wing, the 33d at Florida's Eglin AFB, flew the Phantom in a nontraining function.

88. L. Welch, Oral History Interview; L. Welch, Interview, October 1, 2013; *History of the 366th Tactical Fighter Wing, Jan–Jul 1965*, 1:1–2.

89. R. Meyer, "Phantom over Tampa Bay"; *History of the 4453d Combat Crew Training Wing, Jul–Dec 1966*, 1:1–5; *Combat Crew Training Course: F-4C*, 1–2; *USAF Replacement Training Course: F-4C*, 2–3. In December 1963 ground training and simulator practice for previous fighter pilots converting to the F-4 entailed 180 total hours of instruction. By late 1966, the ground instruction requirement alone for transitioning pilots (combat crew training) had grown to more than 250 hours. For those pilots destined for the front seat of the Phantom but who had not yet accumulated appreciable flight time in another fighter, a separate four-month-long RTU (replacement training unit) syllabus was developed in 1966 that included 81 hours of flight time and 240 hours of ground instruction. For comparison, the October 1952 training syllabus for new F-86 fighter pilots entailed 78 hours of academic instruction, no simulator time, and 80 hours of flight training. *Format for the Syllabus of Instruction: F-86*, 2–3. Olds claimed that he completed the F-4 training course in less than a week. Olds, *Fighter Pilot*, 249–52. However, Olds's program was an abbreviated checkout course that was used "to transition commanders and other key personnel in the F-4," and even then, training records indicate that Olds was enrolled in the program from August 18 through September 27. Normally, the program included eleven sorties, and it may be the flying portion of Olds's training that was curtailed and completed in less than a week. *History of the 4453d Combat Crew Training Wing, Jul–Dec 1966*, 1:5, 22.

90. On adverse yaw in the F-4, see Stuck, "Aircraft Control"; *F-4 Aircrew Operational Procedures*, 23–24; Leatham, "Adverse Yaw"; Calvert, "Turning the F-4"; Anderegg, *Sierra Hotel*, 12–13; Cook, *Once a Fighter Pilot*, 192–95; Olds, *Fighter Pilot*, 249–51. Allen's June 1962 assessment of the F-110, in which he extolled the Phantom's flight characteristics and bragged that the aircraft "exhibits no dishonest characteristics when being yanked, banked and kickin' rudder," proved not entirely accurate. Allen, "Meet the F-110A," 21. To avoid the "adverse yaw demon," pilots needed to avoid applying any lateral pressure to the control stick when the aircraft was wallowing at high AOA.

91. *Combat Crew Training Course: F-4C*; Anderegg, *Sierra Hotel*, 20–21; *F-4C Air Combat Tactics*, Supplement no. 1; *History of the 15th Tactical Fighter Wing, Jan–Jun 1966*, 1:8–16; *History of the 15th Tactical Fighter Wing, Jul–Dec 1966*, 2:9–20. Whereas ACM focused primarily on close-range, visual dogfighting, air combat tactics (ACT) encompassed the entire air-to-air combat environment, from long-range intercepts through ACM.

92. *Combat Crew Training Course: F-4C*. See also McCarthy, *Phantom Reflections*, 17.

93. *Radar Academic Training Course: F-4C*; R. Meyer, "Phantom over Tampa Bay"; Sokol, interview; Anderegg, interview, October 2, 2013. On the initial development of the radar course, see Krebs to ATC and TAC, "F-110 Radar Pilot Training"; Norman to USAF, "F-110 Radar Pilot Training"; Heflin to TAC, "Radar Pilot Training (APQ-72)"; Feezel to CSAF, "Bench Sets/Component Lab Sets"; Stewart to HQ USAF (AFPTR-F-N), "Requirements for the Radar Pilot Training"; *History of the Air Training Command, Jul–Dec 1962*, 1:115–16; *History of the Air Training Command, Jan–Jun 1963*, 1:119–20.

94. *F-4 Aircrew Operational Procedures*, 29.

95. *T.O. 1F-4C-34-1-1*, sect. 1, pp. 12–25, 27. See also *T.O. 1F-4C-2-19*. One GIB's assessment of the selectable scope ranges is recorded in Attinello, *Air-to-Air Encounters in Southeast Asia: Account of F-4 and F-8 Events Prior to 1 March 1967*, 196; see also R. Scott, *F-4C Fighter Screen and Escort*, 9. To compute the target's altitude, the GIB mentally solved the equation: target altitude in feet = (radar elevation angle × target range in miles × 100) ± ownship altitude in feet. *4453d CCTW RTU Weapons School: Intercept Techniques*, 28.

96. *4453d CCTW RTU Weapons School: Intercept Techniques*; *F-4C Intercept*.

97. *T.O. 1F-4C-34-1-1*, sect. 1, pp. 12–30. The antenna control stick was also used to control some of the radar's other functions and displays. For example, the azimuth coverage of the radar could be reduced from its normal 120 degrees to only 60 degrees (to reduce scan time) and then slewed as required using the antenna control.

98. Ibid. See also *T.O. 1F-4C-2-19*.

99. *T.O. 1F-4C-34-1-1*, sect. 1, pp. 12–30. See also *T.O. 1F-4C-2-19*. Of note, the radar contained several additional and more complex operating modes that could be used if the radar encountered electronic jamming. The GIB was expected to be completely familiar with these alternate operating modes as well.

100. *F-4C Air Combat Tactics*, 18, 21.

101. Ibid., 21; *4453d CCTW RTU Weapons School: Intercept Techniques*, 2–9; *F-4 Aircrew Operational Procedures*, 61. To avoid confusion, intra-aircraft "commentary" used terms slightly different from the "brevity code" that fighter pilots used when communicating between aircraft. Brevity code terms included, for example, *angels* to indicate altitude in thousands of feet, *no joy* to indicate the target could not be found, and *Clara* to indicate that there were no enemy targets detected by the radar.

102. "Advantage Is Here: Why the Phantom II Has a Two Man Crew."

103. *Combat Crew Training Course: F-4C*, 1–2; *Radar Academic Training Course: F-4C*, 1; *USAF Replacement Training Course: F-4C*, 2–3; *Format for the Syllabus of Instruction: F-86*, 2–3.

104. P. Smith, "Ode to a Fighter Pilot" (emphasis added).

105. *History of the 12th Tactical Fighter Wing, Jul–Dec 1964*, 1:19–21; Thornborough and Davies, *Phantom Story*, 92; P. Davies, *USAF F-4 Phantom II MiG Killers, 1965–68*, 8.

106. *History of the 15th Tactical Fighter Wing, Jul–Dec 1965*, 1:15–17; CSAF to TAC, "TWO BUCK Deployment"; Alden to 9 AF, "TACOP Final Report for 45 TFS," 1, 3.

107. Hone, "Southeast Asia," 521.

108. *History of the 15th Tactical Fighter Wing, Jul–Dec 1965*, 1:15–20; Van Staaveren, *Gradual Failure*, 146–48, 210. On an alternative policy, see Michel, *Clashes*, 166.

109. Van Staaveren, *Gradual Failure*, 48–49. There is an abundance of literature on the Gulf of Tonkin incident, the American's reprisal attack, and Congress's ensuing resolution that granted President Johnson the authority to "take all necessary steps, including the use of armed force," to pursue objectives in Vietnam. See, for example, Goulden, *Truth Is the First Casualty*; Galloway, *Gulf of Tonkin Resolution*; McNamara, *In Retrospect*; Moïse, *Tonkin Gulf*.

110. Van Staaveren, *Gradual Failure*, 50–51, 58–64. The Air Force had been supporting the fledgling Laotian air force's attacks against the same target set for nearly two months prior.

111. Ibid., 9–25, 83; *Flaming Dart Air Strikes*; McNamara, *In Retrospect*, 170–74; Tilford, *Setup*, 102–4; Clodfelter, *Limits of Air Power*; Pape, *Bombing to Win*, chap. 6. For decades, Rolling Thunder would be cited as a model of failed diplomacy and poorly executed military strategy.

112. Van Staaveren, *Gradual Failure*, 69–70, 83–84; Andrews, "To Fly and Fight," 61–63.

113. Van Staaveren, *Gradual Failure*, 84; Andrews, "To Fly and Fight," 61–62.

114. R. Scott, *Soviet Bloc AA Weapons*; Van Staaveren, *Gradual Failure*, 70, 86–89; Andrews, "To Fly and Fight," 61–63; McCrea, *US Navy, Marine Corps, and Air Force Fixed-Wing Aircraft Losses*, 1–15; Michel, *Clashes*, 37; Hone, "Southeast Asia," 521; Werrell, *Archie to SAM*, 117–19; Momyer, "Evolution of Fighter Tactics in SEA," 59. Van Staaveren noted that after the initial Rolling Thunder attacks, Secretary of the Air Force Eugene Zuckert worried that the service might lose its "credibility" if its advanced tactical aircraft continued to be lost to "unsophisticated" enemy weapons. The Air Force initiated several studies trying to discern a pattern in its losses to AAA. See, for example, English, Marlow, and Sandborn, *Reduction of Tactical Fighter Losses*; Marlow and English, *Reduction of Tactical Fighter Losses*.

115. Momyer, *Airpower in Three Wars*, 134–36.

116. Quoted in Andrews, "To Fly and Fight," 132.

117. Van Staaveren, *Gradual Failure*, 113–14, 161–73; Langguth, "US Aides Indicate a Red Missile Hit Jet"; Alden to 9 AF, "TACOP Final Report for 45 TFS," 6; Futrell et al., *Aces and Aerial Victories*, 5.

118. Gaspard, *SA-2 Deployment and Tactics*; McCrea, *US Navy, Marine Corps, and Air Force Fixed-Wing Aircraft Losses*, 1–15; Van Staaveren, *Gradual Failure*, 164; Baldwin, "US Raids 2 Missile Sites"; Hone, "Southeast Asia," 528–34; Andrews, "To Fly and Fight," 128–44; Anderegg, *Sierra Hotel*, 21–22; *8th TFW Tactical Doctrine*, March 1967, 38–43; Handley, *Nickel on the Grass*, 124n64; McCarthy, *Phantom Reflections*, 30–31, 76; Olds, "End of Tour Report," 1–2; Momyer, *Airpower in Three Wars*, 136–39, 150–52; Futrell et al., *Aces and Aerial Victories*, 5–7; Michel, *Clashes*, 28–38; Werrell, *Chasing the Silver Bullet*, 48–50. Anderegg described the SA-2 as "fast but not agile," noting that "if the F-4 or F-105 pilot saw the missile coming, he could evade it easily by first diving to pick up some airspeed. Then, once the missile committed down toward him, the pilot would pull back up into the SAM and cause it to overshoot and miss. Aircraft hit by SAMs were almost always jets whose pilots never saw the missile." SAMs would account for about 7 percent of total Air Force aircraft losses during the war.

119. Robinson to C15, "Report of the 47th TFS Deployment," 13.

120. Michel, *Clashes*, 16–20; Van Staaveren, *Gradual Failure*, 52, 70, 96; Momyer, *Airpower in Three Wars*, 133–34, 153–58; Harr, Muskat, and Williams, *Air Combat Tactics Evaluation Final Report*; Muskat, "Featherduster"; Saunders, "PACAF Tactics/Techniques (Bulletin #8)"; R. Scott, *Counter-Air Tactics*. In the Feather Duster tests, American pilots evaluated the relative effectiveness of their F-100s, F-104s, F-105s, and F-4Cs against the MiG-15/-17, which was simulated by an F-86. It was the second time the Sabre functioned as a proxy for the MiG; recall Boatner, "Letter Report on Relative Aerial Combat Capability."

121. Olds, Oral History Interview, July 12, 1967, 8–9.

122. Futrell et al., *Aces and Aerial Victories*, 10. Futrell calculated that during the period from September to December 1966, "of the 192 strike aircraft actually engaged by MIG's, 107 (or 55.73 percent) jettisoned their ordnance."

123. CSAF to TAC, "TWO BUCK Deployment"; Van Staaveren, *Gradual Failure*, 89–99, 241; Momyer, *Airpower in Three Wars*, 157; Hone, "Southeast Asia," 506–9, 515; Futrell et al., *Aces and Aerial Victories*, 3–5; Futrell, *Ideas, Concepts, Doctrine, 1907–1960*, chap. 7–8; Futrell, *Ideas, Concepts, Doctrine, 1961–1984*, chap. 3–4. The Joint Chiefs of Staff submitted its request to McNamara to move a squadron of Air Force Phantoms to Southeast Asia on

March 19. The preferred doctrine for gaining air superiority was even written into the requirements documentation for the F-4C, which noted, "Ideally, the enemy's air arm should be destroyed on the ground with either conventional or nuclear weapons." *F-4C System Package Program*, August 31, 1962, sect. 5, p. 3.

124. "US Fliers Bomb 2 Bridges"; Langguth, "Hanoi MIG's Down 2 American Jets"; "Bomb Loads Slowed F-105's"; Van Staaveren, *Gradual Failure*, 106–9; Attinello, *Air-to-Air Encounters in Southeast Asia: Account of F-4 and F-8 Events Prior to 1 March 1967*, 33–34; Attinello, *Air-to-Air Encounters in Southeast Asia: F-105 Events Prior to 1 March 1967*, 35–41; Michel, *Clashes*, 8–10.

125. Alden to 9 AF, "TACOP Final Report for 45 TFS," 1, 5–6, 10; Attinello, *Air-to-Air Encounters in Southeast Asia: Account of F-4 and F-8 Events Prior to 1 March 1967*, 57; Futrell et al., *Aces and Aerial Victories*, 22; Michel, *Clashes*, 21.

126. Attinello, *Air-to-Air Encounters in Southeast Asia: Account of F-4 and F-8 Events Prior to 1 March 1967*, 43; Attinello, *Air-to-Air Encounters in Southeast Asia: F-105 Events Prior to 1 March 1967*, 43–47. Navy Phantoms encountered MiGs on June 4 and June 17; everyone escaped the first encounter unscathed, but during the second two of the four MiG-17s were shot down. See Attinello, *Air-to-Air Encounters in Southeast Asia: Account of F-4 and F-8 Events Prior to 1 March 1967*, 45–53. An earlier engagement between four Navy Phantoms and four MiG-17s on April 9 was eventually determined to be an unfortunate mistake—the MiGs were Chinese, not North Vietnamese. In the ensuing confusion, one MiG was shot down and a Phantom lost, the latter to either low fuel or an errant American missile. See Langguth, "One MIG Seen on Fire"; "Peking Reports Hainan Air Fight"; Van Staaveren, *Gradual Failure*, 110–11; Attinello, *Air-to-Air Encounters in Southeast Asia: Account of F-4 and F-8 Events Prior to 1 March 1967*, 35–41.

127. *F-4C System Package Program*, August 31, 1962; *F-4C System Package Program*, September 1963; R. Meyer, "Phantom over Tampa Bay."

128. Alden to 9 AF, "TACOP Final Report for 45 TFS," 1, 5–6. There was a shortage of 600-gallon centerline external fuel tanks when the 45th TFS first arrived in combat, which limited fuel capacity and sortie duration; see references in Attinello, *Air-to-Air Encounters in Southeast Asia: Account of F-4 and F-8 Events Prior to 1 March 1967*, 55–57.

129. Alden to 9 AF, "TACOP Final Report for 45 TFS," 1, 6.

130. Ibid., 6.

131. Ibid., 1, 4, 10; Sokol, interview; L. Welch, interview, October 1, 2013; Hargrove, Oral History Interview, 31–32; Van Staaveren, *Gradual Failure*, 138; Attinello, *Air-to-Air Encounters in Southeast Asia: Account of F-4 and F-8 Events Prior to 1 March 1967*, 43. Van Staaveren stated that the Phantoms first started flying bombing missions at the end of May 1965, but the sorties were likely strike CAP missions instead of dedicated bombing missions. Attinello recorded that the Phantoms' MiG sighting on May 31, 1965, occurred after the Phantoms were "climbing out from air-to-ground delivery after MIGCAP," a typical strike CAP profile. On the mechanics of manual dive bombing in the Phantom, see Anderegg, *Sierra Hotel*, 5–9.

132. Alden to 9 AF, "TACOP Final Report for 45 TFS," 1–2, 4–5, 7, 11–14; Robinson to C15, "Report of the 47th TFS Deployment," 9–12; Raddin, "End of Tour Report," 21. One of the additional missions was Road Recce, which entailed weaving alongside a designated road at high speed and low altitude, usually about 450 knots and 5,000 feet, searching for targets to destroy with an assortment of bombs, rockets, or bullets from the gun pod. The requirement for the Phantom to carry a podded external gun was added by TAC in April 1962. On the

SUU-16, see *History of the Tactical Air Command, Jan–Jun 1962*, 1:251n75; *History of the Tactical Air Command, Jul–Dec 1962*, 1:238–39; Mehserle, *Capability of the F-4C Aircraft/SUU-16/A Gun Pod; F-4C/SUU-16A*; Broadway, "Have Guns"; Fino, "Breaking the Trance."

133. Robinson to C15, "Report of the 47th TFS Deployment," 5.

134. Momyer, *Airpower in Three Wars*, 103–112; Braswell to Yudkin, "Background Material on SEA Command and Control"; Michel, *Clashes*, 38–39; Van Staaveren, *Gradual Failure*, 209–10, 240; Wayne Thompson, *To Hanoi and Back*, 18–19; Cunningham, "William W. Momyer," 171–72; *History of the 366th Tactical Fighter Wing, Oct 1966–Mar 1967*, 1:10; Momyer, "Evolution of Fighter Tactics in SEA," 58.

135. Hone, "Southeast Asia," 524. The 12th TFW moved to Cam Ranh in November 1965, and the 8th TFW set up shop at Ubon in December 1965. The 366th TFW eventually established itself at DaNang in March 1966.

136. Knaack, *Encyclopedia of US Air Force Aircraft*, 1:273–77; Olds, "End of Tour Report," 11–13; McInerney, "F-4D"; Thornborough and Davies, *Phantom Story*, 108. The F-4D also carried the Air Force's IR-guided Falcon missile, the AIM-4D (discussed in more detail in the next subsection), and was equipped with a lead-computing optical sight for external gun pod employment. See McInerney, "LCOS." On the WRCS and dive toss, see also Olds, Oral History Interview, n.d., 78; McCarthy, *Phantom Reflections*, 87–88; Buhrow, Oral History Interview, n.d., 22–24; Anderegg, *Sierra Hotel*, 63–64; McNally, "WRCS"; McInerney, "F-4 Dive Toss."

137. Olds, "End of Tour Report," 14–17; Anderegg, *Sierra Hotel*, 3–4, 14–15; *8th TFW Tactical Doctrine*, December 1967. Cf. *8th TFW Tactical Doctrine*, March 1967, which was almost 230 pages, with more than 50 devoted to air-to-air tactics, MiG CAP, and strike escort operations.

138. Andrews, "To Fly and Fight," 12.

139. Cook, *Once a Fighter Pilot*, 170–72. Cook remembered one "personally devastating mission" in which he was directed to drop napalm on his own troops. Initially rebuffing the forward air controller's direction, Cook was told, "The commander on the ground wants the napalm down on top of him and his troops. He says that they're dug in and the enemy is overrunning them. He says do it now or they're all dead anyway!" Cook dropped the napalm, and he "received word later that day or the next that our mission had been a success," but he said he emerged from that particular sortie "deeply wounded with a scar that would always be there."

140. Wallace, interview.

141. Momyer, "Evolution of Fighter Tactics in SEA," 62.

142. Cook, *Once a Fighter Pilot*, 219.

143. Kimball, Oral History Interview, 24–25; Hendrickson, Oral History Interview, 28. In August 1966, the Phantom crews in the 479th TFS were forced to abandon their "Mig hunt" opportunities when they became the only exclusively "night strike" Phantom squadron. Nash, "Project CORONA HARVEST End of Tour Report," 11, 49.

144. *8th TFW Tactical Doctrine*, March 1967, 97.

145. McCarthy, *Phantom Reflections*, 68–69, 166; Andrews, "To Fly and Fight," 152–62.

146. The description of the engagement is based on the following accounts: Attinello, *Air-to-Air Encounters in Southeast Asia: Account of F-4 and F-8 Events Prior to 1 March 1967*, 55–77; R. Scott, *F-4C—MIG-17 Engagement*; Michel, *Clashes*, 20–24; Futrell et al., *Aces and Aerial Victories*, 22–26; P. Davies, *USAF F-4 Phantom II MiG Killers, 1965–68*, 8–9.

147. "Pilots Describe Downing of MIG's."

148. Alden to 9 AF, "TACOP Final Report for 45 TFS," 5.

149. During the engagement, the lead Phantom climbed to twenty-four thousand feet and orbited the area to monitor the fight. Expecting MiGs to attack as four-ships, the number 2 Phantom extended away from the fight searching for the anticipated accompanying MiGs, but then lost sight of the fight and by the time it had reoriented itself, the engagement was over. The day after the mission, each victorious two-person F-4 crew was awarded a Silver Star; the aircrews from the accompanying F-4s received Distinguished Flying Crosses.

150. *F-4C System Package Program*, August 31, 1962, sect. 5, p. 13. The F-100D aircrews using the GAR-8 came to the same conclusion in 1959, observing that "the greatest limitation . . . in employing the GAR-8 . . . is the inability to positively identify a fighter target as friend or foe at the range from which GAR-8's are normally fired." Giordano, "F-100D GAR-8 Tactics," 43.

151. Momyer, *Airpower in Three Wars*, 175.

152. Pilots Describe Downing of MIG's." See also McCarthy, *Phantom Reflections*, 29; Mohamed, "Case for Recognition."

153. Cook, *Once a Fighter Pilot*, 211–15.

154. Michel, *Clashes*, 46–51; Van Staaveren, *Gradual Failure*, 117–18; R. Scott, *Big Eye Task Force*; R. Scott, *EC-121D Operations*; Futrell et al., *Aces and Aerial Victories*, 4–5, 12–16; Momyer, *Airpower in Three Wars*, 168–75. See also references in Attinello, *Air-to-Air Encounters in Southeast Asia: Account of F-4 and F-8 Events Prior to 1 March 1967*, 58, 95–97. Of note, Cook's near shootdown of the Navy fighters came after the EC-121 warned him of unidentified aircraft nearby. Cook, *Once a Fighter Pilot*, 211–15. There were also frequent communications issues between the EC-121 aircraft and the Phantoms. Olds complained of the generic MiG warnings, noting that the calls applied to a thirty-by-thirty-mile sector of airspace, which equated to "forty-five thousand cubic miles of sky, and you tell me there's a bloody MIG in there. It's stupid-useless information." Olds, "Briefing to the Staff of Project CORONA HARVEST on Operation BOLO," 67; see also Buhrow, Oral History Interview, April 11, 1968, 19–20. Later during Operation Linebacker, the Navy would lend its own radar control capability to the air fight over North Vietnam using the call sign "Red Crown." The aircrews also started receiving signals intelligence under the call sign "Teaball" informing them of the MiGs' activities. Michel, *Clashes*, 250–53.

155. Olds, "Briefing to the Staff of Project CORONA HARVEST on Operation BOLO," 27, 32; Asay, Covington, and Stone, *History of Operations BOLO*, 8–9.

156. Michel, *Clashes*, 181, 194–95, 267–69; Anderegg, *Sierra Hotel*, 33–34.

157. Robinson to C15, "Report of the 47th TFS Deployment," 11, 14; Michel, *Clashes*, 16; Anderegg, *Sierra Hotel*, 32; Blesse, Corona Ace Oral History Interview, 59–60; Cook, *Once a Fighter Pilot*, 209–10; Dilger, Oral History Interview, 24. Dilger noted that "when you get at the R Min [minimum range] of the missile and the R Max [maximum range] of the eyeball, they start coinciding with each other, and you've almost got an impossible situation." Several pilots continued to favor development of a better, more maneuverable air-to-air missile with a shorter minimum range instead of a gun. See, for example, pilot excerpts in Attinello, *Air-to-Air Encounters in Southeast Asia: Account of F-4 and F-8 Events Prior to 1 March 1967*, 142, 161.

158. Surprisingly, the Air Force seemed vaguely uninterested in the performance of the Phantom's missiles when it first started taking delivery of its new F-4Cs. During TAC's "limited evaluation" of the F-4C, conducted by the 558th TFS, 12th TFW, at MacDill AFB in August 1964, air-to-air testing was assigned a "relatively low test project priority." Of the

forty-six scheduled Sparrow shots, only seventeen sorties were flown, and of those only four successfully launched the test missile. All four test launches were later "termed non-productive" due to failure of the telemetry scoring system. No Sidewinder missiles were launched during the test. Despite the inconclusive findings, the evaluation report was optimistic, declaring, "The F-4C [air-to-air] delivery capability is somewhat apparent." *F-4C Limited Evaluation*, 55–59.

159. Michel, *Clashes*, 156–58; Laslie, *Air Force Way of War*, 12.

160. Reported in Attinello, *Air-to-Air Encounters in Southeast Asia: Account of F-4 and F-8 Events Prior to 1 March 1967*, 196.

161. Extracts from the Heat Treat team's SEA trip report were distributed as an attachment to R. Scott, *F-4C Fighter Screen and Escort*, 6, 10. The team found that during the period from April 23 to May 11, 1966, Air Force F-4Cs fired thirteen AIM-7s (and tried to fire an additional three which never left the aircraft) to down a single MiG—a 6 percent hit rate. Whereas some failures could be attributed to faulty missile maintenance and aircraft loading or improper pilot performance, the team noted that "four of the Sparrows launched during the period 23–24 April were fired under ideal conditions and missed" for inexplicable reasons. See also the Navy's investigation of its aircraft armament in Ault, *Report of the Air-to-Air Missile System*.

162. Michel, *Clashes*, 150–53, 182; Laslie, *Air Force Way of War*, 12; *Project Red Baron II*, 13; John J. Burns, Oral History Interview, June 5–8, 1984, 223–25. See also *Southeast Asia Counter-Air Alternatives*, sect. H, p. 12. On the performance of the 8th TFW's AIM-7 missiles, see Olds, "End of Tour Report," 8–9.

163. McCarthy, *Phantom Reflections*, 59. Ritchie remembered the thirty-seven hours being more like fifty to seventy-five man-hours per flight hour. Ritchie, Oral History Interview, 31. Olds claimed that the "break rate" on the F-4 was 92 percent, "meaning that about 92% of the time a guy taxis into the flightline, there is [something wrong with the jet. It] may only take 15 minutes to fix and on the other hand it may take 5 or 6 hours." Olds, Oral History Interview, n.d., 10–11, 68–69.

164. Asay, Covington, and Stone, *History of Operations BOLO*, Annex F; Ow, *Mission Bolo*; Olds, "Briefing to the Staff of Project CORONA HARVEST on Operation BOLO," 12–14.

165. Dilger, Oral History Interview, 23; Olds, "End of Tour Report," 8–9; Kirk, Oral History Interview, 5.

166. Michel, *Clashes*, 43, 153–56; R. Scott, *F-4C Fighter Screen and Escort*, 7, 10; Laslie, *Air Force Way of War*, 12–13; Olds, "End of Tour Report," 9. See also *Southeast Asia Counter-Air Alternatives*, sect. H, p. 13.

167. Dilger, Oral History Interview, 25–26.

168. Kirk, Oral History Interview, 3 (emphasis added). See also Olds, Oral History Interview, n.d., 69–70.

169. Michel, *Clashes*, 108–12, 156; Thornborough and Davies, *Phantom Story*, 110; *Project Red Baron II*, 13; Sean O'Connor, *Arming America's Interceptors*. When the Air Force abandoned the Sidewinder for its Falcon, the Navy pressed on and developed the AIM-9D, which included a new, cryogenically cooled PbS target detector cell to improve the missile's tracking and IR-discrimination capability. Westrum, *Sidewinder*, 177. The AIM-4D Falcon also used a cooled IR-seeker, but whereas the Navy had designed its missile with a four-hour supply of liquid nitrogen, the Air Force's Falcon had only a two-minute supply. Michel summarized the F-4D front seater's dilemma: "[He could] either arm the AIM-4D early in the engagement and hope he

would get a chance to use it within the next two minutes, or wait and try to remember to arm it after the fight began and when there was a target available. In a turning dogfight where shot opportunities were fleeting, such restraints on a missile clearly were unacceptable" (110).

170. Olds, Oral History Interview, n.d., 70–74; Olds, *Fighter Pilot*, 314–16. Because the new F-4Ds were no longer configured to support the Sidewinder (only the AIM-4D Falcon), the aircraft had to be rewired to accept the old AIM-9B missiles.

171. Michel, *Clashes*, 111, 156. On the perceived advantages of the Navy's newer AIM-9D versus the Air Force's AIM-9B, see Hargrove, Oral History Interview, 14–15. From then on, the Air Force and Navy developed separate Sidewinder missiles produced under separate contracts. The disparity between the services' different versions would have implications a decade later during the Air Intercept Missile Evaluation and Air Combat Evaluation. On the different Sidewinder production models, see Beck, "State of the Art."

172. "US Jets Will Reflect Lessons." See also the earlier announcement, "Air Force Orders New Jet Fighter." On the challenges of incorporating the internal gun into the Phantom, see Bugos, *Engineering the F-4*, 158–59.

173. Knaack, *Encyclopedia of US Air Force Aircraft*, 1:277–84; Thornborough and Davies, *Phantom Story*, 113–14; McInerney, "F-4E"; McInerney, "F-4E Cat III"; Bielinski, "F-4E." The Air Force fighter pilots yearned for the arrival of the new F-4E and its internal gun; see, for example, Kirk, Oral History Interview, 7. The F-4E would undergo numerous upgrades during the war as part of the Rivet Haste program, including a new cockpit design, leading edge slats for improved stability at high AOA, and TISEO. Beck, "Super Phantom"; Handley, *Nickel on the Grass*, 70n30; Anderegg, *Sierra Hotel*, 32–34.

174. Michel, *Clashes*, 40–44, 57–59, 63–68, 87–97, 112–13; Futrell et al., *Aces and Aerial Victories*, 26–27, 45, 73, 75, 78; Momyer, *Airpower in Three Wars*, 153–65; Heffron, *Air-to-Air Encounters over North Vietnam, 1 Jan–30 Jun 1967*; Olds, "End of Tour Report," 3–6; *History of the 8th Tactical Fighter Wing, Oct–Dec 1967*, 1:32–35; *Project Red Baron II*, 8–9; Hone, "Southeast Asia"; Aust, "PACAF Tactics and Techniques Program (Bulletin #5)." MiG-19s did not appear in North Vietnam until after Rolling Thunder had ended.

175. Cunningham, "William W. Momyer," 280–81, 285; Futrell et al., *Aces and Aerial Victories*, 64.

176. Michel, *Clashes*, 127–31, 140–2, 161–62, 172–78; Olds, "End of Tour Report," 3–6; Weaver, *Air-to-Air Encounters over North Vietnam, 1 Jul 1967–31 Dec 1968*.

177. On the kill ratios in Vietnam, see Sayers, "Red Baron Reports." During the first half of 1967, American F-4s destroyed thirteen MiG-21s with no losses, but in the last quarter of 1967, North Vietnamese MiG-21s successfully downed twelve American aircraft, five of them Phantoms, while suffering only one loss. Michel, *Clashes*, 139–40.

178. "Pilots Describe Downing of MIG's."

179. Olds, *Fighter Pilot*, 254; Olds, Corona Ace Oral History Interview, 83.

180. "'Ace' Fighter Wing Commanded by 'Aces.'"

181. *8th TFW Tactical Doctrine*, March 1967, 78 (emphasis added). "Close control" occurs when an individual fighter is directed against an individual target using precise vectors provided by either a ground-based or air-based radar operator.

182. Anderegg, *Sierra Hotel*, xii.

183. On the process by which a new pilot declared his preferences for an aircraft type, and ATC's attempt to be more accommodating, see *History of the Air Training Command, Jan–Jun*

1963, 1:98–99; Connor to Pilot Training Wings, "Undergraduate Pilot Training Assignment Procedures"; Tips to Pilot Training Wings, "Undergraduate Pilot Training Assignment Procedures"; Wood to ATCCS, "Final Report—UPT Assignments."

184. Clifton, "There I Was . . . 2"; Clifton, "There I Was . . . 4." Clifton's pilot training class was one of the first to fly the new Northrop T-38 Talon advanced jet trainer. Graduating from pilot training a year later, Sokol had a similar experience, "If you wanted to go to a fighter and you weren't the number one guy in your class, you went to the back seat of an F-4." Sokol, interview. See also Buhrow, Oral History Interview, n.d., 43–44; Andrews, "To Fly and Fight," 39–40.

185. Hill, Oral History Interview, iv, 20–22.

186. Everest, Oral History Interview, 13.

187. Olds, Corona Ace Oral History Interview, 13.

188. W. H. Hughes, *Probability of Tour Completion for Pilots*, 13. Cf. earlier analysis in S. J. Scott, *Probability of Southeast Asia Tour Completion*.

189. Hill, Oral History Interview, 37. See also McCarthy, *Phantom Reflections*, 13.

190. Olds, Corona Ace Oral History Interview, 53–54; L. Welch, interview, October 1, 2013; Cook, *Once a Fighter Pilot*, 195–97; Andrews, "To Fly and Fight," 55. On the F-105 experience in Vietnam, see Broughton, *Thud Ridge*; Rasimus, *When Thunder Rolled*; Bell, *100 Missions North*; J. Sherwood, *Fast Movers*, chap. 2. Cliver, a young pilot who was able to finagle an F-105, remarked that "the best way" to start flying fighters was in a single-seater: "You never count on anybody else. It's all you, baby, so figure it out." Cliver, interview.

191. Cook, *Once a Fighter Pilot*, 195.

192. Scholin, "New Phlyers for the Phantom," 48.

193. Sokol, interview; Anderegg, interview, October 2, 2013; Hill, Oral History Interview, 20–22; Buhrow, Oral History Interview, n.d., 43–44; McCarthy, *Phantom Reflections*, 32. Wallace, on the other hand, actually opted for a right seat in the C-141 over the back seat of an F-4, explaining that at the time he was "thinking about going to the airlines." After being in the C-141 for eighteen months and not getting the flying time he was once promised, Wallace said he "got bored . . . and was able to wrangle an F-100 assignment . . . to Southeast Asia." Wallace, interview.

194. Hendrickson, Oral History Interview, 61. Of course, there were exceptions. See, for example, accounts in Hargrove, Oral History Interview, 29–36; Buhrow, Oral History Interview, n.d., 45; Skogerbee, Oral History Interview, 5. Another commander reported that, for the "night strike" mission, "it was preferable to have a pilot in the rear cockpit. Often the aircraft commander flew with much of his cockpit blacked out in order to see terrain and targets. The aircrew in the aft cockpit monitored flight instruments, provided delivery parameters, and operated the radar and elements of the weapons delivery station." Nash, "Project CORONOA HARVEST End of Tour Report," 11.

195. Hendrickson, Oral History Interview, 61–62; Anderegg, interview, October 2, 2013; Sokol, interview.

196. L. Welch, interview, October 1, 2013.

197. John J. Burns, Oral History Interview, June 5–8, 1984, 216–17. See also Robbins, Oral History Interview, 106; McCarthy, *Phantom Reflections*, 154. McCarthy remembered getting a "few orientation rides in the backseat to familiarize us [front seaters] with the systems, but I was not very knowledgeable about all their intricacies." Anderegg recalled the GIBs having a

"very ad-hoc, very informal group called the GIB Protective Association—the GPA. We got together a lot. And we talked about the front seaters, and who could do what well, and who couldn't do what so well, and what guys were like to fly with." Anderegg, interview, October 2, 2013.

198. Anderegg, *Sierra Hotel*, 13–14; Andrews, "To Fly and Fight," 189–90; Olds, Corona Ace Oral History Interview, 79–82; Olds, *Fighter Pilot*, 245–46; Blesse, Corona Ace Oral History Interview, 24–28, 33–36; L. Welch, Oral History Interview, 40–41; Hannah, *Striving for Air Superiority*, 89; Laslie, *Air Force Way of War*, 10–11. According to Andrews, the TAC commander expressed his concern that with TAC crews flying repeated combat tours in Vietnam, an "irreversible gap" might emerge between TAC and SAC that could "fracture and compartment the Air Force." Anderegg declared the "one tour policy" concept "noble and well-intentioned, but its effect was devastating to the combat capability of the tactical forces." Moreover, he thought the notion of a "universally assignable pilot"—that any Air Force pilot could fly any Air Force aircraft with a bare minimum of training—reeked of "institutional arrogance."

199. Anderegg, *Sierra Hotel*, 14; Anderegg, interview, October 2, 2013.

200. Dilger, Oral History Interview, 6–7.

201. Michel, *Clashes*, 162–66. Michel attributed the decline in the performance of Air Force fighter pilots during the war to the "one tour policy." He observed that from April 1965 through June 1967, about 65 percent of the Air Force's fighter pilots in Vietnam had backgrounds in TAC, and the service posted a 3:1 kill-ratio advantage over the MiGs. In contrast, from June 1967 through March 1968, less than 30 percent of Air Force combat pilots came from TAC, and the service's kill ratio dropped to 0.85:1. Much of the F-4 experience necessarily migrated to the Phantom training units as they tried to produce ever-increasing numbers of graduates.

202. Buhrow, Oral History Interview, n.d., 43–45. See also Everest, Oral History Interview, 13; Handley, *Nickel on The Grass*, 159n67.

203. Hill, Oral History Interview, iv, 38–46, 75. See also La Salle, "Project CORONA HARVEST End of Tour Report," 8–9; Osborn, "End of Tour Report," 2. Of note, Hill spent the next five years languishing as a prisoner of war in NVN while his front seater, Major James Low, the same former ace-lieutenant from the Korean War, opted for early release after only eight months—an act considered near-treason among the POWs.

204. Sokol, interview. Sokol told of one occasion in which his front seater, a former T-37 pilot training instructor, became spatially disoriented during a night mission and failed to pull up out of the dive bombing attack. Sokol took control and recovered the plane. Later, the front seater had the gall to tell Sokol that he "really knew what was going on" and didn't need the help. In contrast, some front seaters openly credited their GIB with saving their lives when they became disoriented; see McCarthy, *Phantom Reflections*, 46–50; Handley, *Nickel on The Grass*, 33–36.

205. Buhrow, Oral History Interview, n.d., 44.

206. Momyer to 8 TFW, 12 TFW, and 366 TFW, "F 4 Pilot Upgrading." Momyer's stated goal of upgrading three pilots per month proved overly ambitious.

207. Diefendorf, "End of Tour Report," 5–6.

208. L. Welch, interview, October 1, 2013.

209. Cook, *Once a Fighter Pilot*, 159–61.

210. Doerty, "End of Tour Report," 3; Beisner, "End of Tour Report," 3. Beisner reported that the 12th TFW upgrade program before August 1966 consisted of twenty-one sorties. See also *History of the 8th Tactical Fighter Wing, Oct–Dec 1967*, 1:11–13.

211. Osborn, "End of Tour Report," 2.

212. Buhrow, Oral History Interview, n.d., 44.

213. McCarthy, *Phantom Reflections*, 108; Anderegg, *Sierra Hotel*, 47; Sokol, interview. See also Olds, Oral History Interview, n.d., 82. One commander worried that the Air Force would not properly shepherd the "F-4 back-seaters who upgraded to the front seat and a second SEA tour. These men are the ones we need—the ones who go to war when war happens not when its [*sic*] their turn." Versurah, "Project CORONA HARVEST End of Tour Report," 5.

214. Anderegg, interview, April 9, 2013; Anderegg, interview, October 2, 2013. Although Anderegg volunteered for a second combat tour to earn a front-seat upgrade, he ended up going to Europe because there were no combat slots available.

215. Anderegg, interview, April 9, 2013; Anderegg, interview, October 2, 2013.

216. Sokol, interview.

217. Recorded in Attinello, *Air-to-Air Encounters in Southeast Asia: Account of F-4 and F-8 Events Prior to 1 March 1967*, 161. Attinello documented another front seater complaining after a mission, "Present back seaters spend too much time trying to get to the front seat, and not enough becoming truly proficient with the fire control system" (204). Of course, the corollary was that many front seaters had no interest in learning about or understanding the operations of the radar. There were also differing opinions of the arrangement. For example, Attinello recorded one front seater's preference for having a pilot in the back seat because a "pilot [in the backseat] knows what you are looking for in the front seat" (161). See also Nash, "Project CORONOA HARVEST End of Tour Report," 11; John J. Burns, Oral History Interview, June 5–8, 1984, 216–17; Robbins, Oral History Interview, 106.

218. Recorded in Attinello, *Air-to-Air Encounters in Southeast Asia: Account of F-4 and F-8 Events Prior to 1 March 1967*, 187.

219. Anderegg, interview, October 2, 2013.

220. Anderegg, *Sierra Hotel*, 40–41; Hendrickson, Oral History Interview, 14, 99; L. Welch, Oral History Interview, 89; Tuso, "Navigator's Log"; Robbins, Oral History Interview, 106–8; Ritchie, Oral History Interview, 30–34. Anderegg noted that the decision to put new pilots straight into the Phantom's front seat was made "under pressure from the Department of Defense" after McNamara expressed his opinion that the Air Force was not suffering from a pilot shortage as it claimed. On the later professionalization of the WSO-GIB in the Phantom, see Hendrickson, "GIB and ACM"; Larkins, "WSO in Air-to-Air"; Pratt, "GIB and the F-4."

221. Rasimus, *Palace Cobra*, 19.

222. Anderegg, interview, October 2, 2013.

223. Ibid.

224. Quoted in Scholin, "New Phlyers for the Phantom," 50.

225. Aust, "PACAF Tactics/Techniques Program (Bulletin #1)," Attachment 1:1, 3–4. The PACAF tactics bulletin based its recommendations on a recently conducted Navy test of its F-4Bs in a close-range environment. The Navy's summary report, included in the bulletin, noted that the central question of the brief test was, "Can the F-4B, notwithstanding the fact that someone forget to put guns in it, be utilized as an air superiority fighter and how does it

compare with other contemporary fighter A/C [aircraft] in this role?" The Air Force would conduct its own similar evaluation later as part of the Feather Duster program; see note 120. Visibility out of the Phantom's cockpit was considered atrocious compared to the unrestricted, bubble canopy of the F-86. Cook remarked, "Those fine folks in St. Louis built a great airplane, but obviously didn't think seeing out of it was a priority." Cook, *Once a Fighter Pilot*, 160. See also John J. Burns, Oral History Interview, June 5–8, 1984, 175; Robbins, Oral History Interview, 108; Anderegg, *Sierra Hotel*, 11.

226. Aust, "PACAF Tactics/Techniques Program (Bulletin #1)," Attachment 1:4; *8th TFW Tactical Doctrine*, March 1967, 82–84; *T.O. 1F-4C-2-19*, sect. 2, pp. 22–26; *F-4C Air Combat Tactics*, 20. The 8th TFW's tactics manual differed from the initial tactics bulletin only with regard to the radar's recommended clutter setting. When in "gyro nor" mode, the radar used inputs from the aircraft's inertial navigation system to remain oriented with respect to the earth's horizon.

227. *8th TFW Tactical Doctrine*, March 1967, 83.

228. Ibid., 82–84. The December 1967 edition of the tactics manual added that the front seater could "help accomplish a quicker lock-on by telling the pilot [GIB] the range of the target and if the target will be coming from top to bottom of the scope or vice versa." *8th TFW Tactical Doctrine*, December 1967, 132.

229. *8th TFW Tactical Doctrine*, March 1967, 81–84.

230. *T.O. 1F-4C-34-1-1*, sect. 1, p. 31; Michel, *Clashes*, 151–52.

231. Asay, Covington, and Stone, *History of Operations BOLO*, 40; Attinello, *Air-to-Air Encounters in Southeast Asia: Account of F-4 and F-8 Events Prior to 1 March 1967*, 59.

232. Aust, "PACAF Tactics/Techniques Program (Bulletin #1)," Attachment 1:4; *8th TFW Tactical Doctrine*, March 1967, 80–82; R. Scott, *F-4C Fighter Screen and Escort*, 9; *F-4C Air Combat Tactics*, 19–20; Hargrove, Oral History Interview, 30–31.

233. Olds, Oral History Interview, July 12, 1967, 39–40.

234. *8th TFW Tactical Doctrine*, March 1967, 80, 82. See also comments on communications difficulties in John J. Burns, Oral History Interview, June 5–8, 1984, 218; Attinello, *Air-to-Air Encounters in Southeast Asia: Account of F-4 and F-8 Events Prior to 1 March 1967*, 114–15, 134, 178, 204, 332.

235. Anderegg, interview, October 2, 2013; Sokol, interview. Ritchie was a big proponent of formed crews; see Ritchie, Oral History Interview, 25–34; Ritchie, Corona Ace Oral History Interview, 34–36.

236. Hendrickson, Oral History Interview, 42.

237. Recorded in Attinello, *Air-to-Air Encounters in Southeast Asia: Account of F-4 and F-8 Events Prior to 1 March 1967*, 134.

238. R. Scott, *F-4C Fighter Screen and Escort*, 3, 9. Scott's bulletin reminded Phantom aircrews that "maximum effort should be made to obtain long range radar contacts and establish an optimum attack position within the launch envelope for AIM-7 firing," even if "this ideal attack cannot be achieved in many cases and close-in fighting may become necessary." On the flexibility of the radar setup, the same bulletin noted that "the radar scope ranges used [in combat] vary with personal pilot preference."

239. McCarthy, *Phantom Reflections*, 66.

240. *8th TFW Tactical Doctrine*, March 1967, 75. The recommended steps for number 3 to set his radar elevation coverage were more cumbersome: "3-bar radar, 50-mile scope, adjusts

the antenna to obtain ½-inch of ground return on the lowest sweep and selects 1-bar scan, automatically raising the antenna to the desired elevation."

241. *T.O. 1F-4C-34-1-1*, sect. 1, p. 30; Sokol, interview.

242. Anderegg, interview, October 2, 2013.

243. Sokol, interview.

244. L. Welch, interview, October 1, 2013.

245. One Phantom instructor commented of the requirement for the GIB and his pilot to be able to "almost read each other's mind when they've got to work fast," but he failed to see the similar, figurative connection between the GIB and his radar. Quoted in Scholin, "New Phlyers for the Phantom," 50.

246. John J. Burns, Oral History Interview, June 5–8, 1984, 218.

247. Bakke, Oral History Interview, 11.

248. John J. Burns, Oral History Interview, June 5–8, 1984, 175.

249. Buhrow, Oral History Interview, April 11, 1968, 28–29. Olds commented, "I can't remember a single pilot that flew the F-4 that doesn't love it." Olds, Oral History Interview, n.d., 77.

250. Sokol, interview. Sokol was in the back seat reading a magazine when he heard the "noise change." He looked outside to discover that they were in "some big-ass dive." He then asked the front seater "what the hell are you doing?" only to discover that "he was asleep."

251. Anderegg, interview, October 2, 2013. See also John J. Burns, Oral History Interview, June 5–8, 1984, 189–90; Olds, *Fighter Pilot*, 250–51.

252. Sokol, interview; Hendrickson, Oral History Interview, 40–41. Hendrickson recalled that one of his duties as a GIB was to alert the pilot of the approaching "Mach tuck" so that he could reduce the pressure on the control stick and avoid overstressing the aircraft.

253. Buhrow, Oral History Interview, April 11, 1968, 28–29.

254. L. Welch, interview, October 1, 2013.

255. John J. Burns, Oral History Interview, June 5–8, 1984, 169. For a contrarian opinion, see Olds, Oral History Interview, n.d., 77–78.

256. The RADAR switch (located on the far-left side of the missile control panel) had three positions. The middle, STBY position "tuned" the Phantom's four Sparrow missiles to the aircraft's radar frequency. The up, PWR ON position activated the Phantom's special continuous wave (CW) radar illuminator, which was needed to help illuminate the enemy target and guide the Sparrow missiles. The 8th TFW guidance for using the top two positions of the switch was elaborate. After starting the engines, the front seater selected STBY to allow the CW transmitter to warm up. During taxi, the front seater then pulled the "forward missile bay firing control" circuit breaker and momentarily selected PWR ON to check if the Sparrow missiles would tune properly. After take-off, he again selected PWR ON for at least five minutes to prevent a "false tune" before returning the switch to STBY. At that point, he could reset the firing control circuit breaker. Then, if the mission was to a "high MIG threat" area, he was directed to position the switch back to PWR ON for the remainder of the flight "to prevent the possibility of the missiles de-tuning." *8th TFW Tactical Doctrine*, March 1967, 59–60.

The far right "interlock" (INTLK) switch on the panel was also used for the Sparrow missiles: with interlocks IN, the pilot was prevented from launching an AIM-7 if the Phantom's fire control system judged the aircraft not in an acceptable firing position, for example, if the radar range to the target was not within the computed range of the missile (between Ra and

Rmin) and/or if the aim dot was not within the allowable steering error (ASE) circle on the radar display. The 8th TFW tactics manual specified times to switch between interlocks IN and OUT. The interlock switch had no effect on Sidewinder missile launches. *8th TFW Tactical Doctrine*, March 1967, 80–81, 88–90. The ARM—SAFE switch in the center right of the panel activated the firing circuits for the Phantom's air-to-air weapons. All functions of the missile control panel are described in *T.O. 1F-4C-34-1-1*, sect. 1, pp. 19–20, 34; *T.O. 1F-4C-2-19*, sect. 2, pp. 35–36.

257. *T.O. 1F-4C-34-1-1*, sect. 1, pp. 20, 37–38; *T.O. 1F-4C-2-19*, sect. 2, pp. 35–36; Anderegg, *Sierra Hotel*, 11–12. The pilot could not control the firing sequence of the Sparrow missiles.

258. Quoted in Attinello, *Air-to-Air Encounters in Southeast Asia: Account of F-4 and F-8 Events Prior to 1 March 1967*, 142–44; Futrell et al., *Aces and Aerial Victories*, 27.

259. Anderegg, *Sierra Hotel*, 11–12; Sokol, interview; *8th TFW Tactical Doctrine*, March 1967, 61.

260. R. Scott, *F-4C Fighter Screen and Escort*, 10; *T.O. 1F-4C-34-1-1*, sect. 1, p. 33.

261. Blood, "End of Tour Report," 89.

262. Olds, Oral History Interview, July 12, 1967, 37, 40; *8th TFW Tactical Doctrine*, March 1967, 89.

263. Boyd, "Aerial Attack Study," 42. Boyd explained that "when employing [an] AIM-9B against a maneuvering target, the cone not only diminishes in size, it also changes in shape." On Boyd's "Aerial Attack Study," see Hammond, *Mind of War*, 46–48; Coram, *Boyd*, 103–5, 112–16, 119.

264. Michel, *Clashes*, 154–55; Dilger, Oral History Interview, 25–26; Kirk, Oral History Interview, 3; Olds, Oral History Interview, July 12, 1967, 37–38, 41; Olds, Oral History Interview, n.d., 69–70, 72–73; *8th TFW Tactical Doctrine*, March 1967, 84–88; Heffron, *Air-to-Air Encounters over North Vietnam, 1 Jan–30 Jun 1967*, 43–44. Attinello recorded one Phantom pilot's complaint that "having to fight with two missile envelopes (Sidewinder and Sparrow) complicates the fighter pilot's problem. A good fighter pilot thinks in terms of available envelopes." Attinello, *Air-to-Air Encounters in Southeast Asia: Account of F-4 and F-8 Events Prior to 1 March 1967*, 161.

265. *8th TFW Tactical Doctrine*, March 1967, 86.

266. Ibid., 88.

267. Greene to Commanding General, Continental Air Command, January 31, 1950.

268. Michel, *Clashes*, 105–6; Blesse, *Check Six*, 121.

269. Alden to 9 AF, "TACOP Final Report for 45 TFS," 5; Cook, *Once a Fighter Pilot*, 145.

270. Mehserle, *Capability of the F-4C Aircraft/SUU-16/A Gun Pod*, 36. TAC's November 1966 test was more pessimistic, declaring the SUU-16 "tactically unsatisfactory" for use in "an air-to-air role." *F-4C/SUU-16A*, 1.

271. Bakke, interview. Bakke remembered a demonstration flight for "a civilian official of influence" that was designed to impress upon him the undesirability of the external gun because of its excess drag and maneuvering limitations. The Air Force was pushing hard at this point to gain approval for the F-4E with its internal gun.

272. Blesse, *Check Six*, 113–14, 119–20; Blesse, Corona Ace Oral History Interview, 58–60. Blesse viewed missiles and guns as "complementary weapons," but he was unable to persuade "the missile people" who occupied the Pentagon in the late 1950s. One was Momyer, who reportedly told Blesse, "You goddamn fighter pilots are all alike. You get a couple of kills with a

gun and you think the gun is going to be here forever. Why can't you look into the future and see that the missile is here and the guns are out?"

273. Bakke, interview.

274. Blesse, *Check Six*, 121. Blesse lamented, "After the extremely capable A-1C radar computing gunsight we used 15 years earlier in Korea, it was difficult to understand how we could find ourselves in this situation."

275. Olds, Oral History Interview, July 12, 1967, 42.

276. 366 TFW to 7 AF, "AFR 57-4 Mod I Proposal"; Bakke, interview; *History of the 390th Tactical Fighter Squadron, May 1967*, 2; Bolt, Oral History Interview, 195–97; Olds, Oral History Interview, July 12, 1967, 42–44; Olds, Oral History Interview, n.d., 76. Blesse's wing commander, Bolt, took credit for the invention in his oral history interview, but Bakke in his interview attributed full credit to Blesse. Olds failed to attribute credit for the modification to either Blesse or the 366th TFW, but the 8th TFW may have also developed the modification independently; see additional references in Maloy to Burns; Kirk, Oral History Interview, 8–10. See also Fino, "Breaking the Trance."

277. 366 TFW to 7 AF, "F-4C Air to Air Modification Program."

278. *History of the 366th Tactical Fighter Wing, Apr–Sep 1967*, 1:3–4; Blesse, *Check Six*, 123; Michel, *Clashes*, 104.

279. Hargrove, Oral History Interview, 22–23. See also Futrell et al., *Aces and Aerial Victories*, 55.

280. Recorded in Attinello, *Air-to-Air Encounters in Southeast Asia: Events from 1 March 1967 to 1 August 1967*, 285. See also Futrell et al., *Aces and Aerial Victories*, 56–57.

281. 366 TFW to NMCC et al., "Change 1 to 366TFW OPREP-3, Number 010, FASTEL 443." See also 366 TFW to NMCC et al., "Change 1 to 366TFW OPREP-3 Number 011, FASTEL 442"; *History of the 366th Tactical Fighter Wing, Apr–Sep 1967*, 1:4–5; Attinello, *Air-to-Air Encounters in Southeast Asia: Events from 1 March 1967 to 1 August 1967*, 284–89.

282. Futrell et al., *Aces and Aerial Victories*, 68, 120–21.

283. 366 TACFTRWG to CSAF, "Presentation to Sec Def."

284. Olds, Oral History Interview, July 12, 1967, 42–43.

285. Kirk, Oral History Interview, 6–7; Olds, Corona Ace Oral History Interview, 25. Kirk noted: "Eventually we're going to have the E-model airplanes with the internal gun. That's the answer. . . . I think the Air Force has learned its lesson. We'll never build another fighter without an internal gun. I'm convinced of that, or at least I hope to G—we don't."

286. Hendrickson, Oral History Interview, 50. See also Payne, "Project CORONA HARVEST End of Tour Report," 4–5.

287. Dilger, Oral History Interview, 12–13.

288. Quoted in P. Davies, *USAF F-4 Phantom II MiG Killers, 1965–68*, 31.

289. Olds, *Fighter Pilot*, 317. See also Olds, Oral History Interview, n.d., 76–77. Indicative of Olds's initial concerns, Blesse remembered that when he was pitching his SUU-16 plan to General Momyer, Olds happened to be in the room. When Momyer asked for Olds's appraisal of the idea, Olds reportedly quipped, "General, I wouldn't touch that with a ten-foot pole!" Blesse, *Check Six*, 123, 125.

290. Futrell et al., *Aces and Aerial Victories*, 155–57. An F-105F and an F-4D shared credit for a MiG kill; hence, the ½ victory credit in the Phantom's final tally.

291. John J. Burns, Oral History Interview, March 22, 1973, 17.

292. Bugos, *Engineering the F-4*, 159.

293. Blesse, *Check Six*, 126.

294. Olds, Oral History Interview, n.d., 24 (ellipsis in original). Olds was specifically referring to his unit's practice of using the last fifteen minutes of each flight to rehearse various tactics.

295. *F-4C Air Combat Tactics*, Supplement No. 1. See also note 91.

296. Bedke, Oral History Interview, 9. Bedke continued, "We just went over [to Vietnam] and freelanced it. . . . We had been using old equipment; we had been using tactics that were acceptable with that equipment; and we just didn't have the foresight to look ahead." Anderegg and Wallace echoed Bedke's assessment in their interviews. Anderegg, interview, October 2, 2013; Wallace, interview.

297. Alden to 9 AF, "TACOP Final Report for 45 TFS," 5; *8th TFW Tactical Doctrine*, March 1967, 72–74. The 45th TFS initially had the two elements fly line-abreast at 6,000-feet distance. The wingmen were instructed to fly 1,500 to 2,000 feet away from their respective leader, slightly higher and at least ten degrees back from the line-abreast position. See also Anderegg, *Sierra Hotel*, 20–21; Michel, *Clashes*, 169–71.

298. *8th TFW Tactical Doctrine*, March 1967, 73. In the December 1967 version of the 8th's tactics manual, the "fighting wing" position was refined to be only a sixty-degree cone, and the altitude difference between the two Phantom elements in "fluid four" was increased from 1,000 feet to between 2,000 and 3,000 feet. *8th TFW Tactical Doctrine*, December 1967, 119–21.

299. Anderegg, *Sierra Hotel*, 21.

300. Michel, *Clashes*, 170–71; McCarthy, *Phantom Reflections*, 67; Anderegg, *Sierra Hotel*, 21; Anderegg, interview, April 9, 2013; Anderegg, interview, October 2, 2013; Sokol, interview; Ritchie, Oral History Interview, 14.

301. Michel, *Clashes*, 171–72; Anderegg, interview, April 9, 2013; Anderegg, interview, October 2, 2013. Anderegg noted in his October 2013 interview that determining the proper spacing between aircraft was essentially a simple trigonometry problem: "It's obvious that if you're out fighting in an environment where [an enemy] missile has a range of 6,000 feet, then you need to be 6,000 feet apart to see 6,000 feet behind the other guy [a calculation based on the Phantom's ability to only see forty-five degrees aft of line abreast]."

302. Anderegg, *Sierra Hotel*, 21. Regarding the difficulty of flying as a wingman in the fighting wing position, one Phantom wingman admitted following a successful engagement on April 29, 1966, that he became "completely disoriented during the various maneuvers and concentrated on staying on Lead's wing." During another engagement on May 8, 1966, two Phantoms collided, belly-to-belly: "Both aircraft were damaged and placed momentarily out of control, but they recovered." Both engagements are documented in Attinello, *Air-to-Air Encounters in Southeast Asia: Account of F-4 and F-8 Events Prior to 1 March 1967*, 187, 204.

303. Sokol, interview. Wallace echoed Sokol's assessment, explaining that in fighting wing, "you were . . . providing a shield between the lead and the guy [sneaking up from behind]." Wallace, interview.

304. Michel, *Clashes*, 171–72; Anderegg, interview, October 2, 2013; Sokol, interview.

305. Bedke, Oral History Interview, 41.

306. Riccioni, "New Method of Employing the F-100C"; Riccioni, "Double Attack System"; Anderegg, *Sierra Hotel*, 59–60.

307. Riccioni to Westover, "Double Attack System." For more on Riccioni, see Coram, *Boyd*, 237–38.

308. Riccioni, "Double Attack System."

309. Aerial Attack Section, 414 Fighter Weapons Squadron, "Anything Else Is Rubbish!"; Anderegg, interview, April 9, 2013; Anderegg, interview, October 2, 2013.

310. Aust, "PACAF Tactics/Techniques Program (Bulletin #1)," Attachment 1:7–8. See also Michel, *Clashes*, 169, 172. Two Navy F-4Bs were flying in "loose deuce" when they achieved the first Phantom victory of the Vietnam War on June 17, 1965, as recorded in Attinello, *Air-to-Air Encounters in Southeast Asia: Account of F-4 and F-8 Events Prior to 1 March 1967*, 47–49. Riccioni noted in his article on double attack that "the Navy uses the term 'loose deuce' for their version" of his formation. Riccioni, "Double Attack System," 27.

311. Bedke, Oral History Interview, 41–42. A July 1966 tactics bulletin on F-4C escort tactics suggested using "loose deuce" if flying over water and under control of ground-based (or sea-based) radar controllers. For all other situations, the bulletin declared "the normal formation is the 'fluid four.'" The difference could be attributed to the smaller chance of a surprise MiG attack when flying over the water. R. Scott, *F-4C Fighter Screen and Escort*, 1–2.

312. Sokol, interview.

313. Anderegg, interview, April 9, 2013; Anderegg, interview, October 2, 2013.

314. Bedke, Oral History Interview, 42.

315. Ritchie, Oral History Interview, 12–14, 55.

316. The USAF's Fighter Weapons School located at Nellis AFB was considered the "graduate" school of tactical air warfare and its instructors some of the top fighter pilots in the Air Force. Hence, the instructors there exercised considerable influence over the development of fighter hardware and tactics. Wilson, "PhD for Fighter Pilots."

317. Roy, "Double Attack Revisited"; Skaret, "Double Attack Revisited One More Time"; Der Baron, 414 Fighter Weapons Squadron, "Anything Else Is Rubbish!," Fall 1971.

318. Aerial Attack Section, 414 Fighter Weapons Squadron, "Anything Else Is Rubbish!," 34.

319. Michel, *Clashes*, 171; Anderegg, interview, October 2, 2013; P. Davies, *USAF F-4 Phantom II MiG Killers, 1972–73*, 8–9. Years later in a subsequent interview, Ritchie backtracked, claiming that he "would have preferred two airplanes—but the book said four [to a formation], and we couldn't get around that. So we used the elements as they were one airplane. An element, we used it as one, really. The wingmen were nothing but extra baggage most of the time." He also complained that he "spent 90 percent of my time over Hanoi trying to keep my wingmen from getting shot down." Ritchie, Corona Ace Oral History Interview, 18, 28–29.

320. *8th TFW Tactical Doctrine*, March 1967, 74; *8th TFW Tactical Doctrine*, December 1967, 121.

321. *8th TFW Tactical Doctrine*, March 1967, 97; *8th TFW Tactical Doctrine*, December 1967, 150; Kirk, Oral History Interview, 16–18; Olds, "End of Tour Report," 7; Olds, Oral History Interview, July 12, 1967, 28–29, 37–38.

322. *8th TFW Tactical Doctrine*, March 1967, 96; *8th TFW Tactical Doctrine*, December 1967, 149–50; McCarthy, *Phantom Reflections*, 67. See, for example, the description of the April 26, 1966, engagement in Attinello, *Air-to-Air Encounters in Southeast Asia: Account of F-4 and F-8 Events Prior to 1 March 1967*, 177–84.

323. Michel, *Clashes*, 172 (emphasis in original).

324. Anderegg, interview, October 2, 2013.

325. *8th TFW Tactical Doctrine*, March 1967, 28–37; *8th TFW Tactical Doctrine*, December 1967, 58–63; Kimball, Oral History Interview, 3–4; Olds, "Briefing to the Staff of Project CORONA HARVEST on Operation BOLO," 20–21; Anderegg, *Sierra Hotel*, 21–22; Anderegg, interview, October 2, 2013.

326. Aerial Attack Section, 414 Fighter Weapons Squadron, "Anything Else Is Rubbish!" The instructors would continue to use the pseudonym and title for their editorial column for several years. See, for example, Der Baron, 414 Fighter Weapons Squadron, "Anything Else Is Rubbish!," Fall 1971; Winter 1971; Fall 1972; and Summer 1975.

327. Ritchie, Corona Ace Oral History Interview, 69; Ritchie, Oral History Interview, 8–9, 41–44. Unfortunately, when the radio malfunctioned, maintenance crews had to remove the GIB's ejection seat to reach several critical radio components.

328. *T.O. 1F-4C-1*, sect. 1, pp. 40–52. Later radios included an option for the pilot to reprogram the preset radio frequencies in the jet. They also offered a wider range of manually tunable frequencies.

329. Kimball, Oral History Interview, 13; Olds, Oral History Interview, n.d., 72–74; Sokol, interview. Olds was discussing the AIM-4D Falcon "growl" during his interview, but there were similar difficulties hearing the Sidewinder over all the radio traffic.

330. Morris, Oral History Interview, 16.

331. Kimball, Oral History Interview, 11.

332. Bedke, Oral History Interview, 26–27. See also Olds, Oral History Interview, n.d., 33, 60–61.

333. Bedke, Oral History Interview, 28.

334. Kimball, Oral History Interview, 12.

335. Ritchie, Oral History Interview, 26; Attinello, *Air-to-Air Encounters in Southeast Asia: Account of F-4 and F-8 Events Prior to 1 March 1967*, 161, 309, 324.

336. Harmer and Anderegg, *Shootdown of Trigger 4*, 30–39. The transcript of Cadillac 1's mission on July 29, 1972, is included in Harmer's and Anderegg's report. The engagement was noteworthy because there was concern that Cadillac 1 might have shot down a friendly Phantom. Harmer and Anderegg concluded that the aircrew did not and instead shot a MiG. A more accessible mission transcript of Ritchie's fifth kill can be found in Futrell et al., *Aces and Aerial Victories*, 20–22. Unfortunately, Futrell's transcription includes only the transmissions relevant to the victory and excludes the numerous other extraneous transmissions that vied for the aircrew's attention.

337. Kimball, Oral History Interview, 13.

338. Ritchie, Corona Ace Oral History Interview, 31.

339. Harmer and Anderegg, *Shootdown of Trigger 4*, 3, 5; Sokol, interview.

340. Futrell et al., *Aces and Aerial Victories*, 110.

341. Hendrickson, Oral History Interview, 57.

342. Ibid., 46–47; Futrell et al., *Aces and Aerial Victories*, 110.

343. Futrell et al., *Aces and Aerial Victories*, 110.

344. Hendrickson, Oral History Interview, 44–47.

345. Anderegg, *Sierra Hotel*, 103; *T.O. 1F-4C-34-1-1*, sect. 1, pp. 70–72; *F-4C Intercept*, 91–94.

346. "Advantage Is Here: How the Phantom Records What the Radar Sees."

347. Ibid.

348. Anderegg, interview, October 2, 2013; Sokol, interview.

349. Olds, "Briefing to the Staff of Project CORONA HARVEST on Operation BOLO," 62. Olds also complained that the Phantom had no adequate capability to record its bombing attacks. Olds, "End of Tour Report," 31–32.

350. Sexton and Jameson, "New F-4 Gunsight Camera," 13.

351. Anderegg, *Sierra Hotel*, 103–4.

352. Futrell et al., *Aces and Aerial Victories*, 121, 139.

353. Ibid., v.

354. Olds, "Briefing to the Staff of Project CORONA HARVEST on Operation BOLO," 62.

355. Unlike in Korea, the Air Force had no tolerance for border violations. A wayward EB-66 aircraft, escorted by pairs of Phantoms, twice violated Chinese airspace in mid-1966. On one of the occasions, the Phantoms shot down a Chinese MiG that had come to intercept them. The Air Force, under heavy pressure from "high Washington levels," implemented measures to minimize the chances of future border excursions. See Van Staaveren, *Gradual Failure*, 274–77. One of the more notorious breakdowns in aircrew discipline occurred when a flight of F-105s "accidentally" strafed a Russian ship in Haiphong harbor. The situation was exacerbated when the aircrews' base leadership tried to conceal the incident by destroying the Thud pilots' gun camera film. When an investigation by General John Ryan uncovered the malfeasance, the base's acting wing commander, Colonel Jack Broughton, was summarily sacked and court-martialed. For more details, see Ryan, Oral History Interview, Tape 2:62–64; Olds, Corona Ace Oral History Interview, 86–90; Olds, *Fighter Pilot*, 320–22; Correll, "Man from Thud Ridge."

356. Sokol, interview.

357. Vandenberg, Oral History Interview, 145.

358. J. Sherwood, *Fast Movers*, 216. On the liberal application of the "ace" moniker, see, for example, "Air Force Phantom Crews Get Two More"; "'Ace' Fighter Wing Commanded by 'Aces'"; *History of the 366th Tactical Fighter Wing, Oct 1966–Mar 1967*, 1:vii.

359. Futrell et al., *Aces and Aerial Victories*, 52, 59–60, 134.

360. Olds, Corona Ace Oral History Interview, 45–47; Olds, *Fighter Pilot*, 318–20.

361. Michel, *Clashes*, 214; Blumenthal, "First 2 Aces of Vietnam War."

362. Ritchie, Oral History Interview, 45–46, 49–58; Ritchie, Corona Ace Oral History Interview, 90–97; Futrell et al., *Aces and Aerial Victories*, 93, 96, 98–99, 102. DeBellevue was away on leave when Ritchie scored his second kill.

363. Quoted in Futrell et al., *Aces and Aerial Victories*, 20–22, 103. See also Ritchie, Oral History Interview, 47, 60–62; Deen, "How Steve Ritchie Got His Fifth MIG."

364. Olds, Corona Ace Oral History Interview, 118–19. Olds added that Ritchie's only flaw was that "he thinks he is god's gift to [fighter aviation]. . . . Steve has a monstrous ego."

365. Ritchie, Corona Ace Oral History Interview, 1–11; J. Sherwood, *Fast Movers*, 217–19.

366. Olds, Corona Ace Oral History Interview, 44; Olds, *Fighter Pilot*, 379.

367. Ritchie, Corona Ace Oral History Interview, 62.

368. Ritchie, Oral History Interview, 4, 31–32.

369. Ritchie, Corona Ace Oral History Interview, 15–16, J. Sherwood, *Fast Movers*, 243–45.

370. Ritchie, Corona Ace Oral History Interview, 35, 38; Futrell et al., *Aces and Aerial Victories*, 103.

371. Deen, "How Steve Ritchie Got His Fifth MIG," 28.

372. Quoted in ibid.

373. "US Air Force Academy Jabara Award"; "National Aeronautic Association Mackay Trophy"; Futrell et al., *Aces and Aerial Victories*, 103–5, 109.

374. Futrell et al., *Aces and Aerial Victories*, vi. Anderegg suggested that the impetus for the policy change can be traced to GIB 1st Lt. John D. Ryan Jr., the general's son. Apparently, when John Jr.'s front seater used an AIM-4D Falcon to down a MiG on December 17, 1967, General Ryan, then serving as the commander of PACAF, called to congratulate his son on becoming a "MiG killer." However, John Jr. reminded his father of the policy and explained that he wouldn't receive full credit for the kill. It is also possible that the policy change was made simply to mirror that used by the Navy. Anderegg, interview, October 2, 2013.

375. L. Welch, Oral History Interview, 90.

376. Everest, Oral History Interview, 14.

377. John J. Burns, Oral History Interview, June 5–8, 1984, 219. Creech claimed that he didn't "have any problem" with the arrangement. Creech, Oral History Interview, 54.

378. Blumenthal, "First 2 Aces of Vietnam War."

379. "First Air Force Ace in Indochina."

380. "Air Force Captain Gets 5th MIG."

381. "Saigon Reports Sharp Fighting."

382. "American Fighter Aces."

383. Olds, Corona Ace Oral History Interview, 22; Blesse, Corona Ace Oral History Interview, 17.

384. *8th TFW Tactical Doctrine*, March 1967, v.

385. Ibid., 72.

386. Olds, Oral History Interview, July 12, 1967, 46.

Chapter 5. Eagles over Nellis: "A Fighter Pilot's Fighter"

1. "No-Win War at Dogbone Lake"; Kaylor, "Mock Dogfights Test Latest Jet Fighters"; Everson, *AIMVAL Final Report*, 5, 7; Anderegg, *Sierra Hotel*, 158–59. McDonnell Aircraft merged with Douglas Aircraft in April 1967, creating McDonnell Douglas.

2. Mathis, "F-15 Air Superiority Fighter"; D. Wells, "Sparrow Missile Program Review," 5000–13.

3. "No-Win War at Dogbone Lake"; Kaylor, "Mock Dogfights Test Latest Jet Fighters."

4. Everson, *AIMVAL Final Report*, 13–14; Tissot and Hildreth, *AIMVAL Final Report*, 1: sect. 3, pp. 1–2.

5. Everson, *AIMVAL Final Report*, 7–9, 13–14; Anderegg, *Sierra Hotel*, 108–9. ACMI is discussed in detail later in this chapter.

6. Glasser, "Research, Development, Test, and Evaluation," 357–60; Currie, "Statement of the Director of Defense Research and Engineering," 455–56; W. Evans, "Lightweight Fighter Prototype," 1021–23.

7. "New Planes Designated for Nellis"; "Nellis AFB Squadron Get First F-5E Tiger Fighters." The Aggressor squadrons were comprised of specially trained Air Force pilots who mimicked Soviet tactics during training events with other Air Force units. See Burns, "Aggressively Speaking"; Nabors, "Aggressor Squadron Briefing," 4907–20; Anderegg, *Sierra Hotel*, 76–88; Laslie, *Air Force Way of War*, 44–46; Michel, "Revolt of the Majors," 146–52; "'Operation Red Flag' Studies"; Hillinger, "'Russian Squadrons' Fly Mock Battles"; Galatz,

"Old West Scene." Knight deemed the Aggressor squadrons "the most significant item in the development of modern air-to-air combat capability." Knight, Oral History Interview, 208.

8. Anderegg, *Sierra Hotel*, 150; *F-15 Eagle*, 1–2; *USAF F-15 Fighter Summary*, 4, 10; *SSAC Proposal Analysis*, 8.

9. *F-15 Eagle*, 1–2; *USAF F-15 Fighter Summary*, 12. The Eagle's "complex inlet" mechanism drew a wary eye from Air Force engineers when it was first proposed; see *SSAC Proposal Analysis*, 8. The F-15's F100 engines experienced significant problems early in the program. Sokol remembered, "They'd bang. They'd pop. They'd stall. They'd stag[nate]." Sokol, interview. The engine problems were generally not related to the inlets but to the fuel control systems; see information on the early engine problems in E. Lewis, *History of the F-15 Eagle*, 94–101; Sullivan, *History of the 58th Tactical Fighter Training Wing, Apr–Jun 1975*, 36–37, 65–71; Werrell, *Chasing the Silver Bullet*, 68–69.

10. Cost figures for aircraft are always troublesome due to changing requirements, procurement schedules, and other variables. The F-15 and F-5E costs cited here are based on Air Force congressional testimony delivered March 13, 1974: at that time, the F-15 unit program cost was $10.5 million; the F-5E had a unit flyaway cost closer to $2.5 million. Bellis, "F-15 Program Review," 4267–68; Stringer, "F-5 Program Review," 4296.

11. Hoag, "Superfighter," 21. The "fighter pilot's fighter" was an oft-used expression for the F-15. For example, Meyer referred to the F-15 as "a fighter pilot's fighter—a modern-day version of the F-86 MiG-killer." J. Meyer, "Air Superiority into the 1980s," 43. See also Bowman, "To Fly and Fight," 10; Osterhout, Berdan, and Gordon, "F-15 Briefing Script," slide 3; "F-15: A Fighter Pilot's Fighter"; "USAF F-15 Eagle: The Fighter Pilot's Fighter"; "Four Reasons Why the F-15 Is a Fighter Pilot's Fighter"; K. Brown, "Fighter Pilot's Fighter Plane."

12. L. Welch, interview, October 1, 2013.

13. *F-15 Eagle Fact Sheet*, February 1973, 2; Werrell, *Chasing the Silver Bullet*, 69–70.

14. Osterhout, Berdan, and Gordon, "F-15 Briefing Script," slide 9.

15. D. Evans, *Sabre Jets over Korea*, 234.

16. "No-Win War at Dogbone Lake," 56.

17. Ibid.; Dvorchak, interview; Sparks, interview; Anderegg, *Sierra Hotel*, 159; Guillermin, *Towering Inferno*. Some news reports mistakenly termed the infamous four-against-four engagement the "Towering Infernal"; see Kaylor, "Mock Dogfights Test Latest Jet Fighters"; "Air Force Tried Out Controversial Planes."

18. "No-Win War at Dogbone Lake," 56. The conclusions reached in "No-Win War at Dogbone Lake" became a topic of discussion during congressional testimony; see Fay, "AIMVAL-ACEVAL Program Review," 1978, 5221–23.

19. Blesse, "Changing World of Air Combat," 34, 37.

20. Ibid. Blesse further noted, "I, myself, would not like to try to become a jet ace again without those capabilities." See also his similar assessments in "No-Win War at Dogbone Lake," 57; Blesse, Corona Ace Oral History Interview, 72–74.

21. B. Watts, "Changing World of Air Combat," 35 (emphasis in original). Earlier in the article, Watts argued that "*the capability to fight in close, relying on aircraft performance and the traditional fighter maneuvers, is still not necessarily antiquated*. . . . Even if you yourself can shoot opponents twenty or thirty miles away, in crowded skies you are still going to have to be able to handle close-in attacks from two or three" (34, emphasis in original).

22. In assessing the F-15's impact on the changing patterns of human-machine interaction within fighter aviation, I break from the pattern of the prior chapters and focus on the AIMVAL-ACEVAL *tests* rather than a wartime experience. This shift is necessary because the first combat test of the American Eagles, by then F-15Cs, did not come until 1991 during Desert Storm. Since the F-15C is still in active service today, many of its manuals remain classified. After the F-15C was fielded in 1979, the original F-15 model became known as the F-15A, and the original two-seat TF-15 was renamed the F-15B.

23. Werrell, *Chasing the Silver Bullet*, 60; Neufeld, *F-15 Eagle*, 6; "Lieutenant General Arthur C. Agan." Neufeld's original study of the F-15 was classified. It has since been declassified, but an unclassified "sanitized copy" was also produced in 1974. All references herein are to the original classified version.

24. Agan, Oral History Interview, February 1970, 11; Agan, Oral History Interview, October 2, 1973, 14–16, 32; Michel, "Revolt of the Majors," 70–72. The "F" designation was reserved for fighters; the "A" designation was used for ground-attack aircraft. Myers, a government consultant on tactical air requirements, had a similar assessment of the F-111; see Myers, Oral History Interview, 9–11. On Myers's contributions to the F-X discussions, see Neufeld, *F-15 Eagle*, 8.

25. Agan, Oral History Interview, October 2, 1973, 7, 15–17, 22, 36; J. Burns, Oral History Interview, March 22, 1973, 1–4; Georgi, Oral History Interview, 14–15; Myers, Oral History Interview, 31–32; Neufeld, *F-15 Eagle*, 7–8; Werrell, *Chasing the Silver Bullet*, 60–61.

26. "General John P. McConnell"; Gillespie, "Mission Emphasis," 167–68.

27. Neufeld, *F-15 Eagle*, 8–10; Werrell, *Chasing the Silver Bullet*, 60–61. Bohn was specifically tasked with assessing ground attack aircraft, but he openly questioned his guidance to "assume" air superiority in the target areas. In his conclusions, as paraphrased by Neufeld, Bohn recommended "the Air Force acquire a mix of high- and low-cost aircraft as the most economical way to strengthen the tactical force." Bohn's high-low construct would define the Air Force's procurement strategy for the next several decades; see J. Burns, Oral History Interview, March 22, 1973, 24–25.

28. Myers, Oral History Interview, 25–26; Neufeld, *F-15 Eagle*, 8, 11–12.

29. Neufeld, *F-15 Eagle*, 11, 17; Werrell, *Chasing the Silver Bullet*, 62; Michel, "Revolt of the Majors," 72; Slay, "Tactical Air Programs," 4811–12. For years, the service had emphasized destroying enemy fighters "on the ground and in [their] nest" using tactical nuclear bombs; see Futrell, *Ideas, Concepts, Doctrine, 1907–1960*, 301.

30. O'Donohue, "Development Concept Paper," 1. On the writing of the "Development Concept Paper," see Rhodarmer, Oral History Interview.

31. J. Burns, Oral History Interview, March 22, 1973, 33–36; Kent, Oral History Interview, 6–9; Neufeld, *F-15 Eagle*, 12–13, 15–17; Werrell, *Chasing the Silver Bullet*, 62–66; Gillespie, "Mission Emphasis," 182–90.

32. The announcement was broadcast servicewide via message traffic: USAF to ALMAJCOM, December 24, 1969. See also Neufeld, *F-15 Eagle*, 48; Werrell, *Chasing the Silver Bullet*, 66. McDonnell Douglas won the competition handily, besting its nearest rival by more than 10 percent in the scoring metrics; see *SSAC Proposal Analysis*, 17. On McDonnell Douglas's submission, initially identified as its "Model 199," see *F-15 Eagle*.

33. Middleton, "Air Force Unveils Fighter," 1; E. Lewis, *History of the F-15 Eagle*, 35. Video of Iris Hansen's christening is included in "Boeing F-15 Eagle."

34. Quoted in "F-15 Rolls Out at St. Louis," 11.

35. Werrell, *Chasing the Silver Bullet*, 63; Gillespie, "Mission Emphasis," 214.

36. Hildreth, Oral History Interview, 59–63; J. Burns, Oral History Interview, March 22, 1973, 24–26; Kent, Oral History Interview, 3, 6; Agan, Oral History Interview, October 2, 1973, 12–13, 18–21; Neufeld, *F-15 Eagle*, 10, 13–14; Werrell, *Chasing the Silver Bullet*, 62. Hildreth was TAC's project officer for the A-7. Neufeld cited Agan's explanation that he endorsed the A-7 decision so that the "F-X could now be justified as a 'more sophisticated, higher performance aircraft, . . . an air superiority replacement for the F-4.'"

37. McGough cited both criticisms in his Oral History Interview, 6–7.

38. Neufeld, *F-15 Eagle*, 66–68; Werrell, *Chasing the Silver Bullet*, 65; Rhodarmer, Oral History Interview, 33–36. Countering the Soviet Foxbat's impressive altitude and top-speed capabilities became a major emphasis for F-15 designers and tacticians. See discussions in Osterhout, Berdan, and Gordon, "F-15 Briefing Script," slides 29–33; *F-15 Eagle*, vi–vii. During testing, the F-15 practiced intercepts against an unidentified "target flying at Mach 2.8 to 3.0 at 80,000 feet" to validate its expected performance and tactics against the Foxbat; see Flood and Rodero, *F-15 Initial Operational Test and Evaluation—Air-to-Air*, Annex A:94–101. The fact that the F-15 was the only American fighter deemed capable of tangling with the Foxbat became a source of pride for the engineers at McDonnell Douglas; one company advertisement noted: "The USAF F-15, the fighter pilot's fighter, won't be out-foxed in the air battle arena."

39. Quoted in Neufeld, *F-15 Eagle*, 25–26. See also J. Burns, Oral History Interview, March 22, 1973, 34–35; Rhodarmer, Oral History Interview, 32–33.

40. O'Donohue, "Development Concept Paper," 5. Another report from August 1970 noted that "many F-15 aircraft characteristics which are essential to air-to-air combat are equally as important to air-to-ground combat." Holbury, *Concept of Operations for the F-15*, 13–14.

41. Kent, Oral History Interview, 6–7. See other references to the F-15's ground attack potential in *USAF F-15 Fighter Summary*, 15; *F-15 Eagle Fact Sheet*, 1976, 5.

42. *F-15 Eagle*, 2, 9 (emphasis in original).

43. Flood and Rodero, *F-15 Initial Operational Test and Evaluation—Air-to-Ground*, v.

44. Mathis to AFSC/SDNJ, "F-15 Attack Version"; Chain to TAC/DOO and CSAF/XOO, "TAC 1 TFW DOC"; L. Welch, interview, July 10, 2012; E. Lewis, *History of the F-15 Eagle*, 72, 83–84. The first F-15 pilots converting from other TAC fighters all received training in ground attack: sixteen hours of academics and three sorties. In contrast, they received forty-two hours of academics and thirteen sorties on "aerial attack." *USAF Conversion Course, F-15*, 1–2, 21, 47–49.

45. On interservice rivalry as a source of innovation within the military, see Coté, "Politics of Innovative Military Doctrine."

46. O'Donohue, "Development Concept Paper," 5–7; Neufeld, *F-15 Eagle*, 24, 26, 33, 67–68; Rhodarmer, Oral History Interview, 19–20. The F-14's variable geometry wings facilitated its long loiter–quick sprint capability. The Tomcat could carry up to six Phoenix missiles, but with each pair of missiles weighing substantially more than a Volkswagen Beetle, aircrews usually carried only four Phoenixes under the fuselage. The F-14-Phoenix combination was advertised as possessing a "multi-shot" capability against distant targets, whereby a single F-14 could launch multiple Phoenix missiles against multiple targets, but the targets all needed to be flying within a small, prescribed volume of airspace. It was understood that a dedicated RIO would be required to orchestrate the "multi-shot" targeting, especially when encountering the

ECM that was expected to be used by the likely Soviet bomber targets. Grumman won the contract for the F-14 in February 1969.

47. "Talking Paper on 'Dr. DuBridge Letter to Mr. Packard,'" 1.

48. Neufeld, *F-15 Eagle*, 23–24, 68; Rhodarmer, Oral History Interview, 5, 17–19; Flax, Oral History Interview, 26–28. Flax served as the chief scientist for the Air Force and later as president of the Institute for Defense Analyses (IDA).

49. "Talking Paper on 'Dr. DuBridge Letter to Mr. Packard'"; "Suitability of the F-14B for the F-15 Mission"; "Why TAC Cannot Use the Navy F-14B"; "F-14B versus F-15"; "F-15: 'F-14A/B with Phoenix Missile.'" See also Rhodarmer, Oral History Interview, 19–24. The F-X concept paper declared, "There is no substitute for fighter maneuverability, when evading a tail attack by the enemy, or when trying to gain an advantageous position for a kill, in close combat," and it referenced Boyd's energy-maneuverability (E-M) theory and Ps (specific power) curves to illustrate the anticipated performance advantage of the F-X relative to current and future Soviet threats. O'Donohue, "Development Concept Paper," 5. Neufeld described Boyd's E-M theory as "a revolutionary analytical technique that permitted fighter 'jocks' to communicate with engineers"; Boyd himself described it as a "concept" that "provide[d] a capability for describing aircraft performance in general, and maneuverability in particular, with more precision." Neufeld, *F-15 Eagle*, 18–19; Boyd and Christie, "Expanded Maneuverability Theory." On Boyd's E-M theory and its relationship to the F-15, see also Hammond, *Mind of War*, 51–82; Coram, *Boyd*, 135–53, 188–231; Werrell, *Chasing the Silver Bullet*, 60–61, 63–65; Gillespie, "Mission Emphasis," 208–13; Agan, Oral History Interview, October 2, 1973, 10; Boyd, Oral History Interview, 1–24; Flax, Oral History Interview, 19–20; Georgi, Oral History Interview, 11–12; Myers, Oral History Interview, 15–16, 22–24, 29–30, 38; Sprey, Oral History Interview; L. Welch, Oral History Interview, 171. Welch, working in the Air Force's Studies and Analyses branch at the time, was able to incorporate Boyd's E-M calculations into his TAC Avenger air combat simulation program to help quantify the F-X's advantages. L. Welch, interview, July 10, 2012. On TAC Avenger, see Witze, "TAC AVENGER Is Out to Win"; Kent, Oral History Interview, 11–15; L. Welch, Oral History Interview, 127–28.

50. "Talking Paper on 'Dr. DuBridge Letter to Mr. Packard,'" 2–3.

51. Osterhout, Berdan, and Gordon, "F-15 Briefing Script," slide 37.

52. Ibid. The F-15 "spread the word" briefing is referenced in E. Lewis, *History of the F-15 Eagle*, 30. The "unjammable" reference in the briefing alluded to the propensity for radar-guided missiles to be decoyed by ECM, not necessarily the gun's mechanical functionality and reliability. One official explained in an interview, "The general philosophy on the F-15 is that it's primarily a gun-fighter. Its mission is to get the enemy with the gun, but incidentally while you're getting into position for a gun pass, if it's convenient, fire a missile." D. Davis, Oral History Interview, May 22, 1973, 9.

53. E. Lewis, *History of the F-15 Eagle*, 15.

54. Ibid., 15, 45–46; O'Donohue, "Development Concept Paper," 13; D. Davis, Oral History Interview, May 22, 1973, 10–12; J. Burns, Oral History Interview, March 22, 1973, 12–13.

55. E. Lewis, *History of the F-15 Eagle*, 71–72; Osterhout, Berdan, and Gordon, "F-15 Briefing Script," slide 36; Ulsamer, "Accent Is on Flying," 29–30; *F-15 Eagle Fact Sheet*, February 1973.

56. *F-15 Eagle*, 8; *USAF F-15 Fighter Summary*, 14. The up-cant on the Eagle's gun made it more difficult to employ in an air-to-ground role since the aircraft needed to use a steeper dive angle to shoot a ground target.

57. O'Donohue, "Development Concept Paper," 10.

58. E. Lewis, *History of the F-15 Eagle*, 9–11. Others were also starting to question the wisdom of the F-15 and its complex avionics, notably defense analyst Pierre Sprey and Riccioni, who was then working in the Pentagon on a new lightweight fighter design. The criticisms would eventually contribute to the AIMVAL-ACEVAL tests. See also Kent, Oral History Interview, 15–17; L. Welch, Oral History Interview, 147–49, 155; McIntosh, Oral History Interview, 33–34; Rhodarmer, Oral History Interview, 15, 28–29; Flax, Oral History Interview, 16; Sprey, Oral History Interview; Neufeld, *F-15 Eagle*, 64–66; Coram, *Boyd*, 231–56; Hammond, *Mind of War*, 83–91.

59. *F-15 Avionics and Armament Mission Requirements*, v, xv; E. Lewis, *History of the F-15 Eagle*, 10; McIntosh, Oral History Interview, 20–22; Rogers, Oral History Interview, 35–36. The F-15's planned EOIATS, which the pilots referred to in their report as TISEO based on its predecessor in the F-4E, had earned praise during the source selection competition: "[It] significantly exceeds the minimum requirements and will increase the effectiveness of this [aircraft] system." *SSAC Proposal Analysis*, 11. Blesse was not a fan of scrapping EOIATS on the Eagle, complaining that the Air Force's "fuzzy thinkers" were willing to "pay 15 million goddamn dollars for the F-15" but wouldn't cough up "another $170,000 per airplane to put the TISEO in it." Blesse, Corona Ace Oral History Interview, 82.

60. *Examination of the Utility of the F-15 Attack Radar*, 1, 10–11. A study of the F-15's concept of operations used similar reasoning to justify the AIM-7F on the Eagle, arguing that the radar-guided missile "will allow the F-15 to extract a penalty from the enemy . . . who is reluctant to close for maneuvering combat with the superior performing F-15." Holbury, *Concept of Operations for the F-15*, 6.

61. Osterhout, Berdan, and Gordon, "F-15 Briefing Script," slide 18, 41.

62. *F-15 Eagle Fact Sheet*, 1976, 3–6. On the cockpit displays, see also Osterhout, Berdan, and Gordon, "F-15 Briefing Script," slide 43.

63. Sullivan, *History of the 58th Tactical Fighter Training Wing, Oct–Dec 1974*, 10–11. President Gerald Ford was on hand at Luke AFB to welcome the F-15 into active Air Force service, proclaiming: "This is the month of the pioneer in America. It is the month of the *Mayflower* and our earliest settlers; and this is the day of a new pioneer—a pioneer of peace—the F-15 fighter. There were 102 passengers on the *Mayflower*. The crossing from England to the new world took more than two months. And the end of the journey is freedom. The F-15 can fly across that same Atlantic route today in a matter of hours. The purpose of its journey is still that of the *Mayflower* more than 350 years ago: freedom. That is what really matters—the purpose of a journey. And I am here today to underscore to you and the world that this great aircraft is constructed by the American people in pursuit of peace. Our only aim—with all of this aircraft's new maneuverability, speed and power—is the defense of freedom." TAC/OII to AIG/OI, "AFNS Release, 'President Welcomes.'"

64. *History of the 1st Tactical Fighter Wing, Oct–Dec 1975*, 1:9–18; Bloomcamp to 1 TFW/CCH, "History: 1 October–31 December 1975." The Air Force had decided to relocate the 1st TFW from MacDill AFB to Langley AFB, the site of TAC headquarters and also much closer to the politicians in Washington DC, before the 1st's first Eagles arrived. On the 1st TFW move, which was effective July 1, 1975, see Knight, Oral History Interview, 213; *History of the Ninth Air Force, Jul–Dec 1975*, 1:57–65; *History of the Ninth Air Force, Jan–Dec 1976*, 1:14–19. On the lineage of the 1st TFW as an air superiority wing, see *History of the*

1st Tactical Fighter Wing, Jul–Sep 1975, 1:2–3; "1st TFW to Nest Eagles"; "Building an F-15 Wing."

65. Bloomcamp to 9 AF/DOST, "Helmet Color in the F-15"; Leming to 9 AF/DOST, "Helmet Color in the F-15." Welch affirmed that the Eagle pilots' focus while he was the commander at Langley was on "maneuvering combat. . . . The reason for the airplane was superiority in maneuvering combat." L. Welch, interview, October 1, 2013.

66. Osterhout, Berdan, and Gordon, "F-15 Briefing Script," slide 2.

67. Gillespie, "Mission Emphasis," 225.

68. O'Donohue, "Development Concept Paper," 7.

69. Glasser, "Research, Development, Test, and Evaluation," 366–67. Glasser further noted during his testimony that because the F-15 had only a single air-to-air mission, "it does not have all of the complexities in the avionic equipment that one associates with the multipurpose aircraft that we have become familiar with."

70. O'Donohue, "Development Concept Paper," 7.

71. Ibid.

72. *F-15 Eagle*, viii, 16.

73. *USAF F-15 Fighter Summary*, 22. Flax took credit for encouraging the service to build a two-seat version of the F-15. Flax, Oral History Interview, 17–18, 32–33.

74. Olds, "Lessons of Clobber College," 22.

75. Ritchie, Oral History Interview, 29 (emphasis in original).

76. O'Donohue, "Development Concept Paper," 7. Flax acknowledged the "philosophical arguments" that were often mustered in the debate over the Eagle's single- versus two-seat cockpit. Flax, Oral History Interview, 17, 32.

77. Cited in Gillespie, "Mission Emphasis," 224. Risner acknowledged the pilots' preference for a single-seat cockpit in the F-15: "A typical fighter pilot wants to do everything himself, make his own decisions, and be responsible for his actions. He doesn't want any more crew." Risner, Oral History Interview, 158.

78. Hendrickson, Oral History Interview, 51.

79. W. Flanagan, "The Fighter Force." In his article, Flanagan urged "military decision-makers" to "give serious consideration to converting the F-15s to two-seat aircraft"; the service eventually would with the air-to-ground focused F-15E Strike Eagle, but the service's existing air superiority F-15A/C aircraft remained single seat.

80. Hendrickson, Oral History Interview, 51; Anderegg, *Sierra Hotel*, 152, 163. Hendrickson continued to work diligently to improve the professionalization of the WSO-GIB in the Phantom; see, for example, Hendrickson, "GIB and ACM."

81. *F-15 Avionics and Armament Mission Requirements*, iv, vi.

82. Ibid., 125–26. The enthusiasm is also apparent in Osterhout, Berdan, and Gordon, "F-15 Briefing Script," slide 10. Hughes Aircraft advertisements noted that the APG-63 radar achieved its relatively clutter-free display through a combination of "a low sidelobe antenna, a guard receiver, and frequency rejection of both ground clutter and ground moving targets." *Eyes of the Eagle*, 2. Of course, the radar still occasionally displayed false targets and lost real ones in clutter or the Doppler notch; see assessments in Flood and Rodero, *F-15 Initial Operational Test and Evaluation—Air-to-Air*, Annex A:39, 83. On the advantages of a PD radar, see Stimson, *Introduction to Airborne Radar*, 15–30. Welch recalled that the F-15's PD radar was championed by "this little lowly major [who] argued vehemently for a . . . look-down,

shoot-down pulse-Doppler radar. We [fighter pilots] were aghast because that meant a thirty-six-inch flat plate dish in the nose of an F-15. You know what a thirty-six-inch dish in the nose does to acceleration? He won. He was persistent, he was analytic, his logic prevailed, and he won." L. Welch, interview, July 10, 2012; L. Welch, interview, October 1, 2013; L. Welch, Oral History interview, 238.

83. E. Lewis, *History of the F-15 Eagle*, 11.

84. Neufeld, *F-15 Eagle*, 61–62.

85. *Eyes of the Eagle*, 2. The brochure also touted the air-to-ground modes of the APG-63, including "target ranging for automated bomb releases in visual attacks" and "a mapping mode for navigation."

86. Olds, Oral History Interview, July 12, 1967, 39–40; Sokol, interview; Anderegg, *Sierra Hotel*, 157. Anderegg noted of the F-15: "Some [pilots] called the constant movement of switches on the stick and throttle 'playing the piccolo.' Others joked that before one could become an Eagle pilot, one had to take piccolo lessons."

87. Hoag, "Superfighter," 22, 25.

88. Carson, "Flying the F-15," 26.

89. Anderegg, *Sierra Hotel*, 157.

90. *F-15 Eagle*, 6–7; *USAF F-15 Fighter Summary*, 7.

91. *T.O. 1F-15A-1*, sect. 1, p. 28. The hydromechanical system incorporated two additional advancements. The first was a mechanical pitch trim compensator that, as described by the flight manual (*T.O. 1F-15A-1*), automatically trimmed the stabilators "without affecting stick position to compensate for . . . changing speed, operating flaps or speed brake, or [weapons] separation." McDonnell Douglas noted that the aircraft's auto-trim capability "greatly enhances gunnery tracking capability and helps inhibit pilot induced oscillation." *F-15 Eagle*, 7. The hydromechanical system also used a "mechanical aileron-rudder interconnect," which adjusted the commanded rudder deflection based on "longitudinal stick position. With the control stick aft of neutral, lateral stick motion causes rudder deflection in the same direction as stick motion. With the control stick forward of neutral, lateral stick motion causes rudder deflection in the opposite direction of the stick motion." *T.O. 1F-15A-1*, sect. 1, p. 29. See also Toliver, *F-15A Follow-On Operational Test Final Report*, Annex A:5.

92. *T.O. 1F-15A-1*, sect. 1, p. 30. CAS did not control the ailerons; any roll commands generated by the CAS were achieved through differential displacement of the stabilators. CAS also ensured that "a given stick displacement produce[d] a corresponding constant load factor [or 'g'-force], regardless of altitude or speed," thereby eliminating the Mach tuck phenomenon common in earlier supersonic fighters. *F-15 Eagle*, 6–7. See also Toliver, *F-15A Follow-On Operational Test Final Report*, Annex A:5.

93. *F-15 Eagle*, 7; *USAF F-15 Fighter Summary*, 7.

94. Hoag, "Superfighter," 22–23. Describing the F-15, Welch recalled that it was "almost impossible to get into trouble flying the airplane . . . [it] was so honest." L. Welch, interview, October 1, 2013. Sokol, Wallace, and Anderegg echoed the assessments. Sokol, interview; Wallace, interview; Anderegg, interview, October 2, 2013. Harkening back to Mahurin's description of his Sabre, Welch also likened his F-15 to a "Cadillac." Although Welch was specifically referencing the ample physical space in its cockpit, the Eagle's accoutrements also contributed to its Cadillac-feel. L. Welch, Oral History Interview, 221.

95. *USAF F-15 Fighter Summary*, 16; *F-15 Eagle*, 12.

96. Osterhout, Berdan, and Gordon, "F-15 Briefing Script," slide 13; Hoag, "Superfighter," 22. The radar also had the option of presenting targets in a "velocity versus azimuth display," which, Toliver explained in his test report, "provided the capability for detecting high closing rate targets at ranges greater than [long-range search] mode." The "VS" or velocity search mode was used when facing very fast aircraft like the MiG-25 Foxbat. On long-range search (LRS) and VS, see Toliver, *F-15A Follow-On Operational Test Final Report*, Annex A:11–12, 17, 32; *T.O. 1F-15A-34-1-1*, sect. 1, pp. 52, 66–69; *T.O. 1F-15A-2-25*, sect. 1, pp. 63–70.

97. Sullivan, *History of the 58th Tactical Fighter Training Wing, Jan–Mar 1975*, 75.

98. Sokol, interview. Carson declared, "Anyone who could not pick out the target on this scope needs his eyes checked." Carson, "Flying the F-15," 24.

99. Hoag, "Superfighter," 22; Anderegg, *Sierra Hotel*, 154; *T.O. 1F-15A-34-1-1*, sect. 1, pp. 40–44. With the radar mode control set to AUTO, "whenever the pilot selected a new weapon, . . . the radar operating mode, display range scale, and the az[imuth]- and el[evation]-scan were initialized to the parameters programmed in the CC. For example, the selection of MRM automatically commanded the 80-[mile] range scope, 120-degree azimuth scan, 2-bar elevation coverage, and long-range search mode. The pilot could override any of the initialized conditions . . . [using] the radar control panel." Toliver, *F-15A Follow-On Operational Test Final Report*, Annex A:17–18. During the initial F-15 testing, the CC was programmed to initialize to a 4-bar, 120-degree azimuth, 20-mile range scope whenever SRM was selected; and a 10-mile scope range and the supersearch auto acquisition mode whenever GUN was selected with the weapon select switch. Flood and Rodero, *F-15 Initial Operational Test and Evaluation—Air-to-Air*, Annex A:4, 7–8 and sect. 3, p. 2.

100. *Eyes of the Eagle*, 2, 4; Toliver, *F-15A Follow-On Operational Test Final Report*, Annex A:11–12; Flood and Rodero, *F-15 Initial Operational Test and Evaluation—Air-to-Air*, Annex A: sect. 3, pp. 2–14; *T.O. 1F-15A-34-1-1*, sect. 1, pp. 25–33, 45, 52–54, 66–67. Hughes claimed in its *Eyes of the Eagle* brochure that the APG-63 was the "only production radar to employ both high and medium PRF." The altitude coverage cues earned praise from the F-15 test pilots; Flood and Rodero noted in their report that the "F-15 allowed precise elevation coverage without excess pilot work load" (Annex A:41–42). On the advantages of the different PRFs, see Stimson, *Introduction to Airborne Radar*, 325–34.

101. Hoag, "Superfighter," 25.

102. Osterhout, Berdan, and Gordon, "F-15 Briefing Script," slide 14; *T.O. 1F-15A-34-1-1*, sect. 1, pp. 45, 56, 69; Flood and Rodero, *F-15 Initial Operational Test and Evaluation—Air-to-Air*, Annex A:24. Anderegg declared of the Eagle's APG-63, "No Air Force pilot had ever flown with a radar so capable and so easy to use." Anderegg, *Sierra Hotel*, 156. Toliver explained in his report that when the pilot commanded lock-on with the TDC, the radar would correlate the position of the acq symbol with the synthetically generated target symbol and the stored radar return and then command the antenna to the "elevation bar, azimuth position, and PRF on which the target was detected." The radar would then initiate a "rapid, 2-bar elevation and ±3 degrees azimuth acquisition raster" around the target's last detected position. After the radar received two subsequent returns, it would enter track. Toliver, *F-15A Follow-On Operational Test Final Report*, Annex A:18–19.

103. Sokol, interview. See also Anderegg, *Sierra Hotel*, 154.

104. Flood and Rodero, *F-15 Initial Operational Test and Evaluation—Air-to-Air*, Annex A:30–31 and sect. 3, pp. 2–14; *T.O. 1F-15A-34-1-1*, sect. 1, pp. 25–33, 75–77; Anderegg, *Sierra*

Hotel, 154. Aspect angle measures the position of one fighter relative to the flight path of another. For example, if an F-15 was directly behind the tail of another aircraft, the F-15 would be at a "tail" or 0-degree aspect angle. As the F-15's position moved forward toward the other aircraft's nose, the aspect angle would increase. Directly abeam the other aircraft would be measured as 90 degrees of aspect, with "left" or "right" describing which side the F-15 was on relative to the other aircraft. If the F-15 was directly in front of the other aircraft, the aspect angle would be 180 degrees or a "head" aspect. In the F-15, aspect angle was displayed in tens of degrees. Hence, a "15R" radar-displayed aspect angle would indicate that the F-15 was 150 degrees off the target's tail (or 30 degrees off its nose) and on the target's right side. See also Shaw, *Fighter Combat*, 347–48; *1 TFW Aerial Attack Manual*, 1:ix. On the weapons estimations, Toliver explained in his report: "Rmax1 was the computed maximum aerodynamic range of the AIM-7F for a nonmaneuvering, $2m^2$ target. Rmax2 was computed for a target performing evasive maneuvers during the terminal phase of the missile flight. A missile launched at Rmax2 had a good probability of success. . . . Rmin represented the minimum range required for missile guidance and arming." Toliver, *F-15A Follow-On Operational Test Final Report*, Annex A:11–12. Later, the F-15 CC would be reprogrammed to display the target's magnetic heading instead of its G-loading. On using the displayed information to execute a proper radar intercept, see *1 TFW Aerial Attack Manual*, 1:9–42.

105. Anderegg, *Sierra Hotel*, 153–54. The A-7 was the first Air Force fighter equipped with a HUD, although its performance paled in comparison to the HUD on the F-15.

106. L. Welch, interview, October 1, 2013; Wallace, interview; Sokol, interview. Selecting symbols reject (SYM REJ) on the HUD control panel removed the scales. *T.O. 1F-15A-2-24*, sect. 1, p. 20.

107. Anderegg, *Sierra Hotel*, 153–54; Toliver, *F-15A Follow-On Operational Test Final Report*, Annex A:13, 15, 21–26; Flood and Rodero, *F-15 Initial Operational Test and Evaluation—Air-to-Air*, Annex A:2, 89 and sect. 2, pp. 1–11; *T.O. 1F-15A-34-1-1*, sect. 1, pp. 15–24.

108. Anderegg, *Sierra Hotel*, 154. The other pilot's assessment is cited by Anderegg. Sokol likewise echoed the praise for the TD box. Sokol, interview.

109. Hoag, "Superfighter," 27. Hoag continued, "Knowing where to look is half the battle; early indications are that visual acquisition will improve at least 20% using this line of sight designation."

110. Ibid., 26. See also Anderegg, *Sierra Hotel*, 154–55; Toliver, *F-15A Follow-On Operational Test Final Report*, Annex A:11; Flood and Rodero, *F-15 Initial Operational Test and Evaluation—Air-to-Air*, Annex A:2; *T.O. 1F-15A-1*, sect. 1, p. 76; *T.O. 1F-15A-34-1-1*, sect. 1, p. 2.

111. Hoag, "Superfighter," 26 (emphasis in original); Flood and Rodero, *F-15 Initial Operational Test and Evaluation—Air-to-Air*, Annex A: sect. 2, p. 5 and sect. 3, pp. 7–8; Anderegg, *Sierra Hotel*, 155. In later CC-software versions, a small triangular "shoot cue" would also appear in the HUD under the TD box whenever the F-15 was in proper weapons parameters.

112. *T.O. 1F-15A-34-1-1*, sect. 3, p. 7.

113. L. Welch, interview, October 1, 2013. The initial test pilots agreed, commenting that "pilot workload in the [long-range intercept] mission was low due to the accessibility, operability, and ease of interpretation and performance of the F-15 [fire control system]." Flood and Rodero, *F-15 Initial Operational Test and Evaluation—Air-to-Air*, Annex A:26.

114. Hoag, "Superfighter," 26, 28; *T.O. 1F-15A-34-1-1*, sect. 1, pp. 75–79. Unlike the Phantom, the Eagle did not have a special continuous-wave illuminator to help guide its Sparrow missiles; the Eagle used the APG-63's high PRF waveform instead. This meant that before launching a Sparrow, the F-15 CC had to ensure that the radar was tracking the target using high PRF, and if not, the CC needed to command the radar to make the switch.

115. *USAF F-15 Fighter Summary*, 4.

116. Hoag, "Superfighter," 25–26; Osterhout, Berdan, and Gordon, "F-15 Briefing Script," slide 16; Toliver, *F-15A Follow-On Operational Test Final Report*, Annex A:18; Flood and Rodero, *F-15 Initial Operational Test and Evaluation—Air-to-Air*, Annex A:7–8, 78; *T.O. 1F-15A-34-1-1*, sect. 1, pp. 57, 70.

117. Osterhout, Berdan, and Gordon, "F-15 Briefing Script," slide 15; Toliver, *F-15A Follow-On Operational Test Final Report*, Annex A:18; Flood and Rodero, *F-15 Initial Operational Test and Evaluation—Air-to-Air*, Annex A:7–8, 78, 91; *T.O. 1F-15A-34-1-1*, sect. 1, pp. 57, 70. The pilots in the 1st TFW observed that "supersearch is degraded during high 'G' maneuvering" and therefore recommended using boresight in those situations. *1 TFW Aerial Attack Manual*, 1:66.

118. Toliver, *F-15A Follow-On Operational Test Final Report*, Annex A:18. More simply stated, supersearch would return to the "operating mode from which [it] was entered." While the Eagle advocates boasted of the two-second lock-resolution time in their "spread the word" briefing, two seconds could still seem like an eternity during a stressful engagement. Osterhout, Berdan, and Gordon, "F-15 Briefing Script," slide 15.

119. Flood and Rodero, *F-15 Initial Operational Test and Evaluation—Air-to-Air*, Annex A:24–25, 73. Flood and Rodero noted that some of the early test pilots complained of the "low position" of the weapon select switch, which "required the pilot to abandon radar antenna elevation control. . . . This hand movement resulted in inadvertent displacement of the antenna elevation control." The only weapon that required a radar lock to function was the Sparrow. However, a lock provided valuable target range data and allowed the CC to compute and display helpful weapons employment information in the HUD whenever the pilot was trying to employ a Sidewinder or shoot the gun. See also Toliver, *F-15A Follow-On Operational Test Final Report*, Annex A:18.

120. Hoag, "Superfighter," 27–28; Toliver, *F-15A Follow-On Operational Test Final Report*, Annex A:12; *T.O. 1F-15A-34-1-1*, sect. 1, pp. 80–88. The HUD and VSD provided similar CC-computed weapons employment displays as with the Sparrow, except that only a single Rmax was displayed with the Sidewinder.

121. Hoag, "Superfighter," 26–27; Osterhout, Berdan, and Gordon, "F-15 Briefing Script," slide 36.

122. *T.O. 1F-15A-2-24*, sect. 1, pp. 38–40. Of note, the F-15 weapons employment manual included a stadiametric ranging reference chart for a variety of target wingspans. *T.O. 1F-15A-34-1-1*, sect. 1, p. 94.

123. Hoag, "Superfighter," 27; Toliver, *F-15A Follow-On Operational Test Final Report*, Annex A:24, 35; Flood and Rodero, *F-15 Initial Operational Test and Evaluation—Air-to-Air*, Annex A:3; *T.O. 1F-15A-2-24*, sect. 1, pp. 20, 36–40. On gun employment, see *T.O. 1F-15A-34-1-1*, sect. 1, pp. 89–96.

124. Toliver, *F-15A Follow-On Operational Test Final Report*, Annex A:30–31; Flood and Rodero, *F-15 Initial Operational Test and Evaluation—Air-to-Air*, Annex A:118–19; *T.O. 1F-15A-34-1-1*, sect. 4, p. 1–4.

125. Shepard, *Tactical Weapons Employment Effectiveness*. See also L. Welch, *Weapons Delivery Image Recording System*; Romer and Braeden, "Letter of Agreement"; *History of the 27th Tactical Fighter Squadron, Jan–Mar 1976*, 10. Early attempts to use the HUD film to validate pilots' weapons employment eventually culminated in unit-level "Top Gun" evaluation programs; see *History of the 1st Tactical Fighter Wing, Jan–Mar 1978*, 61, 66.

126. Flood and Rodero, *F-15 Initial Operational Test and Evaluation—Air-to-Air*, Annex A:119; *History of the 27th Tactical Fighter Squadron, Apr–Jun 1976*, chap. 3, "Imagery Recording"; Sokol, interview; Wallace, interview. The 1st TFW's *Aerial Attack Manual* asked pilots to "attempt to record the range, angle-off, 'G' loading, and closure rate of the firing aircraft, on the [audio] tape recorder. This information, coupled with the gun camera film, will give an estimation of kill probability and will be useful in the debriefing." *1 TFW Aerial Attack Manual*, 1:6, 9.

127. Colonel Robert Titus is quoted in Ulsamer, "F-15: Tomorrow's Champion," 31, 33. Titus was working in the office of the deputy chief of staff for research and development.

128. Quoted in Anderegg, *Sierra Hotel*, 157.

129. Anderegg, interview, October 2, 2013.

130. Many of the Eagle's test pilots experienced these issues firsthand. For example, one report noted that, despite the distinct HUD symbology, the pilots sometimes had a difficult time recognizing what mode they were in or which weapon they had selected. Consequently, they "frequently cycled" the weapon select switch "to reaffirm" its position, which had the unfortunate effect of causing "undesirable automatic changes in radar modes and search coverages." Flood and Rodero, *F-15 Initial Operational Test and Evaluation—Air-to-Air*, Annex A:24, 73.

131. L. Welch, interview, October 1, 2013.

132. Coram, *Boyd*, 231.

133. The AIMVAL-ACEVAL Preliminary Test Plan message estimated the tests' total costs at $45 million. By the time of the official AIMVAL-ACEVAL public news release, cost estimates had risen to $64 million. A 1981 review of the tests pegged the final cost figure at $150 million. Curiously, test officials' statements to Congress on AIMVAL-ACEVAL repeatedly claimed a $42 million price tag, but they sometimes caveated their expense figures as R&D (research and development) costs. Sparks, one of the tests' designers and analysts, suggested that tests' operations costs, to include ancillary aircrew training sorties, were likely included in the Air Force's internal cost estimates but excluded from the official reports. Sparks, interview; see also references to AIMVAL-ACEVAL training support sorties in Sullivan, *History of the 58th Tactical Fighter Training Wing, Jul–Sep 1976*, 24–25; Laudise to 58 TFTW DO, "ACEVAL/AIMVAL Training"; ACEVAL AIMVAL JTF to DDR&E, "ACEVAL-AIMVAL Oct Progress Report." On the tests' costs, see AFSC/CS to ADTC/CC et al., "ACEVAL/AIMVAL Test Plan"; Lake, "ACEVAL-AIMVAL Program Review," 5044; DSAF/OIP to AIG, "ACEVAL/AIMVAL News Release"; "Point Paper on the ACEVAL Flight Trials"; Fay, "AIMVAL-ACEVAL Final Report Briefing," 420. The test sortie totals also experienced an increase of nearly a third over initial estimates; see *History of the Air Force Test and Evaluation Center, 1 Jan–31 Dec 1976*, 1:214; Tissot and Hildreth, *AIMVAL Final Report*, 1: sect. 3, pp. 2–3; McKenzie and Hildreth, *ACEVAL Final Report*, 1: sect. 3, pp. 1, 7; Everson, *ACEVAL Final Report*, 10, 138; Fay, "AIMVAL-ACEVAL Final Report Briefing," 420.

134. Tissot, the JTF director during AIMVAL, is quoted in *History of the Air Force Test and Evaluation Center, 1 Jan–31 Dec 1977*, 1:271.

135. On the "quantity" versus "quality" debate, see Fallows, *National Defense*; Kross, *Military Reform*; Merritt and Sprey, "Quality, Quantity, or Training"; Deming, "Force Structure Decision"; "Fightergram: The Quality-Quantity Controversy"; Vandevender, "Quality-Quantity Rebuttal"; Anderegg, "Meeting the Threat"; Correll, "Reformers"; Michel, "Revolt of the Majors."

136. Anderegg, *Sierra Hotel*, 159; Fino, "Doubling Down."

137. Blesse, "Changing World of Air Combat," 34.

138. The exchange occurred during Lake's March 11, 1976, testimony. The staffer specifically wanted to know if the tests would provide "the rock hard requirements for the advanced missile in the 1980's." Lake, "ACEVAL-AIMVAL Program Review," 5050.

139. Later, the commander of the tests' F-15s and F-14s agreed with the staffer's assessment, remarking, "The test objective of quantifying the influence of 'numbers' on engagement outcomes had not only not been achieved, but was 'probably an impossible task.'" Paraphrased in B. Watts, *Six Decades of Guided Munitions*, 50.

140. Quoted in "No-Win War at Dogbone Lake," 56.

141. Sokol, interview. There were initially six Air Force F-15 pilots assigned to support the AIMVAL-ACEVAL tests. When the program fell behind schedule, two more pilots were added; one was Sokol, who happened to be at Nellis as an instructor at the Weapons School preparing for the inaugural F-15 class. Another AIMVAL-ACEVAL F-15 pilot told Congress after the tests, "We learned in a matter of hours or days, perhaps, what normally might take 6 or 8 months." Recorded in Fay, "AIMVAL-ACEVAL Program Review," 1978, 5220.

142. L. Welch, interview, July 10, 2012. In a joint interview, Sparks, Cliver, and Dvorchak all expressed similar sentiments. See also Amlie, *Non-statistical Look at AIMVAL/ACEVAL*, 1.

143. First to receive the Eagle, the 58th Tactical Fighter Training Wing at Luke AFB continued to provide initial training for all F-15 pilots. On the training program, see: Sullivan, *History of the 58th Tactical Fighter Training Wing, Oct–Dec 1974*, 31–34; Sullivan, *History of the 58th Tactical Fighter Training Wing, Apr–Jun 1975*, 35–38; Sullivan, *History of the 58th Tactical Fighter Training Wing, Jul–Sep 1975*, 25–27; Sullivan, *History of the 58th Tactical Fighter Training Wing, Jan–Mar 1976*, 23–25; Sullivan, *History of the 58th Tactical Fighter Training Wing, Apr–Jun 1976*, 40–44; Sullivan, *History of the 58th Tactical Fighter Training Wing, Jul–Sep 1976*, 35–39. The Eagle also continued to undergo testing at Edwards and Luke AFBs; see, for example, Toliver, *F-15A Follow-On Operational Test Final Report.*

144. *1 TFW Aerial Attack Manual*, 1:i, 1, 67.

145. Ibid., 54–55.

146. L. Welch, interview, July 10, 2012. Differing from Welch's recollection, the 64th Aggressor Squadron at the time of their spring 1976 deployment was still flying the underpowered, two-seat T-38; its F-5Es would not arrive until later that year. *History of the Ninth Air Force, Jan–Dec 1976*, 1:172. While at Langley, one of the Aggressor's T-38s crashed in the Atlantic during a training mission. See also "'Aggressors' Organized to Reduce Casualties"; "Eagles Tangle with Aggressors."

147. "DACT Lessons Learned," Attachment 1:2.

148. Ibid.; *1 TFW Aerial Attack Manual*, 1:1, 67. The F-15 pilots' reluctance to use their radars and Sparrow missiles in training also stemmed from the traditions of the F-4 Phantom community and its practice of only assessing simulated kills based on Sidewinder or gun employment. See Anderegg, *Sierra Hotel*, 162–63.

149. "DACT Lessons Learned," Attachment 1:2.

150. *History of the Ninth Air Force, Jan–Dec 1976*, 1:173–75; Arnold, "Red Flag"; Volloy, "Red Flag in Perspective"; "Dummy War at Nellis"; "'Operation Red Flag' Studies"; Hillinger, "'Russian Squadrons' Fly Mock Battles"; Anderegg, *Sierra Hotel*, chap. 9; Michel, "Revolt of the Majors," chap. 9; Laslie, *Air Force Way of War*, chap. 3. The significance of the new exercise was recognized eighteen months later when TAC and its commander Dixon received the prestigious Collier Trophy, awarded annually "for the greatest achievement in aeronautics or astronautics in America," for their role in "developing and implementing Red Flag, an unprecedented combat simulated flight training program for aircrews, . . . a significant contribution to national defense." "National Aeronautic Association Collier Trophy"; Michel, "Revolt of the Majors," 222; Laslie, *Air Force Way of War*, 67. Michel argued that Red Flag and the corresponding focus on "new, innovative program[s] of realistic, combat-oriented training" were as critical to America's future airpower successes as the advanced technology in the aircrews' aircraft. Michel, "Revolt of the Majors," 13. Laslie presents a similar argument. Michel noted that the realism of Red Flag "combat" was reflected in its accident rate: thirty-two major accidents the first full year in 1976; more than four times the accident rate in TAC (220).

151. Quoted in *History of the Ninth Air Force, Jan–Dec 1976*, 1:174.

152. Initially, the Red Flag exercises were numbered sequentially up through Red Flag VIII. Thereafter, the numbering scheme was changed to reflect the fiscal year, which began on October 1st. Hence, Red Flags 77-1 and 77-2 were both conducted during calendar year 1976.

153. *History of the Ninth Air Force, Jan–Dec 1976*, 1:177–81; "Air Force Eagle Joins Nellis Fighters"; Bloomcamp, "Red Flag 77-4 After Action Report"; Miller to TAC/DOOW, "Red Flag 77-7 After Action Report"; 1TFW/DO to TAC/DOOW, "Red Flag 77-8 After Action Report"; 1TFW/DO to TAC/DOOW, "Red Flag 77-9 (Nellis) After Action Report"; 1TFW/DO to TAC/DOOW, "Red Flag 77-9 (Michael AAF) After Action Report"; 1TFW/DO to TAC/DOOW, "RED FLAG 78-1 After Action Report." Welch later commended the "unofficial testing" his pilots conducted at Red Flag and around Langley. L. Welch, Oral History Interview, 238–39; L. Welch, interview, July 10, 2012. Langley's participation in Red Flag 77-9 was unique because the unit was split between two locations—one group enjoyed life at Nellis AFB near Las Vegas; the other was deployed to the remote airstrip at Michael Army Air Field in Utah in order to demonstrate the Eagle's ability to operate out of a "bare base" location for thirty days. See also Bolinger, "F-15 Operates Out of Desert." While some Eagle pilots preferred flying at low altitude during the Red Flag missions as a way to minimize their chance of being detected by the adversaries, others found that the "problems of navigation, formation flying, visual search, and radar search in the low-altitude, high speed arena taxes the pilot considerably." 1TFW/DO to TAC/DOOW, "Red Flag 77-10 After Action Report," 5. On the Eagle pilots' training experiences flying against other units around Langley AFB, see Jaquish to 1 TFW/DO, "DACT OPS Summary"; Fanelli to 1 TFW/DOW, "Beaufort MCAS DACT Operations Summary."

154. Flood and Rodero, *F-15 Initial Operational Test and Evaluation—Air-to-Air*, Annex A:25, 85.

155. *F-15 Conversion Training Manual*, 27–29; Anderegg, *Sierra Hotel*, 59–60. The conversion manual explained that the "prebriefed procedures" included "the types of turns to be made, the parameters (G/AOA, power) at which the turns will be made, and the methods (radio calls, signals) by which the turns will be initiated." Procedures were later developed that would

allow a line-abreast formation to execute a turn without relying on a radio call from the flight leader; see Jumper, "COMM-OUT Turns."

156. Sokol, interview. Sokol recalled that when the Air Force started soliciting applicants for the initial F-15 cadre, he was working fourteen-hour days and flying two weekends a month as an F-4 instructor at George AFB, California. He scribbled out his application but got frustrated midway through when it asked for his nationality and religion, information that the Air Force obviously already had on file. "I grabbed a big red marker and I wrote Polish for nationality. And I wrote Druid for religion." After he was selected for the Eagle assignment, Sokol's commander told him, "You will never know how much good that did you."

157. On the "death of fighting wing" in the Phantom community, see Anderegg, *Sierra Hotel*, 59–60, 83–85; Gish, "F-4 Air to Air Training."

158. L. Welch, interview, October 1, 2013.

159. Sokol, interview.

160. Bloomcamp, "Red Flag 77-4 After Action Report," 4. See also 1TFW/DO to TAC/DOOW, "Red Flag 77-9 (Michael AAF) After Action Report," 10.

161. Jaquish to 1 TFW/DO, "DACT OPS Summary," 3.

162. Anderegg, *Sierra Hotel*, 164.

163. Partin, "Program Management Directive for Project 670E," 2. CLAW supposedly stood for "Concept of a Lightweight Aerial Weapon." Griffith, "AIMVAL-ACEVAL," 24. For more on the Air Force's short-lived CLAW program, see Dilger, "CLAW Missile Program Review," 4705–25. Explaining the difference between the AIM-9 Sidewinder and the CLAW concept weapon, one Air Force general during Dilger's testimony offered: "You might be able to consider some Sidewinders as missiles; we are hoping the CLAW will be a 'hitile'" (4723).

164. Quinn, "Agile Program Review," 4695–4705; Waterman, "Research, Development, Test, and Evaluation, Navy," 794–97. Because it incorporated so many radical technologies, the Navy's Agile missile was clearly going to be expensive. Waterman pinned the expected cost at $298 million. In his 1973 testimony to the House Subcommittee on Appropriations, Malcolm Currie, the director of DDR&E, expressed his concerns with the Agile program, stating, "Estimates indicate that at least a 20-percent reduction (preferably 40 percent) would be necessary to enable procurement of the missile in large quantities." Currie, "Statement of the Director of Defense Research and Engineering," 456. Welch explained that the Navy designed the Agile to be highly maneuverable to compensate for the restricted turning performance of its F-14 Tomcat. L. Welch, interview, July 10, 2012. See also Hildreth, Oral History Interview, 160–61.

165. Haff, "Sidewinder Program Review," 4619–34; Van Dyke, "Sidewinder Program Review," 5013–38; Westrum, *Sidewinder*, 192–95; Beck, "State of the Art." The AIM-9L was the first Air Force-Navy collaborative Sidewinder program since the two services parted ways after the AIM-9B during Vietnam.

166. Griffith, "AIMVAL-ACEVAL," 24–25; Lake, "ACEVAL-AIMVAL Program Review," 5040, 5048–49; "Advanced Short Range Air-to-Air Missile Technology," 29370. An Air Force message explained, "AIMVAL is a direct result of FY 75 CLAW/Agile development program denials by Congress." CSAF/XO to TAC/CV et al., "Air Force ACEVAL/AIMVAL Test Directive." The SRAAM (short-range air-to-air missile) is not to be confused with the Air Force's AGM-69 SRAM (short-range attack missile), the latter "a [nuclear] air-launched, air-to-surface missile" developed for the B-52G/H and FB-111. On SRAAM, see, for example, Kirkconnell

"SRAAM Program Review," 4634–50; on SRAM, see, for example, Glasser, "Statement," 3613–16.

167. Griffith, "AIMVAL-ACEVAL," 24–25; *History of the Air Force Test and Evaluation Center, 1 Jan–31 Dec 1976*, 1:213; Fay, "AIMVAL-ACEVAL Program Review," 1977, 4569; Anderegg, *Sierra Hotel*, 159; Lake, "ACEVAL-AIMVAL Program Review," 5038–39.

168. COMNAVELECSYSCOM to CSAF and CNO, "ACEVAL/AIMVAL Joint Test."

169. ADTC/XR to AFSC/TEV, "ACEVAL-AIMVAL Test Plans"; "F-15s, F-5Es Catching Up with F-14"; Knight, Oral History Interview, 232; Hildreth, Oral History Interview, 161. On the declared importance of casting the F-15 in the best light possible, see CSAF/XO to TAC/CV, "AIMVAL/ACEVAL F-15s"; Swarts, "Interest Paper on AIMVAL/ACEVAL," 3.

170. COMNAVELECSYSCOM to CSAF and CNO, "ACEVAL/AIMVAL Joint Test"; Hildreth, Oral History Interview, 161–62.

171. Lake, "ACEVAL-AIMVAL Program Review," 5056. See also Fay, "AIMVAL-ACEVAL Program Review," 1977, 4579.

172. L. Welch, interview, October 1, 2013. Alluding to a fighter pilot's propensity to use his hands to reconstruct the motion of opposing fighter aircraft during a fight, Welch explained, "If you could get your hand back [farther than the other guy's], then you could turn tighter."

173. Wallace, interview.

174. Anderegg, *Sierra Hotel*, 104, 108–9.

175. Quick, "Cleared in Hot"; Turner, "ACMI Update"; Anderegg, *Sierra Hotel*, 108–9; Everson, *AIMVAL Final Report*, 7–10; Taylor, "Air Combat Maneuvering Instrumentation." The actual ACMI pods mounted to each aircraft were officially termed the AIS (Airborne Instrumentation System). Heeding the recommendations of the Ault report, the Navy initiated development of ACMI—its initial version was called ACMR (Air Combat Maneuvering Range)—as a tool to help pilots visualize weapons employment zones. On the Air Force's initial ACMI requirements, see Momyer, *Air Combat Maneuvering Range*. On the public rollout of ACMI at Nellis AFB, see "Dogfight Pilot Training at Nellis"; "Nellis Pilots Given New Battle Training." The Air Force was also in the process of constructing an East Coast ACMI range for use by both Navy and Air Force aircraft; it became operational in early 1977. See Husted, "1st TFW Inherits Pilot Training System"; Waddell, "ACMI Improves Pilot Combat Training"; Husted to 1 TFW/DO, "Langley AFB ACMI Operations."

176. Quick, "Cleared in Hot," 11, 14–15; Turner, "ACMI Update," 21–22; Anderegg, *Sierra Hotel*, 108–9; Everson, *AIMVAL Final Report*, 7–10; Fay, "AIMVAL-ACEVAL Program Review," 1977, 4574; Risner, Oral History Interview, 182; Dvorchak, interview.

177. Fay, "AIMVAL-ACEVAL Program Review," 1977, 4575. There were lingering concerns, however. To accurately predict the weapons' probability of kill, the vulnerability of the target aircraft had to be incorporated into the ACMI model. However, this threatened to introduce service bias into the ideally objective Pk (probability of kill) tables used during AIMVAL-ACEVAL: Air Force officers argued that the F-15 was less vulnerable than the Navy's F-14; the Navy obviously argued the reverse. To resolve the dispute, a neutral vulnerability model based on an outdated F-4E aircraft was used for both the F-15 and F-14. Understandably, the conservative decision did not placate Air Force officers. See TAC/DR to AFTEC/CC, "ACEVAL/AIMVAL Preliminary Test Plan Review"; Marsh to AFTEC/CC, "AFSC Review of ACEVAL/AIMVAL Preliminary Test Plan." The pilots also complained that the Red forces' missiles were

uncharacteristically effective—Sokol said the Aggressors were given "hit-iles" instead of "miss-iles." Sokol, interview.

178. ACEVAL AIMVAL JTF to DDR&E, "ACEVAL-AIMVAL Monthly Progress Report (as of 6 Jan 76)"; COMNAVELECSYSCOM to DDR&E, "ACEVAL AIMVAL Feb Progress Report"; ACEVAL AIMVAL JTF to DDR&E, "ACEVAL-AIMVAL Oct Progress Report"; Fay, "AIMVAL-ACEVAL Program Review," 1977, 4571; Taylor, "Air Combat Maneuvering Instrumentation"; *History of the Air Force Test and Evaluation Center, 1 Jan–31 Dec 1976*, 1:218–19.

179. Lake, "ACEVAL-AIMVAL Program Review," 5042–43; Fay, "AIMVAL-ACEVAL Program Review," 1977, 4571–12; Dvorchak, interview; Sparks, interview. To enforce aircrew adherence to the VID restriction during the tests, Lake told Congress that he planned to have "random nonexercise [friendly] aircraft in the exercise area." Shooting down one of the "nonplayers" would, according to Fay, result in "a bad day" for the offending aircrew. On implementation, see McKenzie and Hildreth, *ACEVAL Final Report*, 1: sect. 7, p. 8.

180. See, for example, Momyer to Ellis, "CORONA HARVEST," 4, 24.

181. *History of the Air Force Test and Evaluation Center, 1 Jan–31 Dec 1976*, 1:216–17; TAC/DR to CSAF/XOO, "ACEVAL/AIMVAL Test Constraints." See also Blesse, Corona Ace Oral History Interview, 82–87. A similar interservice disagreement emerged over the use of RHAW (radar homing and warning) equipment during the tests. The Air Force's F-15 did not have its RHAW (or later RWR [radar warning receiver]) system installed yet; the Navy's F-14 did. The Aggressors also wanted a RHAW capability that mimicked the system used by the Soviet MiG-21. In November 1976, just prior to the tests, JTF officials declared that no one could use their RHAW, simulated or actual. However, the Aggressors were eventually provided with a pseudo-RHAW capability. ACEVAL AIMVAL JTF to CNO and CSAF/XOO, "ACEVAL-AIMVAL Initial Starting Conditions." See also *History of the Air Force Test and Evaluation Center, 1 Jan–31 Dec 1976*, 1:215–16; ACEVAL AIMVAL JTF to CNO, "F-14/F-15 Use of RHAW in ACEVAL-AIMVAL"; ACEVAL AIMVAL JTF to AIG, "AIM-7F Kill Removal/F-14 RHAW"; TAC/DR to AFTEC/JT, "F-14/F-15 Use of RHAW in ACEVAL/AIMVAL"; Anderegg, *Sierra Hotel*, 160–61.

182. Everson, *ACEVAL Final Report*, 4; McKenzie and Hildreth, *ACEVAL Final Report*, 1: sect. 3, p. 6; O'Keefe, "Eagle Eye"; Laslie, *Air Force Way of War*, 75; Anderegg, *Sierra Hotel*, 161. Blesse was not a fan of the "Eagle Eye" solution, remarking, "The [F-15] is a single-seater fighter with a million different things to do, so now you are going to give the pilot a monocular and let him try that too. . . . That is like giving him 15 different things and then sticking a broom in his ass and telling him to sweep the floor while he does those other 15 things. It is just terrible!" Blesse, Corona Ace Oral History Interview, 84–85. More than twenty years later, F-15C pilots from Langley AFB were still flying with the Eagle Eye on their missions patrolling the No-Fly Zones over Iraq—I was one of them.

183. See Lake, "ACEVAL-AIMVAL Program Review," 5049; SAF/SA to AFTEC/CC, "Review of ACEVAL-AIMVAL Preliminary Test Plan"; ADTC/XR to AFSC/TEV, "ACEVAL-AIMVAL Test Plans"; TAC/DR to AFTEC/CC, "ACEVAL/AIMVAL Preliminary Test Plan Review." For a post-test summary of acknowledged constraints, see Fay, "AIMVAL-ACEVAL Program Review," 1978, 5202–3, 5220–21; Kross, *Military Reform*, 104–5. The inability to see a missile smoke trail is cited in the SAF/SA message and later by Fay during his program review testimony.

184. Haering, "Aim/Ace," 15.

185. TAC/DR to AFTEC/CC, "ACEVAL/AIMVAL Preliminary Test Plan Review."

186. SAF/SA to AFTEC/CC, "Review of ACEVAL-AIMVAL Preliminary Test Plan." One Air Force officer succinctly captured the coming dilemma: "The large scale of the test itself, number of trial repetitions, and bounty of data tend to create the impression that the test results can be taken at face value. Characteristic of such a notion is the attitude that what came out of the test must be right, since we did so much of it and the results did not change." J. Welch to AF/XO, "AIMVAL Analysis and Evaluation," 2.

187. Recorded in Fay, "AIMVAL-ACEVAL Program Review," 1977, 4571. The joint test director tried to assuage the services' concerns, explaining shortly after AIMVAL began, "The majority of the test results and relative effectiveness conclusions will be inferred . . . and should not be treated in terms of specific system absolutes." Quoted in *History of the Air Force Test and Evaluation Center, 1 Jan–31 Dec 1977*, 1:271.

188. Fay, "AIMVAL-ACEVAL Program Review," 1977, 4565. An off-boresight acquisition capability allows the pilot to cue his missile, usually using the radar, to a target that is not directly in front of the aircraft.

189. Everson, *AIMVAL Final Report*, 5, 7, 12–14; Tissot and Hildreth, *AIMVAL Final Report*, 1: sect. 3, pp. 1–3. Each trial required multiple Blue fighters (F-15s or F-14s) and Red fighters, but multiple trials could also be accomplished on a single sortie. Hence, there is no easy correlation between trials and sorties required. AIMVAL required 2,678 total sorties.

190. *T.O. 1F-15A-34-1-1*, sect. 1, p. 58. To make room for vertical scan in the HOTAS scheme, the engineers moved supersearch to the forward position of the auto acquisition switch—subsequent forward switch actuations would toggle between supersearch and boresight. Vertical scan took the old supersearch position, aft on the auto acquisition switch (see figure 5.12). *T.O. 1F-15A-1*, sect. 1, p. 27. Additionally, in preparation for AIMVAL, the F-15's CC also had to be purged of all air-to-ground weapons information to free up memory and computing power to calculate the new weapons parameters for the various missile concepts, including the new AIM-9L.

191. Lake, "ACEVAL-AIMVAL Program Review," 5050, 5052; Fay, "AIMVAL-ACEVAL Program Review," 1977, 4573; Everson, *AIMVAL Final Report*, 5, 7.

192. Tissot and Hildreth, *AIMVAL Final Report*, 1: sect. 3, pp. 3, 6. Of note, an F-15 and an F-5E collided during AIMVAL. The F-15 was destroyed; the F-5E managed to land safely; both pilots survived. The mishap was subsequently cited as evidence of the test's realism; see "Nellis Fighters Collide"; "Two Jets Collide"; Hildreth, Oral History Interview, 172; "No-Win War at Dogbone Lake"; Kaylor, "Mock Dogfights Test Latest Jet Fighters."

193. *History of the Air Force Test and Evaluation Center, 1 Jan–31 Dec 1977*, 1:273.

194. Recorded in Fay, "AIMVAL-ACEVAL Program Review," 1977, 4578.

195. Fay, "AIMVAL-ACEVAL Program Review," 1978, 5206–10, 5225; Newell, "AIMVAL-ACEVAL Results," 1; J. Welch, interview, June 8, 2012; Tissot and Hildreth, *AIMVAL Final Report*, 1: sect. 5, p. 1. In his 1976 testimony, Rear Admiral R. P. McKenzie predicted that the "all aspect, close in" AIM-9L missile would "revolutionize fighter tactics"—AIMVAL seemed to prove him right. Recorded in Van Dyke, "Sidewinder Program Review," 5028. See also Blesse, Corona Ace Oral History Interview, 68–74.

196. Fay, "AIMVAL-ACEVAL Program Review," 1977, 4565.

197. Everson, *ACEVAL Final Report*, 4–10, 28; McKenzie and Hildreth, *ACEVAL Final Report*, 1: sect. 3, pp. 1–8. During ACEVAL, the F-14s retained their helmet-mounted cueing system and their TVSU equipment from AIMVAL; the F-15s did not use the helmet cueing

system but added their "Eagle Eyes" to their aircraft to assist with long-range VIDs. ACEVAL required 3,222 total sorties.

198. Dvorchak to Fino. Dvorchak's assessment is confirmed in Fay, "AIMVAL-ACEVAL Program Review," 1978, 5211–13; Newell, "AIMVAL-ACEVAL Results," 2. Drawing laughter from his congressional audience, Fay commented during his testimony, "I would like to point out, sir, that we spent 2 years getting prepared for ACEVAL, 6 months running the test, 2 days to reduce this data, and 2 months to figure out how to say it, sir" (5211).

199. Quoted in B. Watts, "Doctrine, Technology, and War," n65.

200. Fay, "AIMVAL-ACEVAL Program Review," 1978, 5217. While the tests yielded a great deal of information about short-range air combat, the test officials realized that they had analyzed "perhaps [only] 2 inches on the yardstick of air superiority." Fay, "AIMVAL-ACEVAL Final Report Briefing," 435. Additionally, although concluding that "numbers are the determining factor in the outcome of air combat," they still could not explain, to the Air Force's satisfaction, "what caused numbers to be so important." J. Welch to AF/XO, "AIMVAL Analysis and Evaluation," 2–3. Later analysis revealed that in fact "quantifiable variables such as numbers only accounted for about 10–20 percent of the variation in outcomes, whereas human factors [such as situational awareness] had 'more than five times the effect on results.'" B. Watts, *Six Decades of Guided Munitions*, 50. See also Fay, "AIMVAL-ACEVAL Program Review," 1978, 5212–13. On the importance of situational awareness in air combat, see Spick, *Ace Factor*; B. Watts, "Doctrine, Technology, and War"; B. Watts, *Six Decades of Guided Munitions*, 49–54.

201. The Defense Reform Movement gained significant momentum on Capitol Hill following a hearing organized by Senator Sam Nunn (D-GA) in December 1980. A few months later, a formal Military Reform Caucus formed under the leadership of senator and future presidential-hopeful Gary Hart (D-CO); its members focused their efforts on uncovering waste and corruption in the US military. See Nunn, *Impact of Technology on Military Manpower*; Hart and Lind, *America Can Win*; Wheeler and Korb, *Military Reform*; Fallows, *National Defense*; Locher, *Victory on the Potomac*.

202. Coates and Neikirk, "Cheaper Jets Shoot Down Claims for New Models." Chuck Spinney, one of the Pentagon-analyst Reformers, appeared on the cover of the March 7, 1983, issue of *Time* with the question, "US Defense Spending: Are Billions Being Wasted?" looming in the foreground. Isaacson et al., "Winds of Reform." Sprey and Riccioni were colleagues of Spinney and prominent participants in the public reform debates as well.

203. Spinney, *Defense Facts of Life*. Based on his analysis, Spinney concluded: "Our strategy of pursuing ever increasing technical complexity and sophistication has made high technology solutions and combat readiness mutually exclusive." *Impact of Technology on Military Manpower*, 196. The Reformers' tactical airpower arguments, including Spinney's data-intensive brief, were popularized by award-winning columnist James Fallows. See Fallows, "Muscle-Bound Superpower"; Fallows, "America's High-Tech Weaponry"; Fallows, "I Fly with the Eagles"; Fallows, *National Defense*. See also L. Welch, Oral History Interview, 147–49, 155; Michel, "Revolt of the Majors," 80–83; Correll, "Reformers"; Laslie, *Air Force Way of War*, 89–92.

204. The term "mutually devastating" was used to describe the all-aspect missile environment in McKenzie and Hildreth, *ACEVAL Final Report*, 1: sect. 4, p. 30. The same term also appeared in one anonymous pilot's satirical summary of AIMVAL-ACEVAL that eventually found its way onto the TAC commander's desk. In it, the pilot likened the tests to "a

shootout in an isolated tennis court between giants armed with rifles, pistols, and knives and midgets armed with only pistols and knives. . . . Shootout Results: More or less as expected: Mutually Devastating." J. Welch to McMullen, March 8, 1977; T. McMullen to Dixon, March 15, 1977.

205. Fay, "AIMVAL-ACEVAL Program Review," 1978, 5208. Referencing the quality versus quantity debate, one enthusiastic Air Force general reported, "ACEVAL . . . strongly inferred that more quality and quantity are required." TFWC/CC to TAC/CV, "Recap of ACEVAL-AIMVAL Briefings." Several officials also identified some of the issues with the Reformers' favored "quantity" arguments, not least of which was the burdensome personnel requirements, in both pilots and ground crew, that accompanied the favored fourfold increase in Air Force combat aircraft. See arguments in Perry, *Impact of Technology on Military Manpower*, 9, 40, 48; Anderegg, "Meeting the Threat," 2, 6; Kross, *Military Reform*, 143–45; Kross, "Military Reform: Past and Present."

206. L. Welch, interview, July 10, 2012.

207. Blesse, Corona Ace Oral History Interview, 71–72; Sparks, interview.

208. Sparks, interview.

209. Munger, "Advanced Medium Range Air-to-Air Missile," 436–54; Fay, "AIMVAL-ACEVAL Final Report Briefing," 430–31; R. Bond, "Advanced Medium Range Air to Air Missile," 348–49, 358; O'Hara, "Briefing," 1018–27; Perry, *Impact of Technology on Military Manpower*, 5, 41; Hildreth to TAC/CV, "AIMVAL Final Report Recommendations"; Hildreth, Oral History Interview, 165–66. Following a set of 1978 congressional briefings, one Air Force officer reported that the AIMVAL-ACEVAL findings were "well received by the Congressmen and staffers" and "substantially justified the AMRAAM program in FY79 [fiscal year 1979]." TFWC/CC to TAC/CV, "Recap of ACEVAL-AIMVAL Briefings." Whereas AMRAAM flourished post AIMVAL-ACEVAL, the short-range, IR-guided equivalent ASRAAM (advanced short-range air-to-air missile) never materialized.

There was another important component to long-range aerial combat that was not widely acknowledged at the time—the need for better EID capabilities or clearer rules of engagement (ROE), or better yet, both. See, for example, Blesse, "Changing World of Air Combat," 36–37; Brown, "Technology, Military Equipment, and National Security," 19; Martin to Under Secretary of Defense for Research and Engineering, "AIMVAL Test Results and IFF." The incoming commander of TAC, General Wilbur Creech, recognized the daunting problem. After hearing about the AIMVAL-ACEVAL results, Creech reportedly "seemed to feel that the JTF underestimated [the identification problem] by an order of magnitude. . . . Further he was adamant that present IFF equipment, procedures, reliability and development were inadequate." TFWC/CC to TAC/CV, "Recap of ACEVAL-AIMVAL Briefings." On later EID developments and their use in Desert Storm, see Werrell, *Chasing the Silver Bullet*, 70–71; Putney, *Airpower Advantage*, 289; Carpenter, "Joint Operations in the Gulf War," chap. 3.

210. Sokol, interview.

211. Wallace explained that in the Phantom, "as soon as somebody got a contact, the general tendency was everybody else glommed [locked] on to that." However, he noted that having multiple Phantoms lock and then launch their Sparrow missiles at a single target was not necessarily frowned upon since the missile's reliability was so low. Wallace, interview.

212. L. Welch, Oral History Interview, 239.

213. Sparks, interview.

214. This situation might develop even during a VID-required scenario like that in AIMVAL-ACEVAL. For example, using the combination of the TD box and the Eagle Eye, five- to ten-mile VIDs were not considered abnormal, but the pilot would have an extremely difficult time maintaining a visual track on an additional aircraft for which he had no cueing.

215. Decades later, active electronically scanned array (AESA) radars would eliminate the single-target-tracking limitation of mechanically scanned radars. In the interim, TWS modes allowed a pilot to simultaneously track one target and scan a limited volume of airspace around that target. The Sparrow, however, could not be supported in TWS. When fielded, the different TWS modes were entered from radar track using forward or aft actuations of the auto acquisition switch, further complicating the HOTAS switchology.

Another improvement that emerged after AIMVAL-ACEVAL was the pilot's capability to manually boresight an AIM-9L, which offered a rudimentary form of multitargeting. To multitarget with the Sidewinder after a Sparrow had already been launched, the pilot would need to continue tracking the one target with the radar in order to support the in-flight AIM-7, then visually detect the other target nearby, maneuver his aircraft to place the second target in the middle of the HUD while depressing the AIM-9 boresight button on the right throttle (he could not exceed the sixty-degree radar gimbal on the tracked target during the maneuver, though), uncage the Sidewinder using the nose gear steering button on the control stick, listen for a good tone, and then fire the Sidewinder. An Eagle pilot that could reliably accomplish the cumbersome procedure was considered "shit hot." Finally, ACEVAL also demonstrated the importance of decoy flares for the F-15 to enhance its survivability against any all-aspect IR-guided missiles that were already in flight. Wallace, interview; Cliver, interview; Dvorchak, interview; Sparks, interview; Sokol, interview; L. Welch, interview, October 1, 2013.

216. Sokol, interview.

217. Wallace, interview.

218. Anderegg, *Sierra Hotel*, 160; Anderegg, interview, October 2, 2013; Sokol, interview. The "I've got mine" sort was also occasionally called the "Colonel's sort" or the "field-grade sort," Anderegg explained, because senior officers had "a reputation for being unable to work the radar as well as the young studs"; the older pilots would tend to find a target and grab onto it, hoping that the "younger, more proficient pilots" in the formation would find the other targets.

219. Sparks, interview; Wallace, interview. Wallace described the Aggressors' tactics as "effective . . . to prevent you from sorting, but tactically [they] weren't very salient." The Aggressors were using a "pseudo-RHAW" system to alert them to the Blue forces' targeting. Depending on the account, the "pseudo-RHAW" consisted of either modified "fuzz-buster" radar detectors or ground controlled intercept (GCI)-radio calls to the Red forces whenever a Blue aircraft locked an F-5E; the latter is more likely. See Anderegg, *Sierra Hotel*, 160–61.

220. Sparks, interview. Sparks remembered an Israeli Air Force official who came to observe the tests and, after watching a few engagements, declared that the Aggressors were "too good." The problem, Sparks noted, was that, like the Blue aircrews, the Aggressors were flying nothing but AIMVAL-ACEVAL missions for almost a year, "You get pretty goddamn good [flying that much]."

221. Tissot and Hildreth, *AIMVAL Final Report*, 1: sect. 3, p. 6; McKenzie and Hildreth, *ACEVAL Final Report*, 1: sect. 6, pp. 2–3, 30, and sect. 7, p. 18. Everson concluded in his report that the aircrews' ability to develop stylized tactics that maximized their advantages while negating their opponents' advantages rendered any one side's technological advantage

ephemeral. Everson, *ACEVAL Final Report*, 10–11. Watts declared that the net effect was a tactical stalemate with neither side accruing a significant advantage over the other for any appreciable duration. B. Watts, "Doctrine, Technology, and War."

222. Cliver, interview.

223. Ibid., Wallace, interview.

224. Dvorchak, interview.

225. Jumper, "Instincts of the Fighter Pilot," 5. The wingman's standard four radio calls remained a common wisecrack even after he started to assume a greater offensive role. See, for example, Winters, "Two, Bingo, May-Day . . ."

226. Cliver, interview; Sokol, interview.

227. Dvorchak, interview; Sparks, interview.

228. *1 TFW Aerial Attack Manual*, 1:iv–viii; *4453d CCTW RTU Weapons School: Intercept Techniques*, 4–9.

229. Sparks, interview; Cliver, interview; L. Welch, interview, July 10, 2012. Sokol explained that the Israeli Air Force played a role in the F-15 pilots' decision to abandon the "silence is golden" attitude. One of the AIMVAL-ACEVAL pilots participated in a brief exchange tour with the Israelis, and when he returned, he brought the surprise news: "Those guys are communicating all the time. . . . They're talking all the time." So, Sokol noted, "we started talking more amongst ourselves." Sokol, interview. Anderegg suggested that the "evolution of the brevity code was as important as any evolution in tactics, any evolution in formation, any evolution in weaponry as [the Air Force has] had." Anderegg, interview, October 2, 2013. Still, a better method to assign and accomplish targeting besides radio communications would be through an interaircraft datalink. Datalinked fighter aircraft didn't appear in the Air Force until several decades after ACEVAL. On the continued emphasis of "brevity comm" seven-years after AIMVAL-ACEVAL, see Ross, "Comm Discipline."

230. Cliver, interview. Sokol noted that there were inklings of rudimentary radar sorting tactics being developed at Luke AFB in the F-15 training units, but nothing as rigorous as what emerged during AIMVAL-ACEVAL. Sokol, interview.

231. Sokol, interview. The first F-15 Weapons School class arrived just after ACEVAL. Sokol recalled that the topic of targeting frequently came up in discussions around the bar at the Nellis Officer's Club, a historic clearinghouse for air tactics as fighter pilots rotated in for training or Red Flag exercises. Additionally, Sokol taught the Aggressor pilots the Eagles' intercept and targeting plans (after AIMVAL-ACEVAL) so that when they traveled to unit locations to provide dissimilar air combat training, they could help spread the word on the on the new, preferred targeting tactics. On the F-15 Weapons School, see B. Lamb, "Aerial Attack."

232. Anderegg, *Sierra Hotel*, 160. Similar targeting schemes were developed for various other potentialities, such as two F-15s facing more than two adversaries and four F-15s facing one, two, three, four, or five adversaries.

233. Cliver, interview.

234. Sokol, interview; Wallace, interview.

235. Sokol, interview. Sokol remembered that even in the F-15 Weapons School it was difficult to gain access to the ACMI range at Nellis due to other, higher-priority flight activities and tests.

236. Sparks, interview; Dvorchak, interview; Cliver, interview. Cliver explained his rationale: "Find the rest of the guys. Find the rest of the guys. . . . That's why we begged for search while track or some TWS-kind-of-a-capability in the jet."

237. Wallace, interview; Sokol, interview.

238. Sokol, interview.

239. Wallace, interview. Anderegg recalled that a "big box VTR," housed in a compartment "under the engine intake," arrived in the new C-model of the F-15. "It ran 22-minutes. . . . We did have radar, and we did have HUD—couldn't have them together, one or the other. So at ten miles, there was a . . . 'Go HUD, go HUD' call" to remind the pilots to switch from recording the VSD to recording the HUD. Anderegg, interview, October 2, 2013.

240. Wallace, interview. Wallace noted that "it would have helped out a whole lot more if we had the VCRs" when training the first, new F-15 pilots, "but we didn't."

241. Like many other F-15 instructors, I told my young wingmen that I wanted them to be "robots" working their radar, adhering to the prebriefed search and targeting plan. It was a phrase I had heard growing up as a wingman in the Eagle earlier. I didn't appreciate the philosophical significance of the term at the time.

242. L. Welch, interview, October 1, 2013; Anderegg, interview, October 2, 2013.

243. L. Welch, interview, October 1, 2013. Welch explained, "The F-15 certainly didn't demand any more flying skill. In fact, it demands less." Anderegg affirmed the assessment: "The F-15 is the easiest airplane in the world to fly." Anderegg, interview, October 2, 2013. See also Shower, foreword, 4.

244. Anderegg, *Sierra Hotel*; Michel, "Revolt of the Majors."

245. Blesse, Corona Ace Oral History Interview, 62–65. Contrary to the "fuzzy thinks at TAC," Blesse argued that most fighter pilots understood the value of dissimilar air combat training (e.g. allowing an F-4 pilot to practice fighting against a Navy A-4). By focusing solely on similar combat training, Blesse explained, "the only measure of difference you have now is in the pilots themselves. If one guy cannot get away from another guy, cannot shake him off the rear by doing things that are normal [because they are flying the same aircraft], he will then try things that are abnormal and spin himself in." Thus, similar training was, in Blesse's estimation, "the most dangerous thing you can do." On the lifting of the dissimilar air combat training restriction, see Johnston, "Dissimilar Aircraft Engagement."

246. Michel, "Revolt of the Majors," 219.

247. Quoted in ibid., 186–88, 197–98. Michel also noted that the 1973 Arab-Israeli War influenced Dixon's decision to revamp the Air Force's air combat training programs. See also Laslie, *Air Force Way of War*, 37.

248. Anderegg, *Sierra Hotel*, 47; Michel, "Revolt of the Majors," 144. Michel cited Ron Keys's assessment, "I had been having the time of my life [up to this point] . . . it had never occurred to me I had been poorly trained." Keys eventually became commander of TAC in 2005.

249. Michel, "Revolt of the Majors," 5. See also Anderegg, *Sierra Hotel*, 47–49; Laslie, *Air Force Way of War*. Worden suggested that the fighter pilots' new outlook contributed to their meteoric rise to the upper echelons of Air Force leadership, supplanting the traditional, SAC-minded bomber generals. Worden, *Rise of the Fighter Generals*.

250. Anderegg, *Sierra Hotel*, 42, 46, 55; Anderegg, interview, October 2, 2013. See also Laslie, *Air Force Way of War*, 51–53.

251. V. Mitchell, *Air Force Officers*, 197–99, 201. Mitchell observed that the nationwide push for scientists and engineers following *Sputnik* "dovetailed nicely" with the Air Force's officer-education plans. Both the Aviation Cadet Program and the Officer Candidate School continued to operate for at least another decade. Officer Candidate School (OCS) was different than Officer Training School (OTS); the latter selected only college graduates and continues to operate today. The other commissioning sources included the new Air Force Academy (established in 1954) and Reserve Officer Training Corps (ROTC) programs located on civilian college campuses across the country. Officers who had entered the service without a college degree were often offered the opportunity to pursue one, but the dramatic shift toward college-educated pilots occurred when the majority of World War II and Korean War officers had already retired or separated from the service. See also Anderegg, *Sierra Hotel*, 39, 41–42.

252. For example, General Felix Rogers, himself a product of the Aviation Cadet program who went on to become a World War II ace and, after earning a college degree sponsored by the Air Force, commander of Air University and Air Force Logistics Command, exclaimed in a 1977 interview, "Everybody has got to be a college graduate to fly an airplane. Merde! I don't give a damn what the Human Resources Lab says—everybody doesn't have to be a college graduate to fly an airplane." Rogers, Corona Ace Oral History Interview, 47.

253. Anderegg, *Sierra Hotel*, 55, 74–75, 96–98; Anderegg, interview, October 2, 2013. Despite the additional emphasis it had gained over the years, pilots' knowledge of the information in their flight manuals was still relatively poor—in 1972, only about 10 percent of Phantom pilots and WSOs passed a TAC-sponsored written exam covering their aircraft and weapons systems. Anderegg, *Sierra Hotel*, 74–75; see also Michel, "Revolt of the Majors," 146–48. As fighter aviation became more technically focused, there was an advantage to possessing a technical degree, but the service never excluded officers with nontechnical degrees from its ranks of pilots. A common joke among fighter pilots was the requirement to explain certain topics in simple terms for the history and political science majors in the crowd. See, for example, Snodgrass and New, "BFM for History Majors." (BFM refers to basic fighter maneuvers, aka dogfighting.)

254. L. Welch, interview, October 1, 2013.

255. Sokol, interview.

256. Ibid. Sokol described one premission briefing during his F-4 front-seater training program: "Our whole briefing was at the duty desk. . . . [The instructor] said, 'Have you seen the high-speed yo-yo?' . . .'Have you seen the low-speed yo-yo?' . . .'Have you seen the barrel roll attack?' . . .'Have you seen the butterfly setup?' . . .'We're going to do butterfly setups. Any questions?' 'No, sir.'" The "duty desk" is a dispatch desk in the squadron where pilots learn which specific aircraft they will fly for their mission.

257. Anderegg, *Sierra Hotel*, 50–57. Anderegg explained that the shift toward instructional briefs and debriefs began at the Air Force's Weapons School in 1974 and culminated with a set of articles published in the winter 1976 and spring 1977 issues of *Fighter Weapons Review*. Before that time, the Weapons School course was "nothing more than a rite of passage wherein the students were harassed and hazed rather than instructed"—similar to the classic wingman experience in the operational units (52). It was this Weapons School attitude that had been partly responsible for the resistance to non–fighting wing formations in the Phantom. With the shift in training methods, the Weapons School instructors dedicated themselves to

"improve local training [across the Air Force] and make ourselves more efficient." Keith, "Introduction." The key, according to one instructor, was to introduce a "building block approach" to air combat training, beginning with simple concepts and then gradually working toward more complex employment environments. Jumper, "Air-to-Air Training to Win"; see also Laslie, *Air Force Way of War*, 52–53. Sokol, who was teaching F-15 pilots at the Weapons School at the time, explained: "We're trying to teach the guys how to brief and debrief as much as anything else. I mean, flying the airplane is important. But that whole being an IP's IP [instructor pilot's instructor pilot], that was what we were about. . . . It's not to go out and just prove you beat some guy who wasn't somebody." Sokol, interview.

258. L. Welch, interview, October 1, 2013.

259. Anderegg, interview, October 2, 2013.

260. Ibid. Anderegg explained that it became obvious in the air which pilots had not yet come to this realization, the ones that preferred to spend their time "hanging around the duty desk, drinking coffee," instead of studying. "When you're out there flying with a guy that confronts a situation he hasn't thought about, you can tell it immediately. Because nothing happens. It's quiet on the radio. Airplane's just going straight ahead. You know that this guy—he hasn't seen this before, and he doesn't know what to do about it." Stucky made explicit reference to the fundamental importance of radar employment in the Eagle, likening it to "blocking and tackling" foundations in football. "The bottom line," he wrote, "is that radar skills directly contribute to your SA [situational awareness], and proficiency directly affects how well you can process that information into sound tactical decisions." He also stressed the importance of practicing HOTAS switchology, "If you are *thinking about* hand actions, you're not good enough." Stucky, "Blocking & Tackling," 25 (emphasis in original).

261. L. Welch, interview, October 1, 2013; L. Welch, Oral History Interview, 80–81.

262. Wallace, interview. Absent any remarkable buffoonery from a pilot such as getting lost, Wallace's judgment often came down to, "Were they part of the engagement or were they kind of just tagging along?"

263. Risner, Oral History Interview, 161; D. Martin, "Robinson Risner"; "Brigadier General Robinson Risner." As the vice commander of the USAF Tactical Fighter Weapons Center at Nellis AFB, Risner was the deputy test director for AIMVAL-ACEVAL. After Risner's retirement, Hildreth assumed the post and the deputy director responsibilities.

264. Risner, Oral History Interview, 161.

265. L. Welch, interview, October 1, 2013.

266. Anderegg, interview, October 2, 2013. See also Anderegg, *Sierra Hotel*, 183–84.

267. Davies and Dildy, *F-15 Eagle Engaged*; Shower, foreword, 4.

268. Osterhout, Berdan, and Gordon, "F-15 Briefing Script," slide 4.

269. Hoag, "Superfighter," 25, 28.

270. Dvorchak, interview. Dvorchak continued, "Everything that came out of [the tests] was a byproduct." Sparks and Cliver wholeheartedly agreed with the assessment. Sparks, interview; Cliver, interview. Cliver also offered another perspective on the tests' impact: "We made eight guys pretty proficient, as proficient as you could get. And God only knows how many lives and systems they touched as they proceeded on through their Air Force careers. So that's kind of cool."

271. Anderegg, *Sierra Hotel*, 164; Anderegg, interview, October 2, 2013.

272. Cliver, interview.

Chapter 6. Conclusion

1. Mindell, *War, Technology, and Experience*, 5–6, 41–43, 123–24, 131–32. Mindell noted that Nathaniel Hawthorne also perceived the irony inherent in the *Monitor*, citing Hawthorne's observation: "Even heroism—so deadly a gripe is Science laying on our noble possibilities—will become a quality of very minor importance."

2. Quoted in ibid., 2, 85.

3. Quoted in ibid., 65, 88, 109.

4. Quoted in ibid., 99.

5. Ibid., 148–49.

6. Quoted in ibid., 68. On the *Monitor*'s sailing qualities, see pp. 64–68, 113, 133.

7. Ibid., 148 (emphasis in original). Morrow offered the term *technocratic heroism* to describe the early fighter pilots' circumstances. Morrow, *Great War in the Air*, 26.

8. Mindell, *War, Technology, and Experience*, 146–47.

9. "From Struts to Jets," 62–63.

10. On constructing a cognitive ethnography of the operator, see Hutchins, *Cognition in the Wild*. On our ability to learn about interpersonal relationships through our relationships with our machines, see Sennett, *Craftsman*.

11. L. Welch, interview, October 1, 2013.

12. Olds, *Fighter Pilot*, 293. Olds's romantic passage does not, in my opinion, match the tone of the rest of his memoirs or oral history interviews; the attitudes, however, are consistent. On the Sabre pilot's personal connection, see Salter, *Hunters*, 193. See also Hoover's similar but less eloquent description of "the concept of man and his airplane." Hoover, "Place in the Sun."

13. Sheridan, *Telerobotics, Automation, and Human Supervisory Control*. Sheridan noted that the term *supervisory control* "is derived from the close analogy between the supervisor's interaction with subordinate human staff members in a human organization and a person's interaction with 'intelligent' automated subsystems. . . . In the strictest sense, *supervisory control* means that one or more human operators are intermittently programming and continually receiving information from a computer that itself closes an autonomous control loop through artificial effectors and sensors to the controlled process or task environment" (1). See also Parasuraman and Sheridan, "Model for Types and Levels of Human Interaction." On the application of Sheridan's concepts to military equipment, see *Role of Autonomy in DoD Systems*.

14. For example, Sokol remarked that he had a friend who, after having only flown F-15s (beyond the normal pilot training aircraft), had an opportunity to fly the F-4 while attending the Air Force's Test Pilot School. The friend later told Sokol, "I don't know how you guys could go to war in that airplane, because it's just a full-time job to fly it, let alone do anything else with it." "And it was," Sokol agreed. Sokol, interview.

15. *Joint Strike Fighter*, 18. The F-35 relies on another 14.5 million lines of code for its ground support and maintenance activities.

16. L. Welch, interview, October 1, 2013.

17. The introduction of women into the ranks of fighter pilots in the early 1990s would also have a profound impact on the myth of the fighter pilot, imbued as it was with masculine themes. On some of the controversy and the first female pilots' experiences, see J. Johnson,

"The New Top Guns"; Kantrowitz and Clift, "The Right to Fight"; May, "Women in the Wild Blue Yonder"; Cummings, *Hornet's Nest*; Spears, *Call Sign Revlon*.

18. Originally intended to be the air superiority replacement for the F-15, the F-22 also has the capability to deliver bombs against a set of GPS (Global Positioning System) coordinates. Even foregoing its air-to-ground role, the Raptor still required a dramatic shift in the tasks of the fighter pilot.

19. Fulghum and Fabey, "F-22."

20. "View from the Cockpit," 13; Brower, "Lockheed F-22 Raptor," sect. 2. One recent F-22 pilot noted that in subsequent software versions, the display included additional amplifying details about the target's identification source data; the early test pilots apparently weren't pleased about not having insight into the computer's decision matrix and demanded the change. Laidlaw, interview.

21. Brower, "Lockheed F-22 Raptor," sect. 5. For more detail on the Raptor's "sensor fusion" and its information architecture within the cockpit, see Greeley and Schwartz, "F-22 Cockpit Avionics"; "Holding Four Aces"; "Complete Picture."

22. Fulghum and Fabey, "F-22." Gen. Ronald Keys, commander of ACC, was quoted as remarking, "It's fairly spooky. There's hardly a word spoken among Raptor pilots [inflight]."

23. D. Bond, "Stuck in Hawaii." The infamous Pacific crossing attempt occurred in early 2007; the Raptors were forced to immediately turn around, fly back to Hawaii, and wait for Lockheed to publish a software update for their aircraft. Almost a year earlier, an F-22 pilot was trapped in his jet for over five hours when his canopy failed to open; fire department crews eventually had to cut the canopy open with a chainsaw. Although the failure was eventually determined to be mechanical (a set of too-short screws had backed themselves out of position), I recall Eagle pilots likening the Raptor's canopy problem to the failure of the HAL 9000 in the movie *2001: A Space Odyssey*: "Open the canopy, HAL." "I'm sorry, Dave. I'm afraid I can't do that." Hirschman, "F-22 Fighter Won't Get"; Kubrick, *2001*.

24. Fulghum, "Raptor's Edge." Fulghum quoted David Scott, director of F-22 business development for Lockheed Martin.

25. Fulghum and Fabey, "F-22." The article included an F-22 commander's assessment of his squadron's performance during a recent, large air combat exercise: "We killed [the bandits] rapidly. We didn't do any major turning. It's not that the J-Turn maneuver [a Raptor-specific dogfighting maneuver] isn't fun, but we didn't get a chance to use it." The article reported that the F-22s amassed "144 kills to no losses during the first week of the joint service Northern Edge exercise in Alaska [in 2006], only three air-to-air 'kills' were in the visual arena—two involving AIM-9 Sidewinders and one the F-22's cannon."

26. Blair, "Ten Thousand Feet and Ten Thousand Miles," 62, 65.

27. Goldsmith, "Comments on 'Ten Thousand Feet and Ten Thousand Miles.'"

28. R. Howard, "Comments on 'Ten Thousand Feet and Ten Thousand Miles'"; M. Lamb, "Comments on 'Ten Thousand Feet and Ten Thousand Miles.'" See also Senn, "Ricochets & Replies." Amid the uproar, Blair's implicit accusation that the Air Force was mismanaging the RPA career field garnered congressional attention, with lawmakers quickly calling for a review of RPA officers' promotion rates. "Lawmakers Ask for Review."

29. Panetta, "Distinguished Warfare Medal." The Bronze Star is awarded to an individual who "shall have distinguished himself by heroic or meritorious achievement or service, not

involving participation in aerial flight, in connection with military operations *against an armed enemy*" (emphasis added). "Bronze Star."

30. "Military Order of the Purple Heart Opposes Precedence of New Defense Medal."

31. "VFW Believes Distinguished Warfare Medal Should Not Outrank the Bronze Star." The public outcry spurred Congressman Joe Wilson (R-SC) and forty-eight of his fellow representatives to cosign a letter to Panetta's replacement arguing that the proposed ranking of the DWM above the Bronze Star "is a disservice to our service members and veterans." J. Wilson to Hagel, March 4, 2013.

32. Cited in Holmes, "Why Drone Pilots Deserve Medals."

33. "Medals for Drone Warriors Canceled." On March 20, 2014, Hagel directed a "comprehensive review of the military decorations and awards program," which in part would address "the best way to recognize service members who use remote technology to directly impact combat operations, such as through cyber and remotely piloted aircraft." Garamone, "Hagel Orders Review of Military Decorations."

34. Cited in Cullen, "MQ-9 Reaper Remotely Piloted Aircraft," 19. See also Chapa, "Remotely Piloted Aircraft."

35. Blair, "Ten Thousand Feet and Ten Thousand Miles," 63.

36. Quoted in Michaels, "Drones Change 'Top Gun' Culture." The director, J. D. Wyneken, elaborated, "In the view of many aces, 'Just the very idea of a pilotless aircraft is dishonorable.'" For an organization whose members wouldn't recognize the GIB-aces of the Phantom, such an attitude is hardly surprising. See also Jaffe, "Combat Generation."

37. Quoted in Michaels, "Drones Change 'Top Gun' Culture." Michaels quoted John Hope, the executive director of the Red River Valley Fighter Pilots Association, a popular organization for Vietnam veteran fighter pilots.

38. Cullen, "MQ-9 Reaper Remotely Piloted Aircraft," 43, 168, 272, 277, 287. Cullen's dissertation research provided the first ethnographic account of Air Force RPA operators at work. Although it contained no classified information, Cullen's dissertation was heavily redacted by the Air Force before its release to the MIT archives; officials even redacted Cullen's entire bibliography. Mindell obtained permission to summarize Cullen's work in Mindell, *Our Robots*, chap. 4; see p. 240n113. On the Air Force's early RPAs, see Ehrhard, "Unmanned Aerial Vehicles"; Ehrhard, *Air Force UAVs*. For an alternative perspective of RPA operations, see M. Martin and Sasser, *Predator.*

39. Cullen, "MQ-9 Reaper Remotely Piloted Aircraft," 200, 267. Colonel Case Cunningham, an RPA Wing Commander, noted that in 2016, MQ-9 RPA pilots were "flying CAS [close air support] in an urban environment with great percentages of weapon employment within 'danger close' ranges from friendly forces"—one of the most challenging and dynamic combat environments imaginable. Moreover, Cunningham explained, "[the RPAs are] tasked for some of the exact same theater airpower missions as other fighter aircraft." Cunningham to Fino.

40. "Air Force Approves RPA Initiatives"; Chatterjee, "A Chilling New Post-Traumatic Stress Disorder"; Bumiller, "Drone Pilots"; Ouma, Chappelle, and Salinas, *Facets of Occupational Burnout*; Mindell, *Our Robots*, 152–53.

41. Conflict over the RPA pilots' entitlement to Aviation Career Incentive Pay was one manifestation of the confusion: see Hardison, Mattock, and Lytell, *Incentive Pay*; Buzanowski, "Air Force Officials Approve Incentive Pay"; "ACP Helps Service." The Air Force officially adopted the *pilot* label for its RPA operators during a particularly tumultuous period following

the firings of the secretary of the Air Force and the CSAF in 2008. One senior Air Force official privately noted that the political environment at the time rendered it extremely difficult to call the RPA operators anything other than *pilots*; choosing a different label might fuel the notion that the service was not fully supportive of the RPA mission.

42. Quoted in Eagan, "New ACC Program Begins."

43. Stanley, "COMACC Addresses RPA Health."

44. On the stigma attached to RPA aircrews and their efforts to overcome it, see Blair, "Ten Thousand Feet and Ten Thousand Miles"; Ferran, "Drone 'Stigma' Means 'Less Skilled' Pilots"; *Actions Needed to Strengthen Management*; Jaffe, "Combat Generation"; M. Thompson, "Drone Pilots"; Schogol, "RPA Pilots"; "Lawmakers Ask Review"; Holmes, "Why Drone Pilots Deserve Medals"; Blair and Helms, "Swarm"; Byrnes, "Dark Horizon"; Spinetta, "Glass Ceiling"; Cantwell "Beyond Butterflies"; Cantwell, "Operators of Air Force Unmanned Aircraft Systems"; Mindell, *Our Robots*, 153–57.

45. Lee, "Pilots Needed."

46. Stanley, "COMACC Addresses RPA Health"; Lee, "Pilots Needed." Additional CPIP solutions are outlined in "Air Combat Command Charts Path"; "Air Force Approves RPA Initiatives." In an effort to funnel even more manpower into the RPA enterprise, the Air Force also announced that it would open the RQ-4 Global Hawk, a high-flying and exclusively intelligence-gathering RPA platform, to *enlisted* pilots. As presently envisioned, the enlisted pilots will go through the same RPA-training pipeline as the officer pilots, and upon graduation all will wear the same RPA-pilot wings. See "AF Selects First Enlisted Airmen for Global Hawk." While the enlisted RPA pilot initiative is noble, it will likely only muddy future discussions about RPA pilots' qualifications and expertise.

47. On the RPA flight training program, see Church, "RPA Ramp Up"; "Air Force Officials Announce Remotely Piloted Aircraft Pilot Training"; Fulghum, "Overdrive"; Tirpak, "Putting the Pilot in the RPAs."

48. Mindell identifies three myths of autonomy: the myth of linear progress, the myth of replacement, and the myth of full autonomy. *Our Robots*, 8–10.

49. On the difficulties bridging the cultural divide, see also M. Wells, "Tribal Warfare"; Cantwell, "Operators of Air Force Unmanned Aircraft Systems"; Blair and Helms, "Swarm." Blair and Helms acknowledge that "*pilot* is a term of great prestige in the Air Force," and they extol the service to, "instead of allowing that word to capture us, let us instead capture it and use its gravity to slingshot our service forward." The problem is that the Air Force public relations machine cannot regulate the term's usage; it's bound within the now century-old popular myth of the fighter pilot.

50. Mindell, *Our Robots*, 8–11, 13 (emphasis in original). Mindell offered two follow-on questions: "*How does human experience change? And why does it matter?*"

51. O'Mara observed a similar phenomenon in his study of B-17 bomber aircrews during World War II. With the introduction of a new autopilot-bombsight system in the Flying Fortress, the bomber pilot was stripped of his traditional flying duties approaching the target; in their place, the pilot assumed new tasks and a new professional identity tied to his role as *aircraft commander* in charge of the entire crew's performance. O'Mara, "Socio-Technical Construction of Precision Bombing."

52. Buck, *Pilot's Burden*, 21–22. Buck, a former commercial airline captain, summarized the effects of technology: "When the . . . DC-3 came in, seat-of-the pants flying went out.

While there had always been thinking pilots, until then flying basically had primarily involved fine touch on the controls, a feel for the airplane's limits, and boldness. Flying was an art. The DC-3['s] . . . higher performance was achieved through technology that required the pilot to start down the road to being a technician, a direction that now takes precedence. Flying, however, periodically still needs the fine touch of the art of flying, although hard-nosed scientists believe technology can do it all." See also Mindell, *Our Robots*, chap. 3.

53. McCulloch, "Surgical Professionalism in the 21st Century." McCulloch noted that some senior surgeons worried that increasingly specialized surgical techniques and advanced technologies would challenge "the role and professional ethos of the surgeon." This would result in "a loss of professional values leading to a diminished role for the surgeon as a kind of clinical lathe operator, without any continuing responsibility for his patient—deskilled, disempowered, and easily manipulated by financial decision makers in charge of hospitals" (177).

54. Sibley, *Draft Report of the MITx Subcommittee*. Like most college campuses, MIT in 2014 was struggling to determine the viability of the MOOC (massive open online course) model within higher education. An internal MIT report noted the potential benefits of online education but also emphasized "the values of face-to-face engagement between faculty and students," which included "the opportunities to develop normative sensibilities and moral commitments that will serve as lifelong guides" (10). The chair of the MIT subcommittee therefore recommended the university "go slow, and see what's going to happen." Quoted in Lin, "Group Urges Checks on Online Learning." On higher education, see also Christensen et al., *Disrupting College*.

55. Mindell termed this optimum blend "the perfect 5," a "happy medium" on the scale between no automation (zero) and total automation (ten), that corresponds to "trusted, transparent, reliable, safe autonomy that is fully interactive." Quoted in Dizikes, "Robots and Us."

56. The "everyman a tiger" description occasionally resurfaces; see, for example, Clancy, *Every Man a Tiger*.

57. Roberts, Corona Ace Oral History Interview, 2–3, 24. Meyer offered a similar appraisal: "Nobody ever gets to be an ace by accident." J. Meyer, Oral History Interview, 3.

58. Olds, Corona Ace Oral History Interview, 76.

WORKS CITED

Abbreviations

AFHRA	Air Force Historical Research Agency, Maxwell AFB, AL
AFMC/HO	Air Force Materiel Command History Office, Wright-Patterson AFB, OH
CSDL-HC	Charles Stark Draper Lab–Historical Collection, MIT Museum, Cambridge, MA
DTIC	Defense Technical Information Center, online
MIT	Massachusetts Institute of Technology Library System, Cambridge, MA
MSFRIC	Muir S. Fairchild Research Information Center, Maxwell AFB, AL
NARA	National Archives and Records Administration, online
NHHC	Naval History and Heritage Command, Navy Yard, Washington, DC
NMUSAF	National Museum of the US Air Force, Wright-Patterson AFB, OH

Primary and Secondary Sources

1 TFW Aerial Attack Manual, vol. 1. Langley AFB, VA, January 1976. In *History of the 1st Tactical Fighter Wing, Jan–Mar 1976*, vol. 3. K-WG-1-HI, AFHRA.

"1st TFW to Nest Eagles." *The Flyer.* August 28, 1975. In *History of the 1st Tactical Fighter Wing, Jul–Sep 1975*, vol. 1. K-WG-1-HI, AFHRA.

1TFW/DO. Message to TAC/DOOW. "Red Flag 77-8 After Action Report." 121730Z Aug 77. In *History of the 1st Tactical Fighter Wing, Jul–Sep 1977*, vol. 4. K-WG-1-HI, AFHRA.

———. Message to TAC/DOOW. "Red Flag 77-9 (Michael AAF) After Action Report." 141600Z Sep 77. In *History of the 1st Tactical Fighter Wing, Jul–Sep 1977*, vol. 4. K-WG-1-HI, AFHRA.

———. Message to TAC/DOOW. "Red Flag 77-9 (Nellis) After Action Report." 121930Z Sep 77. In *History of the 1st Tactical Fighter Wing, Jul–Sep 1977*, vol. 4. K-WG-1-HI, AFHRA.

———. Message to TAC/DOOW. "Red Flag 77-10 After Action Report." 201700Z Oct 77. In *History of the 1st Tactical Fighter Wing, Jul–Sep 1977*, vol. 4. K-WG-1-HI, AFHRA.

———. Message to HQ TAC/DOOW. "Red Flag 78-1 After Action Report." 281245Z Nov 77. In *History of the 1st Tactical Fighter Wing, Oct–Dec 1977*, vol. 4. K-WG-1-HI, AFHRA.

4th Fighter-Interceptor Group Tactical Doctrine, July 22, 1951. In *History of the 4th Fighter-Interceptor Group, Jul–Sep 1951*. K-GP-4-HI, AFHRA.

4th Fighter-Interceptor Group Tactical Doctrine, December 28, 1952. In *History of the 4th Fighter-Interceptor Wing, Jul–Dec 1952*. K-WG-4-HI, AFHRA.

4th Fighter-Interceptor Group Tactical Doctrine, June 17, 1953. In *History of the 4th Fighter-Interceptor Wing, Jan–Jun 1953*. K-WG-4-HI, AFHRA.

8th Tactical Fighter Wing Tactical Doctrine, March 1967. In *History of the 8th Tactical Fighter Wing, Jan–Jun 1967*, vol. 2. K-WG-8-HI, AFHRA.

8th Tactical Fighter Wing Tactical Doctrine, December 1967. In *History of the 8th Tactical Fighter Wing, Oct–Dec 1967*, vol. 2. K-WG-8-HI, AFHRA.

"13 Red Jets Shot Down in 2 Battles; Ace Gets 4." *Washington Post*, December 14, 1951.

"208 Rockets Fired at Runaway Plane." *Los Angeles Times*, August 17, 1956.

366 TACFTRWG. Message to CSAF. "Presentation to Sec Def by Col FC Blesse, 366TFW DaNang." 140945Z Jul 67. K717.1622, AFHRA.

366 TFW. Message to 7 AF. "AFR 57-4 Mod I Proposal." May 11, 1967. In *History of the 366th Tactical Fighter Wing, Apr–Sep 1967*, vol. 1. K-WG-366-HI, AFHRA.

———. Message to 7 AF. "F-4C Air to Air Modification Program." May 5, 1967. In *History of the 366th Tactical Fighter Wing, Apr–Sep 1967*, vol. 1. K-WG-366-HI, AFHRA.

———. Message to NMCC, CINCPAC, PACAF CC, and 7AF CC. "Change 1 to 366TFW OPREP-3, Number 010, FASTEL 443." 141800Z May 67. In *PACAF DO Read File, 13-15 May 67*. K717.312, AFHRA.

———. Message to NMCC, CINCPAC, PACAF CC, and 7AF CC. "Change 1 to 366TFW OPREP-3, Number 011, FASTEL 442." 141410Z May 67. In *PACAF DO Read File, 13–15 May 67*. K717.312, AFHRA.

4453d CCTW RTU Weapons School: Intercept Techniques. Course No. 111509F. 1966. In *History of the 4453d Combat Crew Training Wing, Jul–Dec 1966*, vol. 3. K-WG-4453-HI, AFHRA.

A-1 Series (Gun-Bomb-Rocket) Sight. Continental Air Command Manual 50-11, vol. 2, pt. 2. Mitchell AFB, NY: Headquarters, Continental Air Command, 1952. K419.7242050-11, AFHRA.

"The A-1 Sight and Air to Air Gunnery." *Fighter Gunnery*, no. 1 (1951): 3–8. AR.2007.187, NMUSAF.

A-1 Sight for the Control of Gunfire from Fixed Guns, Rocketfire, and Bombing from Aircraft, vol. 1. Cambridge, MA: Instrumentation Laboratory, MIT, April 1946. 146.02-5, AFHRA.

"The A-4 Sight." *Fighter Gunnery*, no. 3 (1951): 15–18. AR.2007.187, NMUSAF.

Accelerated Comparison Test of the K-14 Sight and J-2 Fire Control System in the F-86E for Fighter to Fighter Combat. Project APG/ADB/59-A. Eglin AFB, FL: Air Proving Ground Command, September 26, 1952. AD0004359, DTIC.

"'Ace' Fighter Wing Commanded by 'Aces.'" *USAF News Release*, November 1966. K740.951-2, AFHRA.

"Ace to Tour Jet Bases." *Washington Post*, February 4, 1952.

ACEVAL AIMVAL JTF. Message to AIG. "AIM-7F Kill Removal/F-14 RHAW." 280031Z May 76. In *History of the Air Force Test and Evaluation Center, 1 Jan–31 Dec 1976*, vol. 21. K150.01, AFHRA.

———. Message to CNO. "F-14/F-15 Use of RHAW in ACEVAL-AIMVAL." 222330Z Apr 76. In *History of the Air Force Test and Evaluation Center, 1 Jan–31 Dec 1976*, vol. 21. K150.01, AFHRA.

———. Message to CNO and CSAF/XOO. "ACEVAL-AIMVAL Initial Starting Conditions." 180110Z Nov 76. In *History of the Air Force Test and Evaluation Center, 1 Jan–31 Dec 1976*, vol. 21. K150.01, AFHRA.

———. Message to DDR&E. "ACEVAL-AIMVAL Monthly Progress Report (as of 6 Jan 76)." 061830Z Jan 76. In *History of the Air Force Test and Evaluation Center, 1 Jan–31 Dec 1976*, vol. 21. K150.01, AFHRA.

———. Message to DDR&E. "ACEVAL-AIMVAL Oct Progress Report." 090145Z Nov 76. In *History of the Air Force Test and Evaluation Center, 1 Jan–31 Dec 1976*, vol. 22. K150.01, AFHRA.

"ACP Helps Service Retain Rated, RPA Pilots." Targeted News Service, Washington, DC, January 26, 2011.

Actions Needed to Strengthen Management of Unmanned Aerial System Pilots. GAO-14-316. Report to Congressional Requesters. Washington, DC: US Government Accountability Office, April 2014.

Adas, Michael. *Machines as the Measure of Men: Science, Technology, and Ideologies of Western Dominance*. Ithaca, NY: Cornell University Press, 1989.

ADTC/XR. Message to AFSC/TEV. "ACEVAL-AIMVAL Test Plans REF: AFSC/TEV MSG R161900Z Sep 76." 272017Z Sep 76. In *History of the Air Force Test and Evaluation Center, 1 Jan–31 Dec 1976*, vol. 21. K150.01, AFHRA.

"Advanced Short Range Air-to-Air Missile Technology." *Congressional Record* 121, no. 23 (1975): 29370.

"The Advantage Is Here: How the Phantom Records What the Radar Sees." *Air Force & Space Digest*, November 1962.

"The Advantage Is Here: Why the Phantom II Has a Two Man Crew." *Air Force & Space Digest*, June 1962.

Aerial Attack Section, 414 Fighter Weapons Squadron. "Anything Else Is Rubbish!" *Fighter Weapons Review*, Summer 1971, 33–34. MSFRIC.

"AF Selects First Enlisted Airmen for Global Hawk Pilot Training." Secretary of the Air Force Public Affairs. July 6, 2016. Accessed October 6, 2016. http://www.af.mil/News/ArticleDisplay/tabid/223/Article/828066/af-selects-first-enlisted-airmen-for-global-hawk-pilot-training.aspx.

AFSC/CS. Message to ADTC/CC et al. "ACEVAL/AIMVAL Test Plan." 052310Z Feb 76. In *History of the Air Force Test and Evaluation Center, 1 Jan–31 Dec 1976*, vol. 21. K150.01, AFHRA.

Agan, Arthur C., Lt. Gen., USAF (ret.). Oral History Interview. Interview by Denys Volan, February 1970. K239.0512-899, AFHRA.

———, Lt. Gen., USAF (ret.). Oral History Interview. Interview by Jack Neufeld, October 2, 1973. K239.0512-857, AFHRA.

"'Aggressors' Organized to Reduce Casualties." *Daily Press* (Newport News, VA), May 7, 1976. In *History of the 1st Tactical Fighter Wing, Apr–Jun 1976*, vol. 1. K-WG-1-HI, AFHRA.

"Air Combat Command Charts Path for Remote Aircrews, Units." Air Combat Command Public Affairs. December 11, 2015. Accessed October 5, 2016. http://www.acc.af.mil/News/ArticleDisplay/tabid/5725/Article/660288/air-combat-command-charts-path-for-remote-aircrews-units.aspx.

"Air Force Aces." *Air Force*, May 2011.

"Air Force Approves RPA Initiatives." Secretary of the Air Force Public Affairs. April 11, 2016. Accessed October 5, 2016. http://www.af.mil/News/ArticleDisplay/tabid/223/Article/717598/air-force-approves-rpa-initiatives.aspx.

"Air Force Captain Gets 5th MIG to Become Ace." *New York Times*, September 10, 1972.

"Air Force Eagle Joins Nellis Fighters." *Las Vegas Review Journal*, September 16, 1976.

"Air Force Officials Announce Remotely Piloted Aircraft Pilot Training Pipeline." Targeted News Service, Washington, DC, June 9, 2010.

"Air Force Orders New Jet Fighter." *New York Times*, October 19, 1966.

"Air Force Phantom Crews Get Two More MiG-21s; Squadron Now 'Ace.'" *USAF News Release*, Special Announcement C, November 1966. K740.951-2, AFHRA.

"Air Force Tried Out Controversial Planes." *Beaver County (PA) Times*, May 7, 1978.

Air Service Pursuit and Combat Manual, 1922. Gen. William Mitchell Personal Copy. 248.282-41, AFHRA.

Air-to-Air Guided Missiles. Lecture Manuscript, R-49. Tyndall AFB, FL: Air Tactical School, Air University, April 16, 1948. 239.716721-49, AFHRA.

Alden, William A., Lt. Col., USAF. Memorandum to 9 AF. "TACOP Final Report for 45 TFS from 4 April 1965 to 2 August 1965." October 6, 1965. In *History of the 15th Tactical Fighter Wing, Jul–Dec 1965*, vol. 2. K-WG-15-HI, AFHRA.

Alkire, Darr H., Brig. Gen., USAF. "Gun Camera Brief," n.d. Personal Papers of Charles P. Hollstein. 169.7350, pt. 39, AFHRA.

Allen, R. C., Lt. Col., USAF. "Meet the F-110A." *Fighter Weapons Newsletter*, June 1962, 20–22. MSFRIC.

Allied Drive On in Italy—Planes Smash Foe in Air. Film. United News Newsreels, 1944. 208-UN-103, NARA.

Alper, S. R., and T. E. Caywood. *Weapons for Fighter-to-Fighter Engagement*. Memorandum No. 12. Chicago: University of Chicago, Institute for Air Weapons Research, March 1, 1952. K146.08-3, AFHRA.

"American Fighter Aces." *American Fighter Aces Association*. Accessed June 19, 2013. http://www.americanfighteraces.org/.

"America's Knights of the Sky." *Air Force*, May 1954.

Amlie, Thomas S. *A Non-statistical Look at AIMVAL/ACEVAL*, February 3, 1981. K168.03-2107, AFHRA.

Anderegg, C. R., Col., USAF (ret.). Interview by Steven A. Fino, April 9, 2013.

———, Col., USAF (ret.). Interview by Steven A. Fino, October 2, 2013.

———, Maj., USAF. "Meeting the Threat: Sophistication vs Simplicity." *Fighter Weapons Review*, Fall 1982, 2–6. MSFRIC.

———. *Sierra Hotel: Flying Air Force Fighters in the Decade after Vietnam*. Washington, DC: Air Force History and Museums Program, 2001.

Andrews, William F. "To Fly and Fight: The Experience of American Airmen in Southeast Asia." Ph.D. dissertation, George Mason University, 2011.

Antiaircraft Action Summary: World War II. Information Bulletin No. 29. Washington, DC: Headquarters of the Commander in Chief, United States Fleet, October 8, 1945. Accessed December 19, 2013. http://www.history.navy.mil/library/online/antiaircraft_action_summary_wwii.htm#IV.

Appleman, Roy E. *South to the Naktong, North to the Yalu (June–November 1950)*. United States Army in the Korean War. Washington, DC: Center of Military History, United States Army, 1992.

Arnold, Terry, Maj., USAF. "Red Flag: TAC's Realistic Approach to Readiness." *Air Force*, January 1977.

Asay, C. H., Lt. Col., USAF, Maj. J. D. Covington, USAF, and Capt. J. B. Stone, USAF. *History of Operations BOLO*. In *History of the 8th Tactical Fighter Wing, Jan–Jun 1967*, vol. 2. K-WG-8-HI, AFHRA.

Attinello, John S. *Air-to-Air Encounters in Southeast Asia: Account of F-4 and F-8 Events Prior to 1 March 1967*. Report R-123, vol. 1. Institute for Defense Analyses (IDA), Systems Evaluation Division, October 1967. ADC003627, DTIC.

———. *Air-to-Air Encounters in Southeast Asia: F-105 Events Prior to 1 March 1967*. Report R-123, vol. 2. IDA, Systems Evaluation Division, September 1968. ADC003628, DTIC.

———. *Air-to-Air Encounters in Southeast Asia: Events from 1 March 1967 to 1 August 1967 and Miscellaneous Events*. Report R-123, vol. 3. IDA, Systems Evaluation Division, February 1969. ADC003629, DTIC.

Ault, Frank W., Capt., USN. *Report of the Air-to-Air Missile System Capability Review, Jul–Nov 1968*. Naval Air Systems Command, January 1, 1969. NHHC.

Aust, Abner M., Jr., Col., USAF. "PACAF Tactics and Techniques Program (Bulletin #5)," May 17, 1965. K717.549-1, AFHRA.

———, Col., USAF. "PACAF Tactics/Techniques Program (Bulletin #1)," May 4, 1965. K717.549-1, AFHRA.

Austin, Gordon H., Maj. Gen., USAF (ret.). Oral History Interview. Interview by Hugh N. Ahmann, May 18–20, 1982. K239.0512-1325, AFHRA.

Automatic Range-Only Radar for Airborne Fire Control. Syracuse, NY: General Electric Company, January 16, 1953. ADB963572, DTIC.

"Bailey Pipper." *Fighter Gunnery*, no. 2 (1951): 8–9. AR.2007.187, NMUSAF.

Baker, Royal N., Col., USAF. "Report on F-86 Operations in Korea," April 1, 1953. K-GP-4-SU-OP, AFHRA.

Bakke, Samuel O., USAF (ret.). Interview by Steven A. Fino, April 24, 2010.

———, Maj., USAF. Oral History Interview, July 6, 1967. K239.0512-203, AFHRA.

Baldwin, Hanson W. "US Raids 2 Missile Sites in Vietnam, Wrecking One." *New York Times*, July 28, 1965.

Barthes, Roland. *Mythologies*. Translated by Annette Lavers. New York: Hill and Wang, 1972.

Bates, Matthew, SSgt., USAF. "Retired General Becomes Air Force's Newest Fighter Ace." *Air Force News*, February 13, 2008. Accessed June 17, 2013. http://www.af.mil/news/story.asp?id=123086161.

Bauman, Edward J., 2Lt., USAF. *Test Report on GAR-1, Report #25*. ADJ 56-6984. Holloman AFB, NM: Air Research and Development Command, Holloman Air Development Center, October 29, 1956. Microfilm 31792, AFHRA.

Beck, M. O., Capt., USAF. "State of the Art." *Fighter Weapons Review*, Winter 1974, 28–33. MSFRIC.

———, Capt., USAF. "Super Phantom." *Fighter Weapons Review*, Winter 1972, 18–24. MSFRIC.

Beckwith, Arthur V., Capt., USAF. Memorandum to Commanding General, 5th Air Force. "Wing Fuel-Load in MiG-15." August 28, 1951. In *History of the 4th Fighter-Interceptor Group, Jul–Sep 1951*. K-WG-4-HI, AFHRA.

———, Capt., USAF. Memorandum to Commanding Officer, 4th Fighter Interceptor Wing. "An Analysis of Aerial Engagements over North Korea on 23 October 1951." October 26, 1951. In *History of the 4th Fighter-Interceptor Wing, Oct 1951*. K-WG-4-HI, AFRHA.

———, Capt., USAF. Memorandum to Deputy for Intelligence, 5th Air Force. "Re-evaluation of MiG-15 Tactics." September 19, 1951. In *History of the 4th Fighter-Interceptor Wing, Sep 1951*. K-WG-4-HI, AFHRA.

Bedke, Ernest A., Maj. Gen., USAF (ret.). Oral History Interview. Interview by Douglas G. Lamb, January 25–26, 1988. K239.0512-1787, AFHRA.

Beeson, D. W., Capt., USAAF. "Captain D. W. Beeson, 334th Fighter Squadron, 4th Fighter Group, P-47 and P-51." In *The Long Reach: Deep Fighter Escort Tactics*, 70–72. VIII Fighter Command, May 29, 1944. 524.502A, AFHRA.

Beisner, Gerald J., Lt. Col., USAF. "End of Tour Report," September 14, 1966. K740.131, AFHRA.

Bell, Ken. *100 Missions North: A Fighter Pilot's Story of the Vietnam War.* Dulles, VA: Brassey's, 1993.

Bellis, B. N., Maj. Gen., USAF. "F-15 Program Review." Hearings before the US Senate, Committee on Armed Services, March 13, 1974, pp. 4252–73. Washington, DC: Government Printing Office, 1974. Y4.Ar5/3:P94/6/975/pt8.

Biddle, Tami Davis. *Rhetoric and Reality in Air Warfare: The Evolution of British and American Ideas about Strategic Bombing, 1914–1945.* Princeton, NJ: Princeton University Press, 2002.

Bielinski, Henry E., Maj., USAF. "The F-4E: A Pilot's View." *Air Force*, November 1972.

Bijker, Wiebe E., Thomas P. Hughes, and Trevor Pinch, eds. *The Social Construction of Technological Systems: New Directions in the Sociology and History of Technology.* Cambridge, MA: MIT Press, 1989.

"Black Box to Replace Pilot." *Fighter Weapons Newsletter*, September 1959, 50. MSFRIC.

Blair, David J., Maj., USAF. "Ten Thousand Feet and Ten Thousand Miles: Reconciling Our Air Force Culture to Remotely Piloted Aircraft and the New Nature of Aerial Combat." *Air & Space Power Journal* 26, no. 3 (June 2012): 61–69.

Blair, David J., Maj., USAF, and Capt. Nick Helms, USAF. "The Swarm, the Cloud, and the Importance of Getting There First: What's at Stake in the Remote Aviation Culture Debate." *Air & Space Power Journal* 27, no. 3 (August 2013): 14-38.

Blake, Gordon A., Col., USAF. Memorandum to Maj. Howard E. McCormick, USAF. "Request for Communication and Electronic Equipment Data." October 26, 1949. 239.901-4, AFHRA.

Blesse, Frederick C., Maj., USAF. "The Air Tactics Problem." *Fighter Gunnery Newsletter*, January 1954, 15–18. AR.2007.187, NMUSAF.

———, Maj. Gen., USAF (ret.). "The Changing World of Air Combat." *Air Force*, October 1977.

———, Maj. Gen., USAF (ret.). *Check Six: A Fighter Pilot Looks Back.* Mesa, AZ: Champlin Fighter Museum Press, 1987.

———. Maj. Gen., USAF. Corona Ace Oral History Interview. Interview by Lt. Col. Gordon F. Nelson, USAF, February 14, 1977. K239.0512-1077, AFHRA.

———, Maj., USAF. "No Guts, No Glory." *Fighter Weapons Newsletter*, March 1955, 1–74. MSFRIC.

Blood, Gordon F., Maj. Gen., USAF. "End of Tour Report," 1969. K740.131, AFHRA.

Bloomcamp, Francis W., Col., USAF. Memorandum to 1 TFW/CCH. "History: 1 October–31 December 1975." February 5, 1976. In *History of the 1st Tactical Fighter Wing, Oct–Dec 1975*, vol. 1. K-WG-1-HI, AFHRA.

———, Col., USAF. Memorandum to 9 AF/DOST. "Helmet Color in the F-15." October 19, 1976. In *History of the 1st Tactical Fighter Wing, Oct–Dec 1976*. K-WG-1-HI, AFHRA.

———, Col., USAF. "Red Flag 77-4 After Action Report." March 21, 1977. In *History of the 1st Tactical Fighter Wing, Jan–Mar 1977*, vol. 1. K-WG-1-HI, AFHRA.

Bloss, J. H., Lt. Col., USAF. Memorandum to Commanding Generals, First, Second, Fourth, Tenth, Eleventh, and Fourteenth Air Forces. "Status of Development of New Weapons." September 4, 1947. K410.101-11A, AFHRA.

Blumenthal, Ralph. "First 2 Aces of Vietnam War Are Here for Treat." *New York Times*, May 25, 1972.

Boatner, B. L., Maj. Gen., USAF. "Letter Report on Relative Aerial Combat Capability of F-80C Versus F-86A," n.d. K168.16-18, AFHRA.

"Boeing F-15 Eagle: A Legend is Born," YouTube video, Boeing Company, 2002. Posted by Sascha Körber, April 13, 2009. http://www.youtube.com/watch?v=RRUU28P_Jzg.

Bolinger, Fred, Sgt., USAF. "F-15 Operates out of Desert." *The Flyer*. September 15, 1977. In *History of the 1st Tactical Fighter Wing, Jul–Sep 1977*, vol. 4. K-WG-1-HI, AFRHA.

Bolt, Jones E., Maj. Gen., USAF (ret.). Oral History Interview. Interview by Col. Frederic E. McCoy III, USAF. December 6–7, 1984. K239.0512-1624, AFHRA.

"Bomb Loads Slowed F-105's." *New York Times*, April 5, 1965.

Bond, David. "Stuck in Hawaii: F-22s Stuck for a Week in Hawaii." *Aviation Week and Space Technology*, February 19, 2007.

Bond, Robert M., Brig. Gen., USAF. "Advanced Medium Range Air to Air Missile." Hearings before the US Senate, Subcommittee of the Committee on Appropriations, July 17, 1978, pp. 347–66. Washington, DC: Government Printing Office, 1978. Y4.Ap6/2:D36/4/979/pt6.

———, Capt., USAF. "The Elusive Dart." *Fighter Weapons Newsletter*, September 1959, 29–33. MSFRIC.

Bowen, E. G. "The Tizard Mission to the USA and Canada—Sept 1940." Keynote Address, IEEE Radar Conference. Washington, DC, May 1985. Box 1270, P2 Radar, NMUSAF.

Bowman, Buddy, Maj., USAF. "To Fly and Fight." *Fighter Weapons Newsletter*, Winter 1970, 10–11, 25. MSFRIC.

Boyce, Joseph C., ed. *New Weapons for Air Warfare: Fire-Control Equipment, Proximity Fuses, and Guided Missiles*. Science in World War II, Office of Scientific Research and Development. Boston: Little, Brown, 1947.

Boyd, John R., Capt., USAF. "Aerial Attack Study," August 11, 1964. M-U 43947-5, MSFRIC.

———, Capt., USAF. "Fire Control Problems in the Pursuit Curve Attack." *Fighter Weapons Newsletter*, March 1959, 18–24. MSFRIC.

———, Col., USAF. Oral History Interview. Interview by Jack Neufeld, May 23, 1973. K239.0512-859, AFHRA.

Boyd, John R., Lt. Col., USAF, and Thomas P. Christie. "Expanded Maneuverability Theory." *Fighter Weapons Newsletter*, December 1969, 1. MSFRIC.

Boyne, Walter J. "MiG Sweep." *Air Force*, November 1998.

———. *Phantom in Combat*. Washington, DC: Smithsonian Institution Press, 1985.

Boyne, Walter J., and Philip Handleman, eds. *Brassey's Air Combat Reader*. Washington, DC: Potomac Books, Inc., 2005.

Braswell, Arnold W., Col., USAF. Memorandum to Maj. Gen. Yudkin, USAF. "Background Material on SEA Command and Control Issues." July 15, 1967. In "Background Material for Maj. Gen. Gordon M. Graham's End of Tour Report." K740.131, AFHRA.

"Brigadier General Robinson Risner." *USAF Biography*. Accessed March 28, 2014. http://www.af.mil/AboutUs/Biographies/Display/tabid/225/Article/105823/brigadier-general-robinson-risner.aspx.

Briggs, James E., Lt. Gen., USAF. Memorandum to HQ USAF/AFPDC. "Proposed Changes in Pilot Training." January 25, 1963. In *History of the Air Training Command, Jan–Jun 1963*, vol. 16. K220.01, AFHRA.

Briscoe, Walter A., and H. Russell Stannard. *Captain Ball, VC: The Career of Flight Commander Ball, VC, DSO.* London: Herbert Jenkins Limited, 1918.

Broad. "FEAF Intel Form 1, 21 Dec 1950." In *History of the 4th Fighter-Interceptor Group, Oct–Dec 1950.* K-GP-4-HI, AFHRA.

Broadway, Gary L., Capt., USAF. "Have Guns (Three of the Fastest), Will Travel (High-Low-Fast-Slow)." *Fighter Weapons Newsletter*, December 1965, 4–9. MSFRIC.

"Bronze Star." *Air Force Personnel Center Factsheets*, August 23, 2010. Accessed March 15, 2014. http://www.afpc.af.mil/library/factsheets/factsheet.asp?id=7771.

Broughton, Jack. *Thud Ridge: F-105 Thunderchief Missions over Vietnam.* New York: Lippincott, 1969.

Brower, Ronald W. "Lockheed F-22 Raptor." In *The Avionics Handbook*, edited by Cary R. Spitzer, chap. 32. New York: CRC Press, 2001.

Brown, Harold. "Technology, Military Equipment, and National Security." *Parameters* 13, no. 1 (1983): 15–27.

Brown, Kevin. "A Fighter Pilot's Fighter Plane." *Popular Mechanics*, December 1970.

Bruning, John R., Jr. *Crimson Sky: The Air Battle for Korea.* Washington, DC: Potomac Books Inc., 2005.

Buck, Robert N. *The Pilot's Burden: Flight Safety and the Roots of Pilot Error.* Ames: Iowa State University, 1994.

Buderi, Robert. *The Invention That Changed the World: How a Small Group of Radar Pioneers Won the Second World War and Launched a Technical Revolution.* New York: Simon & Schuster, 1996.

Bugos, Glenn E. *Engineering the F-4 Phantom II: Parts Into Systems.* Annapolis, MD: Naval Institute Press, 1996.

Buhrow, Robert E., Maj., USAF. Oral History Interview. Interview by Lt. Col. Robert Eckert, USAF, and Maj. Harry Shallcross, USAF. n.d. K239.0512-028, AFHRA.

———, Maj., USAF. Oral History Interview. Interview by Grover Jarrett, April 11, 1968. K239.0512-047, AFHRA.

Builder, Carl H. *The Icarus Syndrome: The Role of Air Power Theory in the Evolution and Fate of the U.S. Air Force.* New Brunswick, NJ: Transaction Publishers, 1994.

"Building an F-15 Wing: Interview with Col. Larry D. Welch." *The Flyer.* September 11, 1975. In *History of the 1st Tactical Fighter Wing, Jul–Sep 1975*, vol. 1. K-WG-1-HI, AFHRA.

Bumiller, Elisabeth. "Drone Pilots, Waiting for a Kill Shot 7,000 Miles Away." *New York Times*, July 29, 2012. Accessed April 22, 2014. http://www.nytimes.com/2012/07/30/us/drone-pilots-waiting-for-a-kill-shot-7000-miles-away.html.

Burns, Joe Lee, Maj., USAF. "Aggressively Speaking." *Fighter Weapons Review*, Winter 1974, 20–23. MSFRIC.

Burns, John J., Maj. Gen., USAF. Oral History Interview. Interview by Jack Neufeld, March 22, 1973. K239.0512-961, AFHRA.

———, Lt. Gen., USAF (ret.). Oral History Interview. Interview by Hugh N. Ahmann, June 5–8, 1984. K239.0512-1587, AFHRA.

Burrows, William E. *Richthofen: A True History of the Red Baron.* New York: Harcourt, Brace, and World, 1969.

Bush, V. Letter to Maj. Gen. Carl Spaatz, USAAF, March 11, 1942. 520.87241-2, AFHRA.

Buzanowski, J. G. "Air Force Officials Approve Incentive Pay for Officer, Enlisted RPA Aircrews." Federal Information & News Dispatch, February 24, 2010.

Byrnes, Michael W., Capt., USAF. "Dark Horizon: Airpower Revolution on a Razor's Edge—Part Two of the 'Nightfall' Series." *Air & Space Power Journal* 29, no. 5 (October 2015): 31–56.

Calvert, Donald R., Capt., USAF. "Turning the F-4." *Fighter Weapons Newsletter*, March 1967, 5–9. MSFRIC.

Cameron, Rebecca Hancock. *Training to Fly: Military Flight Training, 1907–1945*. Washington, DC: Air Force History and Museums Program, 1999.

Campbell, Joseph. *The Hero with a Thousand Faces*. 3rd ed. Novato, CA: New World Library, 2008.

Campbell, Joseph, and Bill Moyers. *The Power of Myth*. Edited by Betty Sue Flowers. New York: Anchor Books, 1991.

Campion, Frank. "But Why Are We Losing Air Supremacy over Korea." *Life*, February 4, 1952.

Cantwell, Houston R., Lt. Col., USAF. "Beyond Butterflies: Predator and the Evolution of Unmanned Aerial Vehicles in Air Force Culture." Master's thesis, School of Advanced Air and Space Studies, Air University, 2007.

———, Lt. Col., USAF. "Operators of Air Force Unmanned Aircraft Systems: Breaking Paradigms." *Air & Space Power Journal* 23, no. 2 (Summer 2009): 67–77.

Carpenter, P. Mason, Maj., USAF. "Joint Operations in the Gulf War: An Allison Analysis." Master's thesis, School of Advanced Air and Space Studies, Air University, 1995.

Carson, Don, Capt., USAF. "Flying the F-15." *Air Force*, February 1974.

Catledge, Richard C., Maj. Gen., USAF (ret.). Oral History Interview. Interview by 1Lt. Wayne D. Perry, USAF, September 30, 1987, and December 9, 1987. K239.0512-1768, AFHRA.

CG FEAF. Message to CG AF FIVE. "OP-RQMT GUN-VAL." 130845 Nov 52. In *History of the 5th Air Force, Jul–Dec 1952*, vol. 3, app. 2. K730.01, AFHRA.

———. Message to CG AF FIVE. "OP-RQMT T-160 INSTL in F-86." 220805 Nov 52. In *History of the 5th Air Force, Jul–Dec 1952*, vol. 3, app. 2. K730.01, AFHRA.

———. Message to CG AF FIVE. "Personal from Smart to Barcus." 100815 Sep 52. In *History of the 5th Air Force, Jul–Dec 1952*, vol. 3, app. 2. K730.01, AFHRA.

Chain, John T., Jr., Col., USAF. Message to TAC/DOO and CSAF/XOXX. "TAC 1 TFW DOC." 130041Z Nov 75. In *History of the 1st Tactical Fighter Wing, Jan–Mar 1976*, vol. 1. K-WG-1-HI, AFHRA.

Chapa, Joseph O., Capt., USAF. "Remotely Piloted Aircraft and War in the Public Relations Domain." *Air & Space Power Journal* 28, no. 5 (October 2014): 29–46.

Chatterjee, Pratap. "A Chilling New Post-Traumatic Stress Disorder: Why Drone Pilots are Quitting in Record Numbers." *Salon*. Accessed October 4, 2016. http://www.salon.com/2015/03/06/a_chilling_new_post_traumatic_stress_disorder_why_drone_pilots_are_quitting_in_record_numbers_partner.

Christensen, Clayton M., Michael B. Horn, Louis Caldera, and Louis Soares. *Disrupting College: How Disruptive Innovation Can Deliver Quality and Affordability to Postsecondary Education*. Washington, DC: Center for American Progress, February 2011.

Church, Aaron. "RPA Ramp Up." *Air Force Magazine*, June 2011.

Clancy, Tom. *Every Man a Tiger: The Gulf War Air Campaign*. With Gen. Chuck Horner, USAF (ret.) and Tony Klotz. Updated ed. New York: Berkley Books, 2008.

Cleveland, Charles G., Lt. Gen., USAF. Oral History Interview. Interview by Lt. Col. Judd Katz, USAF, November 1987–January 1988. K239.0512-1784, AFHRA.

Clifton, Chuck. "There I Was . . . Part 1." *Class of 1964, United States Air Force Academy*. Accessed January 4, 2014. http://classof1964usafa.org/Public/Clifton_Charles/story1.htm.

———. "There I Was . . . 2." *Class of 1964, United States Air Force Academy*. Accessed January 4, 2014. http://classof1964usafa.org/Public/Clifton_Charles/story2.htm.

———. "There I Was . . . 4." *Class of 1964, United States Air Force Academy*. Accessed January 4, 2014. http://classof1964usafa.org/Public/Clifton_Charles/story4.htm.

———. "Who Am I." *Class of 1964, United States Air Force Academy*. Accessed January 4, 2014. http://classof1964usafa.org/Public/Clifton_Charles/history.htm.

Cliver, Jeff, Maj. Gen., USAF (ret.). Interview by Steven A. Fino, August 3, 2012.

Clodfelter, Mark. *The Limits of Air Power: The American Bombing of North Vietnam*. Lincoln: University of Nebraska Press, 1989.

Coates, James, and Bill Neikirk. "Cheaper Jets Shoot Down Claims for New Models." *Chicago Tribune*, December 7, 1981.

Combat Crew Training Course: F-4C. TAC Syllabus, Course 111507F. November 1966. In *History of the 4453d Combat Crew Training Wing, Jul–Dec 1966*, vol. 2. K-WG-4453-HI, AFHRA.

Commanding General, 5th Air Force. Memorandum to Commanding General, Far East Air Forces. "Evaluation of the A1CM Gun Sight." July 23, 1952. In *History of the 5th Air Force, Jul–Dec 1952*, vol. 3. K730.01, AFHRA.

COMNAVELECSYSCOM. Message to CSAF and CNO. "ACEVAL/AIMVAL Joint Test." 022324Z May 75. In *History of the Air Force Test and Evaluation Center, 1 Jan 1974–31 Dec 1975*, vol. 24. K150.01, AFHRA.

———. Message to DDR&E. "ACEVAL AIMVAL Progress Report." 130453Z Mar 76. In *History of the Air Force Test and Evaluation Center, 1 Jan–31 Dec 1976*, vol. 22. K150.01, AFHRA.

"A Complete Picture of Who's Out There." *Avionics Magazine: Special Report, Air Dominance with the F-22 Raptor*, 2002: 6–11.

Conference on Plane-to-Plane Fire Control. Washington, DC, August 26, 1941. 520.87241-2, AFHRA.

Connor, Miles A., Col., USAF. Memorandum to Pilot Training Wings. "Undergraduate Pilot Training Assignment Procedures." June 7, 1963. In *History of the Air Training Command, Jan–Jun 1963*, vol. 11. K220.01, AFHRA.

The Control of Plane-to-Plane Fire. National Defense Research Committee, Section D-2 (Fire Control), August 15, 1941. 520.87241-2, AFHRA.

Conway, Erik M. *Blind Landings: Low-Visibility Operations in American Aviation, 1918–1958*. Baltimore: Johns Hopkins University Press, 2006.

Cook, Jerry W., Brig. Gen., USAFR (ret.). *Once a Fighter Pilot . . .* New York: McGraw-Hill, 2002.

Coram, Robert. *Boyd: The Fighter Pilot Who Changed the Art of War*. Boston: Little, Brown, 2002.

Corn, Joseph J. *The Winged Gospel: America's Romance with Aviation*. Baltimore: Johns Hopkins University Press, 1983.

Correll, John T. “The Man from Thud Ridge.” *Air Force*, March 2012.

———. “The Reformers.” *Air Force*, February 2008.

Coté, Owen Reid, Jr. “The Politics of Innovative Military Doctrine: The US Navy and Fleet Ballistic Missiles.” Ph.D. dissertation, MIT, 1996.

Crane, Conrad C. *American Airpower Strategy in Korea, 1950–1953*. Lawrence: University Press of Kansas, 2000.

Creech, Wilbur L., Gen., USAF (ret.). Oral History Interview. Interview by Hugh N. Ahmann, June 1992. K239.0512-2050, AFHRA.

Crouch, Tom D. *Wings: A History of Aviation from Kites to the Space Age*. New York: W. W. Norton, 2003.

CSAF. Message to TAC. “TWO BUCK Deployment.” 202328Z Mar 65. K168.162-10, AFHRA.

CSAF/XO. Message to TAC/CV. “AIMVAL/ACEVAL F-15s.” 312111Z Dec 75. In “Lessons Learned Symposium, 1974 (AFSC).” 03951, AFMC/HO.

———. Message to TAC/CV et. al. “Air Force ACEVAL/AIMVAL Test Directive.” 171237Z Apr 75. In *History of the Air Force Test and Evaluation Center, 1 Jan 1974–31 Dec 1975*, vol. 24. K150.01, AFHRA.

Cullen, Timothy M., Lt. Col., USAF. “The MQ-9 Reaper Remotely Piloted Aircraft: Human and Machines in Action.” Ph.D. dissertation, MIT, 2011.

Cummings, Missy. *Hornet's Nest: The Experiences of One of the Navy's First Female Fighter Pilots*. Lincoln, NE: iUniverse.com, 1999.

Cunningham, Case, Col., USAF. E-Mail to Steven Fino. December 6, 2016.

———, Col., USAF. “William W. Momyer: A Biography of an Airpower Mind.” Ph.D. dissertation, School of Advanced Air and Space Studies, Air University, 2013.

Currie, Malcolm R. “Statement of the Director of Defense Research and Engineering.” Hearings before the US House, Subcommittee of the Committee on Appropriations, September 24, 1973, pp. 407–512. Washington, DC: Government Printing Office, 1973. Y4 Ap6/1:D36/5/974/pt7.

“DACT Lessons Learned.” Memorandum for Record, n.d. In *History of the 27th Tactical Fighter Squadron, Apr–Jun 1976, History of the 1st Tactical Fighter Wing, Apr–Jun 1976*, vol. 1. K-WG-1-HI, AFHRA.

Davies, Peter E. *USAF F-4 Phantom II MiG Killers, 1965–68*. Osprey Combat Aircraft 45. Oxford: Osprey Publishing, 2004.

———. *USAF F-4 Phantom II MiG Killers, 1972–73*. Osprey Combat Aircraft 55. Oxford: Osprey Publishing, 2005.

———. *USAF McDonnell Douglas F-4 Phantom II*. London: Osprey Publishing, 2013.

Davies, Steve. *F-15C Eagle Units in Combat*. Osprey Combat Aircraft 53. Oxford: Osprey Publishing, 2005.

Davies, Steve, and Doug Dildy. *F-15 Eagle Engaged: The World's Most Successful Jet Fighter*. New York: Osprey Publishing, 2007.

Davis, Dale. Oral History Interview. Interview by Jack Neufeld and Maj. Ralph Rowley, USAF, May 22, 1973. K239.0512-963, AFHRA.

Davis, Leighton I., Col., Air Corps. Memorandum to Chief, Armament Laboratory. “Aircraft Sights.” October 20, 1944. A-1 Conferences, CSDL-HC.

———, Lt. Gen., USAF. Oral History Interview. Interview by Maj. Lyn R. Officer, USAF, and Hugh N. Ahmann, April 26, 1973. K239.0512-668, AFHRA.

"The Dawn Patrol." *AFI Catalog of Feature Films*. Accessed December 16, 2013. http://www.afi.com/members/catalog/DetailView.aspx?s=&Movie=3653.

Deen, Sisco, Capt., USAF. "How Steve Ritchie Got His Fifth MIG." *Air Force*, October 1972.

Deming, Wilford E., III, Col., USAF. "A Force Structure Decision." *Fighter Weapons Review*, Winter 1974, 1–8. MSFRIC.

Der Baron, 414 Fighter Weapons Squadron. "Anything Else Is Rubbish!" *Fighter Weapons Review*, Fall 1971, 32–33. MSFRIC.

———. "Anything Else Is Rubbish!" *Fighter Weapons Review*, Winter 1971, 30–31. MSFRIC.

———. "Anything Else Is Rubbish!" *Fighter Weapons Review*, Fall 1972, 16–17. MSFRIC.

———. "Anything Else Is Rubbish." *Fighter Weapons Review*, Summer 1975, 10–12. MSFRIC.

Deullin, Albert B., Lt. "Chasseur Patrols," November 1917. 248.282-42A, AFHRA.

———, Lt. "Pursuit Work in a Single Seater," June 1917. 248.282-42A, AFHRA.

Deur, John M. "Wall of Eagles: Aerial Engagements and Victories in Operation Desert Storm," n.d. NA-399, AFHRA.

Diefendorf, Allen J., Lt. Col., USAF. "End of Tour Report," August 13, 1966. K740.131, AFHRA.

Dilger, Robert G., Lt. Col., USAF. "CLAW Missile Program Review." Hearings before the US Senate, Committee on Armed Services, March 27, 1974, pp. 4705–25. Washington, DC: Government Printing Office, 1974. Y4.Ar5/3:P94/6/975/pt9.

———, Maj., USAF. Oral History Interview, July 6, 1967. K239.0512-202, AFHRA.

Dille, John. "The Jets' First Ace." *Life*, June 4, 1951.

Dizikes, Peter. "Robots and Us." MIT News Office. October 13, 2015. Accessed October 5, 2016. http://news.mit.edu/2015/no-driverless-cars-1013.

Doerty, Walter L., Jr., Lt. Col., USAF. "End of Tour Report," November 5, 1966. K740.131, AFHRA.

"Dogfight Pilot Training at Nellis." *Las Vegas Review Journal*, December 21, 1976.

Donovan, Gen., USAF. Memorandum to HQ USAF. "Configuration of the F4H—USAF Version." December 6, 1961. In *History of the Tactical Air Command, Jul–Dec 1961*, vol. 2. K417.01, AFRHA.

Doolittle, General James H. "Jimmy," and Carroll V. Glines. *I Could Never Be So Lucky Again*. New York: Bantam Books, 1992.

Dorr, Robert F., Jon Lake, and Warren Thompson. *Korean War Aces*. Osprey Aircraft of the Aces 4. London: Osprey Publishing, 1995.

Driggs, Laurence La Tourette. "Aces among Aces." *National Geographic* 33, no. 6 (June 1918): 568–80.

———. Foreword to *Fighting the Flying Circus*, by Edward V. Rickenbacker, v–viii. New York: Frederick A. Stokes, 1919.

DSAF/OIP. Message to AIG. "ACEVAL/AIMVAL News Release." 111943Z Jun 76. In *History of the Air Force Test and Evaluation Center, 1 Jan–31 Dec 1976*, vol. 22. K150.01, AFHRA.

DuBridge, L. A., E. M. Purcell, G. E. Valley, and G. A. Morton. *Radar and Communications: A Report for the AAF Scientific Advisory Group*. Vol. 11 of *Toward New Horizons*, edited by Theodore von Karman. Wright Field, Dayton, OH: Headquarters Air Materiel Command, 1946.

"Dummy War at Nellis Training Combat Crews." *Las Vegas Sun*, December 15, 1975.

Dvorchak, Stephen R., USAF (ret.). E-Mail to Steven Fino. February 9, 2012.

———, USAF (ret.). Interview by Steven Fino, August 3, 2012.

Eagan, Shaun. "New ACC Program Begins, Aimed to Improve MQ-1/9 Community." Air Combat Command Public Affairs. September 1, 2015. Accessed October 5, 2016. http://www.af.mil/News/ArticleDisplay/tabid/223/Article/615577/new-acc-program-begins-aimed-to-improve-mq-19-community.aspx.

"Eagles Tangle with Aggressors in Dogfights." *The Flyer.* May 14, 1976. In *History of the 1st Tactical Fighter Wing, Apr–Jun 1976*, vol. 1. K-WG-1-HI, AFHRA.

Eagleston, Glenn T., Lt. Col., USAF. Memorandum to Commanding General, 4th Fighter Interceptor Wing. "Recommendation concerning Armament Equipment for Present and Future Fighter Type Aircraft." June 28, 1951. In "Test Data and Combat Reports Relative to Vulnerability and Limitations of F-86 Aircraft, Jan–Jul 1951." K-GP-4-SU-OP, AFHRA.

Edwards, Paul N. *The Closed World: Computers and the Politics of Discourse in Cold War America.* Cambridge, MA: MIT Press, 1996.

Ehrenfried, A. D. *General Operating Principles of the A-1 Gun-Bomb-Rocket Sight.* Report R-10. Cambridge, MA: Instrumentation Laboratory, MIT, February 1951. MIT.

Ehrhard, Thomas P. *Air Force UAVs: The Secret History.* Arlington, VA: Mitchell Institute Press, 2010.

———. "Unmanned Aerial Vehicles in the United States Armed Services: A Comparative Study of Weapon System Innovation." Ph.D. dissertation, Johns Hopkins University, 2000.

"Electronics: Maintenance Technique for the APG-30 Radar Set." *Fighter Gunnery Newsletter*, January 1954, 11–12. MSFRIC.

"Enemy Activity in Korea: Air to Air." *FEAF Intelligence Roundup*, no. 113 (October 25–31, 1952): chap. 2. K720.607A, AFHRA.

"Enemy Activity in Korea: Air to Air." *FEAF Intelligence Roundup and Operational Summary*, no. 116 (November 15–21, 1952): chap. 2. K720.607A, AFHRA.

"Enemy Air Activity, 17 Dec 1950." In "Summaries of AF Mission Reports of F-51D, F-80C, F-84E, F-86A, and B-26 Encounters with MiG-15 Aircraft." K730.310-5, AFHRA.

"Enemy Air Activity, 30 Dec 1950." In "Summaries of AF Mission Reports of F-51D, F-80C, F-84E, F-86A, and B-26 Encounters with MiG-15 Aircraft." K730.310-5, AFHRA.

"Enemy Aircraft Destroyed by US Army Air Service." *Air Service Information Circular: Heavier-Than-Air* 1, no. 7 (February 9, 1920). K141.2422, AFHRA.

English, Dave. *Slipping the Surly Bonds: Great Quotations on Flight.* New York: McGraw-Hill, 1998.

English, Peter F., Bobby L. Marlow, and Richard T. Sandborn. *Reduction of Tactical Fighter Losses over North Vietnam.* Working Paper 66/13.1. Directorate of Operations Analysis, Headquarters 7th Air Force, November 23, 1966. K740.31066-13, AFHRA.

Evans, Douglas K., Lt., USAF. "I Fight the Red Jets." *Saturday Evening Post*, July 19, 1952.

———. *Sabre Jets over Korea: A Firsthand Account.* Blue Ridge Summit, PA: Tab Books, 1984.

Evans, William J., Gen., USAF. Letter to HQ USAF/CV. "AIMVAL." March 17, 1976. K168.7339-79, AFHRA.

———, Lt. Gen., USAF. "Lightweight Fighter Prototype." Hearings before the US House, Subcommittee of the Committee on Appropriations, September 27, 1973, pp. 1020–23. Washington, DC: Government Printing Office, 1973. Y4.Ap6/1:D36/5/974/pt7.

Everest, Frank K., Jr., Brig. Gen., USAF (ret.). Oral History Interview. Interview by Hugh N. Ahmann, September 23, 1988. K239.0512-1844, AFHRA.

Everson, David, Col., USAF. *ACEVAL Final Report*. Langley AFB, VA: Tactical Air Command, May 1978. In *History of the Tactical Air Command, Jan–Dec 1977*, vol. 2. K417.0735, AFHRA.

———, Col., USAF. *AIMVAL Final Report*. Langley AFB, VA: Tactical Air Command, October 1977. In *History of the Tactical Air Command, Jan–Dec 1977*, vol. 2. K417.0735, AFHRA.

An Examination of the Utility of the F-15 Attack Radar, 1970. 03952, AFMC/HO.

Eyes of the Eagle: The Hughes APG-63 Radar. Culver City, CA: Radar Avionics-Aerospace Group, Hughes Aircraft Company, 1976. Box 1272, P2 Radar, NMUSAF.

F-4 Aircrew Operational Procedures. TACM 55-4, vol. 1; PACAFM 55-5, vol. 1; USAFEM 55-4, vol. 1. Langley AFB, VA: Headquarters, Tactical Air Command, May 1968. K417.716055-4, AFHRA.

F-4C Air Combat Tactics. Course No. 111509F. MacDill AFB Phase Manual. MacDill AFB, FL, July 1966. In *History of the 15th Tactical Fighter Wing, Jul–Dec 1966*, vol. 2. K-WG-15-HI, AFHRA.

F-4C Intercept. Course No. 111509F. MacDill AFB Phase Manual. MacDill AFB, FL, July 1966. In *History of the 15th Tactical Fighter Wing, Jul–Dec 1966*, vol. 2. K-WG-15-HI, AFHRA.

F-4C Limited Evaluation (DRAFT). TAC Test 63-50. Langley AFB, VA: Headquarters, Tactical Air Command, December 1964. In *History of the 12th Tactical Fighter Wing, Jul–Dec 1964*, vol. 3. K-WG-12-HI, AFHRA.

F-4C System Package Program. 62 ASZM-52. Air Force Systems Command, August 31, 1962. K220.8618-2, AFHRA.

F-4C System Package Program. 62 ASZM-52. Air Force Systems Command, rev. September 1963. 04454, AFMC/HO.

F-4C/SUU-16A Air-to-Air Tactics and Techniques. TAC TR 65-54. Nellis AFB, NV: Headquarters, USAF Tactical Fighter Weapons Center, November 1966. ADA0803446, DTIC.

"F-14B versus F-15," Talking Paper, 1969. 03936, AFMC/HO.

F-15 Avionics and Armament Mission Requirements Analysis: Report of the Ad Hoc Committee. Andrews AFB, MD: Headquarters, Air Force Systems Command, August 17, 1970. 04460, AFMC/HO.

F-15 Conversion Training Manual. USAF Operational Training Course—F-15. Courses F1500B, F1500TX, and F1500I. Luke AFB, AZ, December 1977. Located in *T.O. 1F-15A-1*, NMUSAF.

F-15 Eagle. Report H446. McDonnell Aircraft Company, July 15, 1969. 03950, AFMC/HO.

F-15 Eagle Fact Sheet. 73-2. Washington, DC: Secretary of the Air Force, Office of Information, Internal Information Division, February 1973. 03951, AFMC/HO.

F-15 Eagle Fact Sheet. Andrews AFB, MD: Office of Information, Air Force Systems Command, 1976. 03978, AFMC/HO.

"F-15 Rolls Out at St. Louis." *Flight International*, July 6, 1972.

"F-15: A Fighter Pilot's Fighter." McDonnell Douglas advertisement. *Air Force*, August 1971 and October 1971.

"F-15: 'F-14A/B with Phoenix Missile,'" Talking Paper, n.d. 03936, AFMC/HO.

"F-15s, F-5Es Catching Up with F-14 OR Rates at Nellis, Ryan Says." *Aerospace Daily* 86, no. 23 (August 3, 1977): 179. 168.7339-79, AFHRA.

F-94C Squadron Operational Suitability Test, Appendix V. Project APG/ADA/76-A, n.d. AD027450, DTIC.

"F-110: USAF's New Tactical Fighter." *Air Force & Space Digest*, May 1962.

Fallows, James M. "America's High-Tech Weaponry." *Atlantic Monthly*, May 1980.

———. "I Fly with the Eagles." *Atlantic Monthly*, November 1981.
———. "Muscle-Bound Superpower." *Atlantic Monthly*, October 1979.
———. *National Defense*. New York: Random House, 1981.
Fanelli, William B., Lt. Col., USAF. Memorandum to 1 TFW/DOW. "Beaufort MCAS DACT Operations Summary (20–29 Sep 77)." October 3, 1977. In *History of the 1st Tactical Fighter Wing, Jul–Sep 1977*, vol. 4. K-WG-1-HI, AFHRA.
Fay, Robert H., Lt. Col., USAF. "AIMVAL-ACEVAL Final Report Briefing." Hearings before the US House, Committee on Armed Services, Hearings on Military Posture and H.R. 10929, February 27, 1978, pp. 416–35. Washington, DC: Government Printing Office, 1978. Y4. Ar5/2a:977-78/56/pt3/bk1.
———. "AIMVAL-ACEVAL Program Review." Hearings before the US Senate, Committee on Armed Services, March 9, 1977, pp. 4564–81. Washington, DC: Government Printing Office, 1977. Y4.Ar5/3:P94/6/978/pt7.
———. "AIMVAL-ACEVAL Program Review." Hearings before the US House, Committee on Armed Services, April 5, 1978, pp. 5196–229. Washington, DC: Government Printing Office, 1978. Y4.Ar5/3:D36/7/979/pt7.
"FEAF Intel Form 1, 11 Jul 1951." In *History of the 4th Fighter-Interceptor Wing, Jul 1951*. K-WG-4-HI, AFHRA.
FEAF Report on the Korean War. Vol. 2, 1954. 168.7104-53, AFHRA.
"The Feats of a Phantom." *Air Force & Space Digest*, March 1962.
Feezel, Maj., USAF. Message to CSAF. "Bench Sets/Component Lab Sets for F-4C Fire Control/Navigation/Delivery Systems." October 10, 1962. In *History of the Air Training Command, Jul–Dec 1962*, vol. 11. K220.01, AFHRA.
"Fenway Crowd Sees 1st Jet Ace Awarded DFC." *Daily Boston Globe*, June 26, 1951.
Ferran, Lee. "Drone 'Stigma' Means 'Less Skilled' Pilots at Controls of Deadly Robots." *ABC News*, April 29, 2014. Accessed May 3, 2014. http://abcnews.go.com/Blotter/drone-stigma-means-skilled-pilots-controls-deadly-robots/story?id=23475968.
"Fernandez Bags 13th MIG, Becoming the Top Jet Ace." *New York Times*, May 10, 1953.
"Fightergram: The Quality-Quantity Controversy." *Fighter Weapons Review*, Summer 1974, 34–35. MSFRIC.
The Fighting 51st. Film. United States Air Force, 1952. 342-USAF-20223, NARA.
Final Report on Combat Suitability Test of F-86F-2 Aircraft with T-160 Guns. Project APG/ADA/43-F-1. Eglin AFB, FL: Air Proving Ground Command, August 3, 1953. AD019725, DTIC.
Fino, Steven A. "Breaking the Trance: The Perils of Technological Exuberance in the US Air Force Entering Vietnam." *Journal of Military History* 77, no. 2 (April 2013): 625–55.
———. "Doubling Down: AIMVAL-ACEVAL and US Air Force Investment in TacAir post-Vietnam." *Vulcan: The International Journal of the Social History of Military Technology* 2 (July 2014): 125–61.
"Fire Control Problem." *Fighter Gunnery*, no. 1 (1951): 21. Ar.2007.187, NMUSAF.
"First Air Force Ace in Indochina Is Cited." *New York Times*, August 30, 1972.
"First Jet Pilot Ace Back from Europe, Middle East." *New York Times*, March 3, 1952.
"First Korean Jet Ace Eager to Fight Again." *Los Angeles Times*, November 12, 1952.
"First Triple Jet Ace Helps Bag 9 MIGs." *Chicago Daily Tribune*, May 18, 1953.
"Fixed Price Contract for Services and Supplies for Computing Combination Bomb Sight and Gun Sight," March 26, 1945. A-1 Sight Contract, CSDL-HC.

The Flaming Dart Air Strikes against North Vietnam, 7–11 February 1965: Part I. CI No. 9 Working Paper No. 4, Second Draft, August 2, 1967. Accessed January 10, 2104. http://www.dod.mil/pubs/foi/International_security_affairs/vietnam_and_southeast_asiaDocuments/550.pdf.

Flanagan, John C. "The Selection and Classification Program for Aviation Cadets (Aircrew—Bombardiers, Pilots, and Navigators)." *Journal of Consulting Psychology* 6, no. 5 (September–October 1942): 229–39.

Flanagan, William A., Maj., USAF. "The Fighter Force: How Many Seats?" *Air University Review*, May–June 1981.

Flax, Alexander H. Oral History Interview. Interview by James C. Hasdorff and Jacob Neufeld, November 27–29, 1973. K239.0512-691, AFHRA.

Flood, William G., Maj., USAF, and Maj. Joseph S. Rodero, USAF. *F-15 Initial Operational Test and Evaluation—Air-to-Air (Annex A)*. TAC Project 71C-223W. Final Report (Interim). Nellis AFB, NV: USAF Tactical Fighter Weapons Center, April 1975. 03964, AFMC/HO.

———. *F-15 Initial Operational Test and Evaluation—Air-to-Ground (Annex B)*. TAC Project 71C-223W. Final Report (Interim). Nellis AFB, NV: USAF Tactical Fighter Weapons Center, April 1975. 03966, AFMC/HO.

Format for the Syllabus of Instruction: F-86, Day Fighter (Air-to-Air). Course Number 105902. Nellis AFB, NV: Training Analysis and Development, October 1952. In *History of the Crew Training Air Force, Jul–Dec 1952*, vol. 4. K221.011, AFHRA.

"Four Reasons Why the F-15 Is a Fighter Pilot's Fighter." Sperry advertisement. *Air Force*, September 1972.

Franks, Norman. Introduction to *Knight of Germany: Oswald Boelcke, German Ace*, by Johannes Werner, translated by Claude W. Sykes. Philadelphia: Casemate, 2009.

Fritzsche, Peter. *A Nation of Fliers: German Aviation and the Popular Imagination*. Cambridge, MA: Harvard University Press, 1992.

"From Struts to Jets: Drawings Tell History of US Military Aviation from World War I to the Present." *Life*, February 4, 1952.

Fulghum, David A. "Overdrive." *Aviation Week & Space Technology*, September 26, 2005.

———. "Raptor's Edge." *Aviation Week & Space Technology*, February 9, 2009.

Fulghum, David A., and Michael J. Fabey. "F-22: Unseen and Lethal." *Aviation Week & Space Technology*, January 8, 2007.

Futrell, Robert F. *Ideas, Concepts, Doctrine: Basic Thinking in the United States Air Force, 1907–1960*. 3rd ed. Maxwell AFB, AL: Air University Press, 2001.

———. *Ideas, Concepts, Doctrine: Basic Thinking in the United States Air Force, 1961–1984*. 5th ed. Maxwell AFB, AL: Air University Press, 2004.

———. *The United States Air Force in Korea, 1950–1953*. Rev. ed. Washington, DC: Office of Air Force History, 1983.

———. *United States Air Force Operations in the Korean Conflict, 1 November 1950–30 June 1952*. USAF Historical Study No. 72. Maxwell AFB, AL: USAF Historical Division, Air University, 1953.

———. *United States Air Force Operations in the Korean Conflict, 1 July 1952–27 July 1953*. USAF Historical Study No. 127. Maxwell AFB, AL: USAF Historical Division, Air University, 1956.

Futrell, Robert F., William H. Greenhalgh, Carl Grubb, Gerald E. Hasselwander, Robert F. Jakob, and Charles A. Ravenstein. *Aces and Aerial Victories: The United States Air Force in*

Southeast Asia, 1965–1973. Edited by James N. Eastman Jr., Walter Hanak, and Lawrence J. Paszek. Washington, DC: Office of Air Force History, 1976.

Gabreski, Francis S., Col., USAF. "(End of Tour) Report by Colonel Francis S. Gabreski," July 8, 1952. K720.131-2, AFHRA.

———, Col., USAF. "F-86D Program Deficiencies and Recommendations," n.d. Personal Papers of Francis S. Gabreski. 168.7649-43, AFHRA.

———, Col., USAF. "Our Top Ace Talks Shop." *Air Force*, August 1952.

"Gabreski Tells Truman MIG-15 Is Good as F-86." *New York Times*, June 25, 1952.

Galatz, Karen M. "Old West Scene—Good Guys vs. Bad." *Las Vegas Sun*, October 6, 1977.

"Gallant Feats in the Blue." *Literary Digest*, November 16, 1918.

Galloway, John. *The Gulf of Tonkin Resolution*. Rutherford, NJ: Fairleigh Dickinson University Press, 1970.

Garamone, Jim. "Hagel Orders Review of Military Decorations, Awards Program." American Forces Press Service, March 20, 2014. Accessed on April 18, 2014. http://www.defense.gov/news/newsarticle.aspx?id=121874.

Gardiner, Paul E., Col., USAF. "Effectiveness of F-4H vs A-4D5 in Close Air Support," October 26, 1961. In *History of the Tactical Air Command, Jul–Dec 1961*, vol. 2. K417.01, AFRHA.

Gaspard, Richard E., Col., USAF. *SA-2 Deployment and Tactics in North Vietnam*, June 23, 1966. 168.7571-22, AFHRA.

Gen. MacArthur Leads Attack on Admiralty Islands. Film. United News Newsreels, 1944. 208-UN-94, NARA.

"General John P. McConnell." *USAF Biography*. Accessed February 26, 2014. http://www.af.mil/AboutUs/Biographies/Display/tabid/225/Article/106325/general-john-paul-mcconnell.aspx.

Georgi, William F., Brig. Gen., USAF. Oral History Interview. Interview by Jack Neufeld, June 5, 1973. K239.0512-964, AFHRA.

Getting, I. A. "Conference on Fighter Fire Control." Ad Hoc Group on Airborne Fire Control Systems for Fighter Aircraft, MIT Instrumentation Laboratory, Cambridge, MA, January 16, 1953. AD495903, DTIC.

"GI Gun Cameras and Film." *Fighter Weapons Newsletter*, no. 2 (1955): 61–63. MSFRIC.

Gillespie, Daniel M. "Mission Emphasis and the Determination of Needs for New Weapon Systems." Ph.D. dissertation, MIT, 2009.

Giordano, Bruno A., 1Lt., USAF. "A-4 Sight Check." *Fighter Weapons Newsletter*, May 1956, 10–13. MSFRIC.

———, 1Lt., USAF. "The Dart Target." *Fighter Weapons Newsletter*, November 1956, 17–19. MSFRIC.

———, Capt., USAF. "The Dart Target." *Fighter Weapons Newsletter*, March 1957, 38–53. MSFRIC.

———, Capt., USAF. "F-100D GAR-8 Tactics." *Fighter Weapons Newsletter*, September 1959, 42–43. MSFRIC.

Giraudo, John C., Maj. Gen., USAF (ret.). Oral History Interview. Interview by Lt. Col. Charles M. Heltsley, USAF, January 8–12, 1985. K239.0512-1630, AFHRA.

Gish, Donald L., Maj., USAF. "F-4 Air to Air Training." *Fighter Weapons Review*, Fall 1975, 1–5. MSFRIC.

Glasser, Otto J., Lt. Gen., USAF. "Research, Development, Test, and Evaluation, Air Force." Hearings before the US House, Subcommittee of the Committee on Appropriations, May 27, 1971, pp. 255–380. Washington, DC: Government Printing Office, 1971. Y4.Ap6/1:D36/5/972/pt6.

———, Lt. Gen., USAF. "Statement of Lt. Gen. Otto J. Glasser, Deputy Chief of Staff, Research and Development, HQ, USAF." Hearings before the US House, Committee on Armed Services, May 4, 1971, pp. 3567–616. Washington, DC: Government Printing Office, 1971. Y4.Ar5/2a:971-72/9/pt1.

Glines, C. V. "Charmed Life of Captain Eddie Rickenbacker." *Aviation History* 9, no. 3 (January 1999): 42–50.

Goldsmith, Gary, Maj. "Comments on 'Ten Thousand Feet and Ten Thousand Miles,'" May 7, 2012. Accessed on September 16, 2012. http://www.airpower.au.af.mil/article.asp?id=72.

Goulden, Joseph C. *Truth Is the First Casualty: The Gulf of Tonkin Affair—Illusion and Reality.* Chicago: Rand McNally, 1969.

Gray, Bert, Capt., USAF. Memorandum to Commanding General, 5th Air Force. "Transmittal of Pilot's Statements." June 9, 1951. In "Test Data and Combat Reports Relative to Vulnerability and Limitations of F-86 Aircraft, Jan–Jul 1951." K-GP-4-SU-OP, AFHRA.

Greeley, K. W., and R. J. Schwartz. "F-22 Cockpit Avionics: A Systems Integration Success Story." In *Proceedings of the SPIE* 4022, Cockpit Displays VII: Displays for Defense Applications: 52–62.

Greene, Frederick H., Jr. Letter to Commanding General, Continental Air Command, January 31, 1950. A-1 Sight Correspondence, CSDL-HC.

Griffith, E. J., Lt. Col., USAF. "AIMVAL-ACEVAL: Why/What/How." *Fighter Weapons Review*, Fall 1977, 23–27. MSFRIC.

"The Growth of MiG Alley." In *History of the Fifth Air Force, Jul–Dec 1951*, vol. 3 (Appendix 27):49–59. K730.01, AFHRA.

Guillermin, John. *The Towering Inferno*. Film. Warner Bros., 1974.

"The Gun Camera." *Fighter Weapons Newsletter*, no. 2 (1955): 13. MSFRIC.

Gun Sight: Mark 14. Gunner's Operating Bulletin, No. 2. Headquarters of the Commander in Chief, United States Fleet, n.d. *HyperWar*. Accessed December 22, 2013. http://www.ibiblio.org/hyperwar/USN/ref/Ordnance/GS-Mk14/index.html.

"Gun-Camera Records Kills." *FEAF Intelligence Roundup and Operational Summary*, no. 134 (April 5–18, 1953): chap. 5, p. 1. K720.607A, AFHRA.

"Gun Sight: Mark 18." *HyperWar: Reflector and Computer Sights*. Accessed December 22, 2013. http://www.ibiblio.org/hyperwar/USN/ref/AirGunnery/MARK18GS.html.

Gurelac, Henry E. *Radar in World War II*. Vol. 8 in *The History of Modern Physics, 1800–1950*. Los Angeles: Tomash Publishers; New York: American Institute of Physics, 1987.

Gurney, Gene, Capt., USAF. *Five Down and Glory: A History of the American Air Ace*. Edited by 1Lt. Mark P. Friedlander Jr., USAF. New York: G. P. Putnam's Sons, 1958.

Haering, George. "Aim/Ace." *Topgun Journal* 1, no. 3 (Fall 1977): 14–16. NHHC.

Haff, William, B., Capt., USN. "Sidewinder Program Review." Hearings before the US Senate, Committee on Armed Services, March 14, 1975, pp. 4619–34. Washington, DC: Government Printing Office, 1975. Y4/Ar5/3:P94/6/976/pt9.

Hale, Dudley D., Brig. Gen., USAF. Memorandum to Commanding General, Far East Air Forces. "A1CM Gunsight." September 17, 1952. In *History of the 5th Air Force, Jul–Dec 1952*, vol. 3. K730.01, AFHRA.

Hall, Mordaunt. "The Screen: Sky Battles." *New York Times*, August 16, 1930.

———. "The Screen: The Flying Fighters." *New York Times*, August 13, 1927.

Hallion, Richard P. "Air Dominance: The Essential Achievement." In *Silver Wings, Golden Valor: The USAF Remembers Korea*, edited by Richard P. Hallion, 27–62. Washington, DC: Air Force History and Museums Program, 2006.

———. "Korea: The Forgotten (Air) War." In *Silver Wings, Golden Valor: The USAF Remembers Korea*, edited by Richard P. Hallion, 1–3. Washington, DC: Air Force History and Museums Program, 2006.

Hammond, Grant T. *The Mind of War: John Boyd and American Security*. Washington, DC: Smithsonian Books, 2001.

Handley, Philip W., Col., USAF (ret.). *Nickel on the Grass: Reflections of a U.S. Air Force Pilot*. New York: iUniverse, 2006.

Hannah, Craig C. *Striving for Air Superiority: The Tactical Air Command in Vietnam*. College Station: Texas A&M University Press, 2001.

Hansen, James R. "Aviation History in the Wider View." *Technology and Culture* 30, no. 3 (July 1989): 643–56.

Hardison, Chaitra M., Michael G. Mattock, and Maria Lytell. *Incentive Pay for Remotely Piloted Aircraft Career Fields*. Santa Monica, CA: RAND, 2012.

Hargrove, James A., Jr., Maj., USAF. Oral History Interview. Interview by Lt. Col. Robert Eckert, USAF, and Maj. Harry Shallcross, USAF, September 19, 1967. K239.0512-020, AFHRA.

Harmer, Todd P., Lt. Col., USAF, and C. R. Anderegg. *The Shootdown of Trigger 4: Report of the Project Trigger Study Team*. Washington, DC: Headquarters United States Air Force, April 2001. K146.044-90, AFHRA.

"Harmonization and Firing Techniques for Fighter Aircraft." Armament Laboratory, Air Research and Development Command, March 1952. F2-K, NMUSAF.

Harr, James C., Col., USAF, Capt. Michael S. Muskat, USAF, and Capt. Zeke D. Williams, USAF. *Air Combat Tactics Evaluation Final Report: F-100, F-104, F-105, F-4C vs MiG-15/17 Type A/C (F-86H)*. TAC Mission FF-857. Nellis AFB, NV: USAF Fighter Weapons School, June 17, 1965. AD372500, DTIC.

Hart, Gary, and William S. Lind. *America Can Win: The Case for Military Reform*. Chevy Chase, MD: Adler & Adler, 1986.

Hart, Peter. *Aces Falling: War above the Trenches, 1918*. London: Phoenix, 2008.

Haulman, Daniel L. "Aerial Victory Credits." *USAF Wartime Aerial Victory Credits*, 2000. Accessed June 17, 2013. http://www.au.af.mil/au/afhra/avc.asp.

———. "Introduction: Korean War Aerial Victory Credits." *USAF Wartime Aerial Victory Credits*, 2000. Accessed June 17, 2013. http://www.au.af.mil/au/afhra/avc_intro_korean.html.

"He's a Poor Shot, Says Top Jet Ace: Col. Royal Baker." *New York Times*, April 3, 1953.

Heffron, Charles H., Jr., Lt., USAF. *Air-to-Air Encounters over North Vietnam, 1 Jan–30 Jun 1967*. Project CHECO Report. Headquarters PACAF: Directorate, Tactical Evaluation, November 30, 1967. K717.0413-22, AFHRA.

Heflin, C. J., Col., USAF. Memorandum to TAC. "Radar Pilot Training (APQ-72)." July 20, 1962. In *History of the Air Training Command, Jul–Dec 1962*, vol. 11. K220.01, AFHRA.
"Hell's Angels." *AFI Catalog of Feature Films*. Accessed December 16, 2013. http://www.afi.com/members/catalog/DetailView.aspx?s=&Movie=9655.
Hendrickson, James L., Capt., USAF. "The GIB and ACM." *Fighter Weapons Review*, Summer 1974, 31–33. MSFRIC.
———, Capt., USAF. Oral History Interview. Interview by Maj. Lyn R. Officer, USAF, and Hugh N. Ahmann, January 31, 1973. K239.0512-646, AFHRA.
"Here's a Fighting Man." *Saturday Evening Post*, July 19, 1952.
Hesse, Erwin A., Capt., USAF. "Flying: Continuation Flying Training for Combat Pilots." 4th Fighter Interceptor Group Directive, 4FIG-DIR 60-5, July 21, 1951. In *History of the 4th Fighter-Interceptor Wing, Jul 1951*. K-WG-4-HI, AFHRA.
Hettena, Seth. *Feasting on the Spoils: The Life and Times of Randy "Duke" Cunningham, History's Most Corrupt Congressman*. New York: St. Martin's Press, 2007.
Hildreth, James R., Maj. Gen., USAF. Memorandum to TAC/CV. "AIMVAL Final Report Recommendations." November 28, 1977. 168.7339-79, AFHRA.
———, Maj. Gen., USAF (ret.). Oral History Interview. Interview by James C. Hasdorff, October 27–28, 1987. K239.0512-1772, AFHRA.
Hill, Howard J., Col., USAF (ret.). Oral History Interview. Interview by James C. Hasdorff, July 12, 1991. K239.0512-2221, AFHRA.
Hillinger, Charles. "'Russian Squadrons' Fly Mock Battles at Nellis AFB." *Las Vegas Review Journal*, May 29, 1977.
Hinton, Bruce H., Lt. Col., USAF. "MiG-15 versus F-86A in Korea," July 25, 1951. K-GP-4-SU-OP, AFHRA.
Hinton, Bruce H., Lt. Col., USAF, and Capt. Elmer L. Wingo, USAF. "An Analysis of Operations at Kimpo Air Base: Detachment A, 336th Fighter-Interceptor Squadron, 4th Fighter-Interceptor Group." n.d. In *History of the 4th Fighter-Interceptor Group, Oct–Dec 1950*. K-GP-4-HI, AFHRA.
Hirschman, Dave. "F-22 Fighter Won't Get One Flier's Vote." *Atlanta Journal-Constitution*, June 25, 2006.
Hise, William C., 2Lt., USAF. "An Analysis of the Tactics of Aerial Warfare Used by the MiG-15 Type Enemy Aircraft over North Korea," July 6, 1951. In *History of the 4th Fighter-Interceptor Group, Jul–Sep 1951*. K-GP-4-HI, AFHRA.
———, 1Lt., USAF. "F-86E versus MiG-15 in Korea," January 15, 1952. In *History of the 4th Fighter-Interceptor Group, Jan–Feb 1952*. K-GP-4-HI, AFHRA.
———, 2Lt., USAF. "Markings on MiG-15 Type Enemy Aircraft Over North Korea," September 1, 1951. In *History of the 4th Fighter-Interceptor Group, Jul–Sep 1951*. K-GP-4-HI, AFHRA.
History of ADCOM, 1 Jan–31 Dec 1976. K410.011, AFHRA.
History of the 1st Tactical Fighter Wing, Jul–Sep 1975. Langley AFB, VA. K-WG-1-HI, AFHRA.
History of the 1st Tactical Fighter Wing, Oct–Dec 1975. Langley AFB, VA. K-WG-1-HI, AFHRA.
History of the 1st Tactical Fighter Wing, Jan–Mar 1976. Langley AFB, VA. K-WG-1-HI, AFHRA.
History of the 1st Tactical Fighter Wing, Apr–Jun 1976. Langley AFB, VA. K-WG-1-HI, AFHRA.
History of the 1st Tactical Fighter Wing, Oct–Dec 1976. Langley AFB, VA. K-WG-1-HI, AFHRA.
History of the 1st Tactical Fighter Wing, Jan–Mar 1977. Langley AFB, VA. K-WG-1-HI, AFHRA.
History of the 1st Tactical Fighter Wing, Jul–Sep 1977. Langley AFB, VA. K-WG-1-HI, AFHRA.

History of the 1st Tactical Fighter Wing, Oct–Dec 1977. Langley AFB, VA. K-WG-1-HI, AFHRA.
History of the 1st Tactical Fighter Wing, Jan–Mar 1978. Langley AFB, VA. K-WG-1-HI, AFHRA.
History of the 4th Air Base Group, May 1952. In *History of the 4th Fighter-Interceptor Wing, May 1952*. K-WG-4-HI, AFHRA.
History of the 4th Air Base Group, Jun 1952. In *History of the 4th Fighter-Interceptor Wing, Jun 1952*. K-WG-4-HI, AFHRA.
History of the 4th Fighter-Interceptor Group, Oct–Dec 1950. K-GP-4-HI, AFHRA.
History of the 4th Fighter-Interceptor Group, Jan 1951. K-GP-4-HI, AFHRA.
History of the 4th Fighter-Interceptor Group, Feb–Apr 1951. K-GP-4-HI, AFHRA.
History of the 4th Fighter-Interceptor Group, May–Jun 1951. K-GP-4-HI, AFHRA.
History of the 4th Fighter-Interceptor Group, Jul–Sep 1951. K-GP-4-HI, AFHRA.
History of the 4th Fighter-Interceptor Group, Jan–Feb 1952. K-GP-4-HI, AFHRA.
History of the 4th Fighter-Interceptor Group, Jul 1952. K-GP-4-HI, AFHRA.
History of the 4th Fighter-Interceptor Wing, Jun 1951. K-WG-4-HI, AFHRA.
History of the 4th Fighter-Interceptor Wing, Jul 1951. K-WG-4-HI, AFHRA.
History of the 4th Fighter-Interceptor Wing, Sep 1951. K-WG-4-HI, AFHRA.
History of the 4th Fighter-Interceptor Wing, Oct 1951. K-WG-4-HI, AFHRA.
History of the 4th Fighter-Interceptor Wing, Nov 1951. K-WG-4-HI, AFHRA.
History of the 4th Fighter-Interceptor Wing, Feb 1952. K-WG-4-HI, AFHRA.
History of the 4th Fighter-Interceptor Wing, May 1952. K-WG-4-HI, AFHRA.
History of the 4th Fighter-Interceptor Wing, Jun 1952. K-WG-4-HI, AFHRA.
History of the 4th Fighter-Interceptor Wing, Jul–Dec 1952. K-WG-4-HI, AFHRA.
History of the 4th Fighter-Interceptor Wing, Jan–Jun 1953. K-WG-4-HI, AFHRA.
History of the 8th Tactical Fighter Wing, Jan–Jun 1967. Ubon RTAFB, Thailand. K-WG-8-HI, AFHRA.
History of the 8th Tactical Fighter Wing, Oct–Dec 1967. Ubon RTAFB, Thailand. K-WG-8-HI, AFHRA.
History of the 12th Tactical Fighter Wing, Jul–Dec 1964. MacDill AFB, FL. K-WG-12-HI, AFHRA.
History of the 15th Tactical Fighter Wing, Jan–Jun 1964. MacDill AFB, FL. K-WG-15-HI, AFHRA.
History of the 15th Tactical Fighter Wing, Jul–Dec 1964. MacDill AFB, FL. K-WG-15-HI, AFHRA.
History of the 15th Tactical Fighter Wing, Jul–Dec 1965. MacDill AFB, FL. K-WG-15-HI, AFHRA.
History of the 15th Tactical Fighter Wing, Jan–Jun 1966. MacDill AFB, FL. K-WG-15-HI, AFHRA.
History of the 15th Tactical Fighter Wing, Jul–Dec 1966. MacDill AFB, FL. K-WG-15-HI, AFHRA.
History of the 16th Fighter Interceptor Squadron, Mar 1952. In *History of the 51st Fighter Interceptor Wing, Mar 1952*. K-WG-51-HI, AFHRA.
History of the 27th Tactical Fighter Squadron, Jan–Mar 1976. In *History of the 1st Tactical Fighter Wing, Jan–Mar 1976*, vol. 1. Langley AFB, VA. K-WG-1-HI, AFHRA.
History of the 27th Tactical Fighter Squadron, Apr–Jun 1976. In *History of the 1st Tactical Fighter Wing, Apr–Jun 1976*, vol. 1. Langley AFB, VA. K-WG-1-HI, AFHRA.
History of the 51st Fighter Interceptor Group, Dec 1951. In *History of the 51st Fighter Interceptor Wing, Dec 1951*. K-WG-51-HI, AFHRA.
History of the 51st Fighter Interceptor Wing, Dec 1951. K-WG-51-HI, AFHRA.
History of the 51st Fighter Interceptor Wing, Feb 1952. K-WG-51-HI, AFHRA.
History of the 51st Fighter Interceptor Wing, Mar 1952. K-WG-51-HI, AFHRA.
History of the 51st Fighter Interceptor Wing, Apr 1952. K-WG-51-HI, AFHRA.

History of the 51st Fighter Interceptor Wing, May 1952. K-WG-51-HI, AFHRA.
History of the 51st Fighter Interceptor Wing, Jul–Dec 1952. K-WG-51-HI, AFHRA.
History of the 334th Fighter Interceptor Squadron, Oct 1951. K-SQ-FI-334-HI, AFHRA.
History of the 334th Fighter Interceptor Squadron, Apr 1952. K-SQ-FI-334-HI, AFHRA.
History of the 366th Tactical Fighter Wing, Jan–Jul 1965. Holloman AFB, NM. K-WG-366-HI, AFHRA.
History of the 366th Tactical Fighter Wing, Oct 1966–Mar 1967. Da Nang AB, SVN. K-WG-366-HI, AFHRA.
History of the 366th Tactical Fighter Wing, Apr 1967–Sep 1967. Da Nang AB, SVN. K-WG-366-HI, AFHRA.
History of the 390th Tactical Fighter Squadron, May 1967. In *History of the 366th Tactical Fighter Wing, April 1967–May 1967*. Da Nang AB, SVN. K-WG-366-HI, AFHRA.
History of the 4453d Combat Crew Training Wing, Jul–Dec 1966. Davis-Monthan AFB, AZ. K-WG-4453-HI, AFHRA.
History of the 6400th Air Depot Wing, Feb–Jun 1952. Tachikawa, Japan. K-WG-DEP-6400-HI, AFHRA.
History of the 6400th Air Depot Wing, Jul–Dec 1952. Tachikawa, Japan. K-WG-DEP-6400-HI, AFHRA.
History of the 6400th Air Depot Wing, Jan–Jun 1954. Tachikawa, Japan. K-WG-DEP-6400-HI, AFHRA.
History of the Air Force Test and Evaluation Center, 1 Jan 1974–31 Dec 1975. K150.01, AFHRA.
History of the Air Force Test and Evaluation Center, 1 Jan–31 Dec 1976. K150.01, AFHRA.
History of the Air Force Test and Evaluation Center, 1 Jan–31 Dec 1977. K150.01, AFHRA.
History of the Air Training Command, Jul–Dec 1962. K220.01, AFHRA.
History of the Air Training Command, Jan–Jun 1963. K220.01, AFHRA.
History of the Crew Training Air Force, Jul–Dec 1952. K221.011, AFHRA.
History of the F-15 Eagle. Langley AFB, VA: Tactical Air Command. K417.042-18, AFHRA.
History of the Fifth Air Force, Jul–Dec 1951. K730.01, AFHRA.
History of the Fifth Air Force, Jul–Dec 1952. K730.01, AFHRA.
History of the Ninth Air Force, Jul–Dec 1975. Shaw AFB, SC. K533.01-3, AFHRA.
History of the Ninth Air Force, Jan–Dec 1976. Shaw AFB, SC. K533.01-3, AFHRA.
History of the Tactical Air Command, Jul–Dec 1961. Langley AFB, VA. K417.01, AFHRA.
History of the Tactical Air Command, Jan–Jun 1962. Langley AFB, VA. K417.01, AFHRA.
History of the Tactical Air Command, Jul–Dec 1962. Langley AFB, VA. K417.01, AFHRA.
History of the Tactical Air Command, Jan–Jun 1963. Langley AFB, VA. K417.01, AFHRA.
History of the Tactical Air Command, Jan–Dec 1977. Langley AFB, VA. K417.01, AFHRA.
History of the USAF Tactical Fighter Weapons Center, Jul 1970–Jul 1971. Nellis AFB, NV. K417.0735, AFHRA.
History of the USAF Tactical Fighter Weapons Center, 1977. Nellis AFB, NV. K417.0735, AFHRA.
Hoag, Robert J., Capt., USAF. "Superfighter." *Fighter Weapons Review*, Summer 1974, 18–30. MSFRIC.
Holbury, Robert J., Brig. Gen., USAF. *Concept of Operations for the F-15 Tactical Fighter*. Langley AFB, VA: Headquarters, Tactical Air Command, August 23, 1970. 03927, AFMC/HO.
"Holding Four Aces: Speed, Stealth, Agility, and Revolutionary Avionics." *Avionics Magazine: Special Report, Air Dominance with the F-22 Raptor*, 2002: 3–5.

Holley, I. B., Jr. *Ideas and Weapons: Exploitation of the Aerial Weapon by the United States during World War I; A Study in the Relationship of Technological Advance, Military Doctrine, and the Development of Weapons*. Washington, DC: Office of Air Force History, 1983.

Hollstein, Charles P., Col., USAF. Memorandum to Gen. Banfill. "Comments of Study of Far East Aerial Reconnaissance Problems Submitted by AMC Personnel." February 27, 1951. Personal Papers of Charles P. Hollstein. 168.7350, pt. 40, AFHRA.

Holmes, Jamie. "Why Drone Pilots Deserve Medals." *Slate*, August 2, 2012. Accessed October 8, 2012. http://www.slate.com/articles/technology/future_tense/2012/08/drone_pilots_deserve_military_medals.html.

Holsapple, Joseph R., Maj. Gen., USAF. Memorandum to AFSC and AFLC. "F-4H." November 7, 1961. Included in "F-4C System Package Program," rev. September 1963. 04454, AFMC/HO.

Hone, Thomas. "Korea." In *Case Studies in the Achievement of Air Superiority*, edited by Benjamin Franklin Cooling, 453–504. Washington, DC: Air Force History and Museums Program, 1994.

———. "Southeast Asia." In *Case Studies in the Achievement of Air Superiority*, edited by Benjamin Franklin Cooling, 505–62. Washington, DC: Air Force History and Museums Program, 1994.

Hoover, Bob. "A Place in the Sun." *Fighter Weapons Newsletter*, March 1959, 5–6. MSFRIC.

"How Old Can You Jet?" *Air Force*, April 1951.

"How to Set Up a Film Assessing Room." *Fighter Weapons Newsletter*, no. 2 (1955): 1–9. MSFRIC.

Howard, and William C. Hise. "FEAF Intel Form 1, 18 Nov 1951." In *History of the 4th Fighter-Interceptor Wing, Nov 1951*. K-WG-4-HI, AFHRA.

Howard, William C. Hise, Kerston, and Kenyon. "FEAF Intel Form 1, 16 Oct 1951 (0401)." In *History of the 4th Fighter-Interceptor Wing, Oct 1951*. K-WG-4-HI, AFHRA.

Howard, William C. Hise, Kerston, Ridlen, and Latwas. "FEAF Intel Form 1, 30 Nov 1951." In *History of the 4th Fighter-Interceptor Wing, Nov 1951*. K-WG-4-HI, AFHRA.

Howard, R. T., Lt. Col., USAF. "Comments on 'Ten Thousand Feet and Ten Thousand Miles,'" May 1, 2012. Accessed September 16, 2012. http://www.airpower.au.af.mil/article.asp?id=72.

HQ TAC. Message to ASD. "Proposed System Package for the F4H." 222313Z Dec 61. In *History of the Tactical Air Command, Jul–Dec 1961*, vol. 2. K417.01, AFRHA.

Hughes, Thomas P. *Rescuing Prometheus: Four Monumental Projects That Changed the Modern World*. New York: Vintage Books, 2000.

Hughes, W. H. *Probability of Tour Completion for Pilots in Southeast Asia for 1 January 1966–31 May 1966*. Note 66-4. Headquarters PACAF: Assistant for Operations Analysis, July 15, 1966. 168.7571-19, AFHRA.

"Hughes Falcon: The AF's New Bird of Prey." *Air Force*, May 1955.

Husted, Dave. "1st TFW Inherits Pilot Training System." *The Flyer*. March 3, 1977. In *History of the 1st Tactical Fighter Wing, Jan–Mar 1977*, vol. 2. K-WG-1-HI, AFHRA.

———. Letter to 1 TFW/DO. "Langley AFB ACMI Operations." April 11, 1977. In *History of the 1st Tactical Fighter Wing, Jan–Mar 1977*, vol. 1. K-WG-1-HI, AFHRA.

Hutchins, Edwin. *Cognition in the Wild*. Cambridge, MA: MIT Press, 1995.

Hutchins, Edwin, Nadir Weibel, Colleen Emmenegger, Adam Fouse, and Barbara Holder. "An Integrative Approach to Understanding Flight Crew Activity." *Journal of Cognitive Engineering and Decision Making* 7, no. 4 (December 2013): 353–76.

Impact of Technology on Military Manpower Requirements, Readiness and Operations. Hearing before the US Senate, Subcommittee on Manpower and Personnel of the Committee on Armed Services, December 4, 1980. Washington, DC: Government Printing Office, 1980. Y4.Ar5/3:T22.

Isaacson, Walter, Evan Thomas, Bruce W. Nelan, and Christopher Redman. "The Winds of Reform." *Time*, March 7, 1983.

"It's the Size of the Fight in the Man!" *Air Force*, March 1953.

Jabara, James, Capt., USAF. "We Fly MIG Alley." *Air Force*, June 1951.

"Jabara, Jet Ace, Set for Mid-East Good Will Tour." *Washington Post*, December 29, 1951.

"Jabara to Fight Again." *New York Times*, November 26, 1952.

Jaffe, Greg. "Combat Generation: Drone Operators Climb on Winds of Change in the Air Force." *Washington Post*, February 28, 2010.

James, Clifton E., Maj., USAF. Memorandum to Director of Reconnaissance. "Report of Staff Visit." April 5, 1952. Personal Papers of Charles P. Hollstein. 168.7350, pt. 40, AFHRA.

Jaquish, John E., Lt. Col., USAF. Memorandum to 1 TFW/DO. "DACT OPS Summary." September 7, 1977. In *History of the 1st Tactical Fighter Wing, Jul–Sep 1977*, vol. 4. K-WG-1-HI, AFHRA.

Jenkins, Jack L. "Some Notes on the Behavior of Computing Gunsights." *Fighter Weapons Newsletter*, no.1, February 1956, 47–53. MSFRIC.

Jeon, Chihyung. "Technologies of the Operator: Engineering the Pilot in the US and Japan, 1930–1960." Ph.D. dissertation, MIT, 2010.

"Jet Ace Out of Combat." *New York Times*, May 22, 1951.

"Jet Flier Shy of 'Ace' by 1 in 5 Fights." *Washington Post*, August 17, 1951.

"Jet Plane Explosion Kills Major Bong, Top US Ace." *New York Times*, August 7, 1945.

"Jet 'Superior to' MIG-15: Air Force Reports Production Begun on Improved Plane." *New York Times*, June 4, 1951.

Johnson, David E. *Fast Tanks and Heavy Bombers: Innovation in the U.S. Army, 1917–1945*. Ithaca, NY: Cornell University Press, 1998.

Johnson, J. "The New Top Guns." *Time*, August 12, 1991.

Johnston, Maurice B., Jr., Capt., USAF. "Dissimilar Aircraft Engagement." *Fighter Weapons Newsletter*, March 1968, 26–30. MSFRIC.

Joint Strike Fighter: DOD Actions Needed to Further Enhance Restructuring and Address Affordability Risks. GAO-12-437. Report to Congressional Committees. Washington, DC: US Government Accountability Office, June 2012. Accessed March 14, 2014. http://gao.gov/assets/600/591608.pdf.

Jones, George L., Lt. Col., USAF. Memorandum to Commanding Officer, 4th Fighter Interceptor Wing. "F-86E Combat Evaluation." November 1, 1951. In *History of the 4th Fighter-Interceptor Wing, Oct 1951*. K-WG-4-HI, AFRHA.

Jumper, John, Capt., USAF. "Air-to-Air Training to Win." *Fighter Weapons Review*, Winter 1976, 16. MSFRIC.

———, Capt., USAF. "COMM-OUT Turns: Doing It with Your Mouth Shut." *Fighter Weapons Review*, Spring 1977, 61–64. MSFRIC.

———, Capt., USAF. "Instincts of the Fighter Pilot." *Fighter Weapons Review*, Winter 1976, 5–6. MSFRIC.

"Kansan Becomes History's First Jet Fighter Ace: Wichita Pilot Shoots Down His Fifth and Sixth MIGs in Dogfight over Korean Town." *Los Angeles Times*, May 21, 1951.

"Kansan Downs 2 MIG Planes, Becomes History's 1st Jet Ace." *Washington Post*, May 21, 1951.

Kantrowitz, B., and E. Clift. "The Right to Fight." *Newsweek*, August 5, 1991.

Kaylor, Robert. "Mock Dogfights Test Latest Jet Fighters." *Sarasota Herald-Tribune*, May 7, 1978.

Keith, Larry, Lt. Col., USAF. "Introduction." *Fighter Weapons Review*, Winter 1976, 1. MSFRIC.

Kent, Glenn A., Lt. Gen., USAF. Oral History Interview. Interview by Jacob Neufeld, August 6, 1974. K239.0512-970, AFHRA.

Kenyon, and Beckwith. "FEAF Intel Form 1, 9 Sep 1951." In *History of the 4th Fighter-Interceptor Wing, Sep 1951*. K-WG-4-HI, AFHRA.

Kepner, W. E., Maj. Gen., USAAF. "Major General W.E. Kepner, VIII Fighter Command." In *The Long Reach: Deep Fighter Escort Tactics*, 1–4. VIII Fighter Command, May 29, 1944. 524.502A, AFHRA.

Kilduff, Peter. *Richthofen: Beyond the Legend of the Red Baron*. New York: John Wiley & Sons, 1993.

Kimball, Wesley D., Col., USAF. Oral History Interview. Interview by Capt. Richard Clement, USAF, February 4, 1969. K239.0512-093, AFHRA.

Kirk, William L., Maj., USAF. Oral History Interview, n.d. K239.0512-206, AFHRA.

Kirkconnell, William B., Cmdr., USN. "SRAAM Program Review." Hearings before the US Senate, Committee on Armed Services, March 14, 1975, pp. 4634–50. Washington, DC: Government Printing Office, 1975. Y4/Ar5/3:P94/6/976/pt9.

Knaack, Marcelle Size. *Encyclopedia of US Air Force Aircraft and Missile Systems: Post-World War II Fighters, 1945–1973*. Vol. 1. Washington, DC: Office of Air Force History, 1978.

Knight, James A., Jr., Lt. Gen., USAF (ret.). Oral History Interview. Interview by Col. James C. Posgate, USAF, November 30–December 1, 1983. K239.0512-1561, AFHRA.

Kranzberg, Melvin. "The Newest History: Science and Technology." *Science* 136, no. 3515 (May 11, 1962): 463–68.

Krebs, Frank X., Col., USAF. Memorandum to ATC and TAC. "F-110 Radar Pilot Training." June 15, 1962. In *History of the Air Training Command, Jul–Dec 1962*, vol. 11. K220.01, AFHRA.

Kross, Walter, Lt. Col., USAF. "Military Reform: Past and Present." *Air University Review*, August 1981. Accessed March 29, 2012. http://www.airpower.maxwell.af.mil/airchronicles/aureview/1981/jul-aug/kross.htm.

———. *Military Reform: The High-Tech Debate in Tactical Air Forces*. Washington, DC: National Defense University Press, 1985.

Kubrick, Stanley. *2001: A Space Odyssey*. Film. MGM, 1968.

Kutzscher, Edgar W. "Application of Infrared Radiation in Air to Air Detection Devices." *Fighter Weapons Newsletter*, November 1956, 27–30. MSFRIC.

La Salle, Louis W., Col., USAF. "Project CORONA HARVEST End of Tour Report," November 27, 1968. K740.131, AFHRA.

Laidlaw, Brian, Col., USAF. Interview by Steven A. Fino, January 26, 2017.

Lake, Julian S., R. Adm., USN. "ACEVAL-AIMVAL Program Review." Hearings before the US Senate, Committee on Armed Services, March 11, 1976, pp. 5038–73. Washington, DC: Government Printing Office, 1976. Y4.Ar5/3:P94/6/977/pt9.

Lamb, Bill, Lt. Col., USAF. "Aerial Attack—The F-15 School." *Fighter Weapons Review*, Fall 1984, 16. MSFRIC.

Lamb, M. J., USAF (ret.). "Comments on 'Ten Thousand Feet and Ten Thousand Miles,'" May 1, 2012. Accessed September 16, 2012. http://www.airpower.au.af.mil/article.asp?id=72.

Langguth, Jack. "Hanoi MIG's Down 2 American Jets in First Air Clash." *New York Times*, April 5, 1965.

———. "One MIG Seen on Fire." *New York Times*, April 10, 1965.

———. "US Aides Indicate a Red Missile Hit Jet Near Hanoi." *New York Times*, July 26, 1965.

Larkins, Mark, Lt., USN. "The WSO in Air-to-Air." *Fighter Weapons Review*, Winter 1976, 37–38. MSFRIC.

Laslie, Brian D. *The Air Force Way of War*. Lexington: University Press of Kentucky, 2015.

Laudise, Ted, Col., USAF. Letter to 58 TFTW/DO. "ACEVAL/AIMVAL Testing." August 27, 1976. In *History of the 58th Tactical Fighter Training Wing, Jul–Sep 1976*, by Robert Sullivan. K285.79-42, AFHRA.

"Lawmakers Ask for Review of Unmanned Promotion Rate." *Air Force Times*, October 3, 2012. Accessed October 8, 2012. http://www.airforcetimes.com/news/2012/10/air-force-ask-gao-review-uav-pilots-promotion-rate-100312/.

"Lead Assessing." *Fighter Weapons Newsletter*, no. 2 (1955): 27–28. MSFRIC.

Leatham, Dale W., Capt., USAF. "Adverse Yaw: Recognition, Inducement, and Correction." *Fighter Weapons Newsletter*, December 1962, 4. MSFRIC.

Lee, Caitlin. "Pilots Needed." *Aviation Week & Space Technology*, April 25, 2016.

Leming, Paul J., Jr., Col., USAF. Memorandum to 9 AF/DOST. "Helmet Color in the F-15." November 2, 1976. In *History of the 1st Tactical Fighter Wing, Oct–Dec 1976*. K-WG-1-HI, AFHRA.

"Lessons Learned Symposium, 1974 (AFSC)." 03951, AFMC/HO.

Lewis, Edgar M., Lt. Col., USAF. *History of the F-15 Eagle: A Corporate Memory*. May 18, 1976. K417.042-18, AFHRA.

Lewis, W. David. *Eddie Rickenbacker: An American Hero in the Twentieth Century*. Baltimore: Johns Hopkins University Press, 2005.

"Lieutenant General Arthur C. Agan." *USAF Biography*. Accessed February 26, 2014. http://www.af.mil/AboutUs/Biographies/Display/tabid/225/Article/107889/lieutenant-general-arthur-c-agan.aspx.

"Lieutenant General Leighton I. Davis." *USAF Biography*. Accessed December 19, 2013. http://www.af.mil/AboutUs/Biographies/Display/tabid/225/Article/107302/lieutenant-general-leighton-i-davis.aspx.

Limpantsis, Lt., USAF. "FEAF Intel Form 1, 17 Dec 1950." In *History of the 4th Fighter-Interceptor Group, Oct–Dec 1950*. K-GP-4-HI, AFHRA.

Lin, Leon. "Group Urges Checks on Online Learning at MIT: Social Skills and Credit Integrity at Stake." *The Tech*. April 1, 2014.

Locher, James R. *Victory on the Potomac: The Goldwater-Nichols Act Unifies the Pentagon*. College Station: Texas A&M University Press, 2002.

Long, Charles E., Jr. *USAF Fighter Aircraft Fire Control Systems*. Eglin AFB, FL: Air Proving Ground Command, January 1954. AFMC/HO and ADB192291, DTIC.

The Long Reach: Deep Fighter Escort Tactics. VIII Fighter Command, May 29, 1944. 524.502A, AFHRA.

Luke, Ernest P., Maj., USAF. *Evaluation Report on GAR-1 Weapon System*. ADJ 56-1928. Holloman AFB, NM: Air Research and Development Command, Holloman Air Development Center, March 1956. Microfilm 31792, AFHRA.

Mahurin, Walker M. *Honest John: The Autobiography of Walker M. Mahurin*. New York: G. P. Putnam's Sons, 1962.

———, Maj., USAF. "Major Walker M. Mahurin, 63rd Fighter Squadron, 56th Fighter Group, P-47." In *The Long Reach: Deep Fighter Escort Tactics*, 23–25. VIII Fighter Command, May 29, 1944. 524.502A, AFHRA.

"Major General Frederick C. Blesse." *USAF Biography*. Accessed December 3, 2013. http://www.af.mil/AboutUs/Biographies/Display/tabid/225/Article/107669/major-general-frederick-c-blesse.aspx.

Maloy, Robert W., Col., USAF. Letter to Brig. Gen. Harmon E. Burns, USAF. June 7, 1967. In *History of the 366th Tactical Fighter Wing, Apr–Sep 1967*, vol. 1. K-WG-366-HI, AFHRA.

Marlow, Bobby L., Lt. Col., USAF, and Peter F. English. *Reduction of Tactical Fighter Losses over North Vietnam*. Working Paper 66/13.2. Directorate of Operations Analysis, Headquarters 7th Air Force, December 24, 1966. K740.31066-13, AFHRA.

Marsh, Robert T., Lt. Gen., USAF. Letter to AFTEC/CC. "AFSC Review of ACEVAL/AIMVAL Preliminary Test Plan (Your 272126Z Jan 76)." March 17, 1976. Attachment to Gen. William J. Evans, USAF, letter to HQ USAF/CV, "AIMVAL," March 17, 1976. K168.7339-79, AFHRA.

Martin, Douglas. "Robinson Risner, Ace Fighter Pilot, Dies at 88." *New York Times*, October 28, 2013.

Martin, John J. Memorandum to Under Secretary of Defense for Research and Engineering. "AIMVAL Test Results and IFF." December 21, 1977. 13000, NHHC.

Martin, Matt, and Charles W. Sasser. *Predator: The Remote-Control Air War over Iraq and Afghanistan: A Pilot's Story*. Minneapolis, MN: Zenith Press, 2010.

Mason, Joe L., Lt. Col., USAAF. Memorandum to Commanding Officer, 66th Fighter Wing. "352nd Fighter Group Combat Formations." November 4, 1943. 524.501, AFHRA.

Mass, Warren. "Eddie Rickenbacker: From Flying Ace to Patriot-Entrepreneur." *New American*, July 22, 2013.

Mathis, Robert C., Brig. Gen., USAF. "F-15 Air Superiority Fighter: Presentation for Honorable James R. Schlesinger, Secretary of Defense," May 3, 1974. 03887, AFMC/HO.

———, Brig. Gen., USAF. Memorandum to AFSC/SDNJ. "F-15 Attack Version." April 19, 1974. 03927, AFMC/HO.

Maurer, Maurer. *USAF Credits for the Destruction of Enemy Aircraft: Korean War*. USAF Historical Study No. 81. Maxwell AFB, AL: USAF Historical Division, Air University, 1963.

May, Elaine Tyler. "Women in the Wild Blue Yonder." *New York Times*, August 7, 1991.

McCarthy, Mike. *Phantom Reflections: An American Fighter Pilot in Vietnam*. Mechanicsburg, PA: Stackpole Books, 2007.

McClure, William C., Capt., USAF. Memorandum to Commanding General, Continental Air Command, Commanding General, 14th Air Force, Commanding General, 31st Fighter Wing, Col. H. B. Darling Jr., USAF, and Lt. Col. J. H. Buchert, USAF. "Unit Operations and

Analysis Report on A-1-C-M Gun Sight Test." January 23, 1950. In *History of 31st Fighter Bomber Group, Jan 1950*. K-GP-31-HI, AFHRA.

McCollom, Loren G., Lt. Col., USAAF. Memorandum to Commanding Officer, 66th Fighter Wing. "Narrative Report on Bomber Escort by P-47." November 5, 1943. 524.501, AFHRA.

McCrea, Michael M. *US Navy, Marine Corps, and Air Force Fixed-Wing Aircraft Losses and Damage in Southeast Asia (1962–1973)*. CRC 305. Washington, DC: Center for Naval Analyses, August 1976.

McCulloch, Peter. "Surgical Professionalism in the 21st Century." *The Lancet* 367, no. 9505 (January 14, 2006): 177–81.

McFarland, Stephen L. *Conquering the Night: Army Air Forces Night Fighters at War*. Washington, DC: Air Force History and Museums Program, 1997.

McFarland, Stephen L., and Wesley Phillips Newton. *To Command the Sky: The Battle for Air Superiority over Germany, 1942–1944*. Washington, DC: Smithsonian Institution Press, 1991.

McGehee, James C., Maj. Gen., USAF. Memorandum to ATC (ATXDC). "Flight Indoctrination." May 11, 1963. In *History of the Air Training Command, Jan–Jun 1963*, vol. 16. K220.01, AFHRA.

McGough, Edward A., Maj. Gen., USAF. Oral History Interview. Interview by Jack Neufeld, September 26, 1973. K239.0512-860, AFHRA.

McInerney, Thomas G., Maj., USAF. "F-4 Dive Toss." *Fighter Weapons Newsletter*, March 1968, 7–11. MSFRIC.

———, Capt., USAF. "F-4D." *Fighter Weapons Newsletter*, September 1966, 15. MSFRIC.

———, Capt., USAF. "F-4E." *Fighter Weapons Newsletter*, December 1966, 5. MSFRIC.

———, Maj., USAF. "F-4E Cat III." *Fighter Weapons Newsletter*, March 1969, 30–31. MSFRIC.

———, Capt., USAF. "LCOS: Lead Computing Optical Sight." *Fighter Weapons Newsletter*, September 1966, 10–13. MSFRIC.

McIntosh, Richard K., Col., USAF. Oral History Interview. Interview by Jack Neufeld, March 6, 1973. K239.0512-968, AFHRA.

McKenzie, Robert P., R. Adm., USN, and Maj. Gen. James R. Hildreth, USAF. *ACEVAL Final Report*, vol. 1. Nellis AFB, NV: ACEVAL-AIMVAL JTF, February 1978. In *History of the Air Force Test and Evaluation Center, 1 Jan–31 Dec 1977*, vol. 17. K150.01, AFHRA.

McMillin. "FEAF Intel Form 1, 17 Jun 1951." In *History of the 4th Fighter-Interceptor Group, May–Jun 1951*. K-GP-4-HI, AFHRA.

McMillin, Casolari, Cheshire, and Beckwith. "FEAF Intel Form 1, 18 Jun 1951." In *History of the 4th Fighter-Interceptor Group, May–Jun 1951*. K-GP-4-HI, AFHRA.

McMullen, Richard P. *History of Air Defense Weapons, 1946–1962*. Air Defense Command Historical Study No. 14. Historical Division, Office of Information, HQ Air Defense Command, n.d. K410.041-14, AFHRA.

McMullen, Thomas H., Brig. Gen., USAF. Letter to Gen. Robert J. Dixon, USAF, March 15, 1977. 168.7339-79, AFHRA.

McNally, Ray F., Capt., USAF. "WRCS: Weapons Release Computer Set." *Fighter Weapons Newsletter*, September 1966, 17–27. MSFRIC.

McNamara, Robert S. *In Retrospect: The Tragedy and Lessons of Vietnam*. New York: Times Books, 1995.

McNickle, Melvin F., Col., USAF. DD Form 96 to WCSWF. "Reworking A-4 Sightheads." July 21, 1952. In "AMC, F-86 Correspondence, Documents 201–300." K2021-57, vol. 3, AFHRA.

McVeigh, Donald R., and Robert L. Perry. *The Development of the F-102 Aircraft (through December 1954)*. Wright-Patterson AFB, OH: Historical Branch, Office of Information Services, Wright Air Development Center, April 1957. 04452, AFMC/HO.

"Medals for Drone Warriors Canceled." *New York Times*, April 15, 2013. Accessed September 8, 2013. http://www.nytimes.com/2013/04/16/us/politics/medals-for-drone-warriors-canceled .html.

Mehserle, H. J., Capt., USAF. *Capability of the F-4C Aircraft/SUU-16/A Gun Pod in the Close Support Role*. Eglin AFB, FL: Air Proving Ground Center, Air Force Systems Command, August 1965. K240.04-99, AFHRA.

Melville, Herman. *Battle-Pieces and Aspects of the War*. New York: Harper & Brothers, 1866.

Merlin, Peter W. "The Battle of Palmdale." *China Laker*, Summer 2006. Accessed Jan 6, 2014. http://www.chinalakemuseum.org/secretcitystories/The%20Battle%20of%20Palmdale .pdf.

Merritt, Jack N., and Pierre M. Sprey. "Quality, Quantity, or Training." *Fighter Weapons Review*, Summer 1974, 7–14. MSFRIC.

Meyer, Alan. *Weekend Pilots: Technology, Masculinity, and Private Aviation in Postwar America*. Baltimore: Johns Hopkins University Press, 2015.

Meyer, John C., Gen., USAF. "Air Superiority into the 1980s." *Air Force*, August 1971.

———, Col., USAF. Letter to Commanding General, Air Defense Command. "F-86 Activities Report, 2 January 1951 thru 13 January 1951." February 24, 1951. In *History of the 4th Fighter-Interceptor Group, Jan 1951*. K-GP-4-HI, AFHRA.

———, Col., USAF. Letter to Commanding General, Air Defense Force. "F-86 Activities Report (Special)." April 15, 1951. In *History of the 4th Fighter-Interceptor Group, Feb–Apr 1951*. K-GP-4-HI, AFHRA.

———, Col., USAF. Letter to Commanding General, Air Defense Force. "F-86 Activities Report." May 3, 1951. In *History of the 4th Fighter-Interceptor Group, Feb–Apr 1951*. K-GP-4-HI, AFHRA.

———, Lt. Col., USAF. Memorandum to Commanding Officer, 4th Fighter Interceptor Wing. "F-86-A5 Armament." January 7, 1951. In "Test Data and Combat Reports Relative to Vulnerability and Limitations of F-86 Aircraft, Jan–Jul 1951." K-GP-4-SU-OP, AFHRA.

———, Gen., USAF. Oral History Interview. Interview by Capt. Stephen O. Manning, USAF, July 23, 1975. K239.0512-894, AFHRA.

Meyer, Richard L., Capt., USAF. "Phantom over Tampa Bay." *Fighter Weapons Newsletter*, December 1963, 8–9. MSFRIC.

Michaels, Jim. "Drones Change 'Top Gun' Culture of Air Force." *USA Today*, December 1, 2012. Accessed August 14, 2013. http://www.usatoday.com/story/news/world/2012/11/30/drone -wars/1737991/.

Michel, Marshall L., III. *Clashes: Air Combat over North Vietnam, 1965–1972*. Annapolis, MD: Naval Institute Press, 1997.

———. "The Revolt of the Majors: How the Air Force Changed after Vietnam." Ph.D. dissertation, Auburn University, 2006.

Middleton, Drew. “Air Force Unveils Fighter Designed to Keep Superiority into the ’80’s.” *New York Times*, June 27, 1972.

———. *Air War—Vietnam*. New York: Arno Press, 1978.

“MIG’s Top Sabres, US Jet Ace Says.” *New York Times*, June 2, 1951.

“Military Order of the Purple Heart Opposes Precedence of New Defense Medal.” *Military Order of the Purple Heart*, February 15, 2013. Accessed March 14, 2014. http://www.purpleheart.org/News.aspx?Identity=238.

Miller, Ed Mack. “Thunderbolt Geometry.” *Air Force*, September 1955.

Miller, George H., Col., USAF. Memorandum to TAC/DOOW. “Red Flag 77-7 After Action Report.” July 11, 1977. In *History of the 1st Tactical Fighter Wing, Jul–Sep 1977*, vol. 4. K-WG-1-HI, AFHRA.

Milling, T. D., Maj. *Air Tactics*. Training Regulations No. 440-16. Langley Field, VA: Air Service Field Officers School, 1922. 248.101, AFHRA.

Mindell, David A. *Between Human and Machine: Feedback, Control, and Computing before Cybernetics*. Baltimore: Johns Hopkins University Press, 2002.

———. *Iron Coffin: War, Technology, and Experience Aboard the USS* Monitor. Updated ed. Baltimore: Johns Hopkins University Press, 2012.

———. *Our Robots, Ourselves: Robotics and the Myths of Autonomy*. New York: Viking, 2015.

———. *War, Technology, and Experience Aboard the USS* Monitor. Baltimore: Johns Hopkins University Press, 2000.

Mitchell, A. C. G. “Airborne Fire Control: Specific Systems and Their Performance.” In *Project Vista: A Study of Ground and Air Tactical Warfare with Special Reference to the Defense of Western Europe*, C:App. 205–14. Series B. Pasadena, CA: California Institute of Technology, 1952. K146.07-2B, AFHRA.

Mitchell, Vance O. *Air Force Officers: Personnel Policy Development, 1944–1974*. Washington, DC: Air Force History and Museums Program, 1996.

Mitchell, William “Billy.” *Winged Defense: The Development and Possibilities of Modern Air Power—Economic and Military*. Tuscaloosa: University of Alabama Press, 1925.

Mohamed, Rezk M., Maj., USAF. “A Case for Recognition.” *Fighter Weapons Newsletter*, March 1968, 18–23. MSFRIC.

Moïse, Edwin E. *Tonkin Gulf and the Escalation of the Vietnam War*. Chapel Hill: University of North Carolina Press, 1996.

Momyer, William W., Gen., USAF. *Air Combat Maneuvering Range (ACMR)*. TAC ROC (Required Operational Capability) 11-71. Langley AFB, VA: Headquarters, Tactical Air Command, April 21, 1971. In *History of the USAF Tactical Fighter Weapons Center, Jul 1970–Jul 1971*, vol. 3. K417.0735, AFHRA.

———, Gen., USAF (ret.). *Airpower in Three Wars*. Reprint ed. Maxwell AFB, AL: Air University Press, 2003.

———, Gen., USAF. “The Evolution of Fighter Tactics in SEA.” *Air Force*, July 1973.

———, Lt. Gen., USAF. Memorandum to 8 TFW, 12 TFW, and 366 TFW. “F-4 Pilot Upgrading.” November 21, 1967. In *History of the 8th Tactical Fighter Wing, Oct–Dec 1967*, vol. 2. K-WG-8-HI, AFHRA.

———, Gen., USAF (ret.). Memorandum for Gen. Ellis. “CORONA HARVEST (Out-Country Air Operations, Southeast Asia, 1 January 1965–31 March 1968). K239.031-98, AFHRA.

Morris, Thomas W., Col., USAF. Oral History Interview. Interview by Capt. Richard Clement, USAF, February 4, 1969. K239.0512-092, AFHRA.

Morrow, John H., Jr. *The Great War in the Air: Military Aviation from 1909 to 1921*. Tuscaloosa: University of Alabama Press, 1993.

Munger, Burt, Capt., USN. "Advanced Medium Range Air-to-Air Missile (AMRAAM)." Hearings before the US House, Committee on Armed Services, Hearings on Military Posture and H.R. 10929, February 27, 1978, pp. 436–54. Washington, DC: Government Printing Office, 1978. Y4.Ar5/2a:977-78/56/pt3/bk1.

Muskat, Michael S., Capt., USAF. "Featherduster." *Fighter Weapons Newsletter*, March 1966, 22–26. MSFRIC.

Myers, Charles E., Jr. Oral History Interview. Interview by Jack Neufeld, July 18, 1973. K239.0512-971, AFHRA.

Nabors, Jerry H., Lt. Col., USAF. "Aggressor Squadron Briefing." Hearings before the US Senate, Committee on Armed Services, March 9, 1976, pp. 4907–19. Washington, DC: Government Printing Office, 1976. Y4.Ar5/3:P94/6/977/pt9.

Nalty, Bernard C., ed. *Winged Shield, Winged Sword: A History of the United States Air Force*, vol. 1, *1907–1950*. Washington, DC: Air Force History and Museums Program, 1997.

Nash, Slade, Col., USAF. "Project CORONA HARVEST End of Tour Report," October 17, 1969. K717.131, AFHRA.

"National Aeronautic Association Collier Trophy." Accessed March 8, 2014. http://naa.aero/html/awards/index.cfm?cmsid=62.

"National Aeronautic Association Mackay Trophy." Accessed February 18, 2014. http://naa.aero/html/awards/index.cfm?cmsid=70.

"Navy Identifies Battleship 'X' That Got 3 Cruisers, 32 Planes: The Battle of Santa Cruz." *Washington Post*, October 3, 1943.

"Nellis AFB Squadron Get First F-5E Tiger Fighters." *Las Vegas Review Journal*, December 14, 1975.

"Nellis Fighters Collide." *Las Vegas Review Journal*, March 1, 1977.

"Nellis Pilots Given New Battle Training." *Las Vegas Sun*, December 29, 1976.

Neufeld, Jacob. *The F-15 Eagle: Origins and Development, 1964–1972*. Declassified Copy. Washington, DC: Office of Air Force History, 1974. In *History of the F-15 Eagle*, vol. 2, pt. 1. K417.042-18, AFHRA.

New Developments. Continental Air Command Manual 50-15, vol. 1, pt. 6. Mitchell AFB, NY: Headquarters, Continental Air Command, 1952. K419.7242050-15, AFHRA.

Newell, Maj., USAF. "AIMVAL-ACEVAL Results / Lessons Learned," September 9, 1983. K168.03-2107, AFHRA.

"New England Flier Becomes 16th Jet Ace." *Washington Post*, May 21, 1952.

"New Missile Ready." *New York Times*, September 24, 1958.

"New Planes Designated for Nellis." *Las Vegas Review Journal*, March 12, 1976.

"New Radar Sight Guides Jets' Guns." *New York Times*, April 3, 1950.

Norman, George G., Col., USAF. Memorandum to USAF. "F-110 Radar Pilot Training." June 1, 1962. Attachment to Col. Frank Krebs, USAF, memorandum to ATC and TAC, 15 Jun 1962, in *History of the Air Training Command, Jul–Dec 1962*, vol.11. K220.01, AFHRA.

"No-Win War at Dogbone Lake." *US News & World Report*, January 9, 1978.

Nuckols, Troy A., 1Lt., USAF. *SNL A-57 Gun Machine Cal .50 M3*. Unsatisfactory Report, AAF Form No. 54. War Department, Army Air Forces, n.d. In "Test Data and Combat Reports Relative to Vulnerability and Limitations of F-86 Aircraft, Jan–Jul 1951." K-GP-4-SU-OP, AFHRA.

Nunn, Sam. *Impact of Technology on Military Manpower Requirements, Readiness and Operations*. Hearing before the US Senate, Subcommittee on Manpower and Personnel of the Committee on Armed Services, December 4, 1980, pp. 1–4. Washington, DC: Government Printing Office, 1980. Y4.Ar5/3:T22.

O'Connor, Seamus. "Board Recognizes Korean War Pilot as Ace Decades Later." *Air Force Times*, February 4, 2008.

O'Connor, Sean. *Arming America's Interceptors: The Hughes Falcon Missile Family*. Technical Report. Air Power Australia, April 2012. Accessed January 6, 2014. http://www.ausairpower.net/Falcon-Evolution.html.

O'Donohue, Lt. Col., USAF. "Development Concept Paper: New Air Force Tactical Counter-Air Fighter (F-X)," September 15, 1968. In *History of the F-15 Eagle*, vol. 2. K417.042-18, AFHRA.

O'Hara, J. F., R. Adm., USN. "Briefing." Hearings before the US House, Committee on Armed Services, March 9, 1978, pp. 1018–27. Washington, DC: Government Printing Office, 1978. Y4.Ar5/2a:977-78/56/pt3/bk1.

O'Keefe, Tim., Jr., Lt. Col., USAF. "Eagle Eye." *Fighter Weapons Review*, Fall 1978, 27–29. MSFRIC.

"The 'Old Pros' of MiG Alley." *Air Force*, April 1952.

Olds, Robin, Brig. Gen., USAF. "Briefing to the Staff of Project CORONA HARVEST on Operation BOLO," September 29, 1969. K239.0512-222, AFHRA.

———. Brig. Gen., USAF (ret.). Corona Ace Oral History Interview. Interview by Lt. Col. John N. Dick Jr., USAF, and Lt. Col. Gordon F. Nelson, USAF, February 17, 1977. K239.0512-1079, AFHRA.

———, Col., USAF. "End of Tour Report," n.d. (received at HQ PACAF on October 27, 1967). K717.131, AFHRA.

———. *Fighter Pilot: The Memoirs of Legendary Ace Robin Olds*. With Christina Olds and Ed Rasimus. New York: St. Martin's Press, 2010.

———, Col., USAF. "How I Got My First MIG." *Air Force & Space Digest*, July 1967.

———, Brig. Gen., USAF. "The Lessons of Clobber College." *Grumman Horizons* 8, no. 1 (1968): 18–22. Box 1097, L3-A, NMUSAF.

———, Col., USAF. Oral History Interview, July 12, 1967. K239.0512-160, AFHRA.

———, Brig. Gen., USAF. Oral History Interview. Interview by Maj. Geffen, USAF, and Maj. Folkman, USAF, n.d. K239.0512-051, AFHRA.

Olsen, Emil M. Letter to Holland, Schwarze, and Rhame. "Evaluation of Results of the 1950 USAF Gunnery Meet," April 13, 1950. A-1 Sight Conferences, CSDL-HC.

O'Mara, Raymond P., Col., USAF. "The Socio-Technical Construction of Precision Bombing: A Study of Shared Control and Cognition by Humans, Machines, and Doctrine during World War II." Ph.D. dissertation, MIT, 2011.

"'Operation Red Flag' Studies How to Beat Soviet Pilots." *Las Vegas Review Journal*, May 29, 1977.

Operational Suitability Test of the A-1CM Gun-Bomb-Rocket Sight (F-84E Phase with AN/APG-30 Radar Ranging). Project APG/ADB/8-A. Eglin AFB, FL: Air Proving Ground Command, December 18, 1950. ADB191068, DTIC.

Operational Suitability Test of the F-86E Aircraft. Project APGC/ADB/18-A. Eglin AFB, FL: Air Proving Ground Command, April 24, 1952. ADB971622, DTIC.

Osborn, Terrell J., 1Lt., USAF. "End of Tour Report," January 2, 1967. K740.131, AFHRA.

Osterhout, David, Capt., USAF, Capt. Richard Berdan, USAF, and Capt. William Gordon III, USAF. "F-15 Briefing Script," November 5, 1971. 03972, AFMC/HO and K417.042-18, AFHRA.

Ouma, Joseph A., Wayne L. Chappelle, and Amber Salinas. *Facets of Occupational Burnout among US Air Force Active Duty and National Guard / Reserve MQ-1 Predator and MQ-9 Reaper Operations.* Report AFRL-SA-WP-TR-2011-0003. Wright Patterson AFB, OH: Air Force Research Laboratory, 711th Human Performance Wing, June 2011. ADA548103, DTIC.

Ow, Gordon Y. W., Maj., USAF. *Mission Bolo, 2 January 1967, F-4C vs MiG 21's.* Working Paper 67/3. Directorate, Tactical Air Analysis Center, Headquarters, 7th Air Force, February 13, 1967. K740.31067-3, AFHRA.

PACAF DO Read File, 13-15 May 67. K717.312, AFHRA.

Panetta, Leon. Memorandum. "Distinguished Warfare Medal." February 13, 2013. Accessed March 14, 2014. http://www.defense.gov/news/distinguishedwarfaremedalmemo.pdf.

Pape, Robert A. *Bombing to Win: Air Power and Coercion in War.* Ithaca, NY: Cornell University Press, 1996.

Parasuraman, Raja, and Thomas B. Sheridan. "A Model for Types and Levels of Human Interaction with Automation." *IEEE Transactions on Systems, Man, and Cybernetics—Part A: Systems and Humans* 30, no. 3 (May 2000): 286–97.

Parrott, Lindesay. "Sabres Bag 6 MIG's; Baker Is Top Jet Ace." *New York Times*, March 14, 1953.

Partin, Benton K., Brig. Gen., USAF. "Program Management Directive for Project 670E (Air-to-Air Technology)," September 4, 1973. Microfilm 40233, AFHRA.

Payne, Don H., Col., USAF. "Project CORONA HARVEST End of Tour Report," April 29, 1970. K740.131, AFHRA.

"Peking Reports Hainan Air Fight with US Planes." *New York Times*, April 10, 1965.

Perry, William. *Impact of Technology on Military Manpower Requirements, Readiness and Operations.* Hearings before the US Senate, Subcommittee on Manpower and Personnel of the Committee on Armed Services, December 4, 1980, pp. 4–54. Washington, DC: Government Printing Office, 1980. Y4.Ar5/3:T22.

Persico, Joseph E. *Eleventh Month, Eleventh Day, Eleventh Hour: Armistice Day, 1918 World War I and Its Violent Climax.* New York: Random House, 2007.

Peters, F. Whitten. "The Air Force, Korea, and Kosovo: Past as Prologue." In *Silver Wings, Golden Valor: The USAF Remembers Korea*, edited by Richard P. Hallion, 17–25. Washington, DC: Air Force History and Museums Program, 2006.

Peterson, A. H. "F-86A Operation as a Fighter-Interceptor; K-14 Air Base; 15 December 1950 to 2 January 1951," February 19, 1951. K730.310-6, AFRHA.

"Pilot Is 17th U.S. Jet Ace in Only 6 Weeks in Korea." *New York Times*, June 16, 1952.

"Pilots Describe Downing of MIG's." *New York Times*, July 12, 1965.

A Pilot's Introduction to the E-4 Fire Control System. Report M-40. Research & Development Division, Hughes Aircraft Company, n.d. AD007876, DTIC.

Pisano, Dominick A., Thomas J. Dietz, Joanne M. Gernstein, and Karl S. Schneide. *Legend, Memory, and the Great War in the Air*. Seattle: University of Washington Press, 1992.

"Point Paper on the ACEVAL Flight Trials," February 3, 1981. K168.03-2107, AFHRA.

Porter, William T., Jr., 1Lt., USAF. Memorandum to Director of Reconnaissance. "Report of Staff Visit." March 13, 1952. Personal Papers of Charles P. Hollstein. 168.7350, pt. 52, AFHRA.

Posen, Barry R. *The Sources of Military Doctrine: France, Britain, and Germany between the World Wars*. Ithaca, NY: Cornell University Press, 1984.

Pratt, Bob, Capt., USAF. "The GIB and the F-4." *Fighter Weapons Review*, Summer 1977, 33–36. MSFRIC.

Preston, Benjamin S., Jr., Col., USAF. Memorandum to Commanding Officer, 4th Fighter Interceptor Wing. "Analysis of F-86 Fighter Encounter with MiG-15 in Korea," September 5, 1951. In "Test Data and Combat Reports Relative to Vulnerability and Limitations of F-86 Aircraft, Jan–Jul 1951." K-GP-4-SU-OP, AFHRA.

Project Gun-Val, 01/15/1953–05/01/1953. Film. United States Air Force, 1953. 342-USAF-22561, NARA.

Project Red Baron II: Air to Air Encounters in Southeast Asia. Vol. 1, *Overview of Report*. Nellis AFB, NV: USAF Tactical Fighter Weapons Center, January 1973. M-U 42339-15a, MSFRIC.

Project Vista: A Study of Ground and Air Tactical Warfare with Special Reference to the Defense of Western Europe. 8 vols. Series B. Pasadena, CA: California Institute of Technology, 1952. K146.07-2B, AFHRA.

Pursuit Text. US Army Air Corps Training Regulation 29-229, February 1929. 248.101-8, AFHRA.

Putney, Diane T. *Airpower Advantage: Planning the Gulf Air Campaign, 1989–1991*. Washington, DC: Air Force History and Museums Program, 2004.

Quick, James M., Capt., USAF. "Cleared in Hot." *Fighter Weapons Review*, Winter 1974, 9–17. MSFRIC.

Quinn, J. H., Cmdr., USN. "Agile Program Review." Hearings before the US Senate, Committee on Armed Services, March 27, 1974, pp. 4695–705. Washington, DC: Government Printing Office, 1974. Y4.Ar5/3:P94/6/975/pt9.

Radar Academic Training Course: F-4C. TAC Syllabus, Course 111107R. October 1966. In *History of the 4453d Combat Crew Training Wing, Jul–Dec 1966*, vol. 2. K-WG-4453-HI, AFHRA.

"Radar Gunsight: A Partial Lifting of the Security Veil from American Fighter Equipment." *Flight*, January 16, 1953.

"Radar Ranging." *Fighter Gunnery*, no. 1 (1951): 9–13. AR.2007.187, NMUSAF.

Raddin, James H., Col., USAF. "End of Tour Report," 1970. K740.131, AFHRA.

Raebel, James B., Lt. Col., USAF. "Communications: Frequency Assignments." 4th Fighter Interceptor Group Directive, 4FIG-DIR 100-1, February 6, 1952. In *History of the 4th Fighter-Interceptor Wing, Feb 1952*. K-WG-4-HI, AFHRA.

"Ramey Holds F-86 Better than MIG: Air Force Chief of Operations Says Our Jet Meets Reds' on Better than Even Terms." *New York Times*, January 18, 1952.

Rasimus, Ed. *Palace Cobra: A Fighter Pilot in the Vietnam Air War*. New York: St. Martin's Press, 2006.

———. *When Thunder Rolled: An F-105 Pilot over North Vietnam*. New York: Presidio Press, 2003.

Redmond, Kent C., and Thomas M. Smith. *From Whirlwind to MITRE: The R&D Story of The SAGE Air Defense Computer*. Cambridge, MA: MIT Press, 2000.

"Report of the Chief of Staff." In *War Department Annual Reports, 1917*, 1:125–54. Washington, DC: Government Printing Office, 1918.

A Resume of Our Tactics—F-86E vs MiG-15: Any Resemblance between the Air-to-Air War over Korea and Any Past Air-to-Air War Anywhere Is Purely Coincidental. 51st Fighter-Interceptor Group, 1952. Personal Papers of Francis S. Gabreski. 168.7649-41, AFHRA.

Rhodarmer, Roger K., Maj. Gen., USAF. Oral History Interview. Interview by Jack Neufeld, March 29, 1973. K239.0512-972, AFHRA.

Riccioni, Everest E., Maj., USAF. "The Double Attack System." *Fighter Weapons Newsletter*, December 1963, 18–27. MSFRIC.

———, Maj., USAF. Memorandum to Maj. Gen. Charles D. Westover, USAF. "The Double Attack System." January 30, 1963. Personal Papers of Gen. Charles Westover. 168.7127-46, AFHRA.

———, Maj., USAF. "A New Method of Employing the F-100C in Combat." Attachment to Maj. Riccioni, USAF, memorandum to Maj. Gen. Westover, USAF, January 30, 1963, 168.7127-46, AFRHA.

Rickenbacker, Edward V. *Fighting the Flying Circus*. New York: Frederick A. Stokes, 1919.

Rippon, T. S., Capt., RAMC, and Lt. E. G. Manuel, RAF. "Report on the Essential Characteristics of Successful and Unsuccessful Aviators, with Special Reference to Temperament." *The Lancet* 192, no. 4961 (September 28, 1918): 411–15.

Risner, Robinson, Brig. Gen., USAF (ret.). Oral History Interview. Interview by Capt. Mark C. Cleary, USAF, March 1–2, 1983. K239.0512-1370, AFHRA.

Ritchie, Richard S. Corona Ace Oral History Interview. Interview by Lt. Col. Gordon F. Nelson, USAF, and Lt. Col. John N. Dick Jr., USAF, February 16, 1977. K239.0512-1078, AFHRA.

———, Capt., USAF. Oral History Interview. Interview by Maj. Lyn R. Officer, USAF, and Hugh N. Ahmann, October 11 and 30, 1972. K239.0512-630, AFHRA.

Robbins, Jay T., Lt. Gen., USAF (ret.). Oral History Interview. Interview by James C. Hasdorff, July 24–25, 1984. K239.0512-1593, AFHRA.

Roberts, John W., Gen., USAF. Corona Ace Oral History Interview. Interview by Lt. Col. John N. Dick Jr., USAF, Lt. Col. Gordon F. Nelson, USAF, and Maj. Jesse L. Green, USAF, February 10, 1977. K239.0512-1076, AFHRA.

Robinson, Jack A., Lt. Col., USAF. Memorandum to C15. "Report of the 47th TFS Deployment to South East Asia (TACOP Final)." December 16, 1965. In *History of the 15th Tactical Fighter Wing, Jul–Dec 1965*, vol. 2. K-WG-15-HI, AFHRA.

Rogers, Felix M., Gen., USAF (ret.). Corona Ace Oral History Interview. Interview by Lt. Col. Gordon F. Nelson, USAF, and Lt. Col. John N. Dick Jr., USAF, February 1, 1977. K239.0512-1069, AFHRA.

———, Lt. Gen., USAF. Oral History Interview. Interview by Jack Neufeld, July 17–18, 1974. K239.0512-965, AFHRA.

Role of Autonomy in DoD Systems. Task Force Report. Washington, DC: Department of Defense, Defense Science Board, July 2012.

Romer, Richard A., Maj., USAF, and Col. Cecil H. Braeden, USAF. "Letter of Agreement between 1st Tactical Fighter Wing and 9th Tactical Intelligence Squadron," January 21, 1976. In *History of the 1st Tactical Fighter Wing, Oct–Dec 1975*, vol. 1. K-WG-1-HI, AFHRA.

Ross, Don, Maj., USAF. "Comm Discipline." *Fighter Weapons Review*, Summer 1985, 13–15. MSFRIC.

Roy, Vincent P., Maj., USAF. "Double Attack Revisited." *Fighter Weapons Review*, Spring 1971, 26–32. MSFRIC.

Ryan, John D., Gen., USAF (ret.). Oral History Interview. Interview by Edgar F. Puryear Jr., August 9, 1979. K239.0512-1491, AFHRA.

SAF/SA. Message to AFTEC/CC. "Review of ACEVAL-AIMVAL Preliminary Test Plan." 222327Z Mar 76. In *History of the Air Force Test and Evaluation Center, 1 Jan–31 Dec 1976*, vol. 21. K150.01, AFHRA.

"Saigon Reports Sharp Fighting in Air War." *New York Times*, October 15, 1972.

Salter, James. *Burning the Days: Recollection*. New York: Vintage, 1998.

———. *The Hunters*. New York: Vintage, 1956.

Saunders, Ralph S., Lt. Col., USAF. "PACAF Tactics/Techniques (Bulletin #8)," June 2, 1965. K717.549-1, AFHRA.

Sayers, William. "The Red Baron Reports: What They Really Said." *Air Power History* 52, no. 3 (Fall 2005): 4–13.

Schaffel, Kenneth. *The Emerging Shield: The Air Force and the Evolution of Continental Air Defense, 1945–1960*. Washington, DC: Office of Air Force History, 1990.

Schatzberg, Eric. *Wings of Wood, Wings of Metal*. Princeton, NJ: Princeton University Press, 1999.

Schinz, Albert W., Col., USAF. "A Brief Resume of the Experiences of Colonel Albert W. Schinz's Combat Tour in Korea," July 10, 1952. K720.131-1, AFHRA.

Schmid, H. A., Col., USAF. DRAFT Memorandum. "Operations of the 4th Fighter-Interceptor Wing in Korea." June 11, 1951. In *History of the 4th Fighter-Interceptor Wing, Jun 1951*. K-WG-4-HI, AFHRA.

Schogol, Jeff. "RPA Pilots: In Demand, but Losing Out." *Air Force Times*, August 26, 2012. Accessed October 8, 2012. http://www.airforcetimes.com/prime/2012/08/PRIMEair-force-rap-pilots-promotions-082112/.

Scholin, Allan R. "New Phlyers for the Phantom." *Air Force & Space Digest*, January 1968.

Scott, Robert R., Col., USAF. *Big Eye Task Force*. Bulletin #20. PACAF Tactics & Techniques, September 7, 1965. K717.549-1, AFHRA.

———, Col., USAF. *Counter-Air Tactics*. Bulletin #45. PACAF Tactics & Techniques, July 26, 1966. K717.549-1, AFHRA.

———, Col., USAF. *EC-121D Operations in SEA*. Bulletin #35. PACAF Tactics & Techniques, February 8, 1966. K717.549-1, AFHRA.

———, Col., USAF. *F-4C—MIG-17 Engagement*. Bulletin #25. PACAF Tactics & Techniques, September 27, 1965. K717.549-1, AFHRA.

———, Col., USAF. *F-4C Fighter Screen and Escort*. Bulletin #44. PACAF Tactics & Techniques, July 14, 1966. K717.549-1, AFHRA.

———, Col., USAF. *Soviet Bloc AA Weapons*. Bulletin #18. PACAF Tactics & Techniques, August 31, 1965. K717.549-1, AFHRA.

Scott, S. J. *Probability of Southeast Asia Tour Completion for Tour Length of 100 Combat Sorties.* Note 66-1. Headquarters PACAF: Assistant for Operations Analysis, April 19, 1966. 168.7571-17, AFHRA.

Senn, Christian A. "Ricochets & Replies: Ten Thousand Feet and Ten Thousand Miles." *Air & Space Power Journal* 26, no. 4 (August 2012): 149–60.

Sennett, Richard. *The Craftsman*. New Haven, CT: Yale University Press, 2008.

Sexton, Mike, Capt., USAF, and Capt. Jerry Jameson, USAF. "The New F-4 Gunsight Camera." *Fighter Weapons Review*, Fall 1971, 8–13. MSFRIC.

Shaw, Robert L. *Fighter Combat: Tactics and Maneuvering.* Annapolis, MD: Naval Institute Press, 1985.

Shepard, Ronald L., Maj., USAF. *Tactical Weapons Employment Effectiveness.* Staff Summary Sheet, January 2, 1976. In *History of the 1st Tactical Fighter Wing, Jan–Mar 1976*, vol. 3. K-WG-1-HI, AFHRA.

Sheridan, Thomas B. *Telerobotics, Automation, and Human Supervisory Control.* Cambridge, MA: MIT Press, 1992.

Sherwood, John Darrell. *Fast Movers: Jet Pilots and the Vietnam Experience.* New York: St. Martin's Paperbacks, 1999.

———. *Officers in Flight Suits: The Story of American Air Force Fighter Pilots in the Korean War.* New York: NYU Press, 1998.

Sherwood, Terence R., Jr. *4th Tactical Fighter Wing Chronology.* Seymour Johnson AFB, NC: 4th Tactical Fighter Wing, Office of Information, 1966.

Shower, Michael A., Lt. Col., USAF. Foreword to *F-15 Eagle Engaged: The World's Most Successful Jet Fighter*, by Steve Davies and Doug Dildy, 4–5. New York: Osprey Publishing, 2007.

Sibley, Susan. *Draft Report of the MITx Subcommittee of the Faculty Policy Committee.* Cambridge, MA: MIT, February 2014.

Singer, P. W. *Wired for War: The Robotics Revolution and Conflict in the 21st Century.* New York: Penguin Press, 2009.

Skaret, Garry, Maj., USAF. "Double Attack Revisited One More Time." *Fighter Weapons Review*, Summer 1971, 14–16. MSFRIC.

Skogerbee, Arvid N., Maj., USAF. Oral History Interview. Interview by Lt. Col. Gordon Crozier, USAF, April 11, 1968. K239.0512-048, AFHRA.

Skolnik, Merrill I. "Fifty Years of Radar." *Proceedings of the IEEE* 73, no. 2 (February 1985): 182–97.

Slay, Alton D., Lt. Gen., USAF. "Tactical Air Programs." Hearings before the US Senate, Committee on Armed Services, March 9, 1976, pp. 4803–26. Washington, DC: Government Printing Office, 1976. Y4.Ar5/3:P94/6/977/pt9.

Smith, George F., Col., USAF. Memorandum to Commanding General, Far East Air Force. "Combat Evaluation of the MiG-15 vs the F-86." December 26, 1950. K146.003-140, AFHRA.

Smith, Merritt Roe, and Leo Marx, eds. *Does Technology Drive History? The Dilemma of Technological Determinism.* Cambridge, MA: MIT Press, 1994.

Smith, Perry M., Capt., USAF. "Ode to a Fighter Pilot." *Fighter Weapons Newsletter*, June 1962, 23. MSFRIC.

Snodgrass, Mike, Capt., USAF, and Maj. Terry New, USAF. "BFM for History Majors." *Fighter Weapons Review*, Spring 1988, 12–17. MSFRIC.

Sokol, Thomas, Col., USAF (ret.). Interview by Steven A. Fino, August 4, 2013.

Southeast Asia Counter-Air Alternatives. Headquarters, PACAF, December 31, 1966. K717.310-1, AFHRA.

Sparks, William R., Lt. Col., USAF (ret.). Interview by Steven A. Fino, August 3, 2012.

Spears, Sally. *Call Sign Revlon: The Life and Death of Navy Fighter Pilot Kara Hultgreen*. Annapolis, MD: Naval Institute Press, 1998.

"Special Subject: Two Years of MiG Activity." *FEAF Intelligence Roundup and Operational Summary*, no. 116 (November 15–21, 1952): chap. 4. K720.607A, AFHRA.

Spick, Mike. *The Ace Factor: Air Combat and the Role of Situational Awareness*. Annapolis, MD: Naval Institute Press, 1988.

Spinetta, Lawrence, Lt. Col., USAF. "The Glass Ceiling for Remotely Piloted Aircraft." *Air & Space Power Journal* 27, no. 3 (August 2013): 101–18.

Spinney, Franklin C. *Defense Facts of Life*. Washington, DC: Office of Secretary of Defense, December 5, 1980. ADA111544, DTIC.

Sprey, Pierre. Oral History Interview. Interview by Jack Neufeld, June 12, 1973. K239.0512-969, AFHRA.

"The Square Corner." *Fighter Gunnery*, no. 1 (1951): 14–15. Ar.2007.187, NMUSAF.

SSAC Proposal Analysis: Advanced Tactical Fighter System, F-15. 69ASK-286. Wright-Patterson AFB, OH: Aeronautical Systems Division, Air Force Systems Command, December 25, 1969. 04458, AFMC/HO.

"Standard Aircraft Characteristics: F-86A." Air Materiel Command, December 8, 1950. K243.861-1, AFHRA.

"Standard Aircraft Characteristics: F-86A." Wright Air Development Center, September 21, 1951. K243.861-1, AFHRA.

"Standard Aircraft Characteristics: F-86E." Wright Air Development Center, May 18, 1951. K243.861-1, AFHRA.

"Standard Aircraft Characteristics: F-86F." Wright Air Development Center, May 22, 1952. K243.861-1, AFHRA.

Stanley, Steve, Tech. Sgt., USAF. "COMACC Addresses RPA Health to Senate Armed Services Committee." Air Combat Command Public Affairs. March 17, 2016. Accessed 5 October 2016. http://www.acc.af.mil/News/ArticleDisplay/tabid/5725/Article/696879/comacc-addresses-rpa-health-to-senate-armed-services-committee.aspx.

Stevens, Bob. "There I Was." *Air Force*, May 1973.

Stewart, George B., Col., USAF. Memorandum to HQ USAF (AFPTR-F-N). "Requirements for the Radar Pilot Training (APQ-100) Courses." November 8, 1962. In *History of the Air Training Command, Jul–Dec 1962*, vol. 11. K220.01, AFHRA.

Stimson, George W. *Introduction to Airborne Radar*. 2nd ed. Mendham, NJ: SciTech Publishing, 1998.

Stocker, Joseph. "A Wing Man Has to Have Eyes in the Back of His Head." *Air Force*, March 1955.

Stratemeyer, George E. *The Three Wars of Lt. Gen. George E. Stratemeyer: His Korean War Diary*. Edited by William T. Y'Blood. Washington, DC: Air Force History and Museums Program, 1999.

Strauss, Walter J. *Analysis of Korean Air-to-Air Combat Film for Period 28 June 1950–30 September 1951*. Memorandum No. 16. Chicago: University of Chicago, Institute for Air Weapons Research, March 1953. K146.08-4, AFHRA.

Stringer, E. M., Col., USAF. "F-5 Program Review." Hearings before the US Senate, Committee on Armed Services, March 13, 1974, pp. 4294–304. Washington, DC: Government Printing Office, 1974. Y4.Ar5/3:P94/6/975/pt8.

Stuart, Harold C. "Salute to Combat Leaders." *Air Force* 35, no. 9, September 1952.

Stuck, Don. "Aircraft Control," February 1966. In *History of the 15th Tactical Fighter Wing, Jul–Dec 1966*, vol. 2. K-WG-15-HI, AFHRA.

Stucky, Paul, Capt., USAF. "Blocking & Tackling." *Fighter Weapons Review*, Winter 1981, 24–28. MSFRIC.

"Suitability of the F-14B for the F-15 Mission," Talking Paper, 1969. 03936, AFMC/HO.

Sullivan, Robert C. *History of the 58th Tactical Fighter Training Wing, Oct–Dec 1974*. Luke AFB, AZ. K285.79-42, AFHRA.

———. *History of the 58th Tactical Fighter Training Wing, Jan–Mar 1975*. Luke AFB, AZ. K285.79-42, AFHRA.

———. *History of the 58th Tactical Fighter Training Wing, Apr–Jun 1975*. Luke AFB, AZ. K285.79-42, AFHRA.

———. *History of the 58th Tactical Fighter Training Wing, Jul–Sep 1975*. Luke AFB, AZ. K285.79-42, AFHRA.

———. *History of the 58th Tactical Fighter Training Wing, Jan–Mar 1976*. Luke AFB, AZ. K285.79-42, AFHRA.

———. *History of the 58th Tactical Fighter Training Wing, Apr–Jun 1976*. Luke AFB, AZ. K285.79-42, AFHRA.

———. *History of the 58th Tactical Fighter Training Wing, Jul–Sep 1976*. Luke AFB, AZ. K285.79-42, AFHRA.

Summary of F-86E Aircraft Accidents: From 1 January 1951 through 31 August 1954. Norton AFB, CA: Office of the Inspector General, USAF, Directorate of Flight Safety Research, n.d. In "F-86 Reports, Histories, Surveys, Summaries, and Comments, 1954–1955." K2021-57, vol. 22, AFHRA.

Swarts, Barry, Maj., USAF. "Interest Paper on AIMVAL/ACEVAL," July 14, 1976. In *History of ADCOM, 1 Jan–31 Dec 1976*, vol. 6. K410.011, AFHRA.

T.O. 1F-4C-1. USAF Series F-4C and F-4D Aircraft, Flight Manual, February 15, 1979. F-4C/Ops, NMUSAF.

T.O. 1F-4C-2-19. USAF Series F-4C Aircraft, Maintenance Instructions: Fire Control System, March 1, 1972. NMUSAF.

T.O. 1F-4C-34-1-1. USAF Series F-4C Aircraft, Aircrew Weapons Delivery Manual (Nonnuclear), February 1, 1976. F-4C/Arm, NMUSAF.

T.O. 1F-15A-1. F-15A/B/C/D Aircraft, Flight Manual, January 15, 1984. ER.2011.088, NMUSAF.

T.O. 1F-15A-2-24. Maintenance Instructions: Head Up Display System, September 15, 1978. NMUSAF.

T.O. 1F-15A-2-25. Maintenance Instructions: Radar System (Radar Set and Indicator Group), December 1, 1978. NMUSAF.

T.O. 1F-15A-34-1-1. F-15A and F-15B Aircraft, Nonnuclear Weapon Delivery Manual (Air-to-Air), September 15, 1985. ER.2011.088, NMUSAF.

T.O. 1F-86A-1. F-86A Aircraft, Flight Handbook, August 30, 1957. A-1, F-86A/Ops, NMUSAF.

T.O. 1F-86E-1. F-86E Aircraft, Flight Handbook, November 26, 1956. A-1, F-86E/Ops NMUSAF.

TAC/DR. Message to AFTEC/CC. "ACEVAL/AIMVAL Preliminary Test Plan Review." 241945Z Mar 76. In *History of the Air Force Test and Evaluation Center, 1 Jan–31 Dec 1976*, vol. 21. K150.01, AFHRA.

———. Message to AFTEC/JT. "F-14/F-15 Use of RHAW in ACEVAL/AIMVAL." 251700Z Jun 76. In *History of the Air Force Test and Evaluation Center, 1 Jan–31 Dec 1976*, vol. 21. K150.01, AFHRA.

———. Message to CSAF/XOO. "ACEVAL/AIMVAL Test Constraints." 050200Z May 76. In *History of the Air Force Test and Evaluation Center, 1 Jan–31 Dec 1976*, vol. 21. K150.01, AFHRA.

TAC/OII. Message to AIG/OI. "AFNS Release, 'President Welcomes F-15 into Air Force.'" 191455Z Nov 74. In *History of the 58th Tactical Fighter Training Wing, Oct–Dec 1974*, by Robert Sullivan. K285.79-42, AFHRA.

"Tactics and Techniques of Long-Range Fighter Escort in VIII Fighter Command." Reproduced by HQ, Army Air Forces, Asst. Chief of Air Staff, Requirements Division, Fighter Section, Washington DC, August 29, 1944. 524.522, AFHRA.

"Tactics: V & XIII Fighter Commands." Seventh Air Force, 1943. 741.548, AFHRA.

"Talking Paper on 'Dr. DuBridge Letter to Mr. Packard,'" December 12, 1969. 03936, AFMC/HO.

Taylor, Lt. Col., USAF. "Air Combat Maneuvering Instrumentation (ACMI)," October 1, 1975. 168.7339-79, AFHRA.

Tedder, Arthur William. *Air Power in War*. London: Hodder and Stoughton, 1954.

Test of A-13B Ball Turret with K-17 Sight and AN/APG-5. Project 2-45-58. Eglin Field, FL: Headquarters, Army Air Forces Proving Ground Command, March 6, 1946. ADB220326, DTIC.

TFWC/CC. Message to TAC/CV. "Recap of ACEVAL-AIMVAL Briefings Week of 27 Feb–3 Mar 78." 070200Z Mar 78. 168.7339-79, AFHRA.

"These Seventeen Americans Are Korean War's Jet Aces." *Life*, July 7, 1952.

Thompson, Mark. "Drone Pilots: We Don't Get No Respect." *Time*, May 2, 2012.

Thompson, Warren, and Doug Dildy. *F-86 Sabre vs MiG-15*. Duel 50. London: Osprey Publishing, 2013.

Thompson, Wayne. *To Hanoi and Back: The USAF and North Vietnam, 1966–1973*. Washington, DC: Air Force History and Museums Program, 2000.

Thompson, Wayne, and Bernard C. Nalty. *Within Limits: The U.S. Air Force and the Korean War*. Washington, DC: Air Force History and Museums Program, 1996.

Thor, Robert M., Capt., USAF. "GAR-8." *Fighter Weapons Newsletter*, June 1958, 29–30. MSFRIC.

Thornborough, Anthony M. *USAF Phantoms*. New York: Arms & Armour Press, 1988.

Thornborough, Anthony M., and Peter Davies. *The Phantom Story*. London: Arms & Armour Press, 1994.

Thorne, James H., Maj., USAAF. "Major James H. Thorne, 359th Fighter Squadron, 356th Fighter Group, P-47." In *The Long Reach: Deep Fighter Escort Tactics*, 21–22. VIII Fighter Command, May 29, 1944. 524.502A, AFHRA.

Thyng, Harrison R., Col., USAF. "The Operation of the 4th Fighter Wing in Korea," 1952. K720.131, AFHRA.

Tilford, Earl H., Jr. *Setup: What the Air Force Did in Vietnam and Why*. Maxwell AFB, AL: Air University Press, 1991.
Tilley, Reade, Capt. *Fighter Tactics*. Special Informational Intelligence Report 43-2. Washington, DC: Intelligence Service, US Army Air Forces, January 31, 1943. 142.034-2, AFHRA.
Tips, George R., Lt. Col., USAF. Memorandum to Pilot Training Wings. "Undergraduate Pilot Training Assignment Procedures." June 21, 1963. In *History of the Air Training Command, Jan–Jun 1963*, vol. 11. K220.01, AFHRA.
Tirpak, John A. "Putting the Pilot in the RPAs." *Air Force*, July 2010.
Tissot, Ernest E., R. Adm., USN, and Maj. Gen. James R. Hildreth, USAF. *AIMVAL Final Report*, vol. 1. Nellis AFB, NV: ACEVAL-AIMVAL JTF, September 1977. In *History of the Air Force Test and Evaluation Center, 1 Jan–31 Dec 1977*, vol. 14. K150.01, AFHRA.
Toliver, Richard, Maj., USAF. *F-15A Follow-On Operational Test and Evaluation Final Report: Annex A, System Description*. Test 76-101. Kirtland AFB, NM: Air Force Test and Evaluation Center, August 1976. In *History of the Air Force Test and Evaluation Center, 1 Jan–31 Dec 1976*, vol. 12. K150.01, AFHRA.
"Top U.S. Jet Ace Lost In Korean Air War." *New York Times*, February 12, 1952.
Trest, Warren. *Once a Fighter Pilot: The Story of Korean War Ace Lt. Gen. Charles G. "Chick" Cleveland*. Montgomery, AL: River City Publishing, 2012.
Trotti, John. *Phantom over Vietnam*. New York: Berkley Books, 1984.
Tucker, Roger D., Capt., USAF. "Fire Control." *Fighter Weapons Newsletter*, December 1959, 14–23. MSFRIC.
Turner, Chuck, Maj., USAF. "ACMI Update." *Fighter Weapons Review*, Summer 1977, 17–22. MSFRIC.
Tuso, Joseph F., Maj., USAF. "Navigator's Log: Jack Armstrong to the Age of Aquarius and Beyond." *Air Force & Space Digest*, June 1970.
"Two Jets Collide; Pilots OK." *Las Vegas Sun*, March 1, 1977.
"Two New 'Kills' Give Yank Title of First Jet Ace." *Daily Boston Globe*, May 21, 1951.
Type A-1C(M) Gyro Computing Sight (Gun-Bomb-Rocket). Handbook: Operation and Service Instructions (AN-70BD-4), February 15, 1951. Box 804, NMUSAF.
"US Sabres Splash Red Jets." *Life*, January 29, 1951.
Ulsamer, Edgar. "The Accent Is on Flying." *Air Force*, June 1971.
———. "The F-15: Tomorrow's Champion of the Dogfight Arena." *Air Force*, July 1969.
Umstead, Stanley M., Jr., Lt. Gen., USAF. Oral History Interview. Interview by Capt. Mark C. Cleary, USAF, July 22 and 25, 1983. K239.0512-1533, AFHRA.
United States Air Force Statistical Digest: Fiscal Year 1953. Washington, DC: Directorate of Statistical Services, Deputy Chief of Staff, Comptroller, 1953. AFHRA.
United States Air Force Statistical Digest: Fiscal Year 1964. 19th ed. Washington, DC: Directorate of Data Automation, Comptroller of the Air Force, 1964. AFHRA.
"US Air Force Academy Jabara Award." Accessed February 18, 2014. https://www.usafa.org/Heritage/Jabara_Award.
"US Fliers Bomb 2 Bridges Linking Hanoi with Bases." *New York Times*, April 4, 1965.
"US Jets Will Reflect Lessons Learned in Vietnam." *New York Times*, November 24, 1966.
USAF. Message to ALMAJCOM, 240046Z Dec 69. In *History of the F-15 Eagle*, vol. 2, pt. 1. K417.042-18, AFHRA.

USAF Conversion Course, F-15. TAC Syllabus F1500C. Langley AFB, VA: Headquarters, Tactical Air Command, October 1974. In *History of the 58th Tactical Fighter Training Wing, Oct–Dec 1974*, by Robert Sullivan. K285.79-42, AFHRA.

USAF F-15 Fighter Summary. Wright-Patterson AFB, OH: Deputy for F-15/JEPO, Headquarters for ASD, May 1, 1971. 03969, AFMC/HO.

"The USAF F-15 Eagle: The Fighter Pilot's Fighter." McDonnell Douglas advertisement. *Air Force*, September 1973.

"The USAF F-15, the Fighter Pilot's Fighter, Won't Be Out-Foxed in the Air Battle Arena." McDonnell Douglas advertisement. *Air Force*, September 1971.

USAF Replacement Training Course: F-4C. TAC Syllabus, Course 111509F-RTU. September 1966. In *History of the 4453d Combat Crew Training Wing, Jul–Dec 1966*, vol. 2. K-WG-4453-HI, AFHRA.

Usselman, Steven W. "From Sputnik to SCOT: The Historiography of American Technology." *Magazine of History* 24, no. 3 (July 2010): 9–14.

Vandevender, J. A., Lt., USN. "Quality-Quantity Rebuttal." *Fighter Weapons Review*, Fall 1974, 37. MSFRIC.

Van Dyke, W. H., Jr., Capt., USN. "Sidewinder Program Review." Hearings before the US Senate, Committee on Armed Services, March 11, 1976, pp. 5013–38. Washington, DC: Government Printing Office, 1976. Y4.Ar5/3:P94/6/977/pt9.

Van Staaveren, Jacob. *Gradual Failure: The Air War over North Vietnam, 1965–1966*. Washington, DC: Air Force History and Museums Program, 2002.

Vandenberg, Hoyt S., Jr., Maj. Gen., USAF (ret.). Oral History Interview. Interview by Lt. Col. James D. Weidman, USAF, December 19–22, 1985. K239.0512-1690, AFHRA.

"Velocity Jump: Trajectory Shift and Angle of Departure." *Fighter Gunnery*, no. 1 (1951): 19–20. Ar.2007.187, NMUSAF.

Versurah, Vincent V., Col., USAF. "Project CORONA HARVEST End of Tour Report," March 3, 1971. K740.131, AFHRA.

"VFW Believes Distinguished Warfare Medal Should Not Outrank the Bronze Star, Purple Heart." *VFW—Veterans of Foreign Wars*. Accessed March 14, 2014. http://www.vfw.org/News-and-Events/Articles/2013-Articles/VFW-WANTS-NEW-MEDAL-RANKING-LOWERED/.

"View from the Cockpit." *Avionics Magazine: Special Report, Air Dominance with the F-22 Raptor*, 2002: 12–14.

Volloy, Gerald R., Maj., USAF. "Red Flag in Perspective." *Fighter Weapons Review*, Spring 1979, 2–5. MSFRIC.

Waddell, Donald E., III, Capt., USAF. "ACMI Improves Pilot Combat Training." *The Flyer*. June 2, 1977. In *History of the 1st Tactical Fighter Wing, Jan–Mar 1977*, vol. 2. K-WG-1-HI, AFHRA.

Wagner, Ray. *The North American Sabre*. MacDonald Aircraft Monographs. Garden City, NY: Doubleday, 1963.

Wallace, Jere, Col., USAF (ret.). Interview by Steven A. Fino, October 3, 2013.

Warnock, A. Timothy, ed. *The USAF in Korea: A Chronology, 1950–1953*. Washington, DC: Air Force History and Museums Program, 2000.

Waterman, Peter. "Research, Development, Test, and Evaluation, Navy." Hearings before the US House, Subcommittee of the Committee on Appropriations, September 25, 1973, pp. 639–848. Washington, DC: Government Printing Office, 1973. Y4.Ap6/1:D36/5/974/pt7.

Watts, Barry D., Maj., USAF. "The Changing World of Air Combat, or *Plus Ça Change, Plus C'est la Même Chose*." *Air Force*, December 1977.

———, "Doctrine, Technology, and War." Maxwell AFB, AL, 1996. Accessed March 29, 2012. http://www.airpower.maxwell.af.mil/airchronicles/cc/watts.html.

———. *Six Decades of Guided Munitions and Battle Networks: Progress and Prospects*. Washington, DC: Center for Strategic and Budgetary Assessments, March 2007.

Watts, James W. *Operational Survey of Type A-1CM Sighting System as Used by 49th and 136th Fighter-Bomber Wings (F-84 Aircraft) in the Korean Theatre*. Operations Analysis Memorandum. HQ Fifth Air Force, Operations Analysis Office, February 24, 1952. K730.3101-25, AFHRA.

"We Must Seek Out the Hun and Destroy Him," 1943. 524.501, AFHRA.

Weaver, Robert B., Lt. Col., USAF. *Air-to-Air Encounters over North Vietnam, 1 Jul 1967–31 Dec 1968*. Project CHECO Report. Headquarters PACAF: Directorate, Tactical Evaluation, August 30, 1969. K717.0413-22, AFHRA.

Welch, Jasper A., Jr., Maj. Gen., USAF (ret.). Interview by Steven A Fino, June 8, 2012.

———, Maj. Gen., USAF. Letter to AF/XO. "AIMVAL Analysis and Evaluation." December 5, 1977. In *History of the USAF Tactical Fighter Weapons Center, 1977*, vol. 2. K417.0735, AFHRA.

———, Maj. Gen., USAF. Letter to Brig. Gen. Thomas H. McMullen, USAF, March 8, 1977. 168.7339-79, AFHRA.

Welch, Larry D., Gen., USAF (ret.). Interview by Steven A. Fino, July 10, 2012.

———, Gen., USAF (ret.). Interview by Steven A. Fino, October 1, 2013.

———, Gen., USAF (ret.). Oral History Interview. Interview by Hugh N. Ahmann and Daniel L. Haulman, November 1993, October 1994, May 1997, and May 2000. K239.0512-2126, AFHRA.

———, Col., USAF. *Weapons Delivery Image Recording System for Tactical Fighter Aircraft*. 1TFW Supplement 1 to TACR 55-53. Langley AFB, VA: Headquarters, 1st Tactical Fighter Wing, January 12, 1976. In *History of the 1st Tactical Fighter Wing, Oct–Dec 1975*, vol. 1. K-WG-1-HI, AFHRA.

Wells, D. V., Capt., USN. "Sparrow Missile Program Review." Hearings before the US Senate, Committee on Armed Services, March 11, 1976, pp. 5000–13. Washington, DC: Government Printing Office, 1976. Y4/Ar5/3:P94/6/977/pt9.

Wells, Mark K., Col., USAF. "Tribal Warfare: The Society of Modern Airman." *Air & Space Power Journal* 29, no. 3 (June 2015): 82–87.

Wells, Robert J. "Tips and Info for AN/APG-30 Radar." *Fighter Weapons Newsletter*, February 1956, 58–64. MSFRIC.

Wentworth, John, Capt. "Lectures by Pursuit Pilots of AEF—Pursuit Pilots: Duties, Training, Etc.," 1919. 248.282-42, AFHRA.

Werner, Johannes. *Knight of Germany: Oswald Boelcke, German Ace*. Translated by Claude W. Sykes. Philadelphia: Casemate, 2009.

Werrell, Kenneth P. *Archie to SAM: A Short History of Ground-Based Air Defense*. 2nd ed. Maxwell AFB, AL: Air University Press, 2005.

———. *Chasing the Silver Bullet: US Air Force Weapons Development from Vietnam to Desert Storm*. Washington, DC: Smithsonian Books, 2003.

———. *Sabres over MiG Alley: The F-86 and the Battle for Air Superiority in Korea*. Annapolis, MD: Naval Institute Press, 2005.

Westrum, Ron. *Sidewinder: Creative Missile Development at China Lake*. Annapolis, MD: Naval Institute Press, 1999.

Wheeler, Winslow T., and Lawrence J. Korb. *Military Reform: A Reference Handbook*. Westport, CT: Praeger Security International, 2007.

Whistler, Ross, Lt. Col., USAFR. "A Brief History of United States Air Force Reflector Gunsights." Research Report, Air War College, January 1973. R1-238.3, AFMC/HO.

Whitehead, Asa, Capt., USAF. "Supersonic Air-to-Air Gunnery." *Fighter Weapons Newsletter*, December 1957, 14–16. MSFRIC.

"Why TAC Cannot Use the Navy F-14B," Talking Paper, 1969. 03936, AFMC/HO.

Wickens, Christopher D. "Situation Awareness: Review of Mica Endsley's 1995 Articles on Situation Awareness Theory and Measurement." *Human Factors* 50, no. 3 (June 2008): 397–403.

Wilcox, Robert K. *Scream of Eagles: The Dramatic Account of the U.S. Navy's Top Gun Fighter Pilots and How They Took Back the Skies over Vietnam*. New York: Simon and Schuster, 2005.

Wildenberg, Thomas. "The A-1C(M) Gunsight: A Case Study of Technological Innovation in the United States Air Force." *Air Power History* 56, no. 2 (Summer 2009): 28–37.

———. "A Visionary Ahead of His Time: Howard Hughes and the US Air Force—Part III: The Falcon Missile and Airborne Fire Control." *Air Power History* 55, no. 1 (Summer 2008): 4–13.

Wilson, Bill, Lt. Col., USAF. "PhD for Fighter Pilots." *Fighter Weapons Review*, Summer 1973, 1–3. MSFRIC.

Wilson, Joe. Letter to Chuck Hagel, March 4, 2013. Accessed March 14, 2014. http://joewilson.house.gov/uploadedfiles/hagel_dwm_letter.pdf.

"Wings." *AFI Catalog of Feature Films*. Accessed December 16, 2013. http://www.afi.com/members/catalog/DetailView.aspx?s=&Movie=13362.

Winkler, John L., Sr., Capt., USAF. "A Look at the -34 Aircrew Non-nuclear Weapons Delivery Manual." *Fighter Weapons Newsletter*, no. 3 (September 1965): 16–18. MSFRIC.

Winters, Deacon, Capt., USAF. "Two, Bingo, May-Day . . ." *Fighter Weapons Review*, Fall 1987, 13–17. MSFRIC.

Witze, Claude. "TAC AVENGER Is Out to Win." *Air Force & Space Digest*, January 1968.

Wohl, Robert. *A Passion for Wings: Aviation and the Western Imagination, 1908–1918*. New Haven, CT: Yale University Press, 1994.

———. *The Spectacle of Flight: Aviation and the Western Imagination, 1920–1950*. New Haven, CT: Yale University Press, 2005.

Wolfe, Tom. *The Right Stuff*. New York: Bantam Books, 1980.

Wood, Frank P., Col., USAF. Memorandum to ATCCS. "Final Report—UPT Assignments." August 14, 1963. In *History of the Air Training Command, Jan–Jun 1963*, vol. 11. K220.01, AFHRA.

Worden, Mike, Col., USAF. *Rise of the Fighter Generals: The Problem of Air Force Leadership, 1945–1982*. Maxwell AFB, AL: Air University Press, 1998.

"World's First Jet Ace Downs 2 MIG's within 10 Minutes: Kansas Captain Gets Fifth and Sixth Victims in Battle over Sinuiju." *New York Times*, May 21, 1951.

Y'Blood, William T. *MiG Alley: The Fight for Air Superiority*. Washington, DC: Air Force History and Museums Program, 2000.

Yeager, Chuck. *Yeager: An Autobiography*. With Leo Janos. New York: Bantam Books, 1985.

"You've Got to Prove It!" *Air Force*, May 1953.
Youngling, Edward W., Sheldon H. Levine, John B. Mocharnuk, and Louise M. Weston. *Feasibility Study to Predict Combat Effectiveness for Selected Military Roles: Fighter Pilot Effectiveness*. Final Report, MDC E1634. St. Louis, MO: McDonnell Douglas Corporation, April 29, 1977. ADA041650, DTIC.
Zhang, Xiaoming. *Red Wings over the Yalu: China, the Soviet Union, and the Air War in Korea*. College Station: Texas A&M University Press, 2002.
Ziemke, Caroline. "In the Shadow of the Giant: USAF Tactical Air Command in the Era of Strategic Bombing, 1945–1955." Ph.D. dissertation, The Ohio State University, 1989.

INDEX

Note: Page numbers in *italics* indicate figures.